# Handbook of Bio-Based Materials for Smart Manufacturing

In recent years, the convergence of biotechnology and manufacturing has ushered in a new era of sustainable innovation and efficiency. Incorporating bio-based materials into smart manufacturing processes has become a crucial strategy to minimize environmental impact, promote circularity, and propel technological advancements across various industries. The complete chapters of this handbook are arranged in four sections. It includes a comprehensive overview and a holistic approach to the principles, applications, challenges, and opportunities associated with integrating bio-based materials into modern manufacturing practices.

The *Handbook of Bio-Based Materials for Smart Manufacturing* begins with an exploration of fundamental characteristics, properties, and processing techniques. From bio-based polymers to composites and blends, readers will gain insights into the diverse array of materials derived from renewable biomass sources and their potential applications in smart manufacturing environments. The handbook illustrates the role of bio-based materials in additive manufacturing, coatings, packaging, energy applications, automotive, aerospace, construction, biomedical, textiles, electronics, and more, showcasing the versatility and adaptability of these materials across various sectors.

By examining emerging trends, future directions, and the vision for sustainable manufacturing practices, readers will gain valuable insights into the changing landscape of bio-based materials and their potential impact on the future of smart manufacturing. This handbook is a valuable resource for researchers, engineers, industry professionals, and students seeking to deepen their understanding of bio-based materials and their role in driving sustainable innovation in manufacturing.

**Innovations in Smart Manufacturing for Long-Term Development and Growth**

**Series Editors:**
**Atul Babbar, Gursel Alici, Yu Dong, Ankit Sharma**

**Fabrication Techniques and Machining Methods of Advanced Composite Materials**
*Vikas Dhawan, Atul Babbar, Inderdeep Singh and Jonathan M. Weaver*

**Advances in Pre- and Post-Additive Manufacturing Processes**
Innovations and Applications
*Naveen Mani Tripathi and Ankit Sharma*

**Handbook of 3D Printing in Pharmaceutics**
Innovations and Applications
*Prakash Katakam, Ranvijay Kumar, Nishant Ranjan and Atul Babbar*

**Intelligent Machinery Fault Diagnostics and Prognostics**
The Future of Smart Manufacturing
*Deepam Goyal, Ankit Sharma and Mohamad Abou Houran*

**Nanomaterials for Smart Manufacturing**
Fundamental Principles, Theory, Case Studies, and Advanced Applications
*Dinesh Deshwal, Anil Kumar Narwal, Vivek Srivastava, Atul Babbar and Jonathan M. Weaver*

**Sensor Technologies and Wireless Communications in Industry 5.0**
A Solution for Smart Manufacturing
*Edited by Ramkumar K.R., Bhupendra Singh, Lei Pan and Ankit Sharma*

**Handbook of Smart Manufacturing Technologies**
Industrial Engineering in the Digital Age
*Edited by Mahendra U. Gaikwad, Ankit Sharma, Muthuramalingam Thangaraj and Sachin Salunkhe*

**Handbook of Bio-Based Materials for Smart Manufacturing**
*Edited by Devendra Singh, Naveen Mani Tripathi, Dharminder Singh and Ankit Sharma*

For more information about this series, please visit: www.routledge.com/Innovations-in-Smart-Manufacturing-for-Long-Term-Development-and-Growth/book-series/CRCISMLDG

# Handbook of Bio-Based Materials for Smart Manufacturing

Edited by
Devendra Singh, Naveen Mani Tripathi,
Dharminder Singh, and Ankit Sharma

CRC Press
Taylor & Francis Group
Boca Raton London New York

CRC Press is an imprint of the
Taylor & Francis Group, an **informa** business

Designed cover image: Shutterstock

First edition published 2026
by CRC Press
2385 NW Executive Center Drive, Suite 320, Boca Raton FL 33431

and by CRC Press
4 Park Square, Milton Park, Abingdon, Oxon, OX14 4RN

*CRC Press is an imprint of Taylor & Francis Group, LLC*



ISBN: 978-1-032-93457-0 (hbk)
ISBN: 978-1-032-93501-0 (pbk)
ISBN: 978-1-003-56620-5 (ebk)

DOI: 10.1201/9781003566205

Typeset in Times
by Apex CoVantage, LLC

*Editors would like to dedicate this book to their Parents, wife, kids and Maa Kamakhya Devi.*

# Contents

# Preface

In recent years, the intersection of biotechnology and manufacturing has given rise to a new era of sustainable innovation and efficiency. The utilization of bio-based materials in smart manufacturing processes has emerged as a key strategy to reduce environmental impact, promote circularity, and drive technological advancements in various industries. This book, *Handbook of Bio-Based Materials for Smart Manufacturing*, aims to provide a comprehensive overview of the principles, applications, challenges, and opportunities associated with the integration of bio-based materials in modern manufacturing practices.

The journey into the world of bio-based materials begins with an exploration of their fundamental characteristics, properties, and processing techniques. From bio-based polymers to composites and blends, readers will gain insights into the diverse range of materials derived from renewable biomass sources and their potential applications in smart manufacturing environments. The book delves into the role of bio-based materials in additive manufacturing, coatings, packaging, energy applications, automotive, aerospace, construction, biomedical, textiles, electronics, and more, showcasing the versatility and adaptability of these materials across various sectors.

As we navigate through the chapters, we will uncover the transformative impact of bio-based materials on sustainability, energy efficiency, and product innovation. From biodegradable packaging solutions to lightweight automotive composites, and from smart coatings to biocompatible medical devices, the book presents real-world examples and case studies that highlight the practical applications and benefits of bio-based materials in today's manufacturing landscape.

Furthermore, the book addresses the challenges and opportunities that come with the adoption of bio-based materials, including scalability, regulatory frameworks, standardization, and market acceptance. By exploring emerging trends, future directions, and the vision for sustainable manufacturing practices, readers will gain valuable insights into the evolving landscape of bio-based materials and their potential to shape the future of smart manufacturing. It is our hope that this book serves as a valuable resource for researchers, engineers, industry professionals, and students seeking to deepen their understanding of bio-based materials and their role in driving sustainable innovation in manufacturing. We invite you to embark on this journey with us as we explore the exciting possibilities and transformative potential of bio-based materials for smart manufacturing.

# Acknowledgment

We express our sincere thanks to all contributors, all those friends and colleagues who have activated our norms to take up a study like this.

The editors greatly remember the contribution and support provided by the Hon'ble Director, RGIPT, Prof. Harish Hirani; Chancellor, SRMU, Shri Pankaj Agarwal; Vice Chancellor, Glasgow Caledonian University, United Kingdom, Prof. Steve Decent and Vice Chancellor, Chitkara University, Punjab, Dr. Sandhir Sharma for providing a vibrant research and academic ecosystem.

We are highly indebted to our parents, wife, kids and in-laws whose support and advice have paved the way for our study and investigations. They have created a congenial atmosphere at home for our studies and writing. They are silent observers of our works.

**Dr. Devendra Singh**
**Dr. Naveen Mani Tripathi**
**Dr. Dharminder Singh**
**Dr. Ankit Sharma**

# About the Editors

**Dr. Devendra Singh** is assistant professor in the Institute of Biosciences and Technology at Shri Ramswaroop Memorial University, Barabanki, Uttar Pradesh, India. With a remarkable academic background and extensive research experience, Dr. Singh has established himself as a prominent figure in the biotechnology field. His academic journey began with a Master of Technology (M.Tech.) from Amity University, where he laid the foundation for his future research endeavors. He further pursued his Doctor of Philosophy (Ph.D.) in biotechnology from the Department of Biotechnology, Motilal Nehru National Institute of Technology, Allahabad, India, where he delved deeper into the intricacies of biotechnology.

Following his Ph.D., Dr. Singh gained valuable research experience through a postdoctoral fellowship in the Department of Biotechnology at the National Institute of Technology, Warangal, Telangana, India. This experience honed his research skills and gave him a deeper understanding of the latest advancements in the field. With over three years of teaching experience, Dr. Singh has developed a unique ability to effectively communicate complex scientific concepts to students, making him an exceptional educator.

Dr. Singh's research output is remarkable, with over 51 research papers published in esteemed journals of national and international repute, boasting a cumulative impact factor of more than 71. His contributions to the biotechnology field are further amplified by his numerous book chapters, with over 33 chapters published to date. Furthermore, Dr. Singh has been granted three patents and has had the technology of two patents transferred, demonstrating his innovative approach to biotechnology. He currently runs one CSTUP project and one in-house project under his name. His research interests span various areas, including additive, environmental biotechnology, microbiology, pharmaceutical biotechnology, intellectual property rights (IPR), and biomedical applications.

**Dr. Naveen Mani Tripathi** is currently working as assistant professor in the Mechanical Engineering Department of the Rajiv Gandhi Institute of Petroleum Technology, Jais, Ministry of Petroleum and Natural Gas, Government of India. He completed his PhD from the Mechanical Engineering Department of Ben-Gurion University of the Negev, Israel. After his Ph.D., he worked as a particle scientist in Belgium. His postdoctoral research is from the chair of Process System Engineering of the Technical University of Munich, Germany. He is awarded with many fellowships, such as GATE, GRE, Kreitman, TUM University Foundation Fellow, and Alexander von Humboldt Fellow. His diversified research areas are powder rheology, additive and smart manufacturing, renewable energy, and thermo-fluid. He has more than 10 years of teaching and research experience, including five patents and more than 50 journal/conference/book chapter publications. He is also an editor of a book titled *Advances in Pre- and Post-Additive Manufacturing Processes*, CRC Press. He voluntarily works as an editorial board member and review editor for more than 20 journals. He is the editor of a special issue of the Scopus-indexed journals, including *Recent Patents on Mechanical Engineering*, Betham Science. He also works as a mentor and advisor for different institutional bodies.

**Dr. Dharminder Singh** is currently working as assistant professor in the Department of Mechanical Engineering at Glasgow Caledonian University, UK. He completed his Ph.D. in 2017 from Glasgow Caledonian University, following which he worked as a research and development engineer on a Knowledge Transfer Partnership between the University of Huddersfield and Trillium Flow Technologies for three years, where he conducted research on the condition monitoring of control valves and the development of smart valves. He has published his research in many conferences and journal papers. Dr. Singh's research interests are in the fields of additive manufacturing, thermo-fluids, CFD modeling of flow handling equipment, and the application of machine learning.

**Dr. Ankit Sharma** presently holds the associate director (research) position at the Centre for Research Impact & Outcome (CRIO), Department of Chitkara University, Punjab, India. He is also the head and publisher of the international peer-reviewed Chitkara journals and Chitkara University Publications. His admiring communication and relationship-building capabilities have always focused on the progress of his organization. He received his doctorate in mechanical engineering from the Thapar Institute of Engineering and Technology, Punjab, India. He is a passionate researcher with diversified research interests in manufacturing, medical devices, modern machining processes, additive manufacturing, tissue engineering, and more. He has published more than 55 scientific articles in various peer-reviewed, reputed, top-notch journals, conferences, and books in manufacturing and materials. Dr. Sharma is the founder and the general chair-convenor of the annual flagship international conference ICEMSMCI (International Conference on Emerging Materials, Smart Manufacturing, Computational Intelligence) 2023, 2024, and 2025. He has also been organizing several other international conferences, such as the International Conference on Advanced Materials Sustainable Future (ICAMSF) 2025. He has chaired various international conferences in the domain of advanced manufacturing and materials. He has delivered several seminars and keynote talks on international (the USA, China, and India) platforms and has been awarded with the best research paper awards. He is currently doing consultancy with the health industry by developing biomedical device manufacturing machines. He has also delivered numerous invited industrial keynotes while addressing various topics related to smart manufacturing. Dr. Sharma is working on research commercialization and has filed more than 30 patents. He has also been granted four international patents, along with over 18 granted patents. He contributes to interdisciplinary research, particularly where engineering and global burning issues related to health and manufacturing sectors intersect. Dr. Sharma edited and authored 5 books, serving as a series editor for Taylor & Francis, Elsevier, and ASME books. He is also serving as a guest editor and on the editorial boards of several reputed SCI, ESCI, Scopus-indexed international peer-reviewed journals.

# Contributors

**R.G. Abaszade**
Azerbaijan State Oil and Industry University
and
Azerbaijan University of Architecture and Construction
Baku, Azerbaijan

**Pooja Adtani**
Department of Basic Medical and Dental Sciences
College of Dentistry
Gulf Medical University
Ajman, United Arab Emirates

**Vishnu Agarwal**
Department of Biotechnology
Motilal Nehru National Institute of Technology Allahabad
Prayagraj, Uttar Pradesh, India

**Mohanned Alhussien**
School of Life Sciences Weihenstephan
Technical University of Munich
Liesel-Beckmann-Str. Freising
Germany

**Khushi Bhardwaj**
Department of Microbiology
Gurukul Kangri University
Haridwar, Uttarakhand, India

**Ravindra Naik Bukke**
School of Mechanical and Materials Engineering
Indian Institute of Technology
Mandi, Himachal Pradesh, India

**Ankush Chauhan**
Faculty of Allied Health Sciences
Chettinad Hospital, and Research Institute
Chettinad Academy of Research and Education
Kelambakkam, Tamil Nadu, India

**Deokrishna Kumar Choudhary**
School of Mechanical and Materials Engineering
Indian Institute of Technology
Mandi, Himachal Pradesh, India

**Sachchidanand Sharan Das**
Faculty of Law, Centre for Juridical Studies, Dibrugarh University
Dibrugarh, Assam, India

**Vinay Kumar Dhiman**
Faculty of Food Science and Technology (Food Microbiology)
Department of Food Science and Technology Khalsa College
Amritsar, Punjab, India

**Vivek Kumar Dhiman**
Centre for Herbal Pharmacology and Environmental Sustainability
Chettinad Hospital and Research Institute
Chettinad Academy of Research and Education
Kelambakkam, Tamil Nadu, India

**Marvellous O. Eyube**
Department of Chemistry
Faculty of Physical Science
University of Benin
Benin City, Nigeria

**C.T. Ezeh**
Università degli Studi di Salerno (UNISA)
Fisciano, Salerno, Campania, Italy; and University of Teramo (UNITE)
Teramo, Abruzzo, Italy

**Juveriya Israr**
Faculty of Biosciences
Institute of Biosciences and Technology
Shri Ramswaroop Memorial University
Lucknow-Deva Road, Barabanki
Uttar Pradesh, India
and
Department of Biotechnology
Era University
Lucknow, Uttar Pradesh, India

**Hanna Sara Johnson**
Cranfield University
Bedford, United Kingdom

**Manmohit Kalia**
Department of Biology
Binghamton University
Binghamton, New York, USA

**E.A. Khanmamadova**
Azerbaijan State Oil and Industry University
Baku, Azerbaijan

**Renu Khare**
Faculty of Biosciences
Institute of Biosciences and Technology
Shri Ramswaroop Memorial University
Barabanki, Uttar Pradesh, India

**Manoj Kumar Khurana**
State University of New York
Plattsburgh New York, USA

**Dhruv Kumar**
Department of Mechanical Engineering
IIMT College of Engineering
Greater Noida, Uttar Pradesh, India

**Aparna Kumari**
Institute of Biosciences and Technology
Shri Ramswaroop Memorial University
Lucknow, Uttar Pradesh, India

**Manju Kumari**
Department of Psychiatry
King George's Medical University
Shah Meena Road
Lucknow, Uttar Pradesh, India

**Ravi Kant Kushwaha**
Maharishi School of Pharmaceutical Sciences
Lucknow, Uttar Pradesh, India

**Nikhil Gupta**
Department of Mechanical Engineering at IIMT College of Engineering
Greater Noida, Uttar Pradesh, India

**F. Marra**
Università degli Studi di Salerno (UNISA), University of Teramo (UNITE)
Teramo, Italy

**Anshu Mishra**
Maharishi School of Pharmaceutical Sciences
Lucknow, Uttar Pradesh, India

**Sankalp Misra**
Faculty of Biosciences
Institute of Biosciences and Technology
Shri Ramswaroop Memorial University
Barabanki, Uttar Pradesh, India

**Nishu Mittal**
Institute of Biosciences and Technology
Shri Ramswaroop Memorial University
Lucknow, Uttar Pradesh, India

**Rahul Nadda**
Department of Chemical Engineering
Loughborough University
Loughborough, United Kingdom

**Swapnil Pandey**
Department of Neurology
McKnight Brain Institute
University of Florida
Gainesville, Florida, USA

**Vikas Pandey**
Fujita Health University
Dengakugakubo, Kutsukake-cho
Toyoake, Aichi, JAPAN

**Bhavna Pandey**
Mechanical Engineering Department
Motilal Nehru National Institute of Technology Allahabad
Prayagraj, Uttar Pradesh, India

**Virendra Nath Pathak**
Faculty of Mathematical and Statistical Sciences
Shri Ramswaroop Memorial University
Lucknow, Uttar Pradesh, India

**Manash K. Paul**
Department of Radiation Biology and Toxicology
Manipal School of Life Sciences
Manipal Academy of Higher Education
Manipal, Karnataka, India
and
Department of Pulmonary and Critical Care Medicine
David Geffen School of Medicine
University of California
Los Angeles, California, USA

**Dmitry Portnikov**
Department of Mechanical Engineering
Ben-Gurion University of the Negev
Beer-Sheva, Israel

**Dube Dheeraj Prakashchand**
School of Mechanical and Materials Engineering
Indian Institute of Technology
Mandi, Himachal Pradesh, India

**Rohit Rampal**
Mechanical Engineering Department
Motilal Nehru National Institute of Technology Allahabad
Prayagraj, Uttar Pradesh, India

**Rahutosh Ranjan**
Department of Industrial Engineering
University of Salerno
Salerno, Italy

**Laxmi Rathor**
Department of Biotechnology
Era University
Lucknow, Uttar Pradesh, India

**Mufaddal Huzefa Shakir**
Department of Automobile Engineering
Chandigarh University
Punjab, India

**Rahul Sharma**
Spintec
Grenoble, France

**Hariom Sharma**
National Institute of Health
Bethesda MD, USA

**Simran Sharma**
Department of Electronics and Communication Engineering
Tezpur University
Tezpur, Assam, India

**Shubham Sharma**
Department of Biotechnology
Motilal Nehru National Institute of Technology Allahabad
Prayagraj, Uttar Pradesh, India

**Ankit Sharma**
Chitkara University Institute of Engineering & Technology
Chitkara University
Punjab, India

**Rupendra Shrestha**
Department of Natural and Applied Sciences
Nexus Institute of Research and Innovation (NIRI)
Lalitpur, Nepal

**S. Simal**
University of the Balearic Island (UIB)
Palma, Illes Balears, Spain

**Dharminder Singh**
Department of Mechanical Engineering
School of Computing, Engineering and Built Environment
Glasgow Caledonian University
Glasgow, United Kingdom

**Palak Singh**
Faculty of Biotechnology
Institute of Biosciences and Technology
Shri Ramswaroop Memorial University
Barabanki, Uttar Pradesh, India

**Virendra Singh**
HR Institute of Pharmacy
HRIT University
Ghaziabad, Uttar Pradesh, India

**Akant Kumar Singh**
Department of Mechanical Engineering
Chandigarh University
Punjab, India

**Devendra Singh**
Faculty of Biotechnology
Institute of Biosciences and Technology
Shri Ramswaroop Memorial University
Lucknow-Deva Road
Barabanki, Uttar Pradesh, India

**Ankita Singh**
Faculty of Biosciences
Institute of Biosciences and Technology
Shri Ramswaroop Memorial University
Lucknow-Deva Road
Barabanki, Uttar Pradesh, India

**Akash Singh**
Maharishi School of Pharmaceutical Sciences
Lucknow, Uttar Pradesh, India

**Ambrish Singh**
Rajkiya Engineering College
Azamgarh, Uttar Pradesh, India

**Anmol Srivastava**
Department of Biotechnology
Motilal Nehru National Institute of Technology Allahabad
Prayagraj, Uttar Pradesh, India

**Urvashi Srivastava**
Ex-Associate Professor
Pranveer Singh Institute of Technology
Bhautipratappur Kanpur
Uttar Pradesh, India

**Rajasekaran Subbarayan**
Centre for Herbal Pharmacology and Environmental Sustainability
Chettinad Hospital and Research Institute
Chettinad Academy of Research and Education
Centre for Advanced Biotherapeutics and Regenerative Medicine
Kelambakkam, India

**Naveen Mani Tripathi**
Department of Mechanical Engineering
Rajiv Gandhi Institute of Petroleum Technology
Jais, MoPNG, Government of India
Uttar Pradesh, India

**Virat Mani Tripathi**
Faculty of Law
University of Delhi
Delhi, India

**Pushpendra Tyagi**
Department of Mechanical Engineering
IIMT College of Engineering
Greater Noida, Uttar Pradesh, India

**Ayushi Verma**
Institute of Biosciences and Technology
Shri Ramswaroop Memorial University
Lucknow, Uttar Pradesh, India

**Sunil Kumar Verma**
Faculty of Biotechnology
Institute of Biosciences and Technology
Shri Ramswaroop Memorial University
Lucknow, Barabanki,
Uttar Pradesh, India

**Nidhi Verma**
Department of Biotechnology
Motilal Nehru National Institute of Technology Allahabad
Prayagraj, Uttar Pradesh, India

**Manvendra Yadav**
Atal Incubation Centre-EMPI
New Delhi, India

**Praneta Yadav**
Faculty of Biotechnology
Institute of Biosciences and Technology
Shri Ramswaroop Memorial University
Barabanki, Uttar Pradesh, India

**Gaurav Yadav**
Natural Product Laboratory
Department of Bioengineering
Integral University
Lucknow, Uttar Pradesh, India

**Adhishree Yadav**
Centre of Bioinformatics and Computational Biology
Central Drug Research Institute (CSIR)
Jankipuram Extension
Lucknow, Uttar Pradesh, India

**Pramod Yadav**
Amity Institute of Neuropsychology and Neurosciences
Amity University
Uttar Pradesh, Noida Campus, India

# 1 Introduction to Bio-Based Materials

*Devendra Singh, Sunil Kumar Verma, Naveen Mani Tripathi, Ankit Sharma, and Dharminder Singh*

## 1.1 INTRODUCTION

These three resources can be used to create man-made materials: vegetal resources, fossil resources, and minerals. The materials obtained from each of them are typically proprietary to their resource. If multiple transformation steps are carried out, it is occasionally possible to produce the same substance from either vegetal or fossil resources. For example, ethylene can be polymerized to produce polyethylene (PE). In the petrochemical sector, light gasoline molecules are cracked by steam to produce this gas. It can also be produced via the Fischer–Tropsch process once cellulose is gasified. As a result, the choice of source and procedure is determined by the resource's cost and availability as well as the process's economics and technical viability (Pacheco-Torgal, 2020). Here is a second way to think about how to comprehend the creation of new materials. Another material that is more readily available and/or less expensive can be used in its place if the current one becomes too expensive or scarce, as long as it still meets the performance standards. During the steel crisis, for example, reinforced plastic parts replaced metal ones in autos since they were less expensive and lighter. A third method to take into account is a material's sustainable and ecological features (see Figure 1.1). For example, the energy and fossil fuels required to produce plastics resulted in unintended carbon dioxide emissions into the atmosphere, which exacerbated global warming. Additionally, if certain plastic materials are not properly disposed of after use, they can cause visual pollution and suffocate animals due to their prolonged persistence in the environment (Wool & Sun, 2011).

As a result, selecting materials and determining its origins have become more difficult in the modern world. The availability of raw materials on our planet is not guaranteed, and our ecological consciousness is increasing. Biomaterials are part of the replacement strategy that has already started to be put into practice. Biomaterials, often known as bio-based materials, are materials derived from biomass, as defined in this chapter (Tyagi et al., 2023).

DOI: 10.1201/9781003566205-1

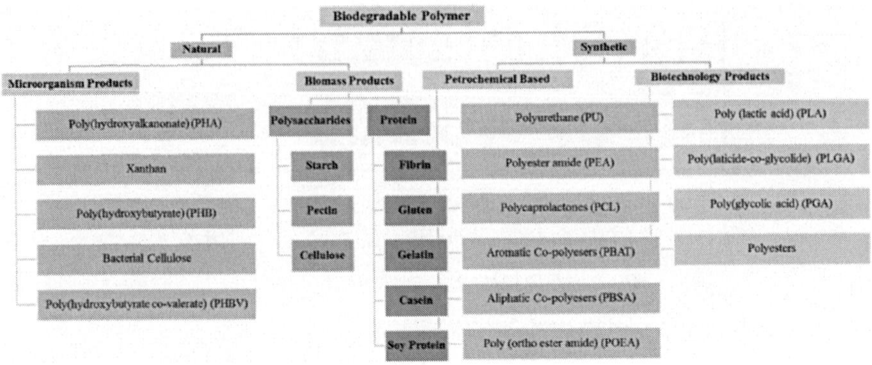


**FIGURE 1.1** Biodegradable polymer classification.

## 1.2 NATURAL FIBERS AND WOOD

Mankind has been endowed by nature with an extensive array of plant-based biomaterials (Chand & Fahim, 2021). It is a common misconception that just because they are natural, their performances must be subpar. Wood, for example, has exceptional hardness, flexibility, and insulating qualities. In the summer, linen serves as an airy and sophisticated textile. The resilience and insulating qualities of stubble explain why they are still preferred for thatched roofs in rural homes and even in historic Japanese shrines (Verma et al., 2024b). Mechanical resistance and insulating qualities are the common features in the above example. These benefits stem mostly from the microscopic structure of these materials, rather than only from their macroscopic shape (such as the hollow cylindrical structure of stubble). Two models can be used to describe most vegetable fibers: wood fibers and cotton fibers (Karimah et al., 2021).

### 1.2.1 Molecular Constitution

Solvent-soluble, water substances, and polysaccharides make up the three main component groups of all natural fibers. Particularly, in cotton fibers, cellulose is the most prevalent component. The repeating unit of this biopolymer is called cellobiose, and it is made up of two glucose-building units that are oriented in the opposite direction relative to the cycle plane. A β-1–4 glucosidic link produces such a structure. According to schematics, cellulose is a "flat" molecule with hydroxyl groups protruding from the "ribbon." The OH groups' accessibility promotes the development of crystalline zones and intermolecular hydrogen bonding. A cotton fiber's cellulose molecule contains almost 7,000 glucose units. The degree of polymerization is lowered to roughly 1,000 units upon extraction. Various pentoses and hexoses make up hemicelluloses. These heteropolysaccharides cannot form crystalline structures because they are frequently branched polymers. Nonetheless, they play a critical role in the composition of natural fibers and form the bonding matrix for the cellulose microfibers along with lignin. Phenylpropane units make up the amorphous, non-polar macromolecule known as lignin. The origin affects the lignin's structure, which is further altered by the extraction process before examination. Recent research has revealed important distinctions in the lignin structures that have been previously reported. Typically, lignin is made up of three fundamental components, which are as follows.

The three main types of units found in hardwoods are the syringyl unit, which is derived from trans-sinapyl alcohol; the p-hydroxyphenyl unit, which is derived from trans-p-cumaryl alcohol; and the guaiacyl unit, which is derived from trans-coniferyl alcohol, which is abundant in softwoods. Lignin and hemicelluloses are connected by covalent bonds made of ether and ester. Known as the lignin–carbohydrate complex, or LCC, it is formed by the two macromolecules. LCC can clump together around cellulose microfibrils because hemicelluloses and cellulose can connect through hydrogen bonding (Jones & Brischke, 2017).

### 1.2.2 The Wood and Timber Hierarchical Structure

Wood has been used as an energy source as well as a structural material. Because of its mechanical qualities and aesthetic appeal, wood is still a popular construction material. Due to the global production levels and species diversity, wood is a plentiful commodity that is found in nearly every country in the world. The ligneous and compact material that makes up a tree's trunk, branches, and roots is known as wood. The word "timber" refers to sawn wood, typically the tree trunk, which is used for building because of its strength. A total of 135 million $m^3$ of sawlogs and veneer logs are produced worldwide. Germany, Sweden, and Finland together make up about half of the EU-28's production in Europe, which totals 154.4 million $m^3$. Wood has a hierarchical structure that extends from the nanoscale to the macroscopic level. The layers of each cell wall in the wood are made up of 40–50% cellulose molecules, which are arranged first in microfibrils and later in macrofibrils (Jones & Brischke, 2017).

The lumen (L) is surrounded by a void, cell wall, which is made up of a main and a secondary wall. The main wall (0.1–0.2 μm) is made up of proteins, pectin, hemicelluloses, cellulose, and lignin, and does not have a specific layout. The secondary wall consists of three oriented cellulose layers, designated S1, S2, and S3. The microfibrils in the S1 layer (0.2–0.3 μm) and S3 layer (0.1 μm) are angled between 50 and 70° with respect to the fiber's axis. The microfibril angle in the S2 layer (1–5 μm), which makes up 90% of the cell wall's weight, ranges from 10 to 30°. The fiber's strength and modulus increase with a smaller angle, while the elongation at break decreases. The middle lamella (M) is made up of lignin and polysaccharides and divides two adjacent cells. It is approximately 0.2 and 1 μm thick. The hollow, cylindrical cells are oriented parallel to the trunk at the mesoscopic level. From the middle of the trunk to the bark are radial sheets of horizontal cells rich in lignin. Because of their darker color, they are more visible in a transversal incision of the trunk. Anisotropy in wood is caused by the orientation of the rays and the cells (Bledzki et al., 2008). The primary mechanical characteristics of several wood species are listed in Table 1.1.

Despite all of its benefits, wood's durability limits its application as a material. Hardwoods are generally more resilient than softwoods, and softwoods have a limited lifespan. Nonetheless, both types of wood provide great food for xylophagous insects and fungi (Dhiman et al., 2024). As a result, it is common practice to

**TABLE 1.1**
**Selective Wood Species' Mechanical Characteristics at a 12% Water Content.**

| Wood species | Fiber length (mm) | Density (ton $m^{-3}$) | Strength at break (MPa) | | | Flexural modulus (GPa) | Reference |
|---|---|---|---|---|---|---|---|
| | | | Flexural | Tensile | Compression | | |
| Mahogany | | | | | | | (Movahedi-Rad et al., 2024; Oladele et al., 2013) |
| Poplar | 1.1 | 0.3–0.5 | 65–85 | 80–100 axial 2 ⊥ | 30–40 axial 7.5–10 ⊥ | 9–11 | |
| Okoume | | | | | | | |
| Azobe | – | 1.1 | 227 | 150–200 axial 5 ⊥ | 90–100 axial >20 ⊥ | – | (van de Kuilen & Blass, 2005) |
| Teck | 0.7–1.2 | 0.3–0.8 | 100–170 | 120–150 axial 4 ⊥ | 50–80 axial 18–20 ⊥ | 11.5–15 | (Jones & Brischke, 2017) |
| Hard oak | | | | | | | |
| Soft oak | 1.1–1.2 | 0.3–0.95 | 75–130 | 100–120 axial 3 ⊥ | 40–60 axial 12–15 ⊥ | 9–12.5 | (Smejda-Krzewicka et al., 2024) |
| Beech | | | | | | | (Chabert et al., 2024) |
| Chestnut | | | | | | | (Wu & Liao, 2013) |
| Maple | | | | | | | (Hanana et al., 2018) |
| Pines | 4 | 0.3–0.85 | 65–85 | 100–120 axial 1.8 ⊥ | 40–50 axial 7.5–8 ⊥ | 9–11 | (Jalali et al., 2023) |
| Red cedar | 3–7 | 0.3–0.9 | 50–70 | 90–100 axial 1.2 ⊥ | 35–45 axial 6–8 ⊥ | 8–10 | (Ülker & Hiziroglu, 2020) |
| Fir | | | | | | | (Jones & Brischke, 2017) |
| Spruce | | | | | | | (Gößwald et al., 2021) |

impregnate wood with substances that act as insecticides and antifungals. The most widely used are the soluble salts of arsenic, chromium, and copper (CCA). Wood treatment chemicals are currently subject to regulations in Europe as a result of the Biocides 98/8/CE ordinance, which aims to prevent serious harm to people, animals, or the environment from these chemicals. New industrial techniques for wood preservation have been created as a result of extensive laboratory and pilot plant research in response to these ecological issues.

### 1.2.3 Plant Fibers

The characteristics of plant fibers vary depending on the species, age, and growth location. Based on where they are located within the plant, they can be divided into different groups (see Figure 1.2). Cotton is the fiber that is produced worldwide by a wide margin, over 30 times as much as jute. It is mostly used in industrial textile applications, which account for the majority of its utilization. As a result, in the manufacture of biomaterials, a few other fibers are more significant than cotton.

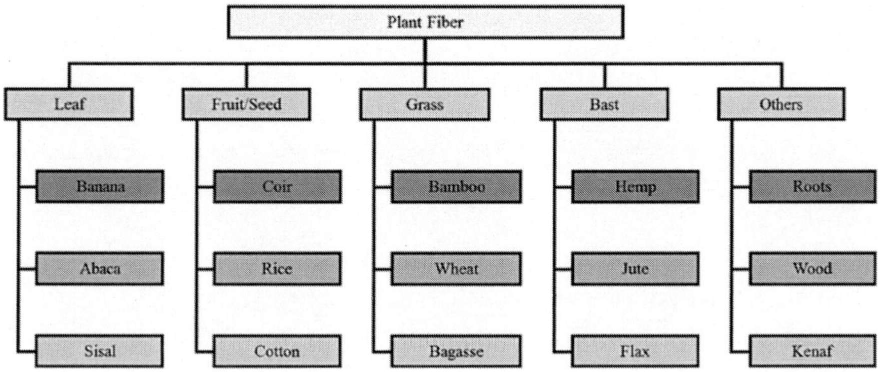


**FIGURE 1.2** Plant fibers are categorized based on where they are located within the plant.

It is interesting to note that developing nations, particularly those in Asia, provide the majority of the natural fibers used in commerce (apart from wood) (Montanari et al., 2021). The source determines the fiber's properties. The primary attributes of commercially available vegetable fibers are listed in Table 1.2. The sections that follow provide a brief explanation of each type of fiber and its use.

#### 1.2.3.1 Cotton

*Gossypium* spp. is the plant that produces cotton, a soft fiber that develops around the plant's seeds. The American species *Gossypium hirsutum*, as well as *Gossypium barbadense*, account for nearly all commercial cotton farmed today in the world. Cotton is the most commonly used natural fiber fabric in clothes because it is frequently spun into thread and employed to create fabrics. A wood cell's secondary walls and cotton fibers have comparable structural features. Since there is no intermediate lamella holding the fibers together, cotton is almost entirely composed of cellulose and contains just 98% lignin. Since it lubricates the fibers' interaction with the machinery and reduces fiber breakage, the natural wax on the fibers' surface makes threading much easier. Longer cotton fibers are employed in textile applications, and shorter cotton fibers, or linters, are used to produce high-grade cellulose for use in biomaterials. For high-purity cellulose chemical transformation, cotton is employed (Cai et al., 2024; Verma & Kumar, 2022).

#### 1.2.3.2 Flax Fiber or Linen

In modern usage, the term "linen" refers to a group of intertwined fabrics used for residential settings. The flax plant fibers, *L. usitatissimum*, were used almost exclusively in the manufacturing of linens. Flax is now a rare, expensive, and prestige fiber that is only produced in small amounts. The characteristic nodes of flax fibers, which give the cloth its flexibility and texture, can be used to identify them (Ma et al., 2023). The irregular polygonal forms that make up the fiber's cross-section add to the fabric's rough feel. Line, when properly prepared, can absorb and release water quickly. It can absorb 20% more moisture without getting wet. There are many applications

**TABLE 1.2**
**The Most Prevalent Plant Fibers' Characteristics.**

| | Plant Fibers | | | | | | |
|---|---|---|---|---|---|---|---|
| **Characteristics** | **Coconut plant** | **Ramie** | **Sisal** | **Jute** | **Hemp** | **Cotton** | **Linen** |
| **Source** | Coconut | Ramie plant | Agave sisalana plant | Jute plant | Hemp Plant | Cotton plant | Flax plant |
| **Density (g cm$^{-3}$** | 1.15 | 1.56 | 1.45 | 1.44 | 1.07 | 1.5–1.6 | 1.54 |
| **Elongation at break (%)** | 15–40 | 1.2–3.8 | 3–7 | 1.5–1.8 | 1.6 | 7–8 | 1–4 |
| **Diameter (µm)** | 12–24 | 16–126 | 7–47 | 25–200 | 10–51 | 10–20 | 5–76 |
| **Length (mm)** | 0.3–1 | 40–250 | 0.8–8 | | 5–55 | 10–55 | 4–77 |
| **Microfibril angle (°)** | 45 | 7.7 | 20 | 8 | 6.2 | | 10 |
| **Modulus of elasticity (GPa)** | 4–6 | 60–130 | 9–21 | 25–30 | 35 | May-13 | 12–85 |
| **Cellulose content (%)** | 43 | 83 | 67–78 | 61–71 | 78 | 87–99 | 64–71 |
| **Shape factor (L/D)** | 35 | 3500 | 100 | 110 | 960 | 2000 | 1700 |
| **Strength at break (MPa)** | 130–175 | 400–1000 | 350–700 | 390–770 | 390 | 290–600 | 600–2000 |
| **Fiber type** | Stiff and rough | Strong, durable, and stiff | Stiff and rough | Smooth and flexible | Strong and durable | Soft and fluffy | Soft and flexible |
| **Thermal stability** | High | Low | Moderate | Low | Moderate | Moderate | Low |
| **Application** | Coarse filling material and upholstery | Industrial sewing thread, packing materials, and fishing nets | Insulation, ropes, and mats | Packaging and textile | Biocomposites and textiles | Textiles and medical supplies | Curtains, tablecloths, pillows, rugs, rope, and medical supplies |

for linen, including fabrics for beds and bathrooms, upholstery, wall coverings, and oil painting supports, as well as industrial goods such as luggage, canvases, and sewing thread. Because of its archival integrity, strength, and longevity, linen is preferred over cotton (Elfaleh et al., 2023).

#### 1.2.3.3 Jute

Jute is a vegetable fiber that is long, smooth, and shiny, which may be spun into sturdy, coarse threads. Plants belonging to the genus Corchorus are used to make it. In terms of quantity produced and range of applications, jute is the second most versatile and affordable natural fiber after cotton, and it takes 4–5 months to grow. Jute hurd, sometimes referred to as the parenchyma or interior woody core of the jute stem, has good potential to combat deforestation since its mechanical qualities are a hybrid of those of wood and textile fiber. Beyond its use in agriculture, it can be made into coarse cloth, which is then used to wrap sacks and bales. Additionally, the fibers are woven into area rugs, carpets, curtains, chair covers, and linoleum backings. While synthetic materials are taking the place of jute in many of these applications, jute's biodegradable nature makes some synthetic alternatives inappropriate. Young tree containers that allow for straight planting without the need to uproot the tree, and geotextiles—a finely knit fabric composed of natural fibers—for controlling soil erosion, protecting seeds, controlling weeds, and numerous other agricultural and landscape applications are a few examples of these applications. Biodegradable jute geotextiles can be left to decompose, which will increase soil fertility (Agumba et al., 2024).

#### 1.2.3.4 Hemp

The standard term for *C. sativa* that is grown for non-drug industrial purposes is hemp. Canada and the European Union both grant licenses for the growing of hemp. Strong fibers are produced from hemp, which grows swiftly. Because of its comparable etymologies, hemp was once frequently employed to make ropes, carpets, and canvas. Compared to linen, the end fiber is more lignified, less regular, and flatter. Hemp is mostly used as a filler in plastic–natural fiber composites in Europe, primarily for the automobile sector. A stronger mechanical attachment using the plastic matrix is represented by the microscopic protuberances on the fiber's surface. Another modern use is the creation of insulating carpets for homes (Zamboulis et al., 2023).

#### 1.2.3.5 Sisal

The *Agave sisalana*, which produces sisal, is used to make rope because of its tough fiber. Sisal plants are unidentified hybrid animals that are sterile. They do not really grow in Yucatan, Mexico, even though they came from the port of Sisal there. Instead, the henequen (*Agave fourcroydes*) is currently grown there. Sisal plants are made up of a 1.5- to 2-m-tall rosette of sword-shaped leaves. The sisal plant usually yields 200–250 economically useful leaves throughout its 7- to 10-year lifespan. There are about 1,000 packs of fiber in each leaf. The fiber component—which makes up just 4% of plant weight—is extracted by decortication. Because of its strength, resilience, stretchability, affinity for specific dyes, and tolerance to saltwater deterioration, sisal is prized for use in cordage. Three sisal grades are employed by the industry: higher-grade fiber is treated and used by the carpet industry; lower-grade fiber, due to its high cellulose and hemicellulose content, is processed by the paper industry; and medium-grade fiber is used in the cordage industry to make ropes. Products constructed of resonant sisal, such as wall tiles and furniture, are developing quickly. The range has recently been extended to include interior auto parts. Recently, sisal has been used in the automotive industry as an environmentally acceptable component and as a strengthening agent in place of glass fiber and asbestos (Lei & Feng, 2020).

#### 1.2.3.6 Coconut Fibers

Growing throughout the tropical world, the coconut palm (*C. nucifera*) is used for ornamentation as well as for a variety of non-culinary and culinary purposes. Almost every portion of the coconut palm has some human use. Fiber from coconut husk, known as coir, is used to make stuffing for brushes, ropes, rugs, and mats. It is also often used to make potting compost in gardening (Ali et al., 2022).

## 1.3 BIOMATERIALS: ISOLATED AND TRANSFORMED BIOPOLYMERS

As amazing chemical reactors, plants are capable of creating incredibly multifaceted macromolecules. These substances, found in the cell walls, act as the plant's energy stores (e.g., starch) or have specialized roles (e.g., proteins). Numerous studies have been conducted and are still being carried out on these biopolymers because the majority of them can be used to create industrial biomaterials. Other researchers have also aimed at the cutting-edge and sustainable development of the bio-based additive manufacturing processes (Agrawal et al., 2024; Atal et al., 2023; Bhadouria et al., 2022; Ranjan et al., 2024a, 2024b; Ranjan & Tripathi, 2024; Tripathi & Sharma, 2024; Satishkumar et al., 2024a, 2024b; Jafar et al., 2024; Singh et al., 2024a; Sharma et al., 2024; Ratsimba et al., 2021; Neveu et al., 2020a, 2020b, 2020c). Figure 1.3 discusses a summary of the most significant bio-based polymers (Verma et al., 2024c). The development of viable applications for these biopolymers requires the development of extraction technologies. One of the easiest processes is starch isolation, which just needs a few basic physical techniques for extraction and purification (the most popular being wet milling). On the other hand, in order to release the cellulose fibrils from the cell wall, chemical treatment is necessary to break the LCC covalent bonds. This can be accomplished using NaOH and $Na_2SO_4$ or solutions of sulfuric acid and hydrogen sulfite. Through the use of gentle mechanical action, these heat treatments turn biomass chips into fibers (Verma et al., 2025). Alkali and oxidizing agents are employed in a multi-stage refining technique to remove residual lignin from the raw pulp in order to further purify it. Ultimately, pentanes and oligosaccharides are eliminated by extractions using either hot or cold alkali. Finally, 99% pure cellulose may be produced. The structure of the separate biopolymers determines their properties. In this chapter, we have restricted our discussion to the primary biopolymers that are being utilized in commercial goods (Isikgor & Becer, 2015).

### 1.3.1 Cellulose

The cellulose molecule's ribbon-like shape facilitates its arrangement into angled packs of roughly 50–100 molecules. Crystallites form when the organization is

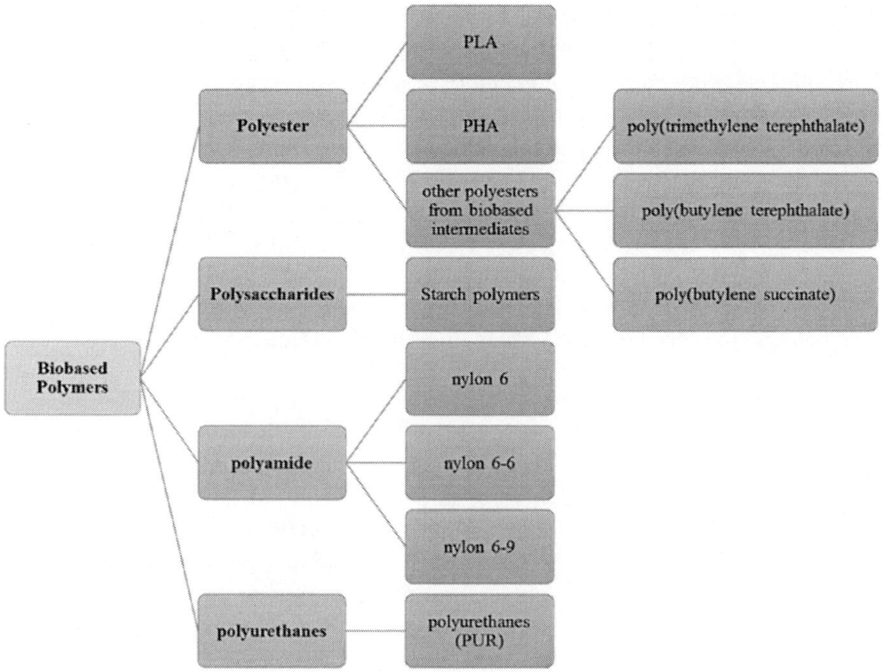


**FIGURE 1.3** A summary of the most significant bio-based polymers.

perfectly regular. In nature, most species (e.g., wood and cotton) have a crystallinity rate of about 50%. Thus, there are crystalline and amorphous regions in the same molecule. Even when cellulose is chemically pulped from wood, its fibrous structure is preserved. In fact, pulping dissolves and assaults the middle lamella or LCC. Consequently, the cell walls—primarily made of cellulose fibers—are recovered without any significant chemical deterioration. Common bulk materials, like paper or absorbent cushions, benefit from having a fiber structure like this (the so-called non-wovens). High-value applications such as hydrogels, chromatography stationary phases, and medicinal formulations also employ pure celluloses (Singh et al., 2025). Following chemical treatment, cellulose exhibits a significant boost in potential. First, we look at one specific instance of cellulose molecule modification: regeneration. It is possible to distinguish between the two forms of regeneration: chemically modified and unmodified (Jing et al., 2024).

#### 1.3.1.1 Chemical Modification for Cellulose Regeneration

$CS_2$ and cellulose react in an alkaline solution during viscose processes to produce a viscous cellulose xanthate solution. Before being extruded to create continuous sheets, threads, or individual beads, the resultant solution is filtered to remove solid particles. Cellulose xanthate is hydrolyzed and transformed again into cellulose when it is submerged in a water solution with sulfuric acid and additional additions. A continuous thread suitable for textile applications is produced when the extrusion profile is shaped like a small cylinder. This fabric is dazzling and has good dyeing qualities;

it is known by its industrial name, Rayon®, or "artificial silk." Variations exist in the viscose process. Special high-value products, such as modal–polynosic fibers, can be obtained depending on the cellulose quality and makeup of the regenerating bath. Cellophane® is the resultant product if the extrusion profile is a sheet. Once more, the hydrolyzing solution is supplemented with certain chemicals to achieve the transparency and plastic-like quality of the sheet. This product has a low shearing resistance yet has a high tensile strength, which makes it a great film for conditioning and packing cookies or sweets. Due to environmental concerns, the viscose process has been gradually (although not completely) abandoned because $CS_2$ is poisonous and easily causes an explosion. Nonetheless, it continues to be significant historically in the field of cellulose chemistry (Magalhães et al., 2023).

#### 1.3.1.2 Cellulose Regeneration in the Absence of Chemical Alteration

Solvents are used to make cellulose solutions, which may be extruded and filtered in the procedures and goods outlined in this section. Precipitation happens in place of chemical regeneration because there is no derivatization of the hydroxyl groups in cellulose. Two of the oldest cellulose solvents are $Cu(NH_3)_4(H_2O)_{22}$ and $C_4H_{12}CuN_{4-2}$; only the latter is utilized on an industrial scale to produce cupro fiber. Subsequently, LiCl systems with an aprotic polar solvent and N-methylmorpholine-N-oxide (NMMO) in water were created as alternative cellulose solvents. However, only the first one has been successful in producing lyocell fiber. Ionic solutions with the ability to dissolve cellulose have been created more recently. 1-N-butyl-3-methylimidazolium chloride (BMIMCl) is the most well-known (Verma et al., 2024a). However, there have not been any noteworthy reports of industrial advancements using this technology to fabricate yarn up to this point. Some of these solvents have not found practical use because of the poor mechanical qualities of the fiber, as well as challenges with solvent recycling and process parameter control (Carolin et al., 2023).

#### 1.3.1.3 Microcrystalline Cellulose

The goal of controlled acid treatment is to break down molecular bonds in the amorphous cellulose zones in order to produce microcrystalline cellulose (MCC). $H_2SO_4$ or HCl is typically applied to native or regenerated cellulose for 15 minutes at 110°C. This results in the formation of thixotropy-displaying colloidal gels. Because of its binding and disintegrating characteristics, MCC is utilized in the production of compressed medicinal tablets. MCC is used in the ice cream industry to increase smoothness by preventing the production of large ice crystals. MCC is utilized in toothpaste and water-based paints as a rheological modifier. MCC is also utilized in chemistry labs as a chromatography (column, thin layer, etc.) support (Trache et al., 2016).

#### 1.3.1.4 Nanocellulose

Because of its hierarchical structure, cellulose's amorphous portion can be dissolved mechanically, chemically, or enzymatically to extract nanoparticles. These nanoparticles are derived from a variety of natural resources and are known as micro-fibrillated cellulose (MFC) or cellulose nanocrystals. Individual microfibrils are defibrillated by mechanical shearing; this is followed by a treatment of high-acid

**TABLE 1.3**
**Cellulose-Based Nanoparticles.**

| Nanoparticle type | Production method | Size range | Crystallinity | Aspect ratio | Key properties | Reference |
|---|---|---|---|---|---|---|
| Cellulose nanocrystals (CNCs) | Acid hydrolysis of cellulose | Length: 50–500 nm and width: 2–20 nm | High (54–88%) | High (10–70) | High aspect ratio, high crystallinity, high strength, and transparent films | (Calvino et al., 2020; Singh et al., 2024b) |
| Cellulose nanofibrils (CNFs) | Mechanical treatment of cellulose fibers | Length: >10 μm and width: 2–20 μm | Moderate (35–50%) | High (100–1000) | High aspect ratio, moderate crystallinity, high strength, and transparent films | (Hossain & Tajvidi, 2024) |
| TEMPO-oxidized cellulose nanofibers (TOCNFs) | TEMPO-mediated oxidation of cellulose followed by mechanical treatment | Length: 500–2000 nm and width: 3–5 nm | Moderate (45–55%) | High (100–700) | High aspect ratio, moderate crystallinity, high surface charge, and high viscosity | (Tang et al., 2024) |
| Bacterial cellulose (BC) | Biosynthesis by bacteria | Length: 201–3000 nm and width: 11–76 nm | High (60–90%) | Moderate (3–40) | high tensile strength, high crystallinity, high elastic modulus, and high water retaining capacity | (Mbituyimana et al., 2021) |

hydrolysis that cuts fibrils longitudinally, which modifies the aspect ratio and size of the cellulose-based nanoparticles (see Table 1.3). Due to their various benefits, these nanoparticles are currently utilized to create bio-nanocomposites. Their hydrophilic nature, which is linked to their natural origin, is their primary disadvantage, which restricts their application in traditional thermoplastic and/or thermoset resins (Khalid et al., 2021; Singh et al., 2023).

#### 1.3.1.5 Bacterial Cellulose

As a byproduct of various bacteria, bacterial cellulose (BC) is yielded in sufficient quantities by *Gluconacetobacter* to be of commercial importance. It is nearly pure cellulose, free of lignin, and other extraneous substances like hemicelluloses and pectins, despite having a gel-like appearance and a solid content of less than 1% (Dhiman et al., 2024). The ability to develop BC to its preferred structure and shape, in addition to its purity, is its second primary advantage. It offers extraordinary physical qualities that are analogous to nanocellulose particles made from plant cellulose (Gregory et al., 2021).

## 1.3.2 Cellulose Derivatives

### 1.3.2.1 Cellulose Inorganic Esters

Cellulose nitrates represent the most important inorganic esters of cellulose. According to the degree of substitution (DS), or the average number of hydroxyl groups modified in a unitary glucose unit, the cellulose nitrates range from a resistant but inflammable polymer in film and photographic applications (celluloid grade, DS = 1.9) to a polymer for lacquers (DS = 2.05–2.35), and to a powerful explosive that burns spontaneously in air (gun cotton, DS = 2.7). In the process of producing vegetable parchment for the bakery business, another cellulose inorganic ester, cellulose sulfate, is formed. A continuous paper sheet is submerged in a concentrated $H_2SO_4$ bath (15–75%) kept at a low temperature for a few seconds as part of this procedure. The cellulose fibers begin to hydrolyze in an acidic solution and produce sulfates. Afterward, the sheet is put through a number of rinse baths to remove the $H_2SO_4$ and hydrolyze the sulfates. The inner mat of cellulose fibers is shielded from the outside by a gelatinized cellulose continuous matrix that forms on both sides of the paper (Carolin et al., 2023).

### 1.3.2.2 Cellulose Ethers

The commonly synthesized cellulose derivatives are cellulose ethers. Viscose solutions can be produced by cellulose, either with organic solvents or water-soluble, depending on the substituent type used to replace the –OH function. Viscosity in water increases to 50 Pa/s when added at quantities of about 1–2%. As a result, they are useful alternatives to xanthan gum and are employed as rheological modifiers in a variety of compositions. The raw material may consist of wood pulp or cotton linters. The latter, with a greater degree of polymerization, are employed in the synthesis of highly viscous cellulose ethers. Carboxymethyl cellulose (CMC) is an odorless, white substance that is safe to use. The more widely used sodium salt is made when alkali cellulose reacts with $CH_2ClCO_2Na$ rather than $C_2H_3ClO_2$. The degree of polymerization and the DS both affect how sodium CMC's rheological characteristics change. Commercial sodium CMC is soluble in water and has DS values ranging from 0.1 to 0.9. Sodium CMC is commonly used as a component in numerous materials rather than as a biomaterial on its own. These include cement (e.g., retarder agent), paper, and textile additives (for ironing, printing, and anti-shrinking) (Lai et al., 2023; Lim et al., 2024).

## 1.3.3 Starch

The primary source of energy in outstanding vegetative plants is starch. It is found in significant amounts in manioc, wheat, potatoes, and corn. A homopolymer of D-anhydroglucopyranose units, starch can have a composition of up to 99% composition. There are two distinct forms of amylose: amylopectin and amylose. Seventy to eighty percent of the polymer is branched. The presence of native starch is manifested as partly crystalline (25–40%) granules, with a diameter of 100 μm, exhibiting a complicated structure that has been the focus of thousands of research publications. At normal temperatures, starch granules are insoluble in water. Starch reversibly absorbs water at 50–10°C, which also lowers hydrogen bonding (Wang et al., 2023).

### 1.3.4 Starch Derivatives

In the polymer chain, carbonyl and carboxyl groups are formed as a result of starch oxidation. Additionally, starch undergoes depolymerization and turns yellow. Beneficially, oxidized starches have lower viscosities and gelatinization temperatures than native starches. They are utilized as coatings and adhesives in the paper industry. Non-ionic starch ethers are employed as coating agents in the paper industry and food business to prevent water from escaping frozen products. In the paper industry, cationic starch ethers are used as a flocculant to aid in the selective separation of negatively charged particles and to improve the cohesiveness and stiffness of cellulose fibers (Liang et al., 2024). Thermoplastic substances known as starch esters have characteristics that are comparable to cellulose esters but are not quite the same. Specifically, starch acetate (DS $< 0.2$) is applied as a food or detergent ingredient and a coating agent in the paper sector (Falua et al., 2022).

### 1.3.5 Chitin and Chitosan

In nature, chitin, a polysaccharide made of $C_8H_{15}NO_6$, is a hard, semi-transparent biomaterial. It is the primary material found in the exoskeletons of crabs, lobsters, and shrimp. Aside from cephalopods like octopuses and squids, chitin can also be found in mushrooms, insects, and beetles. However, crustaceans are by far the primary industrial source of chitin. Chitin can be considered structurally similar to cellulose, except that in each monomer, one hydroxyl group is replaced by an acetylamine group. Because of this, neighboring polymers are able to form more hydrogen bonds, which strengthens the polymer (Lv et al., 2023). Considerable interest exists in chitin as a surgical thread material because of its toughness and robust characteristics. Furthermore, because of its biodegradability, it fades over time as the wound heals. Chitin also possesses certain peculiar qualities that hasten human wound healing. In the industrial setting, chitin is often employed as an ingredient to stabilize and thicken medications and meals. It is also used by the textile industry as it can be formed into fibers; in particular, socks are made from chitin textiles because they are said to have inherent antibacterial and anti-odor properties. Chitin can, in addition, be used in various procedures to size and reinforce paper (Al-Rooqi et al., 2022). Commercial chitosan is produced by deacetylating chitin. $C_8H_{15}NO_6$ and $C_{12}H_{24}N_2O_9$ are arbitrarily distributed in this linear polysaccharide. Commercial chitosans exhibit between 10% and 100% deacetylation. The pKa value of the amino group in chitosan is approximately 1.5. As a result, the charge density of chitosan is dependent on both the degree of deacetylation and pH. It dissolves in acidic-to-neutral liquids and has a positive charge (Kumar et al., 2022). Put differently, chitosan attaches itself easily to negatively charged surfaces like mucosal membranes. It is biocompatible and biodegradable, and can improve the transport of polar medications across epithelial surfaces. Chitosan has refined characteristics and is available for use in biomedical applications. Because of its flocculating qualities, chitosan is employed in the filtration stage of water treatment engineering. It is capable of eliminating oils, heavy minerals, and phosphorus from water. It also serves the same purpose in clarifying beer and wine (in place of egg albumin) (Nasirov et al., 2023).

### 1.3.6 Proteins

#### 1.3.6.1 Oilseed Proteins

Ford's study on using soybean proteins to make lighter cars at the start of the century marked the beginning of major advancements in the field of protein-based materials. Other protein sources, including sunflower or rapeseed, can be synthesized using the same kind of procedures and materials. Any non-covalent alteration in the structure of oilseed proteins, known as denaturation, is essentially what controls how they are textured. This loss of structure may change the molecules' secondary, tertiary, or quaternary structure and is akin to the fusing of crystallites in semi-crystalline polymers (Singh et al., 2024c). At 0–190°C coupled with a moisture level of 0 to about 30%, polypeptides exhibit glass transition behavior like conventional amorphous polymers, provided that thermal denaturation is attained. This phenomenon is highly dependent on moisture concentration. The next step involves shaping proteins via injection molding, hot pressing, or extrusion. For instance, at Iowa State University, Jay-Lin Jane's team spent the 1990s characterizing a wide range of compounds derived from soybean proteins. The Agro-Industrial Chemistry Laboratory characterized a wide range of materials, including sunflower proteins (Garbacz et al., 2023).

#### 1.3.6.2 Animal Proteins

Additionally, a wide variety of proteins are produced by the animal kingdom that can be utilized for a variety of non-food purposes, particularly in the synthesis of materials. A family of proteins known as casein is frequently present in milk from mammals. Reacting with formaldehyde, one of the first polymers to be developed, was used to create Galalith. These days, biodegradable, edible, and water-soluble packaging can be made using innovative methods. The procedure uses sodium caseinates, which are extracted from cow milk. It is a traditional two-step technique that requires compounding transparent pellets that have been plasticized with glycerol and blowing film. In recent decades, other protein types, such as keratin, have also been investigated to create biodegradable biomaterials; examples of these include feathers and blood (Rahman et al., 2024; Talabi et al., 2024).

## 1.4 NATURAL FIBER COMPOSITES

Fibers derived from plants, such as wood, can serve as a suitable substitute for carbon and glass in composites reinforced with a thermoplastic matrix. Vegetable fibers have both financial and environmental benefits. They are less harsh on the processing machinery and are affordable. Because of their low density, the entire composite material is made lighter, and their influence on the environment is minimal (see Figure 1.4).

Wood fibers are the recommended filler for thermoplastic matrices in 75% of situations, which may be due to their high availability. Wood–plastic composite, or WPC, is the term used globally to refer to the final composite material. During the last 10 years, the market has expanded, especially in the US, to 700,000 metric tons at an annual growth rate of 11% (Gomez-Caturla et al., 2022).

The most important aspect of composite materials is the bonding of the fiber to the matrix. The compatibility of the filler and matrix has a significant impact on the mechanical properties of the WPC. A molecule that can create a link between matrix

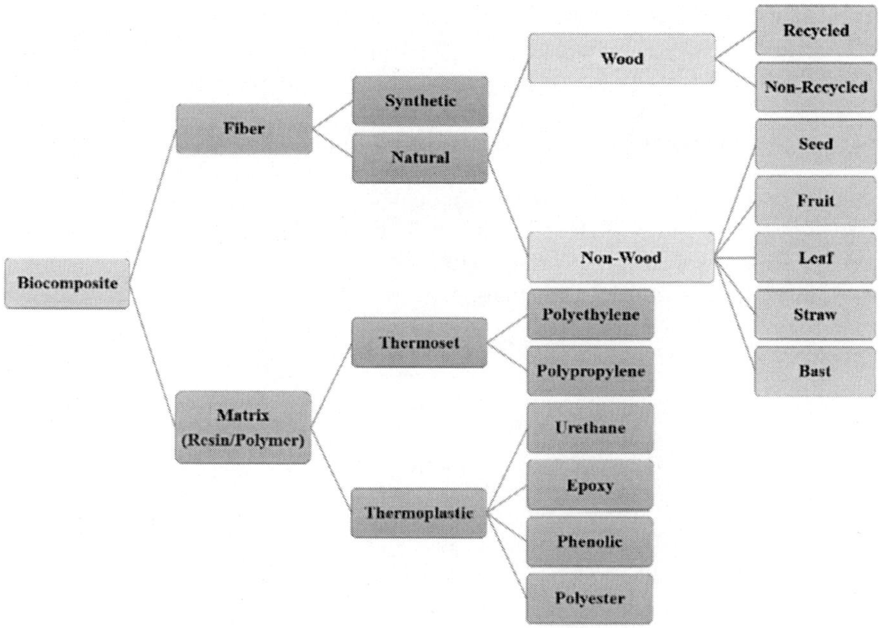


**FIGURE 1.4** Biocomposite structure.

and filler is the only feasible option to increase adhesion. There are three types of bonding at play: non-polar interactions, hydrogen bonds, and covalent bonds. Each of these helps to lessen the innate incompatibility that exists between the hydrophilic fibers and the hydrophobic matrix (Friedrich, 2024).

The beneficial effects of poly (methylene (phenyl isocyanate)) and amino silanes have been proven through some studies. Be aware that the majority of WPCs are not composed of biodegradable materials. Even with natural fibers added, the synthetic matrix is still a fossil resource with little biodegradability. However, these materials also exhibit certain benefits. Indeed, in order to preserve its mechanical qualities during recycling, polypropylene needs to be strengthened with a filler. Thus, WPCs are an ideal way to increase the lifespan of olefins. In terms of WPC recycling, it has been shown that only polyvinyl chloride–wood composites retain their qualities when thermomechanical recycling is carried out. Additionally, the use of wood fibers in WPCs can help reduce the overall carbon footprint of the material compared to traditional plastics. This makes WPCs a more sustainable option for various applications in construction and manufacturing.

## 1.5 CONCLUSION

This chapter provides a technical feasibility perspective on the definition of the biomaterials industry. Nevertheless, it can only be completely understood by a thorough understanding of a wide range of fields, such as ecology, engineering, chemistry, biochemistry, economics, sociology, and medicine. Furthermore, the advancement of these biomaterials will be influenced by political trends and decisions in addition to technical factors. It is anticipated that the promotion of environmentally friendly products will be primarily driven by the will of industrialized societies to advance.

## ACKNOWLEDGMENTS

The authors acknowledge the financial assistance from the Council of Science & Technology, Uttar Pradesh (UPCST) Project (Project ID: 2707). The authors also thank the in-house seed grant project (File No. FY-2023–24/SMG/002) funded by Shri Ramswaroop Memorial University, Lucknow, India.

## REFERENCES

Agrawal, A. P., Ali, S., Tripathi, N. M., Kumar, R., & Srivastava, P. (2024). Optimizing structural steel spur gear design for reduced stress concentrations in Industry 4.0 using finite element analysis. *Recent Patents on Mechanical Engineering*, *17*(1), e050724231667. https://doi.org/10.2174/0122127976309163240620080241

Agumba, D. O., Park, G., Woong Kim, J., & Kim, J. (2024). Biobased natural fiber-reinforced composites derived from lignin-based resin and mercerized jute fibers. *Materials Letters*, *360*, 136055. https://doi.org/10.1016/j.matlet.2024.136055

Ali, B., Hawreen, A., Ben Kahla, N., Talha Amir, M., Azab, M., & Raza, A. (2022). A critical review on the utilization of coir (coconut fiber) in cementitious materials. *Construction and Building Materials*, *351*, 128957. https://doi.org/10.1016/j.conbuildmat.2022.128957

Al-Rooqi, M. M., Hassan, M. M., Moussa, Z., Obaid, R. J., Suman, N. H., Wagner, M. H., Natto, S. S. A., & Ahmed, S. A. (2022). Advancement of chitin and chitosan as promising biomaterials. *Journal of Saudi Chemical Society*, *26*(6), 101561. https://doi.org/10.1016/j.jscs.2022.101561

Atal, S., Dwivedi, D., & Tripathi, N. M. (2023). Corrosion under insulation (CUI) in oil and gas industries. In *Functional Materials for the Oil and Gas Industry* (pp. 127–138). CRC Press. https://doi.org/10.1201/9781003242550-8

Bhadouria, A. S., Kumar, A., Raj, D., Verma, A., Singh, S., Tripathi, P., Kumar, Y., Sinha, A. S. K., Tripathi, N. M., Yeneneh, A. M., & Dwivedi, D. (2022). Corrosion mitigation in oil reservoirs during CO injection. In *Nanotechnology for CO2 Utilization in Oilfield Applications* (pp. 127–145). Elsevier. https://doi.org/10.1016/B978-0-323-90540-4.00014-4

Bledzki, A., Jaszkiewicz, A., Murr, M., Sperber, V. E., Lützendorf, R., & ReußMann, T. (2008). Processing techniques for natural- and wood-fibre composites. In *Properties and Performance of Natural-fibre Composites* (pp. 163–192). https://doi.org/10.1533/9781845694593.1.163

Cai, C., Wang, Z., Ma, L., Xu, Z., Yu, J., & Li, F. (2024). Cotton stalk valorization towards bio-based materials, chemicals, and biofuels: a review. *Renewable and Sustainable Energy Reviews*, *202*, 114651. https://doi.org/10.1016/j.rser.2024.114651

Calvino, C., Macke, N., Kato, R., & Rowan, S. J. (2020). Development, processing and applications of bio-sourced cellulose nanocrystal composites. *Progress in Polymer Science*, *103*, 101221. https://doi.org/10.1016/j.progpolymsci.2020.101221

Carolin C, F., Kamalesh, T., Kumar, P. S., Hemavathy, R. V., & Rangasamy, G. (2023). A critical review on sustainable cellulose materials and its multifaceted applications. *Industrial Crops and Products*, *203*, 117221. https://doi.org/10.1016/j.indcrop.2023.117221

Chabert, A., Fredon, E., Florez, D., Durand, A., & Rémond, R. (2024). Mechanical properties of beech wood treated with malic acid-based polyester. *Holzforschung*. https://doi.org/10.1515/hf-2024-0008

Chand, N., & Fahim, M. (2021). 1—Natural fibers and their composites. In N. Chand & M. Fahim (Eds.), *Tribology of Natural Fiber Polymer Composites, 2nd edn* (pp. 1–59). Woodhead Publishing. https://doi.org/10.1016/B978-0-12-818983-2.00001-3

Dhiman, V., Singh, D., Dhiman, V., Pandey, M., Sharma, A., Pandey, H., & Verma, S. (2024). *Insects as Biotic Stress in Agriculture* (pp. 193–210). https://doi.org/10.1007/978-981-97-4239-4_10

Elfaleh, I., Abbassi, F., Habibi, M., Ahmad, F., Guedri, M., Nasri, M., & Garnier, C. (2023). A comprehensive review of natural fibers and their composites: an eco-friendly alternative to conventional materials. *Results in Engineering*, *19*, 101271. https://doi.org/10.1016/j.rineng.2023.101271

Falua, K. J., Pokharel, A., Babaei-Ghazvini, A., Ai, Y., & Acharya, B. (2022). Valorization of starch to biobased materials: a review. *Polymers*, *14*(11), Article 11. https://doi.org/10.3390/polym14112215

Friedrich, D. (2024). Branding of wood-plastic composites under generic name "biobased-plastic" or specifically as "wood-plastic composite"? A consumer study under compolytics approach. *Journal of Cleaner Production*, *470*, 143276. https://doi.org/10.1016/j.jclepro.2024.143276

Garbacz, K., Wawrzykowski, J., Czelej, M., Czernecki, T., & Waśko, A. (2023). Recent trends in the application of oilseed-derived protein hydrolysates as functional foods. *Foods*, *12*(20), Article 20. https://doi.org/10.3390/foods12203861

Gößwald, J., Barbu, M., Petutschnigg, A., & Tudor, E. (2021). Binderless thermal insulation panels made of spruce bark fibres. *Polymers*, *13*, 1–12. https://doi.org/10.3390/polym13111799

Gomez-Caturla, J., Balart, R., Ivorra-Martinez, J., Garcia-Garcia, D., Dominici, F., Puglia, D., & Torre, L. (2022). Biopolypropylene-based wood plastic composites reinforced with mango peel flour and compatibilized with an environmentally friendly copolymer from itaconic acid. *ACS Applied Polymer Materials*, *4*(6), 4398–4410. https://doi.org/10.1021/acsapm.2c00373

Gregory, D. A., Tripathi, L., Fricker, A. T. R., Asare, E., Orlando, I., Raghavendran, V., & Roy, I. (2021). Bacterial cellulose: a smart biomaterial with diverse applications. *Materials Science and Engineering: R: Reports*, *145*, 100623. https://doi.org/10.1016/j.mser.2021.100623

Hanana, F. E., Désiré, C.-Y., & Rodrigue, D. (2018). Morphology and mechanical properties of maple reinforced LLDPE produced by rotational moulding: effect of fibre content and surface treatment. *Polymers and Polymer Composites*, *26*, 299–307. https://doi.org/10.1177/096739111802600404

Hossain, R., & Tajvidi, M. (2024). Cellulose nanofibrils as a bio-based binder for wood fiber composite insulation panels with enhanced thermo-mechanical properties for structural wall sheathing applications. *Carbohydrate Polymer Technologies and Applications*, *7*, 100429. https://doi.org/10.1016/j.carpta.2024.100429

Isikgor, F. H., & Becer, C. R. (2015). Lignocellulosic biomass: a sustainable platform for the production of bio-based chemicals and polymers. *Polymer Chemistry*, *6*(25), 4497–4559. https://doi.org/10.1039/C5PY00263J

Jafar, M. R., Tripathi, N. M., Yadav, M., & Nasato, D. S. (2024). Additive manufacturing in the age of Industry 4.0 and beyond. In *Advances in Pre-and Post-Additive Manufacturing Processes* (pp. 213–230). CRC Press. https://doi.org/10.1201/9781003428862-11

Jalali, S., Borges, C. da S. P., Carbas, R. J. C., Marques, E. A. de S., Bordado, J. C. M., & da Silva, L. F. M. (2023). Characterization of densified pine wood and a zero-thickness bio-based adhesive for eco-friendly structural applications. *Materials*, *16*(22), Article 22. https://doi.org/10.3390/ma16227147

Jing, L., Shi, T., Chang, Y., Meng, X., He, S., Xu, H., Yang, S., & Liu, J. (2024). Cellulose-based materials in environmental protection: a scientometric and visual analysis review. *Science of the Total Environment*, *929*, 172576. https://doi.org/10.1016/j.scitotenv.2024.172576

Jones, D., & Brischke, C. (Eds.). (2017). 5—Performance of the bio-based materials. In *Performance of Bio-based Building Materials* (pp. 249–333). Woodhead Publishing. https://doi.org/10.1016/B978-0-08-100982-6.00005-7

Karimah, A., Ridho, M. R., Munawar, S. S., Adi, D. S., Ismadi, Damayanti, R., Subiyanto, B., Fatriasari, W., & Fudholi, A. (2021). A review on natural fibers for development of eco-friendly bio-composite: characteristics, and utilizations. *Journal of Materials Research and Technology*, *13*, 2442–2458. https://doi.org/10.1016/j.jmrt.2021.06.014

Khalid, M. Y., Al Rashid, A., Arif, Z. U., Ahmed, W., & Arshad, H. (2021). Recent advances in nanocellulose-based different biomaterials: types, properties, and emerging applications. *Journal of Materials Research and Technology*, *14*, 2601–2623. https://doi.org/10.1016/j.jmrt.2021.07.128

Kumar, P., Verma, S., Rai, S., Prakash, P. S., & Chitara, D. (2022). *Role of Homocysteine Metabolism in Cardiovascular Diseases* (pp. 257–276). https://doi.org/10.1007/978-981-16-6867-8_14

Lai, G., Liu, X., Li, S., Xu, Y., Zheng, Y., Guan, J., Gao, R., Wei, Z., Wang, Z., & Cui, S. (2023). Development of chemical admixtures for green and environmentally friendly concrete: a review. *Journal of Cleaner Production*, *389*, 136116. https://doi.org/10.1016/j.jclepro.2023.136116

Lei, B., & Feng, Y. (2020). Sustainable thermoplastic bio-based materials from sisal fibers. *Journal of Cleaner Production*, *265*, 121631. https://doi.org/10.1016/j.jclepro.2020.121631

Liang, X., Chen, L., McClements, D. J., Peng, X., Xu, Z., Meng, M., & Jin, Z. (2024). Bioactive delivery systems based on starch and its derivatives: assembly and application at different structural levels. *Food Chemistry*, *432*, 137184. https://doi.org/10.1016/j.foodchem.2023.137184

Lim, D.-S., Kim, E.-B., Lee, H.-Y., Kwon, Y.-N., Kim, J.-H., Lee, J.-C., & Myung, S. (2024). Highly miscible bio-based poly(lactic acid)/acetylated cellulose ether ultrafiltration membranes with improved water permeances. *Journal of Membrane Science*, *711*, 123195. https://doi.org/10.1016/j.memsci.2024.123195

Lv, J., Lv, X., Ma, M., Oh, D.-H., Jiang, Z., & Fu, X. (2023). Chitin and chitin-based biomaterials: a review of advances in processing and food applications. *Carbohydrate Polymers*, *299*, 120142. https://doi.org/10.1016/j.carbpol.2022.120142

Ma, H., Guna, V., Raju, T., Murthy, A. N., & Reddy, N. (2023). Converting flax processing waste into value added biocomposites. *Industrial Crops and Products*, *195*, 116434. https://doi.org/10.1016/j.indcrop.2023.116434

Magalhães, S., Fernandes, C., Pedrosa, J. F. S., Alves, L., Medronho, B., Ferreira, P. J. T., & Rasteiro, M. da G. (2023). Eco-friendly methods for extraction and modification of cellulose: an overview. *Polymers*, *15*(14), Article 14. https://doi.org/10.3390/polym15143138

Mbituyimana, B., Liu, L., Ye, W., Ode Boni, B. O., Zhang, K., Chen, J., Thomas, S., Vasilievich, R. V., Shi, Z., & Yang, G. (2021). Bacterial cellulose-based composites for biomedical and cosmetic applications: research progress and existing products. *Carbohydrate Polymers*, *273*, 118565. https://doi.org/10.1016/j.carbpol.2021.118565

Montanari, C., Olsén, P., & Berglund, L. A. (2021). Sustainable wood nanotechnologies for wood composites processed by in-situ polymerization. *Frontiers in Chemistry*, *9*. https://doi.org/10.3389/fchem.2021.682883

Movahedi-Rad, A. V., Ritter, M., Kindler, R. O., Burgert, I., & Panzarasa, G. (2024). Reconstructing poplar wood into a high-performance fiber-reinforced biocomposite. *ACS Sustainable Chemistry & Engineering*, *12*(40), 14694–14703. https://doi.org/10.1021/acssuschemeng.4c04652

Nasirov, F., Abbasov, V., Suleymanova, E., Aslanbeyli, A., Ahmedova, S., & Mammadaliyev, A. (2023). Chitin and Chitosan-biopolymer materials of the 21st century. *PPOR, 24*, 778–824. https://doi.org/10.36719/1726-4685/96/778-824

Neveu, A., Lumay, G., Pillitteri, S., Monsuur, F., Pauly, T., Ribeyre, Q., Francqui, F., Vandewalle, N., & Tripathi, N. M. (2020a). Physical characterization of blends containing mesoporous particles with a focus on electrostatic properties. In *2020 AIChE Virtual Spring Meeting and 16th Global Congress on Process Safety (2020 AIChE Virtual Spring Meeting and 16th Global Congress on Process Safety)*. AIChE. ISBN: 978-0-8169-1113-4.

Neveu, A., Tripathi, N. M., Francqui, F., Lumay, G., & Rigo, O. (2020b). Characterization of powders spreadability in the framework of additive manufacturing processes. *The Proceedings of 2020 Virtual AIChE Annual Meeting (Particle Technology Forum)*, San Francisco, CA, USA. ISBN: 978-0-8169-1114-1.

Neveu, A., Tripathi, N. M., Rigo, O., Francqui, F., & Lumay, G. (2020c). Experimental Investigation of Spreadability of Metal Powders in Recoating Process. *Proceedings—Euro PM2020 Congress and Exhibition (Proceedings—Euro PM2020 Congress and Exhibition)*. European Powder Metallurgy Association (EPMA). ISBN: 978-1-8990-7251-4.

Oladele, I., Omotoyinbo, J. A., & Borisade, M. (2013). Mechanical properties of Mahogany (Swietenia Macrophylla) and Araba (Ceiba Pentandra) dusts reinforced polyester composites. *Leonardo Electronic Journal of Practices and Technologies*, *12*, 1–18.

Pacheco-Torgal, F. (2020). 1—Introduction to biobased materials and biotechnologies for eco-efficient construction. In F. Pacheco-Torgal, V. Ivanov, & D. C. W. Tsang (Eds.), *Bio-Based Materials and Biotechnologies for Eco-Efficient Construction* (pp. 1–16). Woodhead Publishing. https://doi.org/10.1016/B978-0-12-819481-2.00001-5

Rahman, S., Gogoi, J., Dubey, S., & Chowdhury, D. (2024). Animal derived biopolymers for food packaging applications: a review. *International Journal of Biological Macromolecules*, *255*, 128197. https://doi.org/10.1016/j.ijbiomac.2023.128197

Ranjan, R., Singh, D., Pandey, V., Sharma, A., Sinha, A. S. K., & Tripathi, N. M. (2024a). A comprehensive review on the machining process: unconventional as an alternative to conventional machining. *Advances in Pre-and Post-Additive Manufacturing Processes* (pp. 163–195). CRC Press. https://doi.org/10.1201/9781003428862-9

Ranjan, R., & Tripathi, N. M. (2024). Optimizing solar power: engineering efficiency in CdTe/MAsnI3 double absorber solar cell. In *2nd International Conference on Emerging Materials, Smart Manufacturing & Computational Intelligence (ICEMSMCI-2024)*. Punjab: Chitkara University.

Ranjan, R., Verma, S., & Tripathi, N. M. (2024b). The organic HTL in highly efficient germanium-alloyed CsSnI3 perovskite solar cell: a modelling and computational analysis. *Physica Scripta*, *99*(6), 066010. https://doi.org/10.1088/1402-4896/ad4e15

Ratsimba, A., Zerrouki, A., Tessier-Doyen, N., Nait-Ali, B., André, D., Duport, P., Neveu, A., Tripathi, N., Francqui, F., & Delaizir, G. (2021). Densification behaviour and three-dimensional printing of Y2O3 ceramic powder by selective laser sintering. *Ceramics International*, *47*(6), 7465–7474. https://doi.org/10.1016/j.ceramint.2020.11.087

Satishkumar, P., Haldar, B., Tripathi, N. M., Sharma, A., Desai, D. J., Sharma, V. K., Seenivasan., S., & Babbar, A. (2024a). Effect of interlayer temperatures and heat inputs on porosity and hydrogen solubility in wire Arc additive manufactured AA2618 aluminium: a comparative study between pulsed-MIG and CMT methods. In *Advances in Pre-and Post-Additive Manufacturing Processes* (pp. 137–162). CRC Press. https://doi.org/10.1201/9781003428862-8

Satishkumar, P., Kumar, R., Tripathi, N. M., Sharma, A., Desai, D. J., Verma, A., Natarajan., N., & Babbar, A. (2024b). Integrating titanium coating with 3D-printed GFRP panels: an innovative approach to harnessing composite strengths in engineering applications. In *Advances in Pre-and Post-Additive Manufacturing Processes* (pp. 18–38). CRC Press. https://doi.org/10.1201/9781003428862-2

Sharma, A., Babbar, A., Singh, K., Singh, A. K., Tripathi, N. M., & Desai, D. J. K. (2024). Effect of minimum quantity lubrication (MQL) method on machining characteristics for ductile substrates: a future direction. In *Modern Hybrid Machining and Super Finishing Processes* (pp. 35–59). CRC Press. https://doi.org/10.1201/9781003327905-3

Singh, A. K., Shakir, M. H., Yadav, S., Yadav, M., Tripathi, N. M., & Fidan, I. (2024a). Investigations on the mechanical characterization of additively manufactured polymer composite materials fabricated via MEX. In *Advances in Pre-and Post-Additive Manufacturing Processes* (pp. 64–77). CRC Press. https://doi.org/10.1201/9781003428862-4

Singh, D., Sharma, A., Verma, S., Pandey, H., & Pandey, M. (2024b). Impact of nanoparticles on plant physiology, nutrition, and toxicity: a short review. *Next Nanotechnology*, *6*, 100081. https://doi.org/10.1016/j.nxnano.2024.100081

Singh, D., Srivastava, S., Verma, S. K., & Singh, V. (2025). Advances in medicinal plant biotechnology and future directions. In J.-T. Chen (Ed.), *Biotechnology, Multiple Omics, and Precision Breeding in Medicinal Plants, 1st edn* (pp. 258–269). CRC Press. https://doi.org/10.1201/9781003475491-19

Singh, D., Verma, S. K., & Shyam, P. (2024c). Identification and purification of plant secondary metabolite as medicinal raw materials. In *Secondary Metabolites and Biotherapeutics* (pp. 9–38). Elsevier. https://doi.org/10.1016/B978-0-443-16158-2.00003-3

Singh, D., Verma, S., Singh, V., & Perugu, S. (2023). *Green Functional Nanomaterials: Synthesis and Application* (pp. 45–65). https://doi.org/10.1007/978-3-031-31104-8_3

Smejda-Krzewicka, A., Mrozowski, K., & Strzelec, K. (2024). Effect of modified and unmodified oak bark (Quercus Cortex) on the cross-linking process and mechanical, anti-aging, and hydrophobic properties of biocomposites produced from natural rubber (NR). *Materials*, *17*(9), 1968. https://doi.org/10.3390/ma17091968

Talabi, S. I., Ismail, S. O., Akpan, E. I., & Hassen, A. A. (2024). Quest for environmentally sustainable materials: a case for animal-based fillers and fibers in polymeric biocomposites. *Composites Part A: Applied Science and Manufacturing*, *183*, 108216. https://doi.org/10.1016/j.compositesa.2024.108216

Tang, Z., Lin, X., Yu, M., Mondal, A. K., & Wu, H. (2024). Recent advances in TEMPO-oxidized cellulose nanofibers: oxidation mechanism, characterization, properties and applications. *International Journal of Biological Macromolecules*, *259*, 129081. https://doi.org/10.1016/j.ijbiomac.2023.129081

Trache, D., Hussin, M. H., Hui Chuin, C. T., Sabar, S., Fazita, M. R. N., Taiwo, O. F. A., Hassan, T. M., & Haafiz, M. K. M. (2016). Microcrystalline cellulose: isolation, characterization and bio-composites application—a review. *International Journal of Biological Macromolecules*, *93*, 789–804. https://doi.org/10.1016/j.ijbiomac.2016.09.056

Tripathi, N. M., & Sharma, A. (Eds.). (2024). *Advances in Pre-and Post-Additive Manufacturing Processes: Innovations and Applications*. CRC Press. https://doi.org/10.1201/9781003428862

Tyagi, P., Ahmed, S., & Lai, C. W. (2023). Chapter 32—Future prospects of biobased materials. In S. Ahmed & Annu (Eds.), *Advanced Applications of Biobased Materials* (pp. 749–764). Elsevier. https://doi.org/10.1016/B978-0-323-91677-6.00013-1

Ülker, O., & Hiziroglu, S. (2020). Some properties of Eastern Red Cedar (Juniperus Virginiana L.) particleboard panels made using modified starch as BinderSvojstva iverice izrađene od drva crvenog cedra i modificiranog škroba kao veziva. *Drvna Industrija*, *71*, 371–377. https://doi.org/10.5552/drvind.2020.1947

van de Kuilen, J.-W., & Blass, H. (2005). Mechanical properties of azobé (Lophira alata). *Holz Als Roh- Und Werkstoff*, *63*, 1–10. https://doi.org/10.1007/s00107-004-0533-7

Verma, S. K., Khare, R., & Singh, D. (2024a). Current perspectives on neuroendocrine tumors. *hLife*. https://doi.org/10.1016/j.hlife.2024.07.006

Verma, S. K., Pandey, M., Sharma, A., & Singh, D. (2024b). Exploring ayurveda: principles and their application in modern medicine. *Bulletin of the National Research Centre*, *48*. https://doi.org/10.1186/s42269-024-01231-0

Verma, S. K., & Kumar, P. (2022). *Biofuel Policies in India: An Assessment of Policy Barriers* (pp. 44–64). https://doi.org/10.4018/978-1-6684-5269-1.ch004

Verma, S. K., Kumar, P., Mishra, A., Khare, R., & Singh, D. (2024c). Green nanotechnology: illuminating the effects of bio-based nanoparticles on plant physiology. *Biotechnology for Sustainable Materials*, *1*(1), 1. https://doi.org/10.1186/s44316-024-00001-2

Verma, S. K., Singh, D., Sharan, A., Khare, R., Rai, S., Kumar, P., & Singh, V. (2025). Innovations in biomass-derived materials for efficient $CO_2$ sequestration. *Chemistry Africa*. https://doi.org/10.1007/s42250-024-01179-8

Wang, S., Zhang, P., Li, Y., Li, J., Li, X., Yang, J., Ji, M., Li, F., & Zhang, C. (2023). Recent advances and future challenges of the starch-based bio-composites for engineering applications. *Carbohydrate Polymers*, *307*, 120627. https://doi.org/10.1016/j.carbpol.2023.120627

Wool, R., & Sun, X. S. (2011). *Bio-Based Polymers and Composites*. Elsevier.

Wu, C.-S., & Liao, H.-T. (2013). The mechanical properties, biocompatibility and biodegradability of chestnut shell fibre and polyhydroxyalkanoate composites. *Polymer Degradation and Stability*, *99*. https://doi.org/10.1016/j.polymdegradstab.2013.10.019

Zamboulis, A., Xanthopoulou, E., Chrysafi, I., Lorenzo, C., & Bikiaris, D. N. (2023). Poly(ethylene succinate)/hemp fiber composites: fully biobased materials with improved thermal and biodegradation properties. *Sustainable Chemistry for the Environment*, *4*, 100045. https://doi.org/10.1016/j.scenv.2023.100045

# 2 Fundamentals of Bio-Based Polymers

*Ayushi Verma, Aparna Kumari, Nishu Mittal, and Hanna Sara Johnson*

## 2.1 INTRODUCTION

The era of material sustainability faces immense threats from petrochemical resources. Traditional polymers risk health and the environment. It is essential to substitute them with alternatives that are eco-friendly and aid the circular economy. Constant efforts toward producing renewable polymers are a beacon of hope for replacing traditional polymers. Bio-based polymers offer enhanced properties and are cost-effective. Futuristic innovations in ecological polymers have drawn significant attention from industries and markets (Wool & Sun, 2011). Petrochemical resources like peat take thousands of years to form, whereas tropical hardwood is a renewable feed. Preference for natural resources for engineering polymers saves us the hassle of depleting resources. Bio-based polymers are derived from biomass (Verma et al., 2024b). They are biodegradable and can be recycled. Annual waste production globally is estimated to be approximately 1.3 billion tons; this number is expected to triple by 2100. When such waste is incinerated, it releases toxic substances into the environment. The durability of traditional polymers makes them vulnerable to disposal (Babu et al., 2013). These polymers are resistant to microbial degradation and accumulate in the earth as waste. Compared to Petro-based chemicals, bio-based polymers reduce greenhouse emissions and minimize waste load over time. The advantages of bio-based polymers over conventional polymers are summarized in Figure 2.1.

## 2.2 CATEGORIES OF BIO-BASED POLYMERS

The diverse nature of bio-polymers is based on the fact that these polymers are obtained from a plethora of biological resources like microorganisms, agro-waste, and plant resources. These sustainable polymers are divided into three categories. The first category, naturally occurring polymers, which are biomass derivatives. This first generation of biopolymers comprises polysaccharides, proteins, and lignin. Generally found in plants and animals, these polymers require shallow chemical processing and perform structural and storage roles. The second generation of biopolymers is derived from monomers and polymers generated from microorganisms. These polymers are produced by chemical extraction and polymerization of sugars and starches. They also involve fermentation. Examples include PLA, PE, PP, and


DOI: 10.1201/9781003566205-2

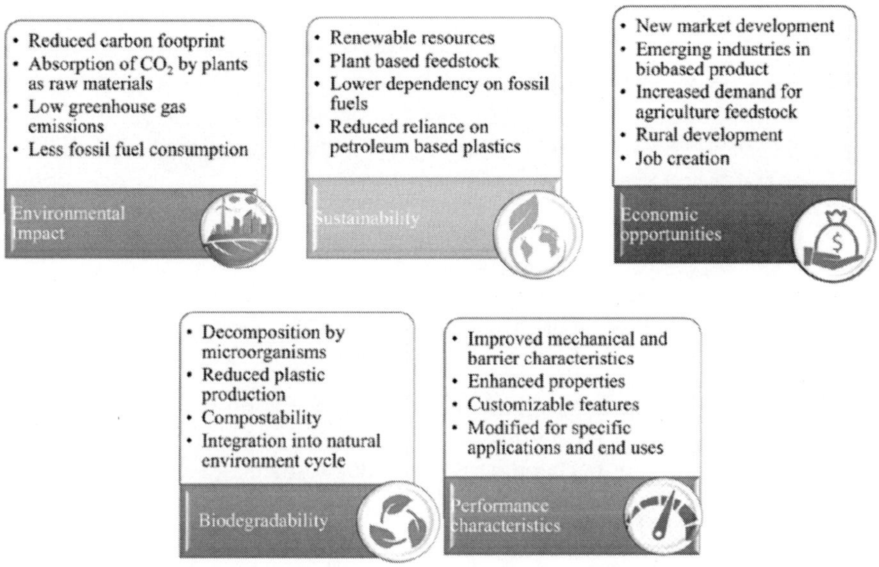


**FIGURE 2.1** Advantages of bio-based polymers.

nylons, as shown in Figure 2.2. Microbes synthesize the third-generation biopolymers via metabolism under highly optimized conditions, such as PHA and bacterial cellulose. Specific polymers like bio-polyethylene terephthalate (PET) partially utilize biological sources and chemical synthesis routes. They reduce the carbon footprint load on the environment compared to traditional plastics (Hassan et al., 2019)

## 2.3 COMMON BIO-BASED POLYMERS

### 2.3.1 Cellulose

Cellulose, a naturally occurring linear polymer, consists of glucose. Glucose subunits are bonded together by β-1,4-glycosidic linkage. It is the most abundantly found organic polymer and is present in the cell walls of algae and plants. It has a wide variety of applications because of its high availability, high tensile strength, and biocompatibility in pharmaceutical and biomedical applications. Cellulose can be chemically modified readily and is highly biodegradable. It is often used in various applications such as packaging materials, textiles, wound dressings, drug delivery, tissue engineering, and cosmetics (Shaghaleh & Wang, 2018)

### 2.3.2 Starch

Starch is a heterogeneous biopolymer composed of linear amylose chains and branched amylopectin chains. It is widely found in corn, potatoes, wheat, and rice. Starch is highly regarded as a versatile material because of its thermoplastic

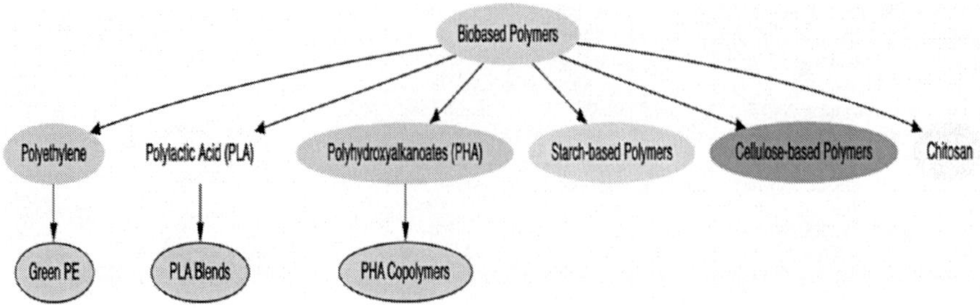


**FIGURE 2.2** Different types of bio-based polymers.

properties, biodegradable nature, and high natural availability (Falua et al., 2022). Starch-based biopolymers create disposable packaging, mulch films, and coatings. They also act as excellent packaging materials due to their flexibility and molding properties.

### 2.3.3 Chitosan

Chitosan is a bio-based polymer composed of chitin, a natural polysaccharide found in insect exoskeletons and fungal cell walls. Chitin deacylation results in the formation of a biocompatible, biodegradable, and non-toxic material known as chitosan. Such a material can form hydrogels and has antimicrobial properties. Chitosan can be used as a natural preservative in the food industry, thus increasing the shelf life of food (Amitaye et al., 2024). In addition to food packaging, chitosan is used as a flocculating agent and is widely applied in heavy metal remediation processes.

### 2.3.4 Polylactic Acid (PLA)

PLA was commercialized in 1990. It is an aliphatic polyester synthesized by bacterial fermentation of corn starch, which provides highly pure lactic acid monomers. Polycondensation of lactide monomer yields low-molecular-weight PLA. NatureWorks LLC, USA, has developed a cost-effective production method for synthesizing high-molecular-weight PLA. In the first step, lactide dimers are created via condensation, followed by ring-opening polymerization using selective catalytic agents. The properties of the resultant polymer depend on the ratio of L- and D-lactic acid forms. PLA is a popular choice due to its extensive properties of good transparency, rigidity, tolerance to processing conditions, and glossy appearance. The mechanical properties of PLA are similar to traditional polymers, while thermal properties require changes in the stereochemistry of the polymer. NatureWorks LLC, with a production capacity of 100,000 tons per year, is the leading supplier of PLA, which is branded as Ingeo (Vink et al., 2007). Other manufacturers, including China, Europe, and Japan, develop PLA for different sectors like medical devices, automobiles, and electronics. Apart from PLA having great mechanical strength and resistance to heat, it is still a challenge to apply it to engineering applications and electronics. Recently, a Japan-based firm, NEC Corporation, has designed PLA with improved heat resistance and thermal stability. Blends of PLA are being utilized to create computer housings, implants for living cells, and textiles, which are on par with their synthetic alternatives. The FDA has also approved PLA for clinical applications.

### 2.3.5 Polyhydroxyalkanoates (PHAs)

PHAs are a group of polyesters synthesized through bacterial fermentation with the ability to substitute traditional polymers. PHAs are naturally occurring in various organisms, but cell production can be induced externally. The general production process involves feedstocks such as municipal waste, vegetable oil, cellulose, or other carbon sources in a fermentation vessel inoculated with seed culture. Fermentation continues until the feed is consumed and PHA is accumulated, which requires about 48 hours. Later, PHA is isolated and purified. Next, solvent extraction is performed, cell debris is removed, and methanol is added to precipitate the dissolved PHA. PHAs have diverse monomers, which makes it a key player suitable for various applications. Thermal characteristics and mechanical properties of PHA mimic polypropene and polystyrene, but slow crystallization makes it suitable to be used as a copolymer. PHAs are used in films, laminates, synthetic paper production, paints, foams, and binders. USA-based venture Metabolix acquired clearance from the Food and Drug Administration for using PHA for food contact applications. Chinese company Tianan Biologic Material Co. is the largest producer of PHB and PHB copolymers (Koller & Mukherjee, 2022). These are popularly used in biomedical applications for creating cardiovascular patches, sutures, and stem cell growth. PHAs are also being employed in drug delivery systems as they favor degradability and biocompatibility.

### 2.3.6 Polybutylene Succinate (PBS)

PEt alternative PBS is an aliphatic ester that is synthesized by condensing 1,4-butanediol and succinic acid. Succinic acid is produced using green routes, which reduces energy consumption, increases yield, and is comparatively a low-cost process. Numerous companies have shown interest in the manufacturing of bio-succinate. A Japanese firm, Mitsubishi Chemicals, has commercialized PBS biopolymer. The primary step in the production of PBS biopolymer is the synthesis of succinic acid from glucose, and later, the resulting compound goes through chemical reduction to produce butanediol. The addition of adipic and sebacic acid monomers produces copolymers of PBS. PBS's thermal and mechanical properties depend highly on configuration and extent of crystallinity (Barletta et al., 2022). PBS has a higher melting point than PLA and is tougher. PBS is commercialized through mulch films, food packaging applications, and plasticizers for polyvinyl chloride (PVC). The only limitation lies in low mechanical flexibility, which can be fixed by creating blends with starch and PVC.

### 2.3.7 Bio-Polyethylene (PE)

PE is a widely used engineered polymer consistently manufactured from fossil fuels. It is synthesized through the polymerization of ethene under STP (standard temperature and pressure) conditions using a catalyst. Ethylene has conventionally been obtained via ethanol dehydration or conversion of naptha and heavy oils into gas and later mixing them with steam. However, rising oil prices have led to the production of "green polyethylene" (Siracusa & Blanco, 2020). This form of PE is made by dehydrating ethanol derived from fermentation by microbes. The idea of the production of PE from bio-ethanol has been introduced previously. In the 1980s, bio-polyethene and bio-PVC from bioethanol were developed by Braskem. At that time, the price of oil and the extent of biotechnological innovations were limited, making the process economically unviable (Castro-Rosas et al., 2016). Today, industrial-scale bio-PE is primarily produced using bioethanol derived from sugarcane. Bioethanol is also obtained from other renewable feed (e.g., sugar beet, wood, starch crops, and waste from plants) through strains of microbes and bio-fermentation. In the general production process, sucrose-rich feedstock undergoes anaerobic fermentation to produce ethanol.

Ethanol distillation is done, yielding an azeotropic hydrous ethanol mixture. Finally, dehydration of ethyl alcohol is performed at rising temperatures using a solid catalyst to generate ethene, which is polymerized to produce PE. Bio-PE possesses the same mechanical and physicochemical properties as conventional polyethene. Braskem, a Brazilian firm, is the leading global producer of bio-PE, commanding 52% of the market and pioneering the introduction of certified bio-PE. In addition to bio-PE, the company also produces other bio-polymers, like bio-PVC, bio-polypropene, and related composites, using identical industrial methods. Present grades of Braskem's bio-polyethene are primarily used in applications such as automobile components, cosmetics, packaging of food, personal care, and toys. The US company Dow Chemicals (Zhang et al., 2016), in collaboration with Crystalsev, ranks as the second leading producer of bio-polyethene with a 12% market share, while Belgium-based Solvay holds a 10% share. Solvay is particularly prominent in the generation of bio-PVC using identical industry techniques. Bio-PE is capable of replacing all applications of traditional fossil-based PE. Due to its affordability and reliable performance, it is widely used across various sectors, including engineering, agriculture, packaging, and everyday commodities.

### 2.3.8 Bio-PET

PET is a versatile, high-performance plastic widely used in products like bottles, fibers, films, and various engineering applications. Despite its importance in the plastic industry, the massive consumption of PET creates significant ecological challenges because of its limited biodegradability and poor sustainability. While material recycling methods for PET have been implemented, they often lead to a noticeable decline in the quality of the recycled material. This degradation is caused by thermal, hydrolytic, and oxidative damage during recycling (Cywar et al., 2022).

Efforts to chemically recycle PET are underway, but the inherent stability of PET makes it resistant to breakdown via standard processes like hydrolysis or alcoholysis. Overcoming these technical barriers is crucial for more effective recycling. A more sustainable approach involves transitioning to bio-based PET (bio-PET) from renewable resources like biomass. A critical step in this direction is replacing petroleum-derived ethylene glycol (EG) with bio-based EG, as depicted in

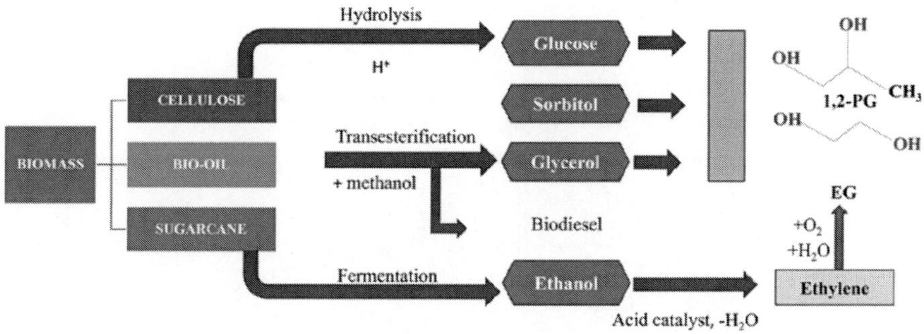


**FIGURE 2.3** Production of Bio-PET from biomass.

Figure 2.3. For instance, the Coca-Cola Company introduced the "Plant Bottle" in 2009 (Ren et al., 2015), made with 30% bio-based materials, including 100% bio-EG and petro-based TPA.

Bio-based bio-terephthalic acid is being produced using natural, renewable feedstocks to enhance PET's sustainability further. The ultimate goal is to make a combination of both bio-EG and bio-terephthalic acid to create PET derived from renewable biomass, achieving a fully sustainable bio-PET.

## 2.4 POLYURETHANE FOAMS: A CASE STUDY

Invented in Germany, polyurethane (PUR) makes up the largest class of cross-linked polymers. Used mainly as foams, polyurethane foams have no alternative as far as their structure is considered. It has been used for almost 90 years now. The chemical structure of polyurethane is highly diverse. It is composed of duromers, thermoplastics, and elastomers (Marson et al., 2021). The feedstock for PUR production is generally petrochemicals and small quantities of biomass. There are several pathways to achieve better and ecologically stable production: (1) utilizing sustainable energy for steam and electricity; (2) sourcing green salt for chlorine synthesis; (3) adopting alternative feedstocks for propene and ethene in polyether backbone synthesis; (4) employing eco-friendly sources of ammonia, toluene, and benzene for the production of methylene diphenyl diisocyanate (MDI) and toluene-2,4-diisocyanate (TDI); and (5) identifying suitable carbon sources for carbon monoxide and formaldehyde. PUR foams are manufactured using polyether and isocyanate. Linkage of the two via urea bonds results in a titrimetric polymer that can decompose on melting, making it unsuitable for mechanical recycling. At present, recycling processes are achieved either through chemical recycling or carbon capture recycling.

### 2.4.1 Chemical Recycling

Components of PUR can be separated by chemical cleavage. Now, these cleaved products are recycled individually. An isolated isocyanate compound can also be converted into a polyol, resulting in a new polyurethane. PET recycling methods can also be applied to PUR due to the resemblance of urethane bonds with ester bonds

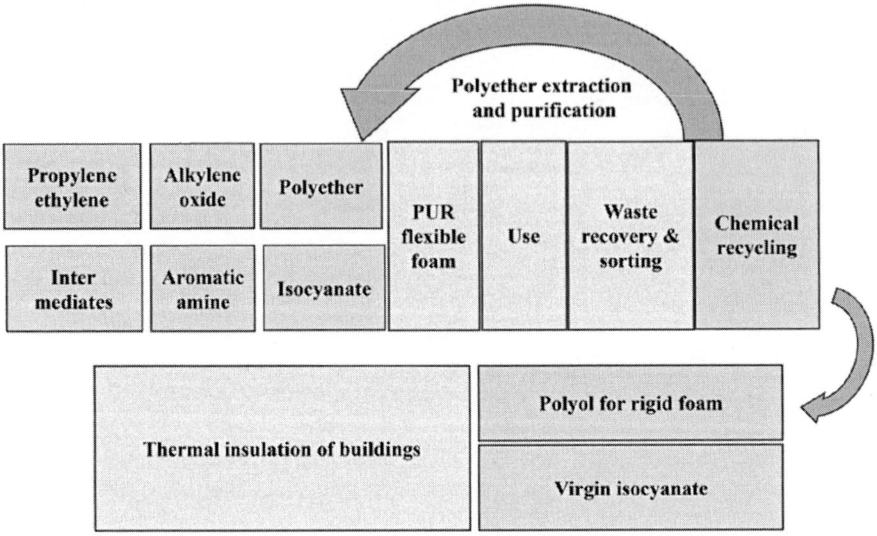


**FIGURE 2.4** Hemolysis technology by Waterlilly® and Renuva®.

in PET. Figure 2.4 shows the hemolysis technology used by Waterlilly® and Renuva® (Thomas et al., 2018). Digesting flexible PUR foam with DEG produced two distinct phases in this process. The upper phase, rich in polyols, could be reused to synthesize new foams that are flexible foams. The other phase contained the isocyanate and was transformed into small carbamates and aromatic amines. Due to the complexity of recovering these compounds, treatment with propene oxide to create N-containing small-chain aromatic polyols is designed explicitly for isoboards used in building thermal insulation. In this process, crude glycerol has been proposed as a "green" alternative to diethylene glycol. This process is popular with the Urbanrec project funded by the EU. A similar concept is applied in the RePURpose project in Denmark. Polyols derived from the chemical breakdown of PET or PUR are marketed in Europe under the brand name Repol® (Johansen et al., 2023; Ranjan et al., 2023, 2024).

### 2.4.2 Carbon Capture and Usage (CCU)

Chemical recycling can be risky; therefore, a second approach is implied, where the carbon dioxide produced during the incineration of polyurethane is recycled. The recycled carbon dioxide is used by plants for photosynthesis, thereby rebuilding natural polymers. This biomimetic route is termed CCU. The ecological impact of carbon dioxide–based chemical manufacturing has also been evaluated (Peres et al., 2022)

### 2.4.3 Bio-based Polymers Approach

Biopolymers are obtained from sustainable sources and can be directly utilized or processed to obtain structures and properties resembling polyurethane chemistry and other traditional polymers (Verma et al., 2024a). Examples of PUR alternatives include polycarbonate polyols, polyethers (Lupranat®), polyesters from castor oil, and pentamethylene diisocyanate (Desmodur® eco N7300 from Covestro).

## 2.5 APPLICATIONS OF BIO-BASED POLYMERS

Considering the gradual decline in environmental quality, more and more attention is being paid to developing greener routes for day-to-day applications. Researchers globally focus on replacing conventional polymers with bio-based polymers for biomedical industries, the food industry, and environmental applications. The vast utilization of these polymers is discussed below (see Figure 2.5).

### 2.5.1 Drug Administration Mechanisms

A drug delivery system is an intelligent approach to achieve precise dosage optimally in a controlled environment. A classic drug delivery system consists of an active biological agent, an excipient as a carrier, and a targeting moiety. Additionally, a few components can increase the precision of drug assimilation and conveniently target mobile cells. Biopolymers often act as components that help increase the drug's therapeutic benefits. Usually, anticancer, antiviral, and contraceptive drugs are administered using such systems. Polymers used in such applications have high biocompatibility and are biodegradable. Approval has been given to natural polymers such as cellulose derivatives, PEG, PLA, PCL, and PLGA. Efforts are being made to utilize PHAs as excipient carriers (Gheorghita et al., 2021).

### 2.5.2 Implants and Scaffolds

Scaffolds are 3D structures designed to support cell growth. They induce cell growth by mimicking the extracellular matrix of the cell. Bio-based polymers are used to create scaffolds as they possess an appropriate architecture suitable for cell adhesion,

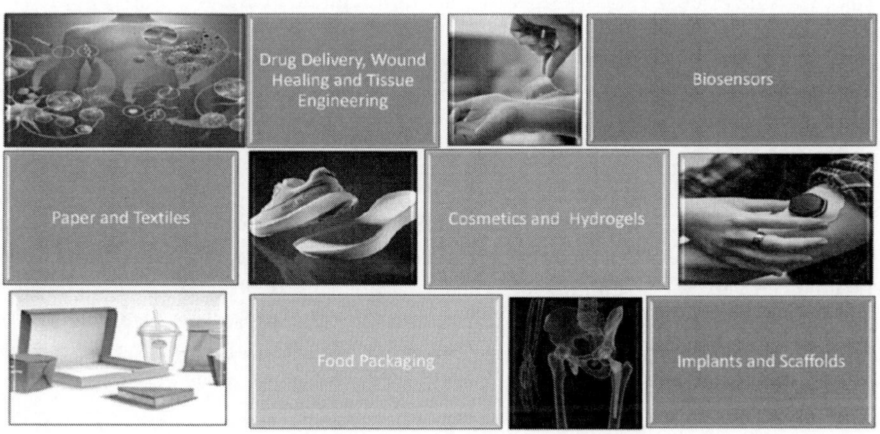


**FIGURE 2.5** Applications of bio-based polymers in the biomedical and pharmaceutical industry.

tissue regeneration, and formation. Other factors, such as pore size and permeability, are also considered when designing a scaffold for proper cell migration and vascularization (Stoppel et al., 2015). Based on tissue function, the scaffold should also degrade and reabsorb at a variable time. The biocompatibility of biopolymers makes them an excellent choice to use as implants. Polymers like PHAs degrade slowly when exposed to surface erosion or hydrolysis and are non-enzymatic. Such polymers are being investigated for nerve repair and bone implants. Being thermoplastic, they are easily molded into desired shapes. To increase hydrophilicity, biopolymers can be either blended or copolymerized with polyethylene glycol (Bhatt & Jaffe, 2015).

### 2.5.3 Hydrogels

Hydrogels are interconnected polymer networks with a dual-phase structure that can absorb water. The interaction between the hydrogel and water is purely electrostatic. This allows for the entrapment of water molecules within the polymer network, giving rise to a swollen morphology. Hydrogels are used to heal wounds, provide moisture, and avoid further lesions. They protect nerve endings from exposure to the environment and provide suitable conditions to accelerate cellular activities (Mahmood et al., 2022).

### 2.5.4 Pharmaceutical Applications

Microcapsules are extravascular systems that efficiently transport glucose and insulin. Investigations are being conducted to design microcapsules with minimal inflammatory reactions, which could be used to treat endocrine problems. Characteristic requirements for microencapsulation are stability of biopolymers, non-interference with cellular activities and the drug to be administered, low toxicity, biosafety, proper drug diffusion and release, and degradability. Conventional polymers have long been used to encapsulate medical compounds (Udayakumar et al., 2021). They have excellent pharmacokinetic properties when compared to small-molecule medications. A significant challenge with conventional polymers is the buildup of toxicity. To reduce concerns over side effects, biopolymers like sodium alginate, cellulose, starch, and agar are usually deployed for oral and intravenous administration. Microspheres and nanocapsules created using these biopolymers are highly stable and are non-immunogenic. Varied applications of bio-based polymers are summarized in Table 2.1.

### 2.5.5 Food Packaging

Food spoilage results from microbial contamination, temperature, humidity, physical damage, and shocks. Proper packaging of food can enhance the shelf life of food, protect it, and make it suitable for transport. Each food product requires a separate strategy of preservation. Fruits and vegetables are preserved by reducing the rate of respiration and transpiration (Agarwal et al., 2023). Dairy products should be prevented from microbial growth, while meat products are subject to spoilage due to discoloration (Lionetto & Esposito Corcione, 2021). Conveying of processing powders through pipelines and their spreading on surface was investigated by the researchers (Santo et al., 2018a, 2018b; Tripathi et al., 2015, 2016, 2017, 2018a, 2018b,

**TABLE 2.1**
**Various Biopolymers and Their Applications.**

| Polymers | Source | Applications | References |
|---|---|---|---|
| Chitosan | Chitin (shellfish and fungi) | Mucoadhesive systems, nanoparticles, hydrogels, and controlled drug release | Sashiwa and Aiba (2004), Sharma et al. (2019) |
| Alginate | Brown seaweed | Microparticles, hydrogels, and sustained drug release | Lee and Mooney (2012), LeRoux et al. (1999) |
| Hyaluronic acid | Animal tissues and bacteria | Targeted drug delivery, hydrogels, and injectable formulations | Lapcık et al., 1998, Tripodo et al., 2015 |
| Cellulose derivatives | Plant biomass | Tablets, films, and hydrogels for oral and transdermal delivery systems | Kamel et al. (2008); Heinze and Liebert (2001) |
| Polylactic acid (PLA) | Corn, sugarcane | Biodegradable nanoparticles and implants for sustained and localized drug delivery | Anderson and Shive (1997) |
| Polyhydroxyalkanoates (PHA) | Bacterial fermentation | Nanoparticles, microspheres, and tissue engineering scaffolds | Chen (2009), Khanna and Srivastava (2005) |
| Gelatin | Animal collagen | Injectable systems, microspheres, and soft tissue delivery applications | Vernon (2011), Coester et al., 2000 |
| Starch derivatives | Plant starch (corn and rice) | Controlled release systems and biodegradable carriers for oral drug delivery | Tharanathan (2003), Han et al., 2018 |
| Pectin | Citrus peels and apples | Colon-targeted delivery, microparticles, and gelling agents for controlled release | Sriamornsak (2011) |
| Soy protein isolates | Soybeans | Nanoparticles and films for controlled drug release and wound healing applications | Kalia and Sharma (2022) (Series, J.S. on Biocatalysis) |

2018c, 2018d, 2018e, 2019a, 2019b, 2019c, 2019d; Kalman et al., 2017, 2018; Naveh et al., 2017; Tripathi & Mallick, 2017; Singh et al., 2015; Sharma et al., 2015; Tripathi & Mallick, 2014). It is found that the optimum velocity is important for the proper distribution of materials. Diverse food packaging strategies like active, intelligent, thoughtful, and green packaging are employed. Bio-based polymers in food packaging offer gas barriers, antimicrobial activities, and thermal packaging. Disposal of such packaging is not detrimental to the environment and is an additional advantage (Grujić et al., 2017).

## 2.6 MODERN BIOPOLYMER SCIENCE

Conventional methods of designing and manufacturing bio-based polymers involve experimentation and characterization of the material. Such results can now be achieved by computational approaches, which are efficient and cost-effective in terms of innovation and scale-up. The combination of machine learning and AI helps in simulating and chemical modeling of new materials (Scacchi et al., 2024). Processing of complex data, prediction of properties, and studying material and mechanism can be attained in less time using computational methods. Integration of current digital tools (Material 4.0) with experimental (Material 1.0), computational (Material 3.0), and theoretical knowledge (Material 1.0) forms the basis of modern biopolymer science. This newer approach is summarized in Figure 2.6

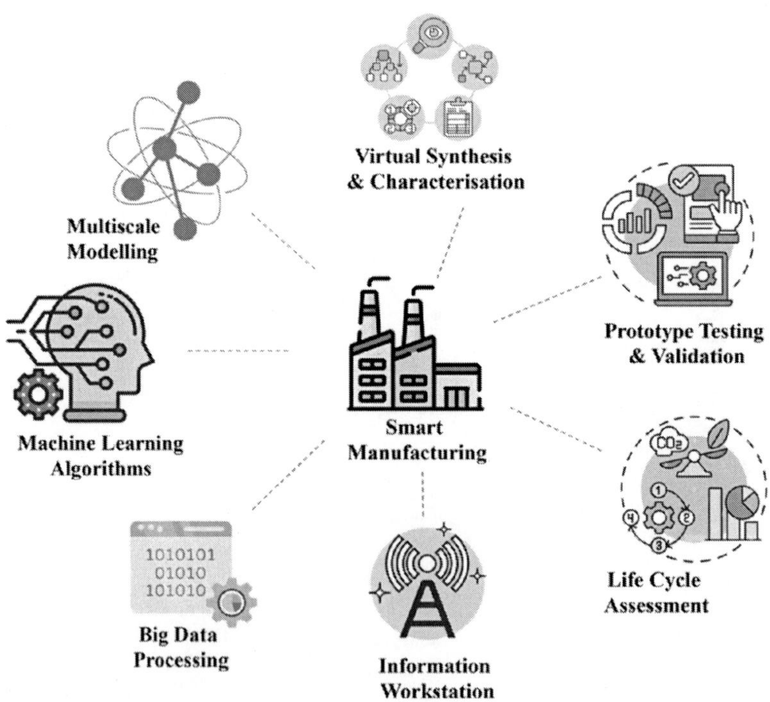


FIGURE 2.6 Computational approaches for polymer designing.

As discussed earlier, PHA synthesis is done via fermentation, and biopolymers are accumulated within microbial cells. A study conducted by Xu et al. (2022) employed mixed microbial cultures along with the activated sludge model dataset. Optimized parameters for PHA synthesis were determined with the help of recurrent neural networks. Apart from synthesis, analysis of bio-based polymers is done using various software that can handle complex data, also termed the materiomics approach (Hufnagl et al., 2021). The degradation of bioplastics with 26 nonlinear regression models and an artificial neural network was studied by Ebrahimzade et al. (2022). Rheological modeling of biopolymers has been established using 3D computational fluid dynamics (Tagliavini et al., 2018).

A multifaceted software tool, "Polymer Expert," with a user-friendly interface, has been implemented for *de novo* polymer designing. It can produce novel polymers with specific properties. It works by creating and expanding a database of repetitive units and performs calculations for quantitative structure, properties, and relationships (QSPR). Searches are conducted to identify potential targets in PEARL, a library containing the full analog database. Other computational tools currently used in synthesis, analysis, and modeling of bio-based polymers are described in Table 2.2

## 2.7 MARKET DYNAMICS OF BIO-BASED POLYMERS AND SWOT ANALYSIS

In 2019, the bio-based polymer industry faced significant challenges: PLA and PHA were sold out despite increased production efforts (Chinthapalli et al., 2019). Bio-based polypropylene (PP) was introduced, and there was a rise in the production of polybutylene adipate-co-terephthalate, polybutylene succinate, bio-based polyethylene, starch-based polymer compounds, and epoxy resins. However, the production of bio-based PET decreased. Policymakers must also adequately support bio-based and biodegradable plastics, treating them like fossil-based plastics under Plastic Policy. According to the latest market report, "*Bio-based Building Blocks and Polymers—Global Capacities, Production and Trends 2019–2024*" by German nova-Institute, 2019 data provide production figures for all bio-based polymers, along with projections through 2024. The global production of bio-based polymers reached 3.8 million tons, accounting for 1% of the total production of petrochemical-based polymers and marking a 3% growth compared to the previous year in 2019. This growth rate is projected to persist through 2024. The primary raw materials for biopolymer synthesis are natural byproducts, which account for 46%, particularly glycerol generated during the production of biodiesel. The key players in the bio-based polymer market are summarized in Table 2.3

The global market for bio-based polymers is expected to surge substantially during the projected duration, fueled by increasing demand for environmentally friendly chemicals worldwide. Additionally, stringent regulatory policies on synthetic polymers are expected to further propel the growth and expansion of the renewable polymer market. In 2021, the global market was valued at USD 7.92 billion and is anticipated to reach USD 12.16 billion by 2029, showing a compound annual growth rate of 5.50% from 2022 to 2029. Analysis of the market, developed by the "Data

**TABLE 2.2**
**Common Tools and Software Used in the Computational Approach of Bio-Based Polymer Designing.**

| Tools | Software | Uses | References |
|---|---|---|---|
| Polymer informatics platform | Polymer Genome and ChemLG | Virtual screening | (Doan Tran et al., 2020) |
| Molecular design software | ChemDraw, Materials Studio, and Schrödinger Materials Science Suite | Designs polymer architecture | (Accelrys, 2013) |
| Machine learning models | Neural Networks and Random Forests; frameworks: TensorFlow and PyTorch | Predicts properties and polymerization pathways | (Raccuglia et al., 2016) |
| Self-consistent field theory (SCFT) | | Studies microphase separation | (Matsen & Schick, 1994) |
| Coarse-grained MD | MARTINI force field, HOOMD-blue | Simulates polymer system and reduces atomistic details | (Marrink et al., 2004) |
| Finite-element analysis (FEA) | Abaqus and ANSYS | Stress–strain analysis of polymer | (Hibbitt et al., 1997) |
| Density functional theory (DFT) | JDFTx and Quantum ESPRESSO | Predicts thermal and mechanical properties | (Hohenberg & Kohn, 1964) |
| Monte Carlo (MC) simulations | GAMS and Crystal Ball Professional | Studies chain confirmation and phase behavior | (Metropolis et al., 1953) |
| Quantum mechanical tools | Gaussian, ORCA, and VASP | Assess the polymerization process, mechanism of reaction, and electronic properties | (Hohenberg & Kohn, 1964) |
| Molecular dynamics | GROMACS, LAMMPS, and NAMD | Simulates self-assembly behavior, material–property relationships, and polymer chain dynamics | (Plimpton, 1995) |

Bridge Market Research," provides detailed expert insight, including a review of import or export trends, cost, manufacturing, and utilization patterns, as well as the impact of climate change, shown in Figure 2.7

### 2.7.1 Packaging Industry Demands

Packaging is a primary sector for bio-based polymers due to their remarkable clarity, gloss, odor, food fats, and oil resistance (Pan et al., 2016). These polymers also enhance stiffness, twist resistance, and print quality in packaging materials. Bio-based polymers are commonly used in various applications, such as supermarket

**TABLE 2.3**
**Major Players in the Bio-Based Polymer Market.**

| Company | Products | References |
|---|---|---|
| NatureWorks LLC | Ingeo (PLA) | NatureWorks official website |
| BASF SE | Ecovio (biopolymer) | BASF – sustainability reports |
| TotalEnergies Corbion | Luminy (PLA) | TotalEnergies Corbion press releases |
| Novamont SpA | Mater-Bi (Bioplastic) | Nova Mont's sustainability reports |
| Braskem | Bio-based polythene | Braskem corporate resources |
| Mitsubishi Chemical Holdings Corporation | Biodegradable and bio-based material | Mitsubishi official reports |
| Corbion | PLA | Corbion sustainability frameworks |
| Arkema Group | Rilsan (Bio-based Nylon) | Arkema group product pages |
| Teijin Limited | Bio-based polycarbonates | Teijin Corporate announcements |
| Biome Bioplastics | Compostable biopolymer | Biome bioplastics information |

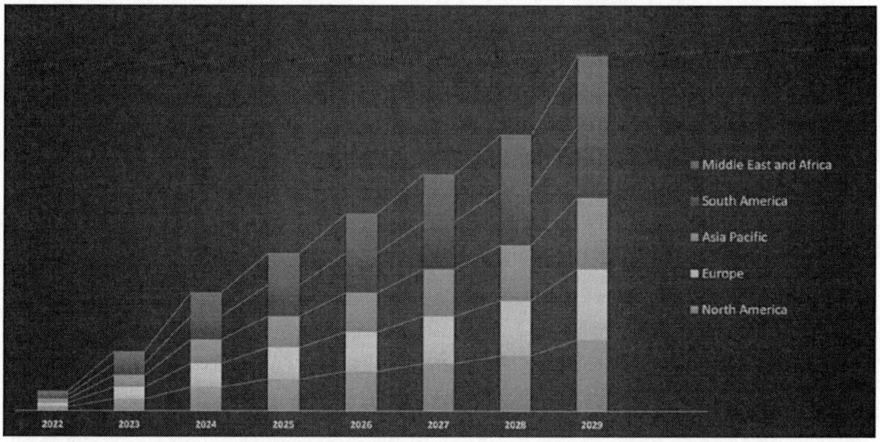


**FIGURE 2.7** Trends in Global Bio-Based Polymer Market (Source: Data Bridge Market Analysis Study, 2022).

produce packaging, bakery boxes, envelopes, display carton windows, bottles, and shopping or carrier bags. As a result, the extensive use of bio-based polymers in packaging is expected to drive market growth (Joseph et al., 2023). In addition, strict regulations on the use of traditional polymers in the future will drive the growth of the bio-based polymer market. Supportive government regulatory actions promoting bio-based polymers, coupled with a reduced reliance on petroleum-based resources, are also expected to contribute to the market's expansion. India took a significant step toward sustainable production by banning single-use plastics in 2002. IIT Guwahati researchers developed a cost-efficient and non-toxic bioplastic to manage solid waste (Choubey et al., 2023)

### 2.7.2 Eco-Friendly Polymers

Moreover, the growing preference for eco-friendly polymers to safeguard the environment, along with regulations banning non-degradable polymers in many countries, presents valuable opportunities for industry participants during the projected timeframe till 2029. Moreover, rising consumer awareness in both developed and emerging nations, coupled with the non-toxic nature of biodegradable polymers, is presumed to further fuel the growth of the bio-based polymer domain in the upcoming years (Jha et al., 2024).

### 2.7.3 Low Awareness and the Cost Factor

The need for more awareness in developing regions poses a significant challenge to the expansion of the biopolymer market, acting as a drawback to its expansion. This factor is anticipated to negatively impact the rate of market dynamics. The biopolymer market report offers a clear idea into notable advancements, influenced by recent innovations, evolving trade regulations, shifts in import–export dynamics, production trends, streamlined value chain, enhanced market presence, and the influence of domestic and regional market players. It also highlights prospects of generating revenue areas, adjustments in market rules, strategic growth analysis, market dimensions, segment growth patterns, application-specific opportunities, product authorizations, launches, regional expansions, and advancements in technology. Patel et al. (2005) stated that the bio-based polymer market's growth will be hindered by its higher costs compared to petroleum-derived polymers.

## 2.8 CONCLUSION

The concerns over petrochemical-based resources highlight the urgent need for sustainable alternatives and innovative solutions. Bio-based polymers represent a promising avenue for reducing environmental impact, enhancing energy security, and promoting a circular economy. Continued research, investment, and policy support are crucial for accelerating the transition to more sustainable and resilient industrial practices.

## REFERENCES

Agarwal, A., Shaida, B., Rastogi, M., & Singh, N. B. (2023). Food packaging materials with special reference to biopolymers-properties and applications. *Chemistry Africa*, *6*(1), 117–144.

Amitaye, A. N., Elemike, E. E., Akpeji, H. B., Amitaye, E., Hossain, I., Mbonu, J. I., & Aziza, A. E. (2024). Chitosan: a sustainable biobased material for diverse applications. *Journal of Environmental Chemical Engineering*, 113208.

Anderson, J. M., & Shive, M. S. (1997). Biodegradation and biocompatibility of PLA and PLGA microspheres. *Advanced Drug Delivery Reviews*, *28*(1), 5–24.

Babu, R. P., O'connor, K., & Seeram, R. (2013). Current progress on bio-based polymers and their future trends. *Progress in Biomaterials*, *2*, 1–16.

Barletta, M., Aversa, C., Ayyoob, M., Gisario, A., Hamad, K., Mehrpouya, M., & Vahabi, H. (2022). Poly (butylene succinate) (PBS): materials, processing, and industrial applications. *Progress in Polymer Science*, *132*, 101579.

Bhatt, R., & Jaffe, M. (2015). Biopolymers in medical implants. In *Excipient Applications in Formulation Design and Drug Delivery* (pp. 311–348). Springer.

Castro-Rosas, J., Cruz-Galvez, A. M., Gomez-Aldapa, C. A., Falfan-Cortes, R. N., Guzman-Ortiz, F. A., & Rodríguez-Marín, M. L. (2016). Biopolymer films and the effects of added lipids, nanoparticles and antimicrobials on their mechanical and barrier properties: a review. *International Journal of Food Science & Technology*, *51*(9), 1967–1978.

Chen, G. Q. (2009). A microbial polyhydroxyalkanoates (PHA) based bio-and materials industry. *Chemical Society Reviews*, *38*(8), 2434–2446.

Chinthapalli, R., Skoczinski, P., Carus, M., Baltus, W., De Guzman, D., Käb, H., . . . & Ravenstijn, J. (2019). Biobased building blocks and polymers—global capacities, production and trends, 2018–2023. *Industrial Biotechnology*, *15*(4), 237–241.

Choubey, V., Govind, M., & Arjune, V. (2023). The emergence of bioplastic as a sustainable polymer in the Indian food packaging industry: a scientometric analysis. *Journal of Data Science, Informetrics, and Citation Studies*, *2*(1), 1–12.

Coester, C. J., Langer, K., Von Briesen, H., & Kreuter, J. (2000). Gelatin nanoparticles by two step desolvation a new preparation method, surface modifications and cell uptake. *Journal of Microencapsulation*, *17*(2), 187–193.

Cywar, R. M., Rorrer, N. A., Hoyt, C. B., Beckham, G. T., & Chen, E. Y. X. (2022). Bio-based polymers with performance-advantaged properties. *Nature Reviews Materials*, *7*(2), 83–103.

Doan Tran, H., Kim, C., Chen, L., Chandrasekaran, A., Batra, R., Venkatram, S., . . . & Ramprasad, R. (2020). Machine-learning predictions of polymer properties with Polymer Genome. *Journal of Applied Physics*, *128*(17).

Ebrahimzade, I., Ebrahimi-Nik, M., Rohani, A., & Tedesco, S. (2022). Towards monitoring biodegradation of starch-based bioplastic in anaerobic condition: finding a proper kinetic model. *Bioresource Technology*, *347*, 126661.

Falua, K. J., Pokharel, A., Babaei-Ghazvini, A., Ai, Y., & Acharya, B. (2022). Valorization of starch to biobased materials: a review. *Polymers*, *14*(11), 2215.

Gheorghita, R., Anchidin-Norocel, L., Filip, R., Dimian, M., & Covasa, M. (2021). Applications of biopolymers for drugs and probiotics delivery. *Polymers*, *13*(16), 2729.

Grujić, R., Vujadinović, D., & Savanović, D. (2017). Biopolymers as food packaging materials. In *Advances in Applications of Industrial Biomaterials* (pp. 139–160). Springer.

Han, Y., Yu, M., & Wang, L. (2018). Physical and antimicrobial properties of sodium alginate/carboxymethyl cellulose films incorporated with cinnamon essential oil. *Food Packaging and Shelf Life*, *15*, 35–42.

Hassan, M. E. S., Bai, J., & Dou, D. Q. (2019). Biopolymers; definition, classification and applications. *Egyptian Journal of Chemistry*, *62*(9), 1725–1737.

Heinze, T., & Liebert, T. (2001). Unconventional methods in cellulose functionalization. *Progress in Polymer Science*, *26*(9), 1689–1762.

Hibbitt, Karlsson, & Sorensen. (1997). *ABAQUS/Standard: User's Manual*, Vol. 1. Hibbitt, Karlsson & Sorensen.

Hohenberg, P., & Kohn, W. (1964). Inhomogeneous electron gas. *Physical Review*, *136*(3B), B864.

Hufnagl, B., Stibi, M., Martirosyan, H., Wilczek, U., Möller, J. N., Löder, M. G., . . . & Lohninger, H. (2021). Computer-assisted analysis of microplastics in environmental samples based on μFTIR imaging in combination with machine learning. *Environmental Science & Technology Letters*, *9*(1), 90–95.

Jha, S., Akula, B., Enyioma, H., Novak, M., Amin, V., & Liang, H. (2024). Biodegradable biobased polymers: a review of the state of the art, challenges, and future directions. *Polymers*, *16*(16), 2262.

Johansen, M. B., Donslund, B. S., Larsen, E., Olsen, M. B., Pedersen, J. A., Boye, M., . . . & Skrydstrup, T. (2023). Closed-loop recycling of polyols from thermoset polyurethanes by tert-amyl alcohol-mediated depolymerization of flexible foams. *ACS Sustainable Chemistry & Engineering*, *11*(29), 10737–10745.

Joseph, T. M., Unni, A. B., Joshy, K. S., Kar Mahapatra, D., Haponiuk, J., & Thomas, S. (2023). Emerging bio-based polymers from lab to market: current strategies. *Market Dynamics and Research Trends C*, *9*(1), 30.

Kalia, S., & Sharma, S. (Eds.). (2022). *Protein-Based Biopolymers: From Source to Biomedical Applications*. Woodhead Publishing.

Kalman, H., Tripathi, N. M., Gabrieli, O. G., & Portnikov, D. (2017). Phase diagrams for pneumatic and hydraulic conveying. *The Proceeding of the 18th International Conferences on Transport and Sedimentation of Solid Particles, T and S 2017*, Prague, Czech Republic (pp. 145–152). ISSN 0867-7964, ISBN 978-83-7717-269-8.

Kalman, H., Tripathi, N. M., Gabrieli, O. G., & Portnikov, D. (2018). Effect of fluid on particles conveying in pipes. *The Conference Proceedings of the 8th World Congress on Particle Technology*, Orlando, FL, USA. ISBN: 978-0-8169-1105-9.

Kamel, S., Ali, N., Jahangir, K., Shah, S. M., & El-Gendy, A. A. (2008). Pharmaceutical significance of cellulose: a review. *Express Polymer Letters*, *2*(11), 758–778.

Khanna, S., & Srivastava, A. K. (2005). Recent advances in microbial polyhydroxyalkanoates. *Process Biochemistry*, *40*(2), 607–619.

Koller, M., & Mukherjee, A. (2022). A new wave of industrialization of PHA biopolyesters. *Bioengineering*, *9*(2), 74.

Lapcık Jr., L., Lapcık, L., De Smedt, S., Demeester, J., & Chabrecek, P. (1998). Hyaluronan: preparation, structure, properties, and applications. *Chemical Reviews*, *98*(8), 2663–2684.

Lee, K. Y., & Mooney, D. J. (2012). Alginate: properties and biomedical applications. *Progress in Polymer Science*, *37*(1), 106–126.

LeRoux, M. A., Guilak, F., & Setton, L. A. (1999). Compressive and shear properties of alginate gel: effects of sodium ions and alginate concentration. *Journal of Biomedical Materials Research: An Official Journal of the Society for Biomaterials, the Japanese Society for Biomaterials, and the Australian Society for Biomaterials and the Korean Society for Biomaterials*, *47*(1), 46–53.

Lionetto, F., & Esposito Corcione, C. (2021). Recent applications of biopolymers derived from fish industry waste in food packaging. *Polymers*, *13*(14), 2337.

Mahmood, A., Patel, D., Hickson, B., DesRochers, J., & Hu, X. (2022). Recent progress in biopolymer-based hydrogel materials for biomedical applications. *International Journal of Molecular Sciences*, *23*(3), 1415.

Marrink, S. J., De Vries, A. H., & Mark, A. E. (2004). Coarse grained model for semiquantitative lipid simulations. *The Journal of Physical Chemistry B*, *108*(2), 750–760.

Marson, A., Masiero, M., Modesti, M., Scipioni, A., & Manzardo, A. (2021). Life cycle assessment of polyurethane foams from polyols obtained through chemical recycling. *ACS Omega*, *6*(2), 1718–1724.

Matsen, M. W., & Schick, M. (1994). Stable and unstable phases of a diblock copolymer melt. *Physical Review Letters*, *72*(16), 2660.

Metropolis, N., Rosenbluth, A. W., Rosenbluth, M. N., Teller, A. H., & Teller, E. (1953). Equation of state calculations by fast computing machines. *The Journal of Chemical Physics*, *21*(6), 1087–1092.

Naveh, R., Tripathi, N. M., & Kalman, H. (2017). Experimental pressure drop analysis for horizontal dilute phase particle-fluid flows. *Powder Technology*, *321*, 355–368. https://doi.org/10.1016/j.powtec.2017.08.029

Pan, Y., Farmahini-Farahani, M., O'Hearn, P., Xiao, H., & Ocampo, H. (2016). An overview of bio-based polymers for packaging materials. *Journal of Bioresources and Bioproducts*, *1*(3), 106–113.

Patel, M., Marscheider-Weidemann, F., Schleich, J., Hüsing, B., Angerer, G., Wolf, O., . . . & Patel, M. (2005). Techno-economic feasibility of large-scale production of bio-based polymers in Europe. *Techncial Report EUR*, 22103.

Peres, C. B., Resende, P. M., Nunes, L. J., & Morais, L. C. D. (2022). Advances in carbon capture and use (CCU) technologies: a comprehensive review and CO2 mitigation potential analysis. *Clean Technologies*, *4*(4), 1193–1207.

Plimpton, S. (1995). Fast parallel algorithms for short-range molecular dynamics. *Journal of Computational Physics*, *117*(1), 1–19.

Raccuglia, P., Elbert, K. C., Adler, P. D., Falk, C., Wenny, M. B., Mollo, A., . . . & Norquist, A. J. (2016). Machine-learning-assisted materials discovery using failed experiments. *Nature*, *533*(7601), 73–76.

Ranjan, R., Tripathi, N. M., Desai, D. J., & Sharma, A. (2023). Performance enhancement of a Ge-Sn Co-alloyed all-inorganic based perovskite solar cell using an organic hole transport layer. *International Conference on Emerging Materials, Smart Manufacturing & Computational Intelligence* (Vol. 13).

Ranjan, R., Verma, S., Kumar, P., & Tripathi, N. M. (2024). 3D-printing approach for fabricating solar cells and piezoelectric devices: a holistic review. *Progress in Engineering Science*, 100016. https://doi.org/10.1016/j.pes.2024.100016

Ren, H., Qiao, F., Shi, Y., Knutzen, M. W., Wang, Z., Du, H., & Zhang, H. (2015). Plantbottle™ packaging program is continuing its journey to pursue bio-mono-ethylene glycol using agricultural waste. *Journal of Renewable and Sustainable Energy*, *7*(4).

Santo, N., Portnikov, D., Tripathi, N. M., & Kalman, H. (2018a). Experimental study on the particle velocity development profile and acceleration length in horizontal dilute phase pneumatic conveying systems. *Powder Technology*, *339*, 368–376. https://doi.org/10.1016/j.powtec.2018.07.074

Santo, N., Tripathi, N. M., & Kalman, H. (2018b). Particle velocity reduction in horizontal-horizontal bends of dilute phase pneumatic conveying. *35th Israeli Conference on Mechanical Engineering (ICME)*, Beer-Sheva, Israel.

Sashiwa, H., & Aiba, S. I. (2004). Chemically modified chitin and chitosan as biomaterials. *Progress in Polymer Science*, *29*(9), 887–908.

Scacchi, A., Vuorte, M., & Sammalkorpi, M. (2024). Multiscale modelling of biopolymers. *Advances in Physics: X*, *9*(1), 2358196.

Series, J. S. on *Biocatalysis*, Vol. 10.

Shaghaleh, H., Xu, X., & Wang, S. (2018). Current progress in the production of biopolymeric materials based on cellulose, cellulose nanofibers, and cellulose derivatives. *RSC Advances*, *8*(2), 825–842.

Sharma, K., Singh, S., Yadav, M., Yadav, S., & Tripathi, N. M. (2015). A review on the performance of the nanofluid based solar collectors–solar energy. *Asian Review of Mechanical Engineering*, *4*(1), 17–26. https://doi.org/10.51983/arme-2015.4.1.2391

Sharma, N., Singh, D., Rani, R., Sharma, D., Pandey, H., & Agarwal, V. (2019). Chitosan and its nanocarriers: applications and opportunities. In *Nanomaterials in Plants, Algae and Microorganisms* (pp. 267–286). Academic Press.

Singh, S., Sharma, K., Lal, K., & Tripathi, N. M. (2015). To study the behaviour of nanorefrigerant in vapour compression cycle-a review. *International Journal of Research in Engineering and Technology*, *4*(04), 648–652.

Siracusa, V., & Blanco, I. (2020). Bio-polyethylene (Bio-PE), Bio-polypropylene (Bio-PP) and Bio-poly (ethylene terephthalate)(Bio-PET): recent developments in bio-based polymers analogous to petroleum-derived ones for packaging and engineering applications. *Polymers*, *12*(8), 1641.

Sriamornsak, P. (2011). Application of pectin in oral drug delivery. *Expert Opinion on Drug Delivery*, *8*(8), 1009–1023.

Stoppel, W. L., Ghezzi, C. E., McNamara, S. L., Iii, L. D. B., & Kaplan, D. L. (2015). Clinical applications of naturally derived biopolymer-based scaffolds for regenerative medicine. *Annals of Biomedical Engineering*, *43*, 657–680.

Studio, D. (2013). *3.5*. San Diego, CA; Accelrys Software Inc.

Tagliavini, G., Solari, F., & Montanari, R. (2018). CFD simulation of a co-rotating twin-screw extruder: validation of a rheological model for a starch-based dough for snack food. *International Journal of Food Engineering*, *14*(2), 20170116.

Tharanathan, R. N. (2003). Biodegradable films and composite coatings: past, present and future. *Trends in Food Science & Technology*, *14*(3), 71–78.

Thomas, S., Rane, A. V., Kanny, K., Abitha, V. K., & Thomas, M. G. (Eds.). (2018). *Recycling of Polyurethane Foams*. Elsevier.

Tripathi, N. M., Francqui, F., & Lumay, G. (2019a). How does the thermal degradation influence the Spreadability, packing dynamics, and electrostatic behaviour of metallic powders?. *Additive Manufacturing Journal*, AMSI, *1*(3), 62–70.

Tripathi, N. M., Francqui, F., Pirenne, T., & Lumay, G. (2019b). Measuring Food Powders Electrical Properties as a result of anti-static content. *The Proceedings of 2019 AIChE Annual Meeting. American Institute of Chemical Engineers*, San Francisco, CA, USA. ISBN: 978-0-8169-1112-7.

Tripathi, N. M., Kalman, H., & Levy, A. (2017). Analysis of acceleration pressure drop in dilute phase pneumatic conveying. *International Symposium of Reliable Flow of Particulate Solidsv, RELPOWFLO V*, Skien, Norway.

Tripathi, N. M., Kalman, H., & Levi, A. (2019c). *Bend Pressure Drop for Pneumatic Conveying= יטמואינפ עונישב םיפופיכ לע ץחל לפמ* (Ph.D. thesis). Ben-Gurion University of the Negev, Israel.

Tripathi, N. M., Levy, A., & Kalman, H. (2016). Initial acceleration pressure drop in dilute phase pneumatic conveying system. *Powder, Granule and Bulk Solids: Innovations and Applications Conference*.

Tripathi, N. M., Levy, A., & Kalman, H. (2018a). Acceleration pressure drop analysis in horizontal dilute phase pneumatic conveying system. *Powder Technology*, *327*, 43–56. https://doi.org/10.1016/j.powtec.2017.12.045

Tripathi, N. M., Levy, A., & Kalman, H. (2018b). Bend pressure drop analysis in dilute phase pneumatic conveying. *The Conference Proceeding of Conveying and Handling of Particulate Solids (CHoPS-2018)*, Greenwich, London, United Kingdom.

Tripathi, N. M., & Mallick, S. S. (2014). *An Investigation into Pressure Drop Across Bends for Fluidised Densephase Pneumatic Conveying Systems* (Doctoral dissertation). Thapar University, India.

Tripathi, N. M., & Mallick, S. S. (2017). Pneumatic conveying of Fly Ash: Bend models investigation. *Advanced Materials Proceedings*, VBRI Press, *2*(8), 526–531. https://doi.org/10.5185/amp.2017/812

Tripathi, N. M., Santo, N., Kalman, H., & Levy, A. (2018c). Analysis of particle flow in vertical dilute phase pneumatic conveying. *35th Israeli Conference on Mechanical Engineering (ICME)*, Beer-Sheva, Israel.

Tripathi, N. M., Santo, N., Kalman, H., & Levy, A. (2018d). Experimental analysis of particle velocity and acceleration in vertical dilute phase pneumatic conveying. *Powder Technology*, *330*, 239–251. https://doi.org/10.1016/j.powtec.2018.02.017

Tripathi, N. M., Santo, N., Levy, A., & Kalman, H. (2018e). Experimental Analysis of Bend effects and secondary acceleration in vertical dilute phase pneumatic conveying. *The Conference Proceeding of Conveying and Handling of Particulate Solids (CHoPS-2018)*, Greenwich, London, United Kingdom.

Tripathi, N. M., Santo, N., Levy, A., & Kalman, H. (2019d). Experimental analysis of velocity reduction in bends related to vertical pipes in dilute phase pneumatic conveying. *Powder Technology*, *345*, 190–202. https://doi.org/10.1016/j.powtec.2019.01.001

Tripathi, N., Sharma, A., Mallick, S. S., & Wypych, P. W. (2015). Energy loss at bends in the pneumatic conveying of fly ash. *Particuology*, *21*, 65–73. https://doi.org/10.1016/j.partic.2014.09.003

Tripodo, G., Trapani, A., Torre, M. L., Giammona, G., Trapani, G., & Mandracchia, D. (2015). Hyaluronic acid and its derivatives in drug delivery and imaging: recent advances and challenges. *European Journal of Pharmaceutics and Biopharmaceutics*, *97*, 40.

Udayakumar, G. P., Muthusamy, S., Selvaganesh, B., Sivarajasekar, N., Rambabu, K., Banat, F., . . . & Show, P. L. (2021). Biopolymers and composites: properties, characterization and their applications in food, medical and pharmaceutical industries. *Journal of Environmental Chemical Engineering*, *9*(4), 105322.

Verma, S. K., Kumar, P., Mishra, A., Khare, R., & Singh, D. (2024a). Green nanotechnology: illuminating the effects of bio-based nanoparticles on plant physiology. *Biotechnology for Sustainable Materials*, *1*(1), 1.

Verma, S. K., Singh, D., & Misra, S. (2024b). Innovative approaches to sustainable waste management and energy generation in developing nations. In *Municipal Solid Waste Management and Recycling Technologies* (pp. 267–300). IGI Global.

Vernon, B. (Ed.). (2011). *Injectable Biomaterials: Science and Applications*. Elsevier.

Vink, E. T., Glassner, D. A., Kolstad, J. J., Wooley, R. J., & O'Connor, R. P. (2007). The eco-profiles for current and near-future NatureWorks® polylactide (PLA) production. *Industrial Biotechnology*, *3*(1), 58–81.

Wool, R., & Sun, X. S. (2011). *Bio-based Polymers and Composites*. Elsevier.

Xu, R. Z., Cao, J. S., Luo, J. Y., Feng, Q., Ni, B. J., & Fang, F. (2022). Integrating mechanistic and deep learning models for accurately predicting the enrichment of polyhydroxyalkanoates accumulating bacteria in mixed microbial cultures. *Bioresource Technology*, *344*, 126276.

Zhang, Z., Hu, S., Chen, D., & Zhu, B. (2016). An analysis of an ethanol-based, whole-crop refinery system in China. *Chinese Journal of Chemical Engineering*, *24*(11), 1609–1618.

# 3 Bio-Based Coating and Surface Treatment

*Deokrishna Kumar Choudhary, Khushi Bhardwaj, Dube Dheeraj Prakashchand, Ravindra Naik Bukke, and Rahul Sharma*

## 3.1 INTRODUCTION

The increasing environmental concerns and regulatory pressures have led to a growing interest in sustainable materials. Bio-based coatings and surface treatments offer a promising solution by reducing reliance on fossil fuels and minimizing greenhouse gas emissions. These coatings, derived from renewable biological sources, provide comparable or superior performance to traditional coatings while being biodegradable and environmentally friendly. This chapter delves into the science behind bio-based coatings, their applications, advantages, and the challenges faced in their widespread adoption. Bio-based coatings are environmentally friendly coatings that are made from renewable biological sources, that is, plant oils, proteins, carbohydrates, and natural polymers (Verma et al., 2024a). They are formulated to mimic or even surpass traditional petrochemical-based coatings in terms of durability, UV stability, adhesion, and resistance to moisture and microbiological growth; as they are biodegradable, they help reduce waste and pollution (Verma et al., 2024b). These days, engineered smart biomaterials exhibit improved characteristics, including cell adhesion and proliferation, and are made to resemble tissue behaviors and qualities. Thus, a variety of polymeric materials, including silicone, polyvinylidene fluoride (PVDF), polypropylene (PP), polyethylene (PE), and poly (methyl methacrylate) (PMMA), can be employed as coatings or as simple medical devices (Wendels et al., 2021). The manufacture of things made from sustainable and renewable resources that are neither dangerous to humans nor the environment is a major concern for our civilization. In this context, the development of sustainable coatings with enhanced and multifunctional properties is required. As a result, the quest for bio-based coatings with high barrier, water resistance, and antibacterial properties is underway. Nowadays, widely utilized materials with excellent moisture barrier qualities for coating manufacture are primarily petroleum based. This must be the obstacle to overcome in research in the coming years.

### 3.1.1 Historical Perspective

The concept of bio-based coatings is not entirely new. Historically, natural materials were used for protective and decorative coatings long before the advent of synthetic

 DOI: 10.1201/9781003566205-3

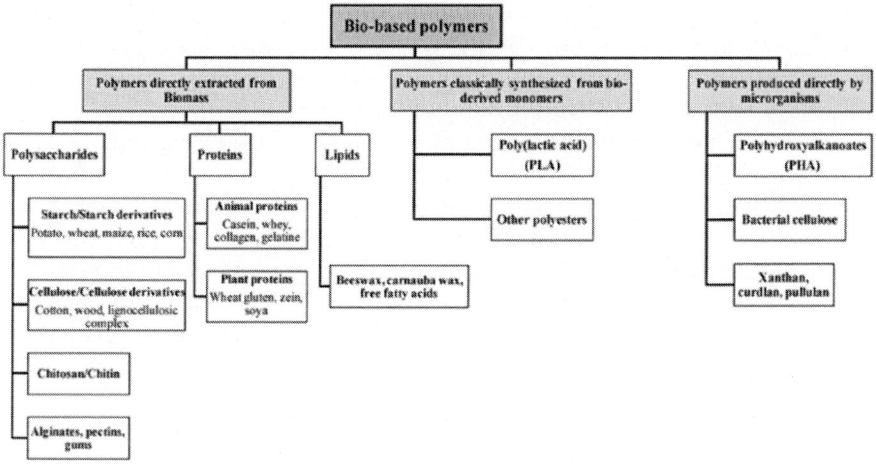


**FIGURE 3.1** Schematic overview of bio-based polymer extraction from biomass, bioderived monomers, and microorganisms, categorized by their origin and production methods.

chemicals. Ancient civilizations utilized natural oils, resins, and waxes for coating applications. For instance, Egyptians used linseed oil to protect their wooden artifacts, while early European painters employed egg tempera, a mixture of egg yolk and pigments, as a durable and vibrant coating for their art. The Industrial Revolution and the development of petrochemical industries in the 19th and 20th centuries shifted the focus toward synthetic coatings, which offered superior performance and consistency. However, the environmental drawbacks of these synthetic materials, including pollution and resource depletion, have rekindled interest in bio-based alternatives. The modern development of bio-based coatings began in earnest in the late 20th century, driven by advancements in green chemistry and growing environmental awareness (Ramezanpour et al., 2024). A bio-based coating with good barrier, water resistance, and antibacterial properties is what is being sought after. These days, materials that are widely employed to produce coatings and have good moisture barrier qualities are essentially petroleum based. This has to be the obstacle that research overcomes in the upcoming years. Prior to delving into the specifics of the review topic, it is vital to emphasize and elucidate the definition of biodegradable and bio-based polymers that are utilized in the manufacture of coatings. The term "biodegradation" describes the breakdown and catabolization of biodegradable polymers into $CO_2$ and $H_2O$ by bacteria and/or enzymes. Alternatively, bio-based polymers can be divided into groups according to the source, method of synthesis, and formulation, as shown in Figure 3.1.

### 3.1.2 Bio-based Coatings Innovations (2020–2024)

Experiments conducted from 2020 to 2024 aimed to enhance bio-based coatings. Key findings include improved barrier properties, with nanocellulose enhancing oxygen and water resistance in polysaccharide-based coatings, making them suitable for packaging, and antibacterial coatings were developed using natural extracts (e.g., essential oils) and zinc oxide (ZnO) nanoparticles. These coatings are used in food packaging to increase the product's shelf life. For UV protection, the UV resistance of coatings is improved by the addition of bio-based UV-absorbing chemicals, like ferulic acid, which is crucial for outdoor applications.

Nowadays, the barrier coatings for paper predominantly rely on fossil-based or synthetic polymers, that is, PE (polyolefins), ethylene vinyl alcohol (EVOH), waxes, and polyvinylidene chloride (PVDC), valued for their cost-effectiveness and accessibility. These materials offer strong water and oxygen barriers for food packaging. However, limitations like finite fossil resources, poor recyclability, and non-biodegradability raise environmental and economic issues, fueling interest in sustainable biopolymer films and coatings with comparable properties (Rastogi & Samyn, 2015). Figure 3.1 shows the side classification of the bio-based polymer. It has three main categories: polymer directly extracted from the biomass, some biological monomers, and microorganisms, which are again divided into polysaccharides, proteins, lipids, polylactic acid, bacterial cellulose, and so on (Wang et al., 2017; Karoki et al., 2024).

### 3.1.3 Bio-based Polyurethane (PU)

Natural processes like metallic structural corrosion can cause significant financial and environmental losses. The most effective way to prevent metallic corrosion and reduce significant corrosion losses is to apply a physical layer, such as an inorganic, organic, or conversion coating (Noreen et al., 2016). Bio-based PU is a versatile polymer derived from renewable resources such as vegetable oils and natural polyols. It serves as an eco-friendly alternative to conventional PUs synthesized from petrochemicals, significantly reducing environmental impact by minimizing reliance on fossil fuels and lowering greenhouse gas emissions. One of the most important issues facing polymer research and engineering today is the creation of ecologically friendly and sustainable polymer products. Because of their adaptable qualities and numerous application areas, bio-based polyesters have garnered more attention as a means of meeting this need. The rapidly expanding bio-based sector has seen the emergence of numerous bio-based polyesters in recent years, thanks to advancements in more sophisticated technology for producing bio-based monomers (Figure 3.2). The bio-based polymers are synthesized through the polymerization of bio-based monomers derived from renewable sources like plants, microorganisms, or algae. It can be the chemical synthesis, such as polycondensation, esterification, and raw material extraction, or we can use some biotechnological approach for the bio-based coating, such as bio-based monomers and dicarboxylic acid (Zhang et al., 2021; Qi et al., 2018; Gupta et al., 2010)

In Figure 3.3, the bio-based PUs and their types show that they have a variety of applications due to their high mechanical characteristics, flexibility, and durability. They are employed in construction as protective coatings and insulation, and in the automotive sector as seats and interior components. Elastomers are essential to producing durable items like tires and seals, and the furniture sector benefits from their use in cushions and coatings (Ortiz et al., 2020).

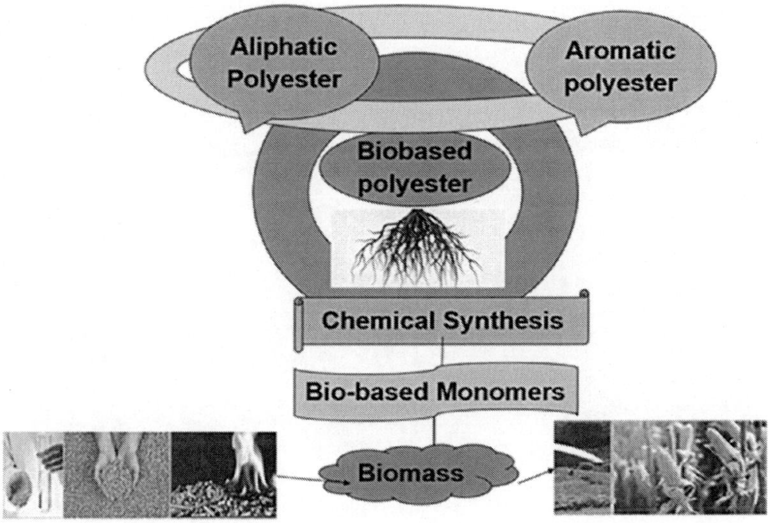


**FIGURE 3.2** A general route for the synthesis process flow.

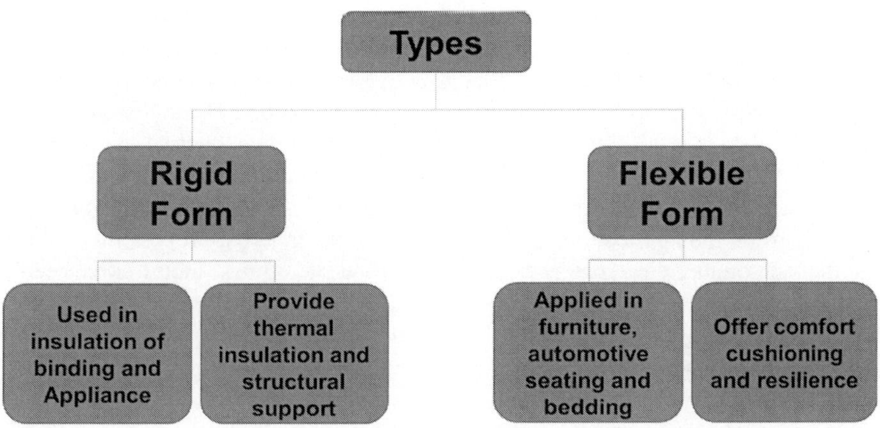


**FIGURE 3.3** Flowchart illustrating types of bio-based polyurethane for rigid insulation, binding applications, and flexible forms for furniture, highlighting their advantages.

### 3.1.4 Bio-based PUs Used in Biomedical Applications

PUs have recently become popular in biomedical applications due to their biocompatibility, excellent chemical stability, and low cytotoxicity. To address challenges related to recycling and degradation, the development of bio-based PUs has emerged as a major focus in current research. PUs, like all other materials in this industry,

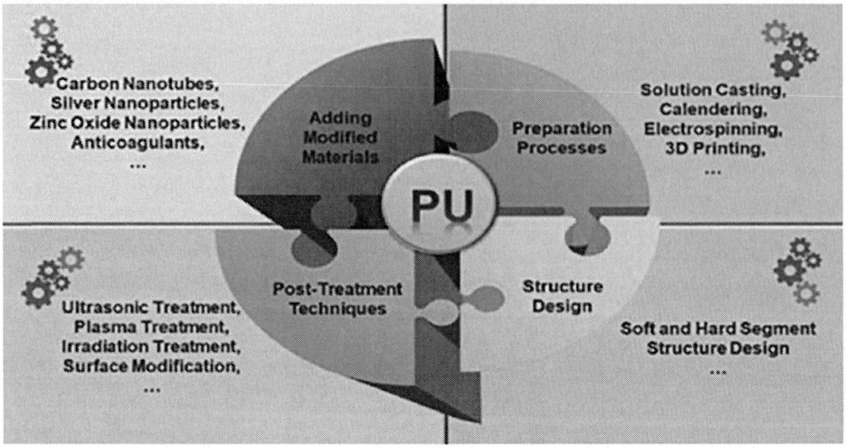


**FIGURE 3.4** Methods for functionalizing polyurethane materials, covering synthesis, design, surface treatments, and material modifications, as outlined.

should go through a long procedure and meet stringent requirements for cytotoxicity, acute and sub-chronic toxicity, carcinogenicity, and hemocompatibility. Requirements depend on the specific living tissue and expected contact duration. These criteria are detailed in regulations like ISO 10993, the global benchmark for the biological assessment of medical devices, recognized by the FDA and European CE marking (Reichert et al., 2020). PU can acquire flexible, conductive, antibacterial, and antithrombotic characteristics through the incorporation of various modifying agents, including conductive materials, antibacterial substances, and anticoagulants, alongside the application of diverse preparative and post-processing methods. To enhance conductivity, materials such as carbon nanotubes, metallic nanowires, and conductive polymers can be integrated, while antimicrobial properties can be introduced through agents like silver nanoparticles, zinc oxide nanoparticles, and chloride-containing polymers. Additionally, the antithrombotic properties can be improved by incorporating modifiers such as natural polysaccharides, anticoagulants, and antiplatelet agents [Cui et al., 2023; Samper et al., 2019].

### 3.1.5 Bio-based Epoxy

Bio-based epoxy resins, derived from renewable sources rather than petroleum-based materials, are attracting attention for their environmental benefits and reduced reliance on fossil fuels. Epoxy resins remain dominant in the coatings market due to their versatility and excellent balance of mechanical strength, adhesive properties, dimensional stability, and resistance to thermal and chemical exposure. These resins are widely used in applications such as adhesives, composites, construction materials, laminates, and electrical insulators. The global market size of epoxy resins reached US$25.8 billion in 2018 and is anticipated to be over US$34 billion by 2022 (da Silva et al., 2022). By 2024, the global market for epoxy resins is expected to

**TABLE 3.1**
**Overview of Bio-based Epoxy Coatings, Including Their Sources, Applications, and Advantages.**

| S. No. | Bio-based epoxy type | Source | Applications | Key benefits | [Ref] |
|---|---|---|---|---|---|
| 11 | Soy-based epoxy | Soybean oil | Protective coatings, adhesives, and composites | Renewable, low VOC emissions | (Zhen et al., 2024) |
| 22 | Linseed oil epoxy | Linseed oil | Industrial coatings and paints | High biodegradability and low toxicity | (Bobade et al., 2016) |
| 33 | Castor oil epoxy | Castor oil | Coatings and electronics encapsulation | High chemical resistance and good mechanical properties | (Lee et al., 2014) |
| 44 | Cellulose-based epoxy | Cellulose (plant fibers) | Structural composites and automotive parts | High strength-to-weight ratio and sustainable | (Colucci et al., 2014) |
| 55 | Lignin-based epoxy | Lignin (wood byproduct) | Coatings, adhesives, and composites | Utilizes waste biomass and cost-effective | (Laurichesse et al., 2014) |

exceed US$37 billion, reflecting continued growth driven by demand in various industrial applications. Table 3.1 shows different types of bio-based epoxy, their source, and their application. The key benefit of bio-based epoxy, such as soya-based, linseed oil-based, castor oil-based, and lignin-based epoxy, is clearly shown in Table 3.1—it has lots of applications in protective coatings, adhesive coatings, industrial coatings, electronics, encapsulation, etc.

### 3.1.6 Bio-based Polyester

Bio-based polyester has emerged as a promising candidate for eco-friendly coating applications, offering a sustainable alternative to conventional petrochemical-based coatings. Derived from renewable resources such as plant oils, sugars, and other biomass, bio-based polyester coatings provide comparable performance characteristics, such as durability, adhesion, and chemical resistance, while significantly reducing the environmental footprint. One of the key advantages of bio-based polyester as a coating material is its potential for reducing volatile organic compound (VOC) emissions. Traditional coatings often rely on solvents that release VOCs, contributing to air pollution and health hazards. In contrast, bio-based polyester coatings can be formulated with low or no VOC content, aligning with stringent environmental regulations and increasing consumer demand for greener products

(Borić et al., 2019). The synthesis of bio-based polyester coatings typically involves the polymerization of bio-derived monomers, such as fatty acids, glycerol, and organic acids, through processes like ring-opening polymerization and polycondensation. These monomers are obtained from renewable feedstocks, including vegetable oils (e.g., soybean and linseed) and sugars (e.g., glucose). The resulting bio-based polyesters can be engineered to achieve desired coating properties, such as hardness, flexibility, and thermal stability (Laurichesse et al., 2014). The bio-based epoxy coatings, including their sources, applications, and advantages, are summarized in Table 3.1. Moreover, bio-based polyester coatings offer the advantage of biodegradability or recyclability, contributing to the circular economy. These coatings can be designed to break down under specific environmental conditions, thereby minimizing the environmental impact at the end of their life cycle. Furthermore, it has the potential for recycling bio-based polyester coatings opens new avenues for sustainable material management (Muñoz-Bonilla et al., 2019). However, despite these benefits, challenges remain in the widespread adoption of bio-based polyester coatings. Coating powder rheology was also tested by different researchers to investigate the angle of repose, flowability, and electrostatic charge (Tripathi et al., 2019a, 2019b, 2019c, 2019d, 2020, 2024; Neveu et al., 2020; Lumay et al., 2019a, 2019b; Kalman et al., 2019a, 2019b, 2019c). One of the primary obstacles is the cost of raw materials and production processes, which are often higher than those of petrochemical-based coatings. The performance of bio-based coatings in extreme environmental conditions, such as high humidity and UV exposure, needs further improvement to match the durability of traditional coatings (Rauta et al., 2019). The various bio-based polyester coatings, their material sources, and key applications are summarized in Table 3.2. Rising sustainability demands are driving R&D in bio-based polyester coatings, with biotech advances lowering costs, enhancing functionality, and industry–academia collaboration aiding global commercialization (Kim & Chung, 2024).

### 3.1.7 Bio-based Alkyd

Bio-based alkyds are a type of sustainable coating made from renewable resources such as vegetable oils, natural fatty acids, and glycerol. These materials have received a lot of interest because they have the potential to replace existing petrochemical-based alkyd resins, which are widely used in paints, varnishes, and other protective coatings. Bio-based alkyd coatings represent a significant innovation in the coatings industry, offering an eco-friendly alternative to traditional petrochemical-based alkyd resins. Derived from renewable resources such as vegetable oils and natural fatty acids, these coatings are increasingly gaining attention for their potential to reduce the environmental impact of industrial and consumer products.

## 3.2 COMPOSITION AND SYNTHESIS

The foundation of bio-based alkyd coatings lies in the use of natural oils and fatty acids as key raw materials. Common sources include linseed oil, soybean oil, sunflower oil, and castor oil. These oils are rich in triglycerides, which are esters formed from glycerol and three fatty acid molecules. The synthesis of alkyd resins involves reacting these triglycerides with polyols, such as glycerol or pentaerythritol, and polybasic acids

**TABLE 3.2**
**Overview of Various Bio-based Polyester Coatings, Their Material Sources, and Key Applications.**

| S. No. | Bio-based polyester coating type | Source of bio-based material | Applications |
|---|---|---|---|
| 11 | Polylactic acid (PLA) coatings | Corn, sugarcane, and beet sugars | 1. Food packaging<br>2. Disposable containers<br>3. Coatings for paper products |
| 22 | Polyhydroxyalkanoates (PHA) coatings | Bacterial fermentation of plant oils and sugars | 1. Biomedical applications<br>2. Biodegradable packaging<br>3. Coatings for medical devices |
| 33 | Poly(butylene succinate) (PBS) coatings | Succinic acid from sugar fermentation and 1, 4-butanediol from biomass | 1. Agricultural films<br>2. Food packaging<br>3. Mulch films |
| 44 | Poly(trimethylene terephthalate) (PTT) coatings | 1,3-Propanediol from corn glucose | 1. Textiles<br>2. Automotive interiors<br>3. Carpets<br>4. Industrial coatings |
| 55 | Soybean oil-based polyester coatings | Soybean oil | 1. Industrial coatings<br>2. Protective wood finishes<br>3. Adhesives |
| 66 | Poly(ethylene furanoate) (PEF) coatings | Furan-2,5-dicarboxylic acid (FDCA) from sugars and ethylene glycol from biomass | 1. Bottles<br>2. Food packaging<br>3. Barrier coatings for containers |

or anhydrides, such as phthalic anhydride. This reaction forms a complex polyester network that gives alkyd resins their desirable properties, such as durability, gloss, and flexibility. One of the critical factors in the performance of bio-based alkyd coatings is the degree of unsaturation in the fatty acids. Unsaturated fatty acids, which contain double bonds, enable the formation of cross-links during the drying process, resulting in a more robust and durable coating. Oils with higher unsaturation levels, such as linseed oil, are preferred for faster drying times and improved film hardness.

### 3.2.1 Environmental Benefits

Bio-based alkyd coatings offer several environmental advantages over traditional coatings. First and foremost, they are derived from renewable resources, which helps reduce the reliance on fossil fuels and the associated carbon footprint. The use of natural oils not only promotes sustainability but also supports the agricultural sector by creating value-added products from crops. In addition to their renewable origin, bio-based alkyds typically emit lower levels of VOCs during application and curing.

VOCs are harmful chemicals that contribute to air pollution and pose health risks to both workers and end-users. Reducing VOC emissions is a significant benefit, aligning with increasingly stringent environmental regulations and growing consumer demand for greener and safer products. Despite their environmental benefits, bio-based alkyd coatings do not compromise performance. They provide excellent adhesion to various substrates, including wood, metal, and plastic. The coatings are also known for their flexibility, allowing them to withstand the expansion and contraction of materials without cracking. This makes them ideal for applications where durability and aesthetic appeal are crucial, such as in architectural coatings, wood finishes, and protective industrial coatings. Moreover, the natural oils used in bio-based alkyd coatings impart hydrophobic properties, enhancing their resistance to water and moisture. This makes them suitable for exterior applications, where weather resistance is essential. The coatings also exhibit good chemical resistance, protecting surfaces from corrosion and degradation caused by exposure to harsh environments.

### 3.2.2 Bio-based Acrylic

Bio-based acrylics have emerged as a game changer in the realm of coatings and surface treatments, providing a more sustainable alternative to traditional petrochemical-based acrylics. These materials are made from renewable resources such as plant-based sugars, starches, and oils, which are converted into acrylic monomers and then polymerized to produce high-performance resins. As businesses prioritize sustainability and environmental responsibility, bio-based acrylics are playing an important role in the development of eco-friendly coatings and surface treatments. The foundation of bio-based alkyd coatings lies in the use of natural oils and fatty acids as key raw materials. Common sources include linseed oil, soybean oil, sunflower oil, and castor oil. These oils are rich in triglycerides, which are esters formed from glycerol and three fatty acid molecules. The synthesis of alkyd resins involves the reaction of these triglycerides with polyols, such as glycerol or pentaerythritol, and polybasic acids or anhydrides, such as phthalic anhydride. This reaction forms a complex polyester network that gives alkyd resins their desirable properties, such as durability, gloss, and flexibility. One of the critical factors in the performance of bio-based alkyd coatings is the degree of unsaturation in the fatty acids. Unsaturated fatty acids, which contain double bonds, enable the formation of cross-links during the drying process, resulting in a more robust and durable coating. Oils with higher unsaturation levels, such as linseed oil, are preferred for faster drying times and improved film hardness (Zia et al., 2016).

### 3.2.3 Bio-based UV Curable

Bio-based UV-curable coatings, made from renewable resources like plant-based monomers, offer a sustainable alternative to conventional UV-curable coatings. These coatings provide rapid curing, high durability, and strong chemical resistance, making them suitable for use in automotive, electronics, and packaging applications. By reducing dependency on fossil fuels and minimizing environmental impact, bio-based UV-curable coatings align with green chemistry principles and meet the growing demand for eco-friendly products. Recent advances in this technology have enhanced their performance, making them increasingly competitive with traditional UV-curable systems.

### 3.2.4 Bio-based Waterborne

Bio-based waterborne coatings are a significant advancement in the coatings industry, offering a sustainable alternative to traditional solvent-based coatings. These coatings are formulated using water as the primary solvent, reducing the reliance on VOCs and minimizing environmental impact. The incorporation of bio-based materials, such as plant-derived resins, polymers, and additives, further enhances their eco-friendliness. One of the key benefits of bio-based waterborne coatings is their reduced environmental footprint. By utilizing renewable resources like vegetable oils, starches, and natural resins, these coatings contribute to the reduction in greenhouse gas emissions and the conservation of non-renewable resources. Additionally, the low VOC content in waterborne formulations makes them safer for both the environment and human health, aligning with stringent environmental regulations and growing consumer demand for green products. In terms of performance, bio-based waterborne coatings offer excellent adhesion, durability, and flexibility. They can be applied to a wide range of substrates, including wood, metal, plastics, and concrete, making them versatile for various industrial and commercial applications. These coatings also provide good resistance to water, chemicals, and UV radiation, which is essential for protecting surfaces in demanding environments (Omonov et al., 2016; Chek & Ang, 2024).

## 3.3 APPLICATIONS OF BIO-BASED COATINGS AND SURFACE TREATMENTS

Bio-based coatings are gaining popularity in a variety of industries due to their environmental benefits and the growing demand for long-term alternatives to traditional petrochemical-based coatings. These renewable resource-based coatings have a wide range of applications, including decorative finishes and industrial protection. Figure 3.5 illustrates the principal applications of bio-based coatings, with main applications in the fields of automotive, architectural, and industrial coatings.

### 3.3.1 Architectural Coatings

*Interior and exterior paints*: Bio-based coatings are used in both interior and exterior paints for residential and commercial buildings. These coatings provide excellent durability, weather resistance, and aesthetic appeal. Their low VOC emissions make them ideal for indoor environments, contributing to healthier indoor air quality. Moreover, they also offer good adhesion, color retention, and gloss, making them a sustainable choice for walls, ceilings, and exteriors.

*Varnishes and wood finishes*: Bio-based varnishes and wood finishes are commonly used to protect and enhance the natural beauty of wood surfaces. These coatings provide a protective layer that guards against moisture, UV radiation, and wear,

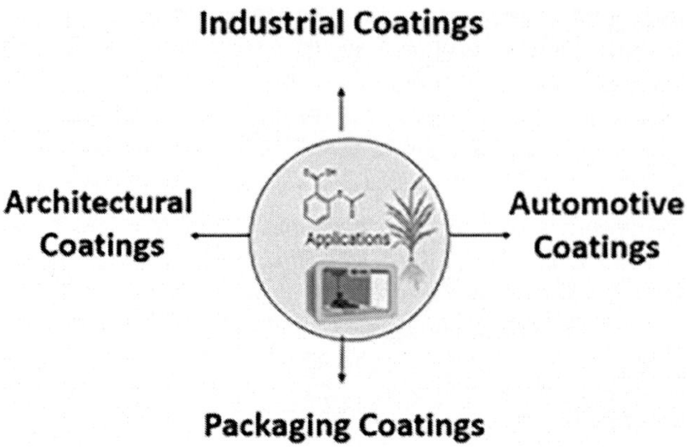


**FIGURE 3.5** Bio-based coatings and its application in automotive, packaging, architectural, and industrial sectors.

extending the life of wooden furniture, flooring, and exterior wooden structures (Stouten et al., 2020).

*Perspective coatings for metals*: In industrial settings, bio-based coatings are applied to metal surfaces to protect against corrosion, wear, and chemical exposure. These coatings are used in manufacturing, machinery, and infrastructure to provide long-lasting protection in harsh environments. Bio-based epoxies and PUs are particularly valued for their high performance in these applications.

*Machinery and equipment*: Bio-based coatings are also used to protect industrial machinery and equipment from environmental factors, including moisture, chemicals, and abrasion. The coatings help reduce maintenance costs and extend the lifespan of equipment, contributing to overall operational efficiency (Panariello et al., 2021).

### 3.3.2 Automotive Coatings

**Exterior and interior coatings:** In the automotive industry, bio-based coatings are applied to both the exterior and interior of vehicles. Exterior coatings provide protection against UV radiation, weathering, and corrosion while maintaining the vehicle's aesthetic appeal with high gloss and color retention. Interior coatings, on the other hand, are used on plastic and metal components to provide durability, scratch resistance, and aesthetic finishes (Fu et al., 2022).

**Underbody coatings:** Bio-based coatings are also used for underbody protection in vehicles, offering resistance to corrosion, impact, and abrasion. These coatings help protect the structural integrity of vehicles, especially in regions with harsh weather conditions (Panariello et al., 2021; Teijido et al., 2023).

**Packaging coatings:** Cinnamaldehyde (CIN), a characteristic bioactive chemical found in the bark and leaves of *Cinnamomum* trees, is both natural and safe. Its high antibacterial action against a variety of foodborne pathogens is gaining popularity as a prospective additive for strengthening and increasing the qualities of bio-based packaging films and coatings (Gigante et al., 2021).

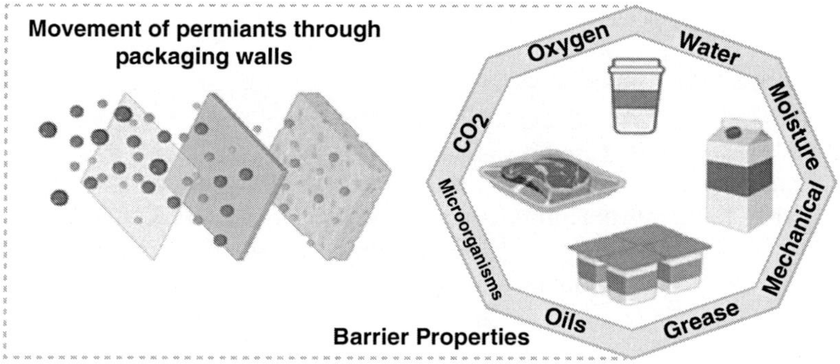


**FIGURE 3.6** Barrier properties of bio-based coatings. Protective films shield packaging from environmental factors (Sun et al., 2024).

**Food packaging:** Bio-based coatings are increasingly used in food packaging to provide barrier properties, such as moisture resistance, grease resistance, and oxygen permeability. These coatings help maintain the freshness and integrity of food products while also contributing to the biodegradability or recyclability of packaging materials (see Figure 3.6).

**Protective films:** Bio-based coatings are also utilized in creating protective films for diverse packaging types beyond food. These films add a layer of defense against environmental factors like humidity and temperature fluctuations, enhancing the safety and stability of goods during transport and storage. As shown in Figure 3.6, bio-based coating materials act as a potential protective barrier, resisting $CO_2$, oxygen, water, moisture, oils, etc. (Zou et al., 2023).

## 3.4 CHALLENGES IN THE DEVELOPMENT AND ADOPTION OF BIO-BASED COATINGS

Despite their advantages, bio-based coatings face several challenges that can hinder their widespread adoption.

**Cost:** The production of bio-based coatings can be more expensive than traditional coatings due to the higher cost of renewable raw materials and the complexity of bio-based synthesis processes. This cost difference can be a barrier to market adoption, particularly in price-sensitive industries.

**Raw material supply:** The availability and quality of bio-based raw materials can vary depending on factors such as crop yields, seasonal variations, and geographic location. This variability can affect the consistency and performance of bio-based coatings, as well as their pricing stability.

**Performance limitations:** While bio-based coatings have made significant strides in performance, there are still some limitations, particularly in demanding applications. For example, bio-based coatings may have longer drying times, lower hardness, or reduced chemical resistance compared to traditional coatings. Ongoing research and development are needed to address these limitations (Sun et al., 2024).

## 3.5 CONCLUSION

An important step toward a more ecologically conscious and sustainable coatings industry is the use of bio-based coatings and surface treatments. These coatings provide an effective substitute for conventional petrochemical-based products by using renewable resources, lowering VOC emissions, and enhancing performance. However, to realize their full potential, issues with cost, performance, and raw material availability must be resolved. Bio-based coatings have the potential to be a significant contributor to sustainable manufacturing and surface treatment technologies in the future through continued research, innovation, and industry collaboration.

## ACKNOWLEDGMENTS

The authors would like to thank SMME, IIT Mandi, for providing the infrastructure facility, and MHRD for the HTRA fellowship.

## REFERENCES

Bobade, S. K., Paluvai, N. R., Mohanty, S., & Nayak, S. K. (2016). Bio-based thermosetting resins for future generation: a review. *Polymer-Plastics Technology and Engineering*, *55*(17), 1863–1896.

Borić, A., Kalendová, A., Urbanek, M., & Pepelnjak, T. (2019). Characterization of polyamide (PA) 12 nanocomposites with montmorillonite (MMT) filler clay used for the incremental forming of sheets. *Polymers*, *11*(8), 1248.

Chek, Y. W., & Ang, D. T. C. (2024). Progress of bio-based coatings in waterborne system: synthesis routes and monomers from renewable resources. *Progress in Organic Coatings*, *188*, 108190.

Colucci, G., Aluigi, A., Tonin, C., & Bongiovanni, R. (2014). Photopolymerization of keratin-based thiol-ene coatings. *Progress in Organic Coatings*, *77*(6), 1104–1110.

Cui, M., Chai, Z., Lu, Y., Zhu, J., & Chen, J. (2023). Developments of polyurethane in biomedical applications: a review. *Resources Chemicals and Materials*, *2*(4), 262–276.

da Silva, L. R., Carvalho, B. A., Pereira, R. C., Diogenes, O. B., Pereira, U. C., da Silva, K. T., . . . & Lomonaco, D. (2022). Bio-based one-component epoxy resin: novel high-performance anticorrosive coating from agro-industrial byproduct. *Progress in Organic Coatings*, *167*, 106861.

Fu, F., Shen, M., Wang, D., Liu, H., Xu, X., Shang, S., . . . & Song, J. (2022). Green preparation of robust hydrophobic bio-based epoxy resin with low dielectric constant. *Industrial Crops and Products*, *188*, 115615.

Gigante, V., Panariello, L., Coltelli, M. B., Danti, S., Obisesan, K. A., Hadrich, A., . . . & Cinelli, P. (2021). Liquid and solid functional bio-based coatings. *Polymers*, *13*(21), 3640.

Gupta, A. P., Ahmad, S., & Dev, A. (2010). Development of novel bio-based soybean oil epoxy resins as a function of hardener stoichiometry. *Polymer-Plastics Technology and Engineering*, *49*(7), 657–661.

Kalman, H., Portnikov, D., Gabrieli, O. G., & Tripathi, N. M. (2019a). What do pneumatic conveying and hydraulic conveying have in common? In J. Sobota & R. Haldenwan (Eds.), *19th International Conference on Transport and Sedimentation of Solid Particles, 2019 (International Conferences on Transport and Sedimentation of Solid Particles; Vol. TS 19)* (pp. 203–210). Wydawnictwo Uniwersytetu Przyrodniczego we Wroclawiu. https://doi.org/10.30825/4.12-23.2019

Kalman, H., Portnikov, D., Gabrieli, O. G., & Tripathi, N. M. (2019b). What do pneumatic conveying and hydraulic conveying have in common?. *Powder Technology*, *354*, 485–495. https://doi.org/10.1016/j.powtec.2019.06.022

Kalman, H., Santo, N., & Tripathi, N. M. (2019c). Particle velocity reduction in horizontal-horizontal bends of dilute phase pneumatic conveying. *Powder Technology*, *356*, 808–817. https://doi.org/10.1016/j.powtec.2019.08.081

Karoki, P. K., Zhang, S., Cai, C. M., Dim, P. E., & Ragauskas, A. J. (2024). Thermally stable and self-healable lignin-based polyester. *Polymer Testing*, *137*, 108515.

Kim, S., & Chung, H. (2024). Biodegradable polymer: from synthesis methods to applications of lignin-graft-polyester. *Green Chemistry*, *26*(21), 10774–10803.

Laurichesse, S., & Avérous, L. (2014). Chemical modification of lignins: towards biobased polymers. *Progress in Polymer Science*, *39*(7), 1266–1290.

Lee, T. J., Kwon, S. H., & Kim, B. K. (2014). Biodegradable sol-gel Coatings of waterborne polyurethane/gelatin chemical hybrids. *Progress in Organic Coatings*, *77*(6), 1111–1116.

Lumay, G., Tripathi, N. M., & Francqui, F. (2019a). How to gain a full understanding of powder flow properties, and the benefits of doing so. *ONdrugDelivery*, *2019*, 42–47.

Muñoz-Bonilla, A., Echeverria, C., Sonseca, Á., Arrieta, M. P., & Fernández-García, M. (2019). Bio-based polymers with antimicrobial properties towards sustainable development. *Materials*, *12*(4), 641.

Neveu, A., Francqui, F., & Tripathi, N. M. (2020). Granular material characterization at the macroscopic scale for the validation and the calibration of numerical models. *The Proceedings of 2020 Virtual AIChE Annual Meeting, (Particle Technology Forum)*, San Francisco, CA, USA. ISBN: 978-0-8169-1114-1.

Noreen, A., Zia, K. M., Zuber, M., Tabasum, S., & Zahoor, A. F. (2016). Bio-based polyurethane: an efficient and environment-friendly coating systems: a review. *Progress in Organic Coatings*, *91*, 25–32.

Omonov, T. S., & Curtis, J. M. (2016). Plant oil-based epoxy intermediates for polymers. In *Bio-Based Plant Oil Polymers and Composites* (pp. 99–125). Elsevier.

Ortiz, P., Vendamme, R., & Eevers, W. (2020). Fully biobased epoxy resins from fatty acids and lignin. *Molecules*, *25*(5), 1158.

Panariello, L., Vannozzi, A., Morganti, P., Coltelli, M. B., & Lazzeri, A. (2021). Biobased and eco-compatible beauty films coated with chitin nanofibrils, nanolignin, and vitamin E. *Cosmetics*, *8*(2), 27.

Qi, M., Xu, Y. J., Rao, W. H., Luo, X., Chen, L., & Wang, Y. Z. (2018). Epoxidized soybean oil cured with tannic acid for fully bio-based epoxy resin. *RSC Advances*, *8*(47), 26948–26958.

Ramezanpour, J., Ramezanzadeh, B., & Samani, N. A. (2024). Progress in bio-based anti-corrosion coatings; A concise overview of the advancements, constraints, and advantages. *Progress in Organic Coatings*, *194*, 108556.

Rastogi, V. K., & Samyn, P. (2015). Bio-based coatings for paper applications. *Coatings*, *5*(4), 887–930.

Rauta, P. R., Mohanta, Y. K., & Nayak, D. (Eds.). (2019). *Nanotechnology in Biology and Medicine: Research Advancements & Future Perspectives*. CRC Press.

Reichert, C. L., Bugnicourt, E., Coltelli, M. B., Cinelli, P., Lazzeri, A., Canesi, I., . . . & Schmid, M. (2020). Bio-based packaging: materials, modifications, industrial applications, and sustainability. *Polymers*, *12*(7), 1558.

Samper, M. D., Ferri, J. M., Carbonell-Verdu, A., Balart, R., & Fenollar, O. (2019). Properties of biobased epoxy resins from epoxidized linseed oil (ELO) crosslinked with a mixture of cyclic anhydride and maleinized linseed oil. *Express Polymer Letters*, *13*(5), 407–418.

Stouten, J., Vanpoucke, D. E., Van Assche, G., & Bernaerts, K. V. (2020). UV-curable biobased polyacrylates based on a multifunctional monomer derived from furfural. *Macromolecules*, *53*(4), 1388–1404.

Sun, J., Leng, X., Zang, J., & Zhao, G. (2024). Bio-based antibacterial food packaging films and coatings containing cinnamaldehyde: a review. *Critical Reviews in Food Science and Nutrition*, *64*(1), 140–152.

Teijido, R., Ruiz-Rubio, L., Lanceros-Méndez, S., Zhang, Q., & Vilas-Vilela, J. L. (2023). Sustainable bio-based epoxy resins with tunable thermal and mechanic properties and superior anti-corrosion performance. *Polymers*, *15*(20), 4180.

Tripathi, N. M., Bocquet, S., Francqui, F., Brochu, M., & Lumay, G. (2019a). Metallic powders thermal degradation: influence on spreadability, packing dynamics and electrostatics. *International Conference on Additive Manufacturing with Powder Metallurgy (AMPM2019)*, Phoenix, AZ, USA (pp. 262–277).

Tripathi, N. M., Francqui, F., & Lumay, G. (2020). Influence of relative air humidity on the flow property of fine Powders. *The Proceeding of Third International Conference on Powder, Granule and Bulk Solids: Innovations and Applications PGBSIA 2020 February 26–28, 2020* (pp. 63–68).

Tripathi, N. M., Kalman, H., & Levy, A. (2019b). Investigation of energy loss in horizontal bends in dilute phase pneumatic conveying systems. *International Congress on Particle Technology, PARTEC-2019*, Nurnberg, Germany.

Tripathi, N. M., Levy, A., & Kalman, H. (2024). Pressure drop through bends at different orientations in dilute phase pneumatic conveying systems. In *Powder Technology Conclave & Exhibition*. Patiala, Punjab, India: PTC.

Tripathi, N. M., Lumay, G., & Francqui, F. (2019c). How to assess cohesion of unconsolidated powders at high flow fields? *9th International Granulation Workshop*, Lausanne, Switzerland.

Tripathi, N. M., Portnikov, D., Levy, A., & Kalman, H. (2019d). Bend pressure drop in horizontal and vertical dilute phase pneumatic conveying systems. *Chemical Engineering Science*, *209*, 115228. https://doi.org/10.1016/j.ces.2019.115228

Verma, S. K., Kumar, P., Mishra, A., Khare, R., & Singh, D. (2024a). Green nanotechnology: illuminating the effects of bio-based nanoparticles on plant physiology. *Biotechnology for Sustainable Materials*, *1*(1), 1.

Verma, S. K., Singh, D., & Misra, S. (2024b). Innovative approaches to sustainable waste management and energy generation in developing nations. In *Municipal Solid Waste Management and Recycling Technologies* (pp. 267–300). IGI Global.

Wang, Z., Yuan, L., Ganewatta, M. S., Lamm, M. E., Rahman, M. A., Wang, J., . . . & Tang, C. (2017). Plant oil-derived epoxy polymers toward sustainable biobased thermosets. *Macromolecular Rapid Communications*, *38*(11), 1700009.

Wendels, S., & Avérous, L. (2021). Biobased polyurethanes for biomedical applications. *Bioactive Materials*, *6*(4), 1083–1106.

Zhang, Q., Song, M., Xu, Y., Wang, W., Wang, Z., & Zhang, L. (2021). Bio-based polyesters: recent progress and prospects. *Progress in Polymer Science*, *120*, 101430.

Zhen, X., Cui, X., Al-Haimi, A. A. N. M., Wang, X., Liang, H., Xu, Z., & Wang, Z. (2024). Fully bio-based epoxy resins from lignin and epoxidized soybean oil: rigid-flexible, tunable properties and high lignin content. *International Journal of Biological Macromolecules*, *254*, 127760.

Zia, K. M., Noreen, A., Zuber, M., Tabasum, S., & Mujahid, M. (2016). Recent developments and future prospects on bio-based polyesters derived from renewable resources: a review. *International Journal of Biological Macromolecules*, *82*, 1028–1040.

Zou, S. L., Xiao, L. P., Li, X. Y., Yin, W. Z., & Sun, R. C. (2023). Lignin-based composites with enhanced mechanical properties by acetone fractionation and epoxidation modification. *Iscience*, *26*(3).

# 4 Bio-Based Materials in Additive Manufacturing

## *Sonication of Sheep Wool Fiber for Feedstock Production for 3D Printing*

*C.T. Ezeh, F. Marra, and S. Simal*

## 4.1 INTRODUCTION

### 4.1.1 3D Ink Fabrication

Three-dimensional printing (3D-P) is the technology of layer-by-layer additive fabrication of CAD-designed parts within the additive manufacturing industry, as opposed to the subtractive (casting) process in extractive manufacturing. The use of wool for advanced processes in additive manufacturing remains limited in the literature, which is attributable to powder conversion difficulties. Ultrasound technology is the application of sound wave propagation, which needs a medium, with water propagating this energy better than air. Sound is energy transferred in a longitudinal wave requiring a medium for travel. The wave travels based on the principle of ultrasound energy absorption in tissues through its conversion to heat (Zhang et al., 2020; Cheng et al., 2007). The study followed an experimental design for sheep wool fiber material processing by sonication; conversion into powder through the dry milling process; characterization of material and technological properties and process impact; and transformation of the resulting wool powder into 3D ink, proffering a solution to the current need for active crystalline polymers for printing 3D functional parts.

Two-way factorial experimentation with pretreatments—mechanical conditioning (cut, milled, and raw) and sonication dosing time (5 and 10 min)—can be deployed for the proposed modification. Sonication introduced a shift in material crystallinity, addressing the yearning for biodegradable functional materials in additive manufacturing technology.

## 4.2 MATERIALS AND METHODS

This section presents the equipment, materials, and methods used in this study. All equipment was available at the laboratory of the University of the Balearic Islands, Palma de Mallorca, Spain.

DOI: 10.1201/9781003566205-4

### 4.2.1 Equipment

The color was measured with a CM-5 colorimeter (Konica Minolta, Japan). An Optika microscope (Olympus, Japan) and Moticam 3 digital camera (Motic, China) were used for image observations, and FESEM Supra 25 (Zeiss, Oberkochen, Germany) was used for the SEM analysis. UP400S (Hielscher Ultrasonics GmbH, Schwabach, Germany) was used for the sonication while cooling with a refrigerating system (Frigiterm, Barcelona, Spain). Particle size reduction (PSR) was performed with a rotor mill (Ultra Centrifugal Mill ZM 200, RETSCH, Haan, Germany). Particle distribution was studied with Microtrac MRB Flow Sync (MTB Tech. Precision, Germany). Mettler Toledo DSC 3 (USA) was used to carry out the differential scanning calorimetric analysis, and thermal analysis (TA) Instrument SDT 2960 Simultaneous DSC-TGA (USA) was used for the thermogravimetric analysis (TGA).

### 4.2.2 Materials

Mallorquin sheep wool was supplied by a local supplier in Mallorca Island, Spain, stored in a dry, cold place, and used as a sample.

### 4.2.3 Methods

#### 4.2.3.1 Structure Evaluation

The sample morphological structure was evaluated under scanning electron microscopy at 100–400 μm magnification. Wool samples—raw, cut, and milled before treatment—were observed as controls, followed by ultrasound samples under investigation. In both cases, micro-images were obtained directly on a scanning electron microscope (FESEM Supra 25, Zeiss, Oberkochen, Germany) at 5 μm resolution, a voltage of acceleration of 10 kV, and a working pressure of 40 Pa. Images were acquired at 100–400× magnification. Spectrum electron microscopy (SEM) was deployed to study the comminution cum ultrasound impact on internal structure, particle size, and material reduction pattern. The resulting powder was subjected to ink formulation for 3D printing using the fused deposition modeling (FDM) process. Thermochemical and thermomechanical analyses, such as thermographic analysis (TGA), dynamic mechanical analysis (DMA), and X-ray diffraction analysis (XDA), were performed on the printed object.

### 4.2.4 Experimental Material Design

To investigate the effect of ultrasound treatment, the samples were submitted to sonication for 5 and 10 min with or without previous mechanical processing. Two mechanical treatments (cutting and milling) before sonication were evaluated. The samples were identified as follows: the first letter indicates the color of the wool—white (W) and red (R); the subsequent letters represent mechanical pretreatment—cut (C) or milled (M); and finally, a number to indicate ultrasonic dosing time, with 0 representing non-sonicated and 5 or 10 (in minutes) dosing time.

### 4.2.5 Mechanical Treatments

The milling process was carried out—30 g of wool was weighed and dropped piecewise (to lower resident time) into the milling cell of a 12-tooth blade, fixed with a 500 μm sieve, operated at 12000 rpm speed before the ultrasound. These samples were named milled (M) for both white and red simples. White milled (WM) and red milled (RM) were obtained as experimental samples for sonification. The same process was followed to obtain controls—white raw, and red raw (RR)—with 0, 5, and 10 indicating ultrasonic dosing time. The cutting of the wool was performed using common scissors, and white cut (WC) and red cut (RC) samples were obtained. The four mechanically prepared samples for sonification were thus WM, RM, WC, and RC.

### 4.2.6 Ultrasound Application

Acoustic energy was applied to WM, RM, WC, and RC, with an ultrasonic generator UP400S (Hielscher Ultrasonics GmbH, Germany) connected to a probe (22 mm diameter). Distilled water (200 mL) was used as a soaking medium. Wool samples (10 g) were immersed in the water within a jacketed glass connected to a thermostatic bath fridge (J.P. Selecta, Frigiterm, Spain), settled at 20°C to maintain the sample operating temperature. The probe was immersed (5 mm) into the water and sonicated for 5 and 10 min to evaluate the effect of sonication times. The cycles and the amplitude were settled at 0.5 and 100%, respectively. The power acoustic density was determined with a calorimetric method by measuring the temperature of distilled water (200 mL), sonicated (without sample) for 5 min, and thereafter the power was calculated from Eq. 4.1.

$$P = mC_p \frac{dT}{dt} \qquad \text{Eq. 4.1}$$

where:
*P* is the power (W), *m* is the mass of the water in (kg), *Cp* is the specific heat of water at 25 °C (J/kg °C), *T* is the temperature (°C), and *t* is the time (s). The acoustic density was calculated as power.

After the ultrasound treatment, the experimental factorial design was summarized as shown in Table 4.1.

Wool samples were dried in a vacuum oven at 30°C—the average final ultrasonic temperature of the sample—air vacuumed at 0.04 bar for the duration of 72–120 hrs, with moisture on the glass cover cleaned intermittently as it condensed noticeably, before being loaded into the chamber in a Petri dish. The wool was milled at varying sieve sizes (0.5, 0.25, and 0.08) using the high-speed retch mill to obtain ultrasound wool powder (UWP). The sieve size variation aided evaluations of material energy releases based on particle sizes, resulting from the pretreatment, according to the postulations of the bond law on PSR, as shown in Eq. 4.2. Understanding this will help the study of the energy gained by the proposed process and the time saving compared to a more complex wet-milling process as reported in the literature, including.

**TABLE 4.1**
**Experimental Design Factors.**

| Name | Material | Mechanical treatment | Sonication time (min) |
|---|---|---|---|
| WR5 | White wool | Raw | 5 |
| WM5 | White wool | Milled | 5 |
| WC5 | White wool | Cut | 5 |
| RR5 | Red wool | Raw | 5 |
| RM5 | Red wool | Milled | 5 |
| RC5 | Red wool | Cut | 5 |
| WR10 | White wool | Raw | 10 |
| WM10 | White wool | Milled | 10 |
| WC10 | White wool | Cut | 10 |
| RR10 | Red wool | Raw | 10 |
| RM10 | Red wool | Milled | 10 |
| RC10 | Red wool | Cut | 10 |

$$\mathbf{W} = \mathbf{K}_{\mathbf{B}}\left(\frac{1}{\sqrt{\mathbf{P}}} - \frac{1}{\sqrt{\mathbf{F}}}\right) \quad \text{Eq. 4.2}$$

where $W$ is the energy required, $P$ is the product diameter, $F$ is the feed diameter, and $K_B$ is the bond constant.

Wool conversion to powder constitutes an industrial constraint for its adoption in 3D technology; it is reported in the literature that wet milling in double and triple operation abides (Marti et al., 2015; Wang et al., 2012).

### 4.2.7 Raw Material Transformation (Bioplastic) Ink Processing

Wool powder in composite with any other powder to form a matrix for printing feedstock requires an adequate understanding of the material properties to obtain a bioplastic. The most technical stage requires understanding the solid mix that will balance the solid–liquid ratios required to formulate a feedstock with the right flow properties for fused filament deposition, hence the first transformation is on the solid matrix composition before the solid–liquid matrix composition.

### 4.2.8 Viscoelastic Characterization—Dynamic Mechanical Advantage (DMA)

Evaluation of the fabricated object's deformation characteristics was necessary to establish the viscoelastic properties of each formulation ratio, with reference to temperature and frequency in compression mode, measuring: storage modulus: E′, G′ (elasticity); loss modulus: E″, G″ (viscosity); and loss tangent: tan $\delta$ (= E″/E′). Figure 4.1 shows the principle of DMA experimental setup. The frequency sweep

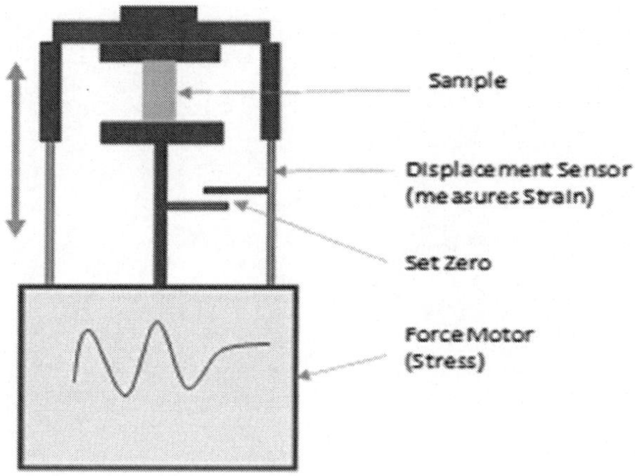


**FIGURE 4.1** DMA investigation setup for wool fiber–cassava composite fused deposition modeling, 3D-printed object.

was 30 Hz at a temperature range of 150°C–200°C for 3–5 hrs, influenced by the combining ratios.

## 4.3 RESULTS AND DISCUSSION

### 4.3.1 Effect of the Treatments Applied to the Wool

To ascertain the physical appearance of the wool, including dust-free and environmental staining, the ultrasound-cleaned wool (UCW) of Mallorquin origin (as shown in Figure 4.2) was subjected to a colorimetric test. $\Delta E$ indicates that non-ultrasound (raw)-treated samples differ in color from the sonicated wool.

#### 4.3.1.1 Raw Material Conversion (Powder Processing)

Drying—the ultrasonically treated wool was dried in a vacuum dryer at the same ultrasonic temperature of 30°C for 72–120 hrs until all surface water was completely removed. Figure 4.3 represents vacuum-dried ultrasonic wool (VUW). The ability of the wool to shrink (raw wool) makes it a necessity that a gradual drying process be followed by vacuuming at intervals to avoid structural distortion to obtain friable glass fiber-like milled samples as presented below.

Figure 4.3 presents the physical effect of sonication on the wool samples with an observable glass fiber-like appearance and feel. The top left is raw red wool sonicated for 10 min; at the center is red wool mechanically treated by cutting before sonication; and the left sample is mechanically treated by single milling with a 500 μm sieve size before sonication.

#### 4.3.1.2 PSR (Comminution)

The wool sample subjected to PSR and comminution unit operation produced a fine particle powder that passed through a 500 μm sieve by dry milling at 12000 rpm. A trial with a smaller sieve size was considered at this stage, taking into account

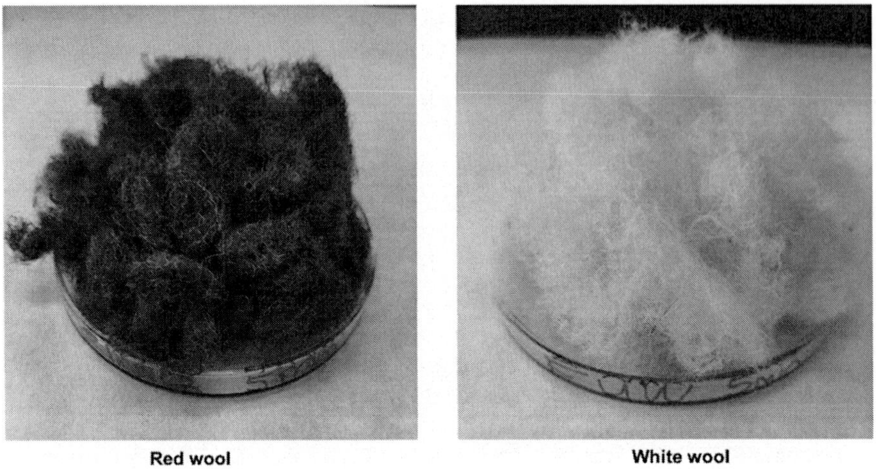


**FIGURE 4.2** Physical appearance of Mallorquin ultrasound-cleaned wool (UCW).

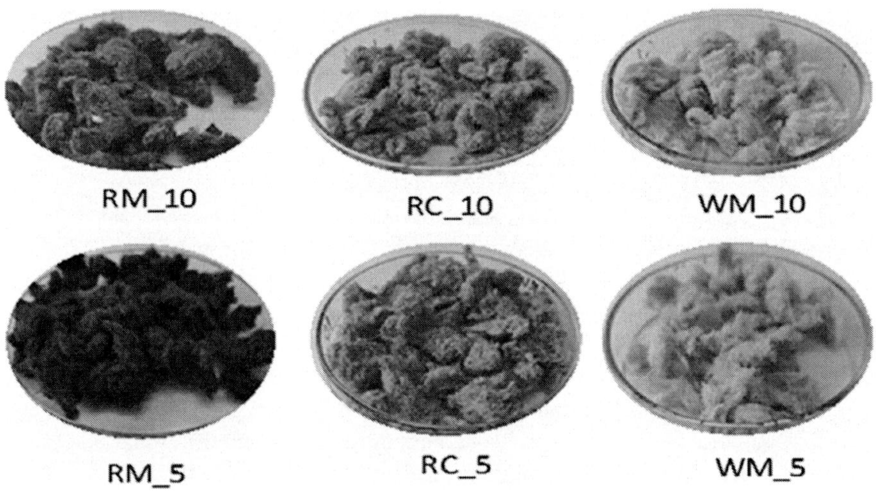


**FIGURE 4.3** Glass fiber-like vacuum-dried ultrasonic wool (VUW). R is raw wool, W is white wool, M means mill mechanical treatment, C is cut mechanical treatment, and 5 and 10 represent ultrasonic dosing time.

the possible twist of the wool by the motor blade, and changing the sieve after the first milling with a 500 µm to overcome this constraint. However, after four milling rounds, the equipment produced a powder with an average particle size of 31.87 µm, which is reduced from φ38.42 µm for the white wool. This represents a 17.05% size reduction. For red Mallorquin wool, particle size reduced from φ48.75 µm to φ38.32 µm, representing a 21.39% size reduction.

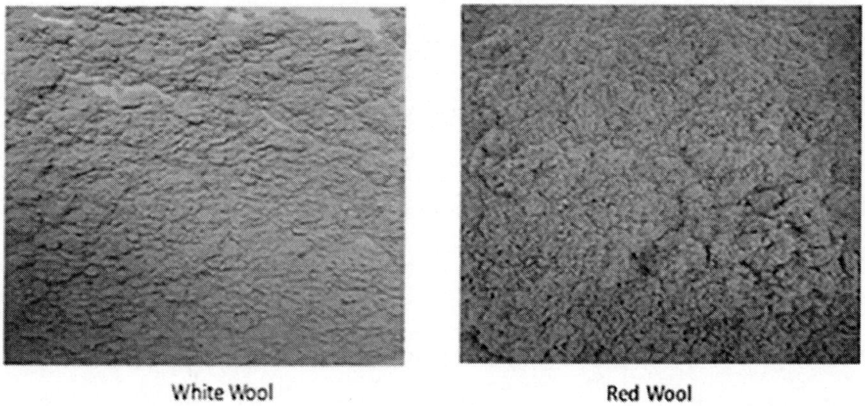


**FIGURE 4.4** White and red ultrasound wool powder.

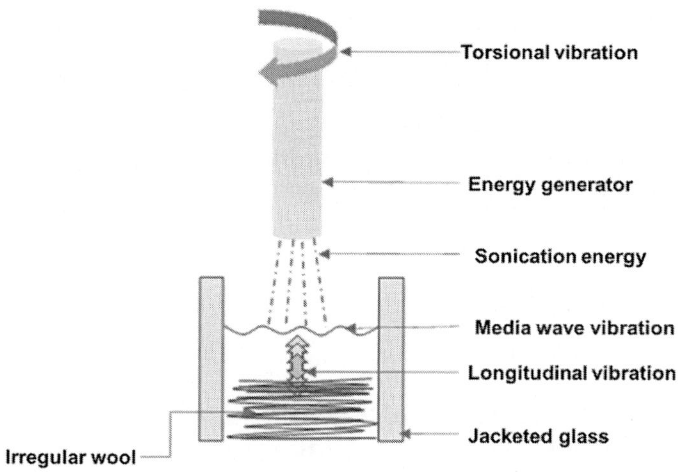


**FIGURE 4.5** Sonication wave energy propagation and milling force mechanism.

### 4.3.1.3 Impact of Sonication on PSR

The median diameter of the enlarged ultrasonic wool sample, as presented in the SEM image and Table 4.5, could indicate enlarged or increased structural porosity as a result of the ultrasonic treatment. It should be noted that the ultrasonic wave propagation travels longitudinally, whereas the rotor mill used in the comminution follows a latitudinal drag-and-tear rotary motion. Yet a fine powder was obtained. Material resident time, reduction, less compaction, and compression from an oscillating hammer bit due to even energy dissipation on the material may yield finer powder.

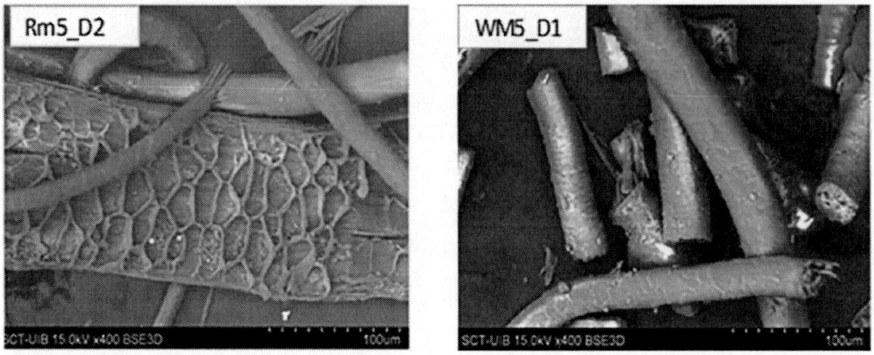


**FIGURE 4.6** Ultrasonic wool internal porosity. R and W are red and white wool, M is milled treatment, 5 implies 5 min dosing time, and D is detailing image (1 and 2) (suspected for enlarged sample diameter).

Lateral splitting of the wool, as presented in Figure 4.6 (Rm5_D2), could be prevented by an oscillatory hammer mill that will enforce break by impaction without lengthy thin wool fiber passage.

A close look at the detailed image (D1) in WM5_D1 shows that the end fragment has similar porosity to that widely observed in the detailed Image 2 (D2) in RM5_D2. The only difference is that D2 fragmented into two lateral equal parts, while D1 was latitudinally cut. One good thing to note is that they are of two different samples (red and white), but dosing time, spatial treatment, and SEM image magnification of 100 μm were the same for milled, and these were never observed in the non-ultrasonic-treated wool.

### 4.3.2 Ultrasonically Engineered Material Analysis

#### 4.3.2.1 Scanning Electron Microscopy on Sonicated and Milled Wool

The SEM image above shows the size reduction pattern with the rotor mill. On the left is the wool before size reduction; after ultrasound treatments, the left presents the milling pattern and the observable changes in the red wool sample as the ultrasound treatment RMD_2 is a clear insight of what is going on in the microdrill of the wool during an ultrasound, whereas RR_D2 and RM10_D2 present an insight on the material's potential PSR under the selected mill when ultrasound exposed. The milled sample from the 500 μm sieve powder explains the internal structure before the final targeted particle size of 80 μm (Figures 4.7 and 4.8).

The microfibril of the Mallorquin shrinks with increasing ultrasonic dosage, which may be attributed to the greasy part of the wool protein susceptible to the given treatment. Although this increase may be statistically insignificant, heat release and temperature define particle size distribution.

#### 4.3.2.2 Particle Size Distribution

The particle size distribution presented finer wool powder of about 20–80 μm (80 μm) in a single milling process, with energy cost and time improvement, when compared

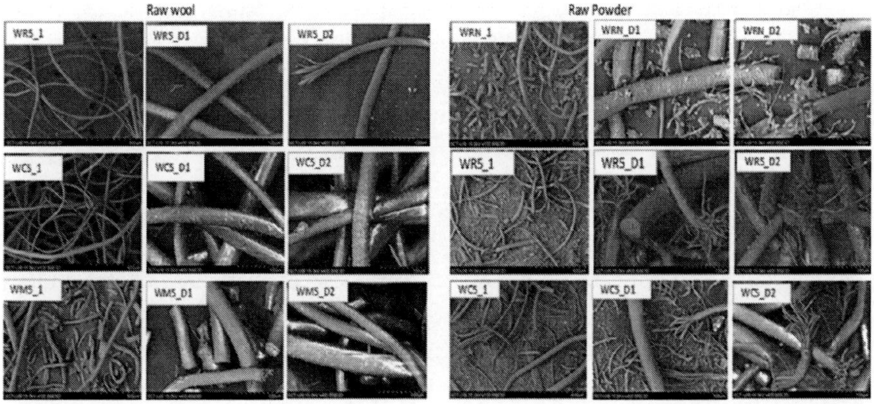


**FIGURE 4.7** White wool (raw and powder) microstructure after 5 min ultrasound treatment. WR5_1: first image showing white raw wool after 5 min ultrasound treatment; WC5_1: white cut wool after 5 min ultrasound treatment; and Dland D2: detail images.

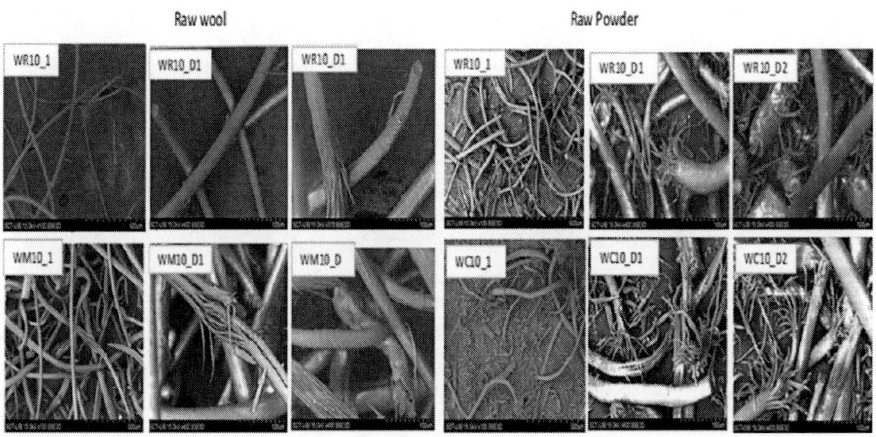


**FIGURE 4.8** White wool (raw and powder) after 10 min ultrasound microstructure. WR10_1: first image showing white raw wool after 10 min ultrasound treatment; WC10_1: white cut wool after 5 min ultrasound treatment; Dland D2: detail images.

with 8–20 μm for a three-stage operation involving wet milling, attrition milling, and air-jet milling, as reported in the literature (Duman et al., 2022; Cheng et al., 2007; Zhang et al., 2020). It could be inferred that sonication improves the milling efficiency. To validate this inference, differential calorimetric results were used to assess the impact of the ultrasound and mechanical treatment of the material and its energy releases (see Figure 4.9 and Table 4.5). This also aided in the verification of the energy requirement for the PSR of the wool samples.

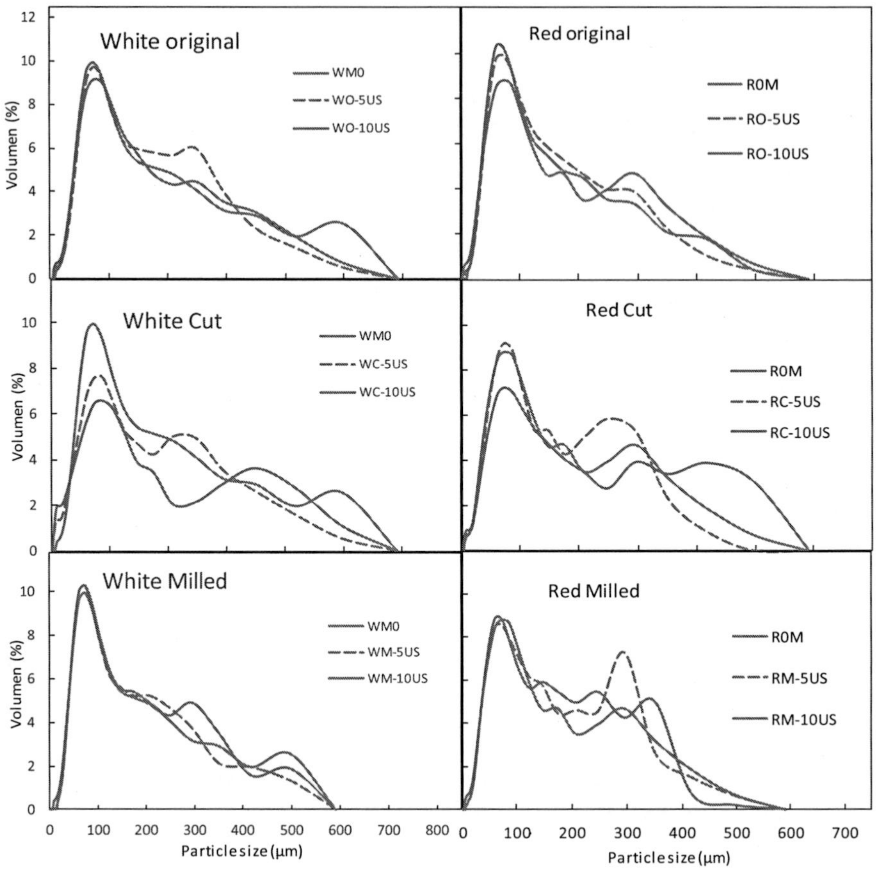


**FIGURE 4.9** Impact of ultrasound on red wool particle size distribution. Pure red, non-treated (PRN_0); treated for 5 min and 10 min (PRN_5 and PRN_10, respectively). The mechanical treatments—milled (M) and cut (C)—replace raw (R) where appropriate. For all white samples, white (W) replaces R in red samples presented.

### 4.3.2.3 X-Ray Diffraction (XRD): Crystallography of Ultrasonic Wool Powder (UWP)

The crystallinity of a material is a unique technological property desirable in 3D additive manufacturing technology for printing functional part objects. The recovery from TA of UWP was further investigated with XRD to check if the ultrasonic pre-treatment impacted a shift from amorphous fiber to a semicrystalline material, considering the observed fiberglass-like behavior exhibited by the milled samples before sonication. The result indicates that semi-crystalline wool powder (SWP) could be possible with ultrasound treatment operations. Figures 4.10–4.14 present the various crystalline compounds observed and the suspected compounds responsible for the material's crystallinity. A predictive analysis of the XRD graph metrics suggests

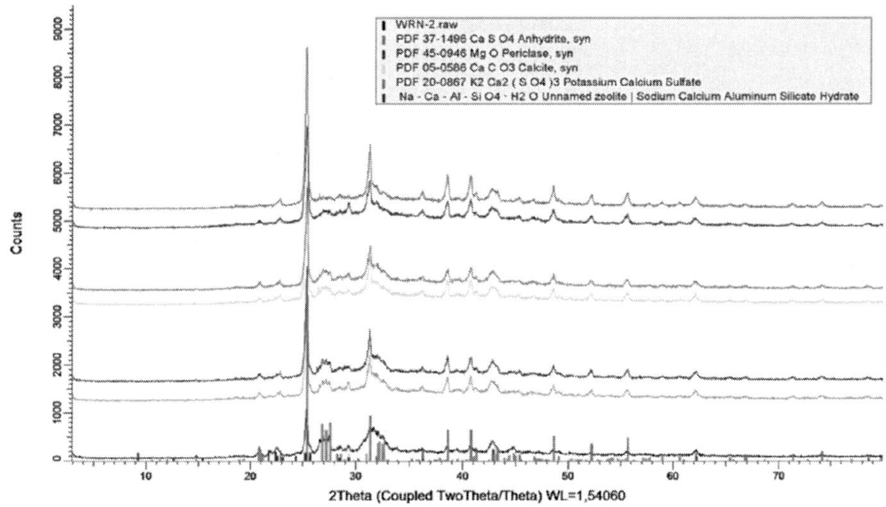


**FIGURE 4.10** X-ray diffraction image of white wool (ultrasound and non-ultrasound). 1. WRN-2.raw (black): white raw non-treated wool; 2. Na-Ca (brown 10 and orange 5): white raw; 3. PDF 05/20 (green 5 and pink 10): white cut (WC); 4. PDF 37/45 (blue 5 and red 10): white milled (WM). Numbers 5 and 10 represent ultrasound dosing time.

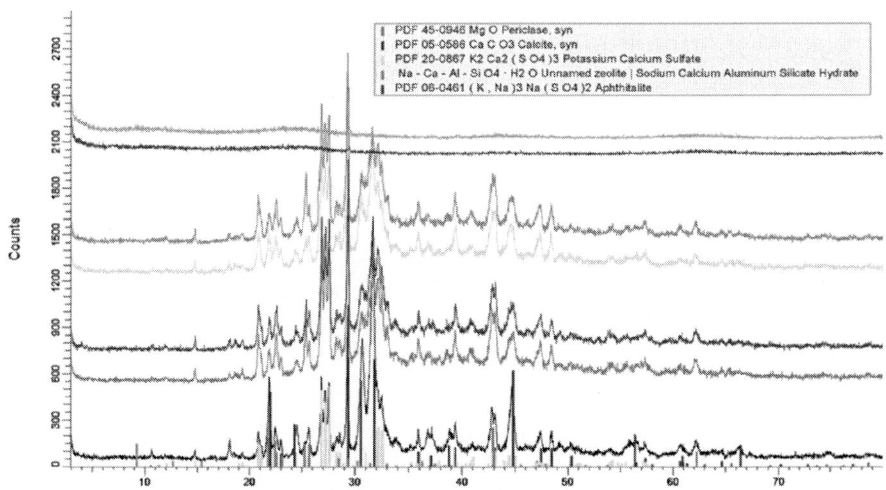


**FIGURE 4.11** X-ray diffraction of red wool (ultrasound and non-ultrasound). 1. RRN-2.raw (black): red raw non-treated (RRN) wool; 2. PDF 05/45 (red 5 and blue 10): red raw (RR); 3. PDF 20/13 (green 5 and pink 10): red cut (RC); and 4. PDF 06 (orange 10 and brown 5): 5 and 10 represent ultrasound dosing time.

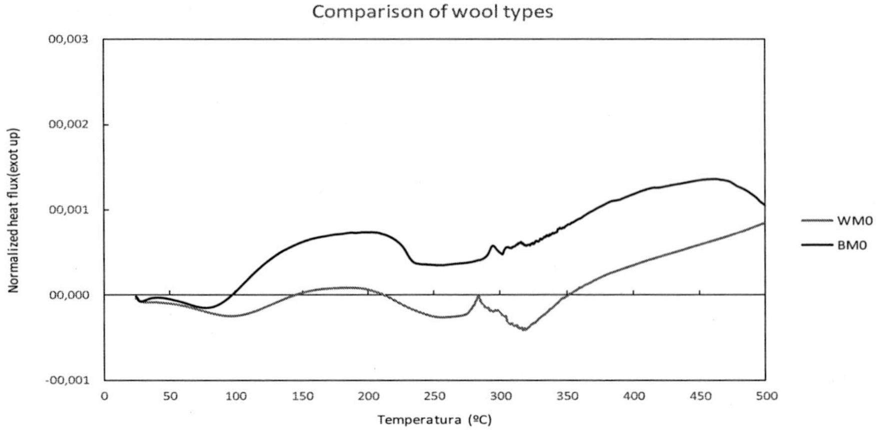


**FIGURE 4.12** Comparison of wool DSC curves (white and brown). White milled non-ultrasound wool (WM0) and red wool milled non-ultrasound wool.

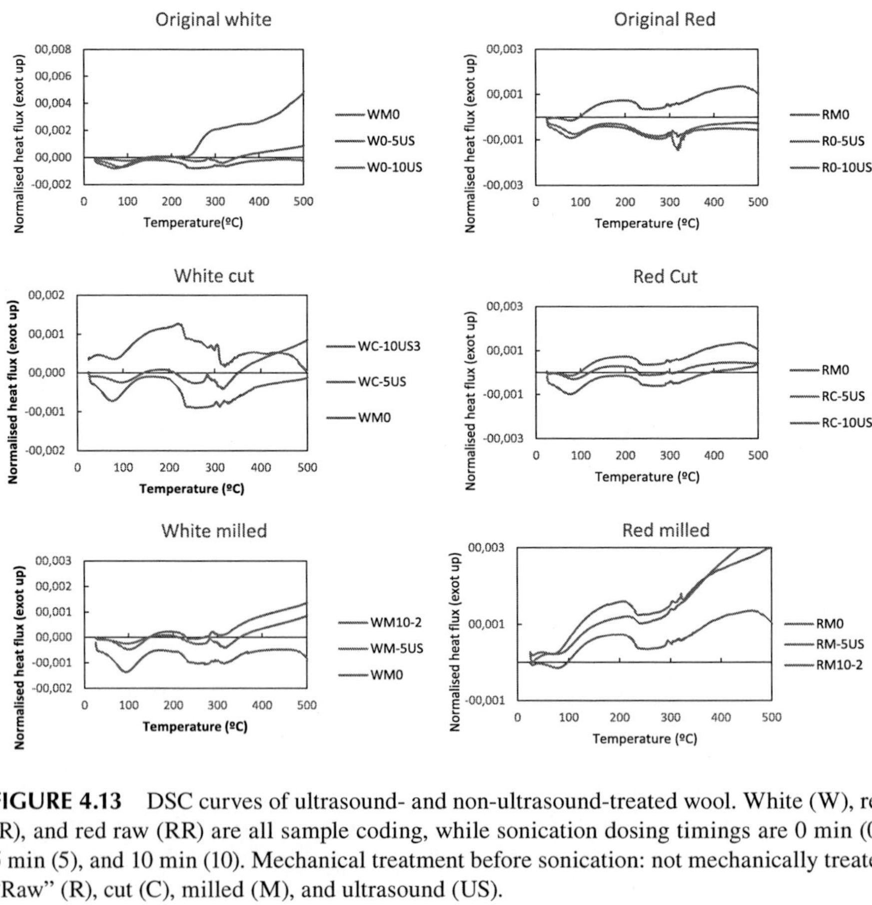


**FIGURE 4.13** DSC curves of ultrasound- and non-ultrasound-treated wool. White (W), red (R), and red raw (RR) are all sample coding, while sonication dosing timings are 0 min (0), 5 min (5), and 10 min (10). Mechanical treatment before sonication: not mechanically treated "Raw" (R), cut (C), milled (M), and ultrasound (US).

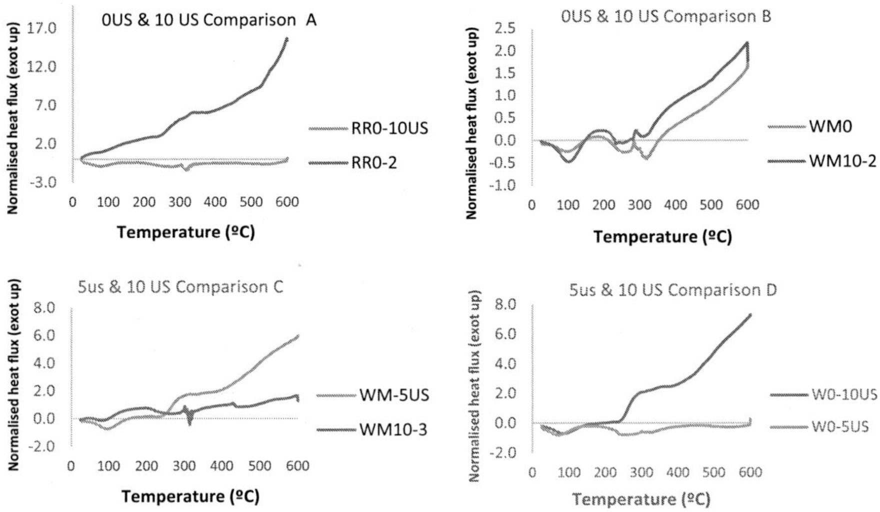


**FIGURE 4.14** Comparison of heat flow and ultrasonic dosing time. White (W), red (R), and RR are all sample coding, while sonication dosing timing are 0 min (0), 5min (5), and 10 min (10), and mechanical treatment before sonication and non-mechanically treated: "Raw" (R), cut (C), milled (M), and ultrasound (US).

anhydrite crystals with calcium sulfate ($CaSO_4$), which forms gypsum crystals in the presence of calcium carbonate—an important industrial raw material. This suggests a possible new product line from wool, usable as fillers, absorbents, drying agents for solids, organic liquids, and gases due to its hygroscopic nature.

The XRD analysis revealed the presence of gypsum crystals, an anhydrite of calcium sulfate ($CaSO_4 \cdot 2H_2O$)—an industrial salt widely used in plaster of Paris (POP), alongside under crystals. The reduction in the sulfur content during sonication may have created a stable oxidation–reduction (Redox) reaction with the more stable calcium in the presence of energized calcium carbonate ($CaCO_3$) to form the anhydrite salt: $SO_2 + 0.5\ O_2 + CaCO_3 \rightarrow CaSO_4 + CO_2$ (Danch and Gadomski, 2000; Gadomski and Luczka, 2000). This reflects the more stable structure reported by Li et al. (2012), but it is not due to the wool's stable fiber; rather, from the shifted crystallinity on ultrasound treatment.

Magnesium oxide (MgO, periclase), calcite ($CaCO_3$), and potassium calcium sulfate ($K_2Ca_2(SO_4)$) identified in the XRD correspond with the chemical analysis reported in Table 4.2, which shows the presence of sulfur, sodium, calcium, magnesium, and manganese as prevent minerals that change with sonication of the wool samples. The sharp peak indicates clear crystallinity, which becomes very refined in the ultrasound-treated sample signals above (colored), and the non-ultrasound-treated sample signal (black) at the base. This discovery agrees with existing research on the binary phase existence of fibers in the amorphous and crystalline state (Danch and Gadomski, 2000).

The presence of zeolite (sodium calcium aluminum silicate hydrate) and aphthitalite in the sonicated samples thus represents the clear impact of acoustic energy on the structure of the wool fiber. Whether these changes in the composition of the material will impact the ease of PSR remains a function of interstitial forces among the formed compound, type of milling machine, and material strength (Faruque et al., 2019; Cheng et al., 2007).

#### 4.3.2.4 Differential Scanning Calorimetry (DSC)

Differential calorimetric curves were obtained from the samples as presented in Figures 4.12–4.14, revealing three distinct phases of thermal processing relatable to the samples, which have been reported to be associated with keratin protein features. While glass transition and water evaporation phases have been reported, this investigation detected glass transition, crystallization, and denaturation (melting) phases, particularly in the treated samples. In principle, glass transition and water evaporation coincide within the peak (Martí et al., 2007), but the mechanical treatment given to the samples before sonication has made it possible to observe refined peaks that suggest crystallinity for the exothermic and endothermic releases. The three-phase transition values are presented in Table 4.4

Figures 4.12–4.14 represent the characterization of DSC curve samples with their values presented in Table 4.2. The glass transition ($T_g$) temperature for both samples was within 58–82°C for UWP, quite different from that reported in the literature

**TABLE 4.2**
**Phase Transition Energies of Wool Samples.**

| | Phase transition temperature | | | Phase transition energy | | |
|---|---|---|---|---|---|---|
| Simple ID | Glass transition $T_g$ (◦C) | Crystallization temp $T_c$ (◦C) | Melting temp $T_m$ (◦C) | Glass transition energy $\Delta H_{eva}$ (◦C) | Crystallization energy $\Delta Hc$ (◦C) | Melting energy $\Delta Hm$ (◦C) |
| Non-ultrasound RR_0 | 275.37 | 298.38 | 483.93 | 2.58 | –223.60 | –26.77 |
| RR_5 | 75.85 | 158.49 | 272.15 | –150.63 | 148.33 | –107.47 |
| RR_10 | 82.39 | 189.51 | 280.44 | –226.23 | 156.71 | –49.02 |
| RC_5 | 75.85 | 208.16 | 237.50 | –150.63 | 81.66 | –10.28 |
| RC_10 | 82.39 | 164.89 | 243.26 | –226.23 | 299.68 | –29.53 |
| RM_5 | 58.13 | 172.45 | 0.00 | –97.42 | 237.09 | 0.00 |
| RM_10 | 77.53 | 167.97 | 207.23 | –407.06 | 241.14 | –88.49 |
| Non-ultrasound WR_0 | 81.03 | 283.30 | 338.84 | -951.23 | 235.70 | –58.03 |
| WR_5 | 80.47 | 153.72 | 241.33 | –439.69 | 108.80 | –118.35 |
| WR_10 | 79.97 | 263.64 | 533.77 | –243.73 | 46.06 | –39.95 |
| wc_0 | NA | NA | NA | –439.69 | –300.74 | 222.82 |
| WC_5 | 55.07 | 256.55 | 382.33 | –44.23 | 79.44 | –402.09 |
| WC_10 | 77.21 | 261.48 | 311.70 | –296.25 | 23.45 | –12.76 |
| WM_5 | 94.83 | 172.45 | 241.48 | –643.90 | 185.68 | –110.87 |
| WM_10 | 69.88 | 167.97 | 341.58 | 30.46 | 479.94 | -253.49 |

(Li et al., 2012), with milled mechanical pretreatment before sonication presenting the lowest $T_g$, suggesting more resistance to heat flow due to large area of sonication which probably experiences intense power density from acoustic energy. The crystallization temperature decreased with increased dosing time and sample surface area. This could be correlatively attributed to the softening of the material due to ultrasound treatment. While DSC evaporation has been attributed to wool fiber internal water (Li et al., 2012), the mechanical treatment should provide a loss gradient with raw samples, in terms of % water loss for different onset temperatures, but this was not the case; onset evaporation temperature was all below 50°C for less than 10% moisture loss. Lipids, as previously reported (Li et al., 2012b), were suspected to be involved; however, variations in ΔHeva were observed under the same dosing time, for the same sample type, but the different mechanical treatment (cutting and milling) before sonication. This affirms that ultrasound dosing time was responsible for the energy differences observed at this transition phase, associated with internal lipid removal. Invariably, the dosing time increased, with energy absorbed altering material strength. Exothermic energy releases and endothermic energy. Table 4.2 presents the phase transition energies of wool samples.

#### 4.3.2.5 Positive Glass Transition Energy

The positive $\Delta H_{ev}$ seen in the redraw samples gives clear insight into the energy requirements of the material. However, the positive value exhibited in the white-milled 10-minute dosing time (Wm10) is subject to validation compared to the negative $\Delta H_{ev}$ observed in the sonicated samples. A close observation of its raw form shows that it was more flexible compared to the red sample, and an increase in its surface area before sonication for 10 min transformed it into a glassier material (fiber glass). This shows the minimum energy requirement to overcome intermolecular forces that bind the materials in glassy states. Three main conditions are responsible for this positive transition of energy, and they include:

- *Low temperature*: It implies that the samples are highly ordered, and any disorder will require energy input to induce the transition into a more flexible state (crystal) needed in 3D printing AM.
- *High viscosity*: Material molecular mobility decreases with increasing viscosity, resulting in more difficulties in flexible state transition, hence a positive transition of energy.
- *Strong intermolecular interaction*: A strong intermolecular interaction of the material could be responsible for these phenomena, implying that energy needs to be supplied to break these interactions and induce flexible state transition.

It is important to note that the sonication treatment greatly impacted the wool crystallization energy as seen in the decreasing $T_g$ energy, with increasing crystallinity. In similitude, crystallization energy $\Delta H_{CRY}$ is considered an important factor in this research, in line with certain crystallinity of polymers—a desired material property in 3D printing—and wool powder adoption in additive manufacturing remains limited by its fabrication process (Duman and Küçük, 2022). The flexible state correlates with material crystallinity, the most desired characteristic of 3D inks sorted for crystalline polymers.

### 4.3.2.6 Thermal Gravimetric Analysis (TGA)

Figures 4.15–4.18 present TGA curves of two wool samples with ultrasonic treatment and mechanical conditioning (cut, milled, and raw) before sonication. The dual mass of the phases was observed with increasing temperature. Moisture loss by evaporation (at around 50°C) was attributed to the first phase, whereas the second pass appears to mark the onset of thermal decomposition of simple (at about 200°C). A correlation between onset temperatures and changes in the mass ratio of the samples obtained is presented in Table 4.3. Powder subjected to TGA obtained from the raw wool samples under XRD is shown in Figure 4.15.

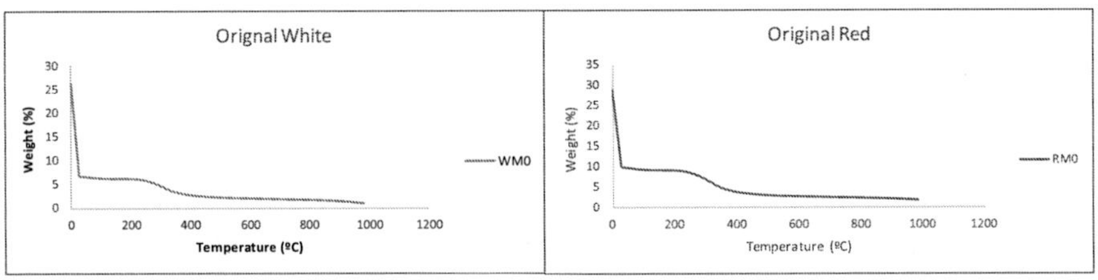


**FIGURE 4.15** Thermogravimetric analytical curve of untreated wool powder samples.

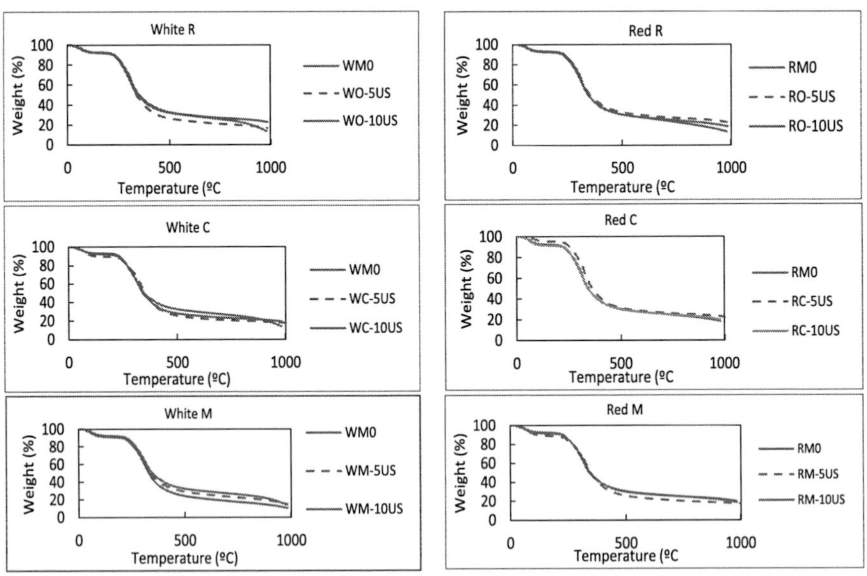


**FIGURE 4.16** Results of the thermogravimetric analysis (TGA). White wool (W) and red wool (R) were subjected or not (O) to a previous treatment of cutting (C) or milling (M), and subsequently treated or not (0) with ultrasound for 5 or 10 min (5US and 10US).

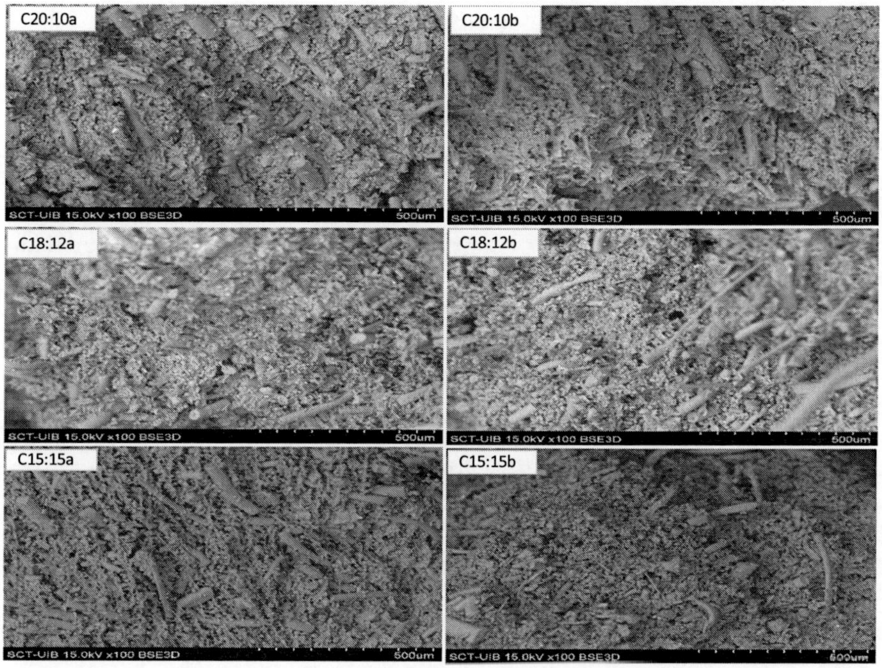


**FIGURE 4.17** Scanning electron microscopy images of mechanically cracked 3D-printed objects from wool ink. Cassava starch (C) and wool powder were used in ratios of 10%, 12%, and 15% for cassava and 20%, 18%, and 15% for wool. (a) First image capture and (b) second image capture of object porosity observation under the microscope.

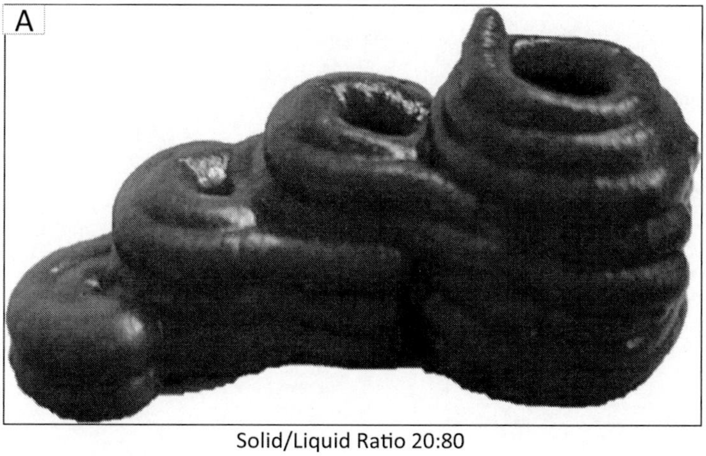


**FIGURE 4.18** Minimum filament solid–liquid ratio for support-free structural stability of functional object printing.

The highest weight loss was observed between 200 and 400°C. Between 150 and 400°C, the primary structures of keratin decomposition are observed, promoted by the breaking of disulfide bonds. This loss was 51% and 54% for white and red wool, respectively. Sonicated white wool witnessed slightly higher weight loss (58%) at temperatures 200–400°C with 10 min dosing time, especially for the samples ground before treatment. Red wool was stable, coincidentally with the thermogram for both ultrasound-treated and untreated samples.

Water evaporation at the unset temperature and structural thermal degradation of the sample were significantly retarded in the ultrasound-treated samples with 5 and 10 min sonification, respectively, compared with the raw wool in both samples. Moreover, cutting pretreatment before sonification appears to retard sample thermal degradation. This is consistent with recent reports (Science Direct 2023; Fourman et al., 2021). More interestingly, the study found that mechanical treatment before sonication delayed thermal degradation even at smaller dosing times and at a higher thermal range of 900°C. While the lower level of loss in mass at increasing temperature has been attributed to the rearrangement of microfibril in the wool protein chains (keratin), further investigation revealed that the sonification shifted the microstructure toward the formation of anhydrite. This phenomenon has been reported as a stronger and more stable fiber structure due to sonication-induced rearrangement in microstructure. This report suspects the formation of a crystalline compound due

**TABLE 4.3**
**TGA Data Set of Ultrasound-Treated and Non-Treated Wool Samples. Mechanical Treatment Cut (c), Milled (m), Non-treated Raw (r), Red Raw (rr), and White (w). Sonication Dosing Time 5 Min (5), 10 Min (10), and 0 Min (0).**

| Simple ID | Water evaporation | | Thermal decomposition | |
|---|---|---|---|---|
| | Onset temp (°C) | Mass loss at 30–900°C (%) | Onset temp (°C) | Mass loss at 30–900°C (%) |
| Non-Ultrasound RR_0 | 37.5 | 9.35 | 350 | 70.395 |
| RR_5 | 48 | 7.67 | 220 | 71.87 |
| RR_10 | 38 | 7.03 | 200 | 71.82 |
| RC_5 | 30 | 11.21 | 230 | 68.35 |
| RC_10 | 48 | 6.76 | 260 | 70.73 |
| RM_5 | 48 | 10.63 | 250 | 67.53 |
| RM_10 | 45 | 6.51 | 200 | 69.15 |
| Non-Ultrasound WR_0 | 38 | 7.71 | 225 | 67.38 |
| WR_5 | 38 | 6.88 | 250 | 72.58 |
| WR_10 | 38 | 6.56 | 210 | 68.80 |
| WC_5 | 50 | 8.38 | 220 | 72.01 |
| WC_10 | 50 | 6.76 | 200 | 70.7 |
| WM_5 | 46 | 10.63 | 200 | 67.53 |
| WM_10 | 50 | 9.98 | 200 | 66.63 |

to sonication, increasing the crystallinity of wool fiber, which has been validated in XRD quantification analysis; see Figure 4.10 and Figure 4.11.

### 4.3.3 Structural Characteristics Evaluation of Printed Object

The structural attributes of 3D-printed objects are a mark of quality. Warping in printed objects remains a concern in developed feedstocks; hence, evaluating the structural competence of the developed wool–cassava composite feedstock was considered in this section.

#### 4.3.3.1 Scanning Electron Microscopy

The 3D-printed geometry remained accurate and consistent with the ink's rheological attributes (see Figures 4.18–4.20). Regarding nozzle size, object height, printing speed, temperature, and pressure agree with previous reports (Ahmadzadeh and Ubeyitogullari, 2023; Baniasadi et al., 2021b). A filament composition (20, 18, and 15%) of wool fiber–cassava starch, at a printing speed of 1200–1500 m/s, at 25–35°C, and nozzle size of 15 and 20 mm produced the desired size, shapes, and designs, which corresponds to the starch result (12.5–15%) reported in the literature. Interestingly, the geometric heights (20–50 mm) remained undistorted in 20% and 18%

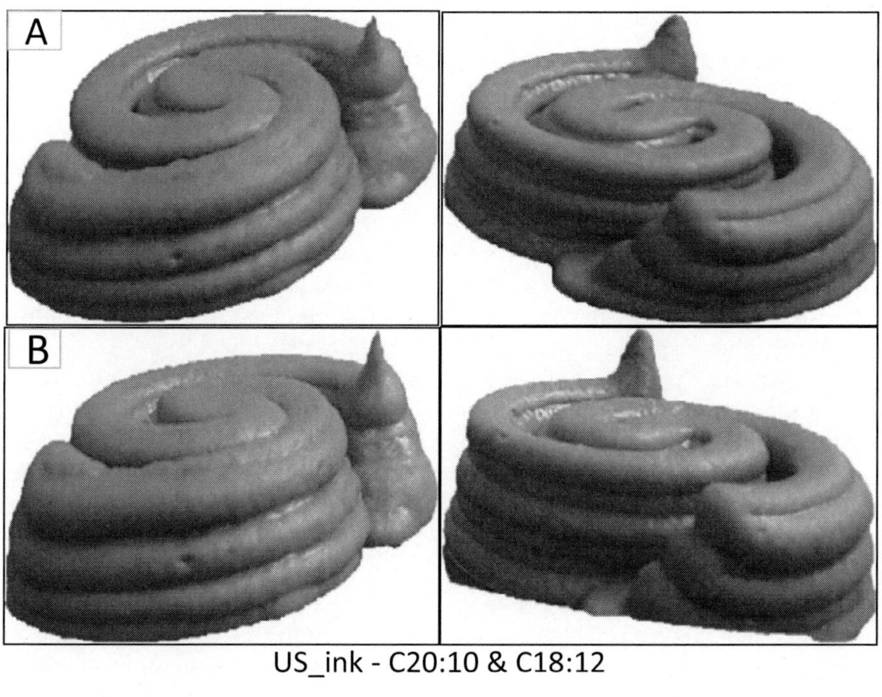


**FIGURE 4.19** Ultrasound enhances filament ink wool–cassava starch ratio. A represents 20% wool, 10% cassava starch, and 70% $H_2O$; and B represents 18% wool, 12% cassava starch, and 70% $H_2O$.

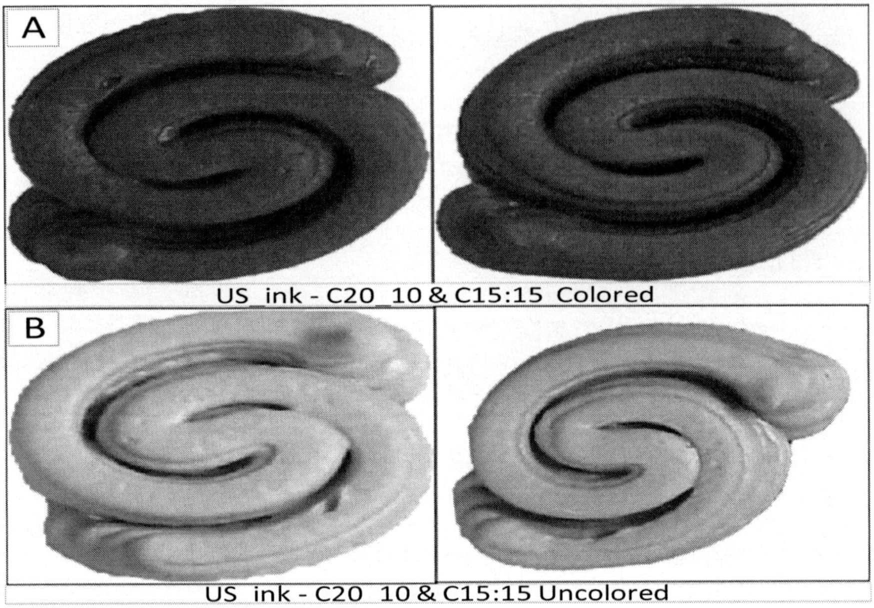


**FIGURE 4.20** Ultrasound enhances filament ink wool–cassava starch ratio. A represents 20% wool, 10% cassava starch, and 70% water, with red colorings, and B represents 15% wool, 15% cassava starch, and 70% $H_2O$ without colors.

of samples. Reducing amylose content improved printed image structure due to its rheological attributes (i.e., viscosity, recovery, G′ and G″ values, and tan $\delta$), contrary to conventional findings (Ahmadzadeh and Ubeyitogullari, 2023). In the wool powder–cassava starch composite filament fabrication experiment, more wool improved the image structure. Figure 4.17 shows SEM images of 3D-printed objects taken from cross-sections of the dry and fractured object obtained from three composite fiber inks. Higher magnification reveals a non-porous structure.

The SEM images obtained of fabricated filament composite inks C20:10, formulated on a magnetic stirrer at 3000 rpm at a constant temperature of 100–150°C, presented visible starch granules as well as wool microfibrils, making it possible to understand the structural interlinkages that occurred in the composite ink, which are consistent with the literature (Ahmadzadeh and Ubeyitogullari, 2023)—presenting their suitability for biodegradation-friendly manufacture.

#### 4.3.3.2 Printed Functional Object Structures

Printability remains a critical factor in the 3D printing of additive manufacturing. It refers to the production of a well-defined structure (Ramesh et al., 2021). Figure 4.18 is poorly defined but gives visible detailing of the designed image printed in a three-dimensional fused deposition modeling (3D-FDM) process. The ability to produce defined graphics in 3D object reality is controlled by ink rheological properties and printing parameters established. The fabricated filament's rheology

**TABLE 4.4**
**Nozzle and Fabricate Filament Printing Parameters.**

| Properties | Value |
|---|---|
| Viscosity | ≥ 12000rpm |
| Filament diameter | ≤ 0.5–20mm |
| Crystallization temperature | 60–120°C |
| Gelling temperature | ≥ 100°C |
| Printing temperature | 30–65°C |
| Ambient temperature for printing | 20–35°C |
| Nozzle diameter | 15–40mm |
| Object height | 2–5cm |

was measured for its viscoelastic properties, with the results discussed in the Results session, while the printing parameters are presented in the Table.4.4. The findings reported by Ramesh et al. (2021) are quite similar to the fabricated PLA filament printing parameters reported in the literature by Yao et al. (2021).

Figure 4.18, printed using the FOODNI 3D printer, maintained the exact design, producing a firm structure without the need for support or pre-heating, consistent with the receiving platform for PLA filament 3D-printed objects reported in the literature (Yao et al., 2021). A total of seven layers from a 40 mm nozzle resulted in a height of ≤ 280 mm object for a FDM-fabricated filament ink of wool fiber–cassava starch composite (FDM-FFI-$W_f/C_{s.}$). The image finishings depict perfect process conditioning and formulation suitability for adoption in additive manufacturing of 3D-printed functional parts. The printing pressure could be observed as perfect, looking at each layer's continuous lines. It has been reported that discontinuous geometry lines (insufficient extrusion) imply low printing pressure. In contrast, high pressure introduces instability in the ink flow (Ramesh et al., 2021), leading to possible jumping or bridged deposition of the inks and missed layers, which was not the case.

### 4.3.4 Fused Filament (Ink) Formulation Printability

The solid–liquid ratio knowledge was most important to understand the right solid matrix balance for a flowing ink composed of wool powder, cassava starch, and water, with 30/70% w/w adopted as the most suitable ratio for obtaining a support-free structure. Wool was observed to contribute more to the stability of the structure, and the percentage it could accommodate a bind (starch) was established as 15% w/w starch. Notably, a sticky deposition unable to stay on the film base with falling height was not observed, suggesting that sheep wool fiber powder (WFP) provides structural support for the 3D object (Figure 4.19).

While Image A of Figure 4.20 represents 10% (w/w) cassava starch (analytical grade) and 20% (w/w) in a solution of 70% (w/w) to formulate the ink, Image B presents 18 and 12% (w/w), respectively, producing a four-layer deposition with smooth, perfect finishing, void of support in both instances. The result was more suitable than 15/15% (w/w) solid matrix composition for the filament reconstitution. It was discovered that the higher the ultrasonic-treated wool powder, the more suitably printable the geometric structure became.

An attempt was made to introduce pigmentation into the ink to ascertain whether there was improvement or alteration, as presented in Figure 4.20. The ink's flowability and the image structure were not affected, but instead, a more appealing smoothness was observable, as shown in the graphic presentation.

## 4.3.5 Drying Printed Object

The 3D-printed object from wool fiber and cassava starch fabricated filament was subjected to drying operations using different techniques (natural air, oven, and freeze).

### 4.3.5.1 Lyophilization and Vacuum Drying

Printed geometry was frozen at −80°C for 1 hr in an ultra-freezer, then withdrawn and transferred into a freeze-dryer at −50°C for 72 hrs. After it was withdrawn, a very nice 3D-printed image was obtained. Another sample was simultaneously placed in a petri dish and kept in the open air at atmospheric conditions for 5 days. A cracked image was observed compared to the other two drying techniques (freeze drying and oven), representing the two outcomes (Figure 4.21).

## 4.3.6 Mechanical Performance of Printed Object

### 4.3.6.1 Textural Analysis

The presented elastic modulus measures resistance to deformation (elasticity), with the stretchy characteristic of low-modulus material when pulled, while high-modulus materials stretch less when pulled. The impact on filament printability in the 3D printing process has been reported as follows: when elasticity is low, nozzle break-off may occur (which is favorable for the FDM process), whereas highly elastic filaments are too soft, likely affecting the geometric height. These properties were evaluated

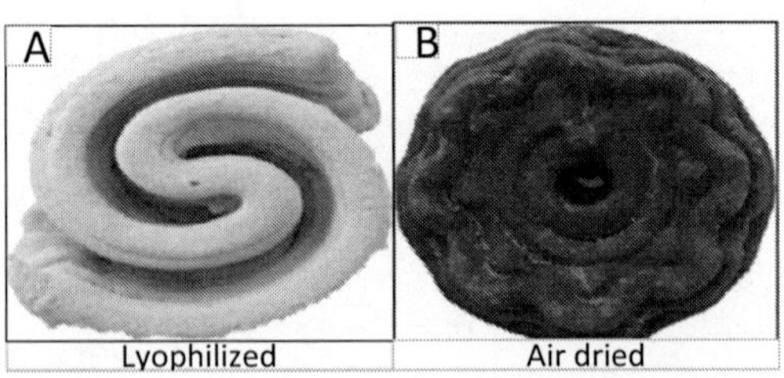


**FIGURE 4.21** Air and freeze-dried, fused deposition model. Three-dimensional (FDM-3D)-printed wool–cassava starch composite filament.

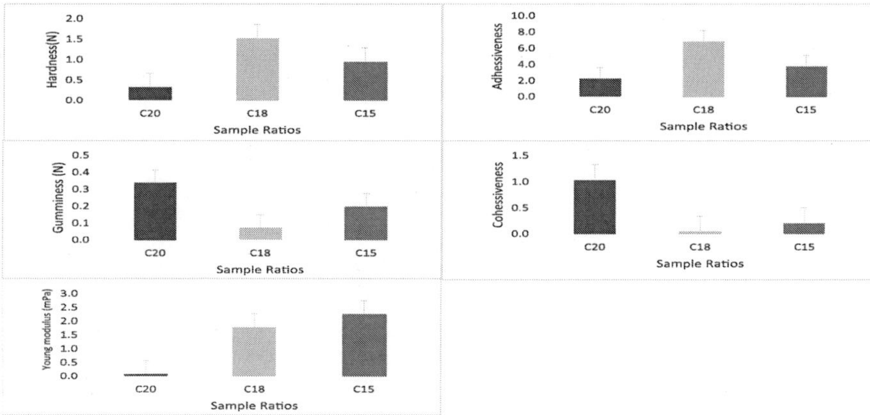


**FIGURE 4.22** Physicochemical properties of fabricated sheep wool and cassava starch ink. C represents the composite ratio of wool powder (20:18:15) and the cassava starch (10:12:15).

on the universal testing machine (UTM), and the results showed a reduced elastic modulus with increasing WFP ratios. This observation is consistent with the literature (Jones and Ashby, 2019; Kimbell and Azad, 2021). Figure 4.22 shows the various textural properties exhibited by the ink for accessing the modulus, obtained from the evaluation of the UTM signals analyzed using Origin software (Origin, 2021b, Origin Lab Corp., Northampton, MA, USA).

Notably, the shape was maintained in all cases compared to what was reported in the literature [Cristina 3D]; hence, all were printable. This difference is attributable to the additive manufacturing potential of WFP due to the treatment (sonication) deployed before conversion to powder.

#### 4.3.6.2 Compressional Strength (Break Analysis)

Mechanical load predicts possible fields of application matching the mechanical properties of parent tissue with fabricated bio-ink for intending artificial tissues is required in tissue engineering (Baniasadi et al., 2021a, 2021b). Mechanical properties could qualify or disqualify 3D ink for use as packing material, automobile parts, or any other functional parts fabrication. Over 200 N was loaded first, and a further 1 ton was added to observe cracking points under compressive force analysis. The signals are captured and reported in the graph presented in Figure 4.23.

The result in Figure 4.23 shows a linear order of deformation, geometrically increasing with load at low strain; the point of densification could be identified as a sharp stress increase occurring on these samples. An additional 1 ton was loaded to observe cracking points under compressive force analysis. The signals were captured and reported in the graph presented in Figure 4.23.

#### 4.3.6.3 Object Porosity and Structural Strength

It was observed that the object porosity increased with increasing wool fiber ratio (C15 < C18 < C20). Uniformity in pore sizes and interconnecting porosity of structures

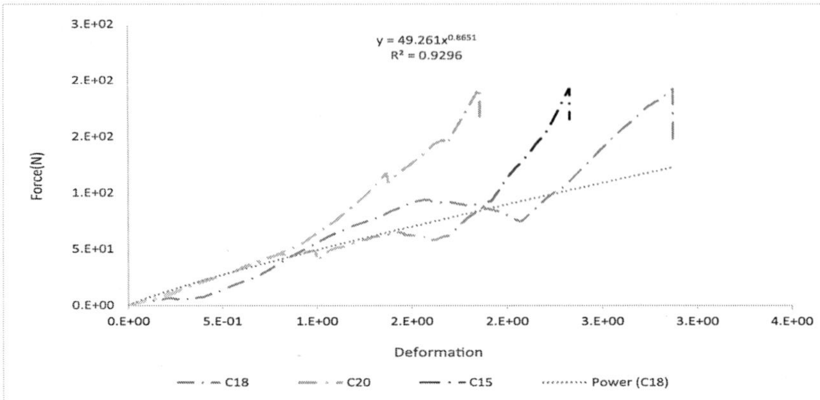


**FIGURE 4.23** Load analysis of wool and cassava starch 3D-printed and dried object. C is the composite ratio (20:18:15) of wool powder and the cassava starch (10:12:15).

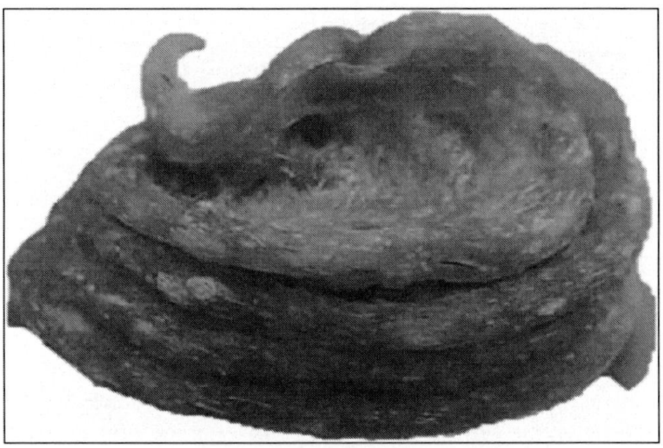

**FIGURE 4.24** Mechanical fracture and porosity.

represents bio-ink suitability for tissue engineering due to proliferation, vascularization, and migration potentials. Figure 4.24 shows that the object is non-porous under scanning electron microscopy upon fracture.

The high porosity observed in higher wool fiber content (C20) correlates with the maintained structural shape (non-shrinkage) compared to lower ratios when dried. The low porosity of the composite could be attributed to the wool microfibrils and possible interconnection between the fragmented fibril and the cassava starch, consistent with the literature (Baniasadi et al., 2021b).

$$porosity(\Phi) = \frac{Msat - Mi}{\rho V} * 100 = \text{saturated mass}$$

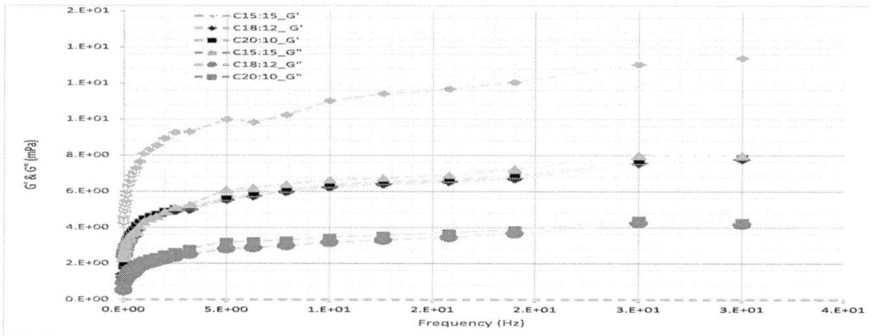


**FIGURE 4.25** Plot of storage modulus, G′ and G″, versus frequency, at different composite ratios of WP/CS blended ratios: cassava starch (10, 12, and 15%) and wool (20, 18, and 15%).

**TABLE 4.5**
**Wool and Cassava Starch Three-Dimensional-Printed (3DP) Object Porosity. *L* Is Length, *H* Means Height, *W* Is Width, *M* Is Mass, and *P* Is Porosities 1 and 2, Representing Initial and Final Values, Respectively.**

| | Height (mm) | | Length (mm) | | Breadth (mm) | | Sample weight (g) | | Porosity (%) |
|---|---|---|---|---|---|---|---|---|---|
| **Samples** | H1 | H2 | L1 | L2 | B1 | B2 | W1 | W2 | |
| **W20:10** | 1.05 | 1.25 | 2.2 | 1.85 | 1.2 | 1.25 | 0.912 | 1.826 | **2.32** |
| **W18:12** | 1.2 | 1 | 2 | 2.2 | 1.3 | 1.15 | 0.849 | 1.73 | **2.23** |
| **W15:15** | 0.8 | 0.5 | 1.8 | 1.9 | 1.1 | 1 | 0.499 | 1.008 | **1.29** |

where $M_i$ = initial Mass, $M_{sat}$ = saturated mass, A = density of absolute ethanol (0.789Kg/L), and V = volume.

### 4.3.7 Viscoelastic Characterization—DMA

The resulting storage modulus G′, loss modulus G″, and tan $\delta$ are extracted as presented in Figures 4.26 and 4.27, respectively. Figures 4.28 and 4.29 show the storage modulus versus loss modulus correlation, as well as the frequency–angular sweep with time of the samples. Analyzing material deformation resistance through the storage modulus G′, while its energy dissipation (mechanical strength) is assessed by the loss modulus, which scholars attribute to material viscous behavior (Wang et al., 2012). The slope in C15:15 shows less suitability for structural fidelity, with the pronounced increase in the slope indicating less strength. This implies that at an equal composite ratio of wool and starch (1:1), the material forms a less stable shape with reduced printability, whereas at a higher wool-to-starch ratio (2:1), that is, C20:10, a more structurally suitable object can be produced. The curve smoothness at any of

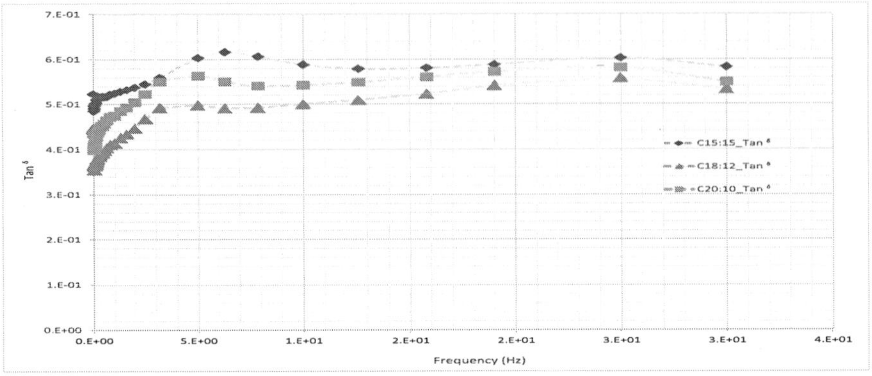


**FIGURE 4.26** Plot of tan $\delta$ versus frequency at different composites for WP/CS-blended ratios: cassava (10, 12, and 15%) and wool (20, 18, and 15%).

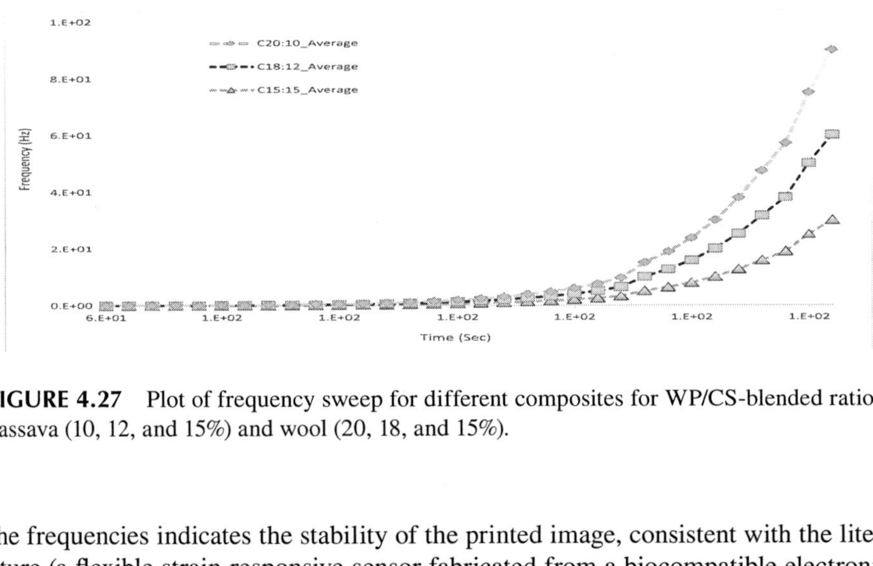


**FIGURE 4.27** Plot of frequency sweep for different composites for WP/CS-blended ratios: cassava (10, 12, and 15%) and wool (20, 18, and 15%).

the frequencies indicates the stability of the printed image, consistent with the literature (a flexible strain-responsive sensor fabricated from a biocompatible electronic ink via an additive-manufacturing process—ScienceDirect, n.d.; Britton et al., 2021), which reports that a suitable ink viscosity for 3D printing is attained (Figure 4.25).

The ratio between the G′ and G″ defines tan $\delta$ used to measure the elasticity of the ink. Tan $\delta < 1$ indicates the existence of elasticity and a solid-like behavior, while tan $\delta > 1$ depicts the viscous property of a more fluid-like behavior. These conditions of elasticity relate directly to the comparison of the storage and loss moduli, such that G′ > G″ indicates high elasticity and a more solid-like ink (He et al., 2020; Vancauwenberghe et al., 2017). For all the fabricated wool and cassava starch ink ratios, G′ > G″, indicating that elastic inks were obtained for all samples with solid-like behavior. Although G′ decreased with increasing wool content, G″ was lower than obtained in C15:15 but remained higher in C20:10 than in C18:12. Notably, tan $\delta$ decreased with

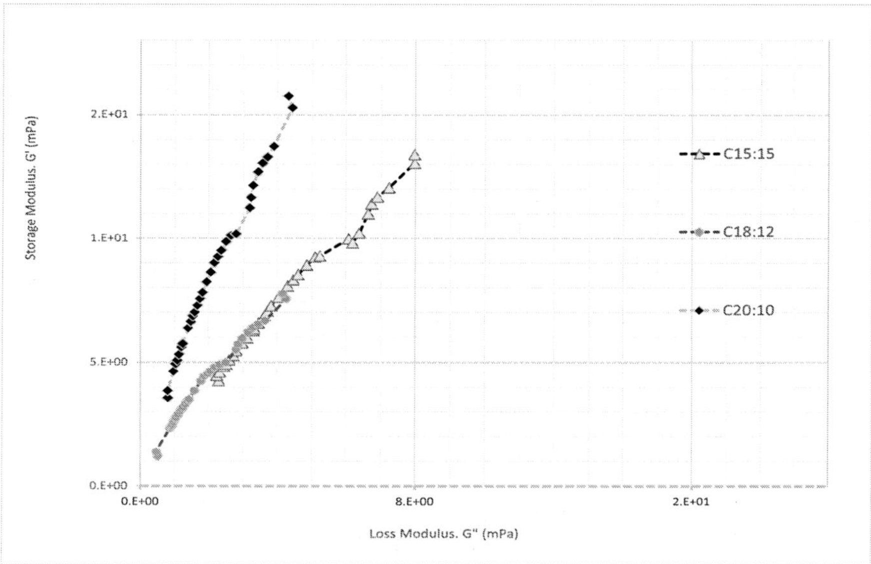


**FIGURE 4.28** Plot of frequency sweep for different composites for WP/CS blends. Cassava starch (C) and wool powder in ratios of cassava (10, 12, and 15%) and wool (20, 18, and 15%).

increasing wool powder content, with C20:10 showing the least value. This indicates low fluidity but mechanical strength, and by inference, high resistance to deformation due to elasticity, resulting in a more stable shape after printing. Tan $\delta$ increased proportionately with increasing frequency; this trend was observed across all the ink ratios. An inferred workability is adopted in applications requiring biocomposite ink for additive manufacturing of functional objects, parts, etc., with well-defined structures.

## 4.4 CONCLUSION

The study observed that sheep wool can be dry-processed into fine powder for 3D additive manufacturing applications, in contrast to the conventional wet milling, attrition milling, drying, and air jet milling (AJM) processes of WFP extraction reported in the literature (Marti et al., 2007). The energy required for PSR appears to be directly proportional to material strength and inversely proportional to the ultrasonic intensity.

This observation could suggest that the sonication-induced structural changes are related to the formation cum shift of wool fiber crystallinity visible in X-ray characterization analysis, contrary to the change in protein structure. Ultrasonic pretreatment impacted PSR and yielded odorless, fine wool powder, with glass transition temperatures of 89.24–55.07°C for white, and 275.37–58.13°C red, while comminution energy was reflected in the phase transition energy, correlating with Bond's law on PSR. The experimental design factored in the material size, uniformity in sonication energy, and sonication dosing time to create a 2 by 3 (5 and 10 min dosing time by cut-milled and milled mechanical size) factorial experimental model conditioning.

All the fabricated ink ratios are printed to design precision with structural stability devoid of warping or shrinkage upon extrusion. This promises relief for bioprinting and the biodegradable potential of wool fiber in 3D technology. With a low porosity of around 1.29–2.32%, the sample porosity increased with increasing WPF. The viscoelastic evidence shows that the ink exhibited a solid-like behavior with storage and loss modulus independent of frequency. A $G' > G''$ obtained, establishes strong water binding within the fabricated ink. The study, hence, unarguably identifies wool powder as a composite for fabricating inks usable in the FDM process of 3D additive manufacturing technology. Conclusively, it is a promising raw material that holds a lot of climatic and innovative advantages for diverse applications in the near future.

## REFERENCES

S. Ahmadzadeh and A. Ubeyitogullari, Generation of porous starch beads via a 3D food printer: the effects of amylose content and drying technique, *Carbohydr Polym*, vol. 301, p. 120296, Feb. 2023, doi: 10.1016/J.CARBPOL.2022.120296. Accessed: Jan. 28, 2023. [Online]. Available: www.sciencedirect.com/science/article/abs/pii/S0144861722012012?via%3Dihub

M. A. al Faruque, R. Remadevi, X. Wang, and M. Naebe, Preparation and characterisation of mechanically milled particles from waste alpaca fibres, *Powder Technol*, vol. 342, pp. 848–855, Jan. 2019, doi: 10.1016/j.powtec.2018.10.049.

H. Baniasadi, R. Ajdary, J. Trifol, O. J. Rojas, and J. Seppälä, Direct ink writing of aloe vera/cellulose nanofibrils bio-hydrogels, *Carbohydr Polym*, vol. 266, p. 118114, Aug. 2021a, doi: 10.1016/J.CARBPOL.2021.118114.

H. Baniasadi *et al.*, 3D printing and properties of cellulose nanofibrils-reinforced quince seed mucilage bio-inks, *Int J Biol Macromol*, vol. 192, pp. 1098–1107, Dec. 2021b, doi: 10.1016/J.IJBIOMAC.2021.10.078.

J. Britton *et al.*, A flexible strain-responsive sensor fabricated from a biocompatible electronic ink via an additive-manufacturing process, *Mater Des*, vol. 206, p. 109700, Aug. 2021, doi: 10.1016/J.MATDES.2021.109700. Accessed: Jan. 28, 2023. [Online]. Available: www.sciencedirect.com/science/article/pii/S0264127521002525

Y. F. Cheng, C. W. M. Yuen, Y. Li, S. K. A. Ku, C. W. Kan, and J. Y. Hu, Characterization of nanoscale wool particles, *J Appl Polym Sci*, vol. 104, no. 2, pp. 803–808, Apr. 2007, doi: 10.1002/app.25631. Accessed: May 29, 2022. [Online]. Available: www.researchgate.net/publication/230353314_Characterization_of_nanoscale_wool_particles

A. Danch and A. Gadomski, On the crystalline-amorphous supermolecular structure of poly(4-methyl-1-pentene) films cast from solution: experimental evidences and theoretical remarks, *J Mol Liq*, vol. 86, no. 1–3, pp. 249–257, Jun. 2000, doi: 10.1016/S0167-7322(99)00146-4.

Ş. Duman and M. Küçük, Production and characterization of keratin microparticles obtained from wool fibers by cryogenic milling method, *Particulate Sci Technol*, 2022, doi: 10.1080/02726351.2022.2028321.

O. Fourmann *et al.*, 3D printing of shape-morphing and antibacterial anisotropic nanocellulose hydrogels, *Carbohydr Polym*, vol. 259, p. 117716, May 2021, doi: 10.1016/J.CARBPOL.2021.117716. Accessed: Jan. 28, 2023. [Online]. Available: www.sciencedirect.com/science/article/pii/S014486172100103X

A. Gadomski and J. Luczka, On the kinetics of polymer crystallization: a possible mechanism, *J Mol Liq*, vol. 86, no. 1–3, pp. 237–247, Jun. 2000, doi: 10.1016/S0167-7322(99)00145-2.

R. He *et al.*, Comparison of the structural characteristics and physicochemical Properties of starches from sixteen cassava germplasms cultivated in China. *Int J Food Properties*, vol. 23, no. 1, pp. 693–707, Jan. 2020, doi: 10.1080/10942912.2020.1752714

D. R. H. Jones and M. F. Ashby, Elastic moduli, *Eng Mater*, vol. 1, pp. 31–47, Jan. 2019, doi: 10.1016/B978-0-08-102051-7.00003-8. Accessed: Jan. 28, 2023. [Online]. Available: www.sciencedirect.com/science/article/pii/B9780081020517000038?via%3Dihub

G. Kimbell and M. A. Azad, 3D printing: bioinspired materials for drug delivery. In *Bioinspired and Biomimetic Materials for Drug Delivery*, pp. 295–318, Jan. 2021, Elsevier. doi: 10.1016/B978-0-12-821352-0.00011-3. Accessed: Jan. 28, 2023. [Online]. Available: www.sciencedirect.com/science/article/pii/B9780128213520000113?via%3Dihub

Q. Li, T. Lin, and X. Wang, Effects of ultrasonic treatment on wool fibre and fabric properties, *J Textile Inst*, vol. 103, no. 6, pp. 662–668, Jun. 2012, doi: 10.1080/00405000.2011.597569. Accessed: Nov. 02, 2022. [Online]. Available: www.researchgate.net/publication/232817561_Effects_of_ultrasonic_treatment_on_wool_fibre_and_fabric_properties

M. Martí, R. Ramírez, A. M. Manich, L. Coderch, and J. L. Parra, Thermal analysis of merino wool fibres without internal lipids, *J Appl Polym Sci*, vol. 104, no. 1, pp. 545–551, Apr. 2007, doi: 10.1002/APP.25586. Accessed: Nov. 02, 2022. [Online]. Available: www.researchgate.net/publication/229956535_Thermal_analysis_of_merino_wool_fibres_without_internal_lipids

S. Ramesh *et al.*, Extrusion bioprinting: recent progress, challenges, and future opportunities, *Bioprinting*, vol. 21, p. e00116, Mar. 2021, doi: 10.1016/J.BPRINT.2020.E00116. Accessed: Jan. 28, 2023. [Online]. Available: www.sciencedirect.com/science/article/pii/S2405886620300439?via%3Dihub

V. Vancauwenberghe *et al.*, Pectin based food-ink formulations for 3-D printing of customizable porous food simulants, *Innov Food Sci Emerg Technol*, vol. 42, pp. 138–150, Aug. 2017, doi: 10.1016/J.IFSET.2017.06.011. Accessed: Jan. 29, 2023. [Online]. Available: www.sciencedirect.com/science/article/abs/pii/S1466856417306793

X. Wang, X. He, G. Huang, and J. Wu, Correlations between alkyl side chain length and dynamic mechanical properties of poly(n-alkyl acrylates) and poly(n-alkyl methacrylates), *Polymer (Guildf)*, vol. 53, no. 2, pp. 665–672, Jan. 2012, doi: 10.1016/J.POLYMER.2011.11.059. Accessed: Jan. 28, 2023. [Online]. Available: www.sciencedirect.com/science/article/abs/pii/S0032386111009931

T. Yao, H. Ouyang, S. Dai, Z. Deng, and K. Zhang, Effects of manufacturing micro-structure on vibration of FFF 3D printing plates: material characterisation, numerical analysis and experimental study, *Compos Struct*, vol. 268, p. 113970, Jul. 2021, doi: 10.1016/J.COMPSTRUCT.2021.113970. Available: www.sciencedirect.com/science/article/abs/pii/S026382232100430X?via%3Dihub

C. Zhang, L. Xia, J. Zhang, X. Liu, and W. Xu, Utilization of waste wool fibers for fabrication of wool powders and keratin: a review, *J Leather Sci Eng*, vol. 2, no. 1, Dec. 2020, doi: 10.1186/S42825-020-00030-3. Accessed: May 05, 2022. [Online]. Available: www.researchgate.net/publication/342659997_Utilization_of_waste_wool_fibers_for_fabrication_of_wool_powders_and_keratin_a_review

# 5 Regulatory Framework and Standards for Bio-Based Materials

*Praneta Yadav, Devendra Singh, Sunil Kumar Verma, Marvellous O. Eyube, and Virendra Singh*

## 5.1 INTRODUCTION

The term "bio" originates from Greek, meaning "life." Bio-based materials are primarily composed of substances derived from living organisms (biomass). These materials should not be confused with biomaterials, which are used in medical applications such as tissue engineering and drug delivery (Katz, 2020). Transitioning from fossil fuel-based resources to plant-derived and renewable materials is critical for a sustainable economy. While bio-based composites are gaining traction in various industries such as food packaging, textiles, and plastics, renewable energy sources like solar and wind are also influencing the shift toward greener solutions (Storz & Vorlop, 2022). Bio-based materials offer alternatives to fossil fuel-based products by utilizing naturally occurring molecules or synthesizing bio-based monomers that undergo polymerization. However, a significant challenge is maintaining consistency in composition and performance. One approach to addressing this challenge is the controlled chemical polymerization of bio-based monomers. For instance, polylactic acid and polybutylene succinate have demonstrated the feasibility of bio-based polymer development. Studies have explored sustainable polymer sources such as cereal grains, which provide raw materials for biodegradable films, fibers, and food packaging (Razzaq et al., 2023). The focus on transforming sustainable feedstocks into viable bio-based materials remains an active area of research.

### 5.1.1 Properties of Bio-based Materials

Bio-based products should be generally clean, recyclable, effective, and efficient to enhance their features and utility. Bio-based materials significantly encourage green production since they appropriately wrap and shield products—two crucial packaging attributes. The packaging system is also made to use materials and energy with the least amount of waste throughout the product life cycle. It is important to consider the properties of food bio-based materials, which may include resistance to water and gas vapor permeability, thermoforming ability, transparency, antifogging ability, printability, availability, UV resistance, water resistance, and resistance to

 DOI: 10.1201/9781003566205-5

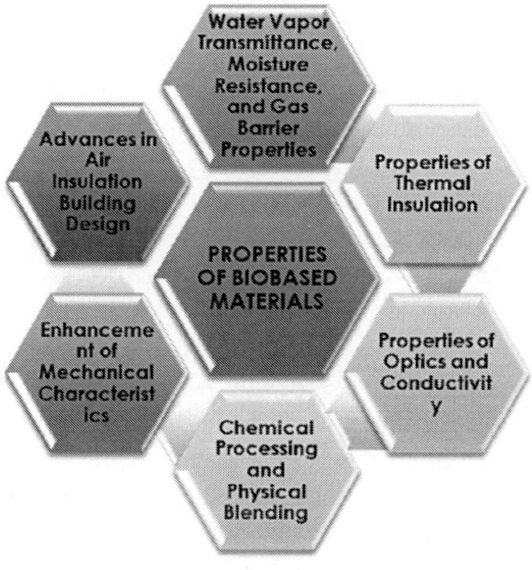


**FIGURE 5.1** Properties of bio-based materials.

acids, alkalis, grease, etc. The price of bio-based products is another important factor (Rahim et al., 2022). Therefore, for bio-based materials to be used in food and other industries, they must have good mechanical, thermal, and gas barriers, as well as water vapor permeability and disposability (see Figure 5.1). Compostable or biodegradable products are achievable if they are created completely from renewable basic materials (Chennouf et al., 2021).

### 5.1.2 Advantages of Bio-based Materials

Bio-based packaging minimizes waste. In 2003, packaging waste accounted for 31.6% of the 78.81 million tons of MSW in the USA, 25% of 56.3 million tons in Europe in 2005, and 10% of 3.3 million tons in Australia in 2004. The USA is implementing a waste disposal strategy that emphasizes recycling, burning, and composting over landfilling. Landfills require large spaces and can emit pollutants and greenhouse gases (Kale et al., 2022). About 14% of the trash stream in the world is composed of plastics. Each year, waste is generated, and only 9% of it gets recycled (Dhiman et al., 2022; Verma et al., 2024a). Leftover waste goes to landfills, requiring more space over time. This creates a waste stream that occupies significant land and can harm urban lifestyles due to limited space and high land costs (Choudhury et al., 2022).

Bio-based plastics help reduce ozone-depleting emissions. Each year, approximately 100 million tons of plastic are produced, generating 500 million tons of carbon dioxide, equivalent to the emissions from 19 million cars. Promoting plastic reuse is crucial to mitigating greenhouse gas emissions. Currently, investment in net carbon reserves stands at 30%, while experts recommend increasing it to 80%. Although the financial costs of adopting biodegradable plastics remain a challenge, their implementation can significantly reduce emissions from the plastics industry. Verma et al. (2024d) state that bio-based materials offer numerous environmental and economic benefits, making them a sustainable alternative to conventional plastics. One of their key advantages is the reduction in greenhouse gas (GHG) emissions. For instance, producing 1 kg of resin using bio-based materials emits only 0.49 kg of carbon dioxide ($CO_2$), significantly lowering GHG emissions. Although the entire production process generates about 2–3 kg of $CO_2$—higher than bioplastic—it remains an eco-friendlier option compared to petroleum-based plastics.

Another major benefit of bio-based materials is their role in mitigating environmental pollution. Shifting from traditional non-biodegradable plastics to bio-based alternatives can significantly reduce plastic waste and its associated ecological impact. Biopolymers, as substitutes for petroleum-based plastics, enhance sustainability while maintaining the necessary durability and functionality for packaging applications. Beyond environmental benefits, bio-based materials contribute to increased profitability for companies. Businesses and NGOs that incorporate bio-based packaging can strengthen consumer trust by demonstrating environmental responsibility. Studies show that consumers are more likely to choose products with bio-based packaging over conventional alternatives, making it an effective indirect marketing strategy. Major corporations such as the Coca-Cola Company and Poland Spring have successfully reduced plastic use by integrating bio-based materials into their packaging.

Furthermore, the adoption of bio-based materials in food packaging could lead to the development of new export commodities, expanding economic opportunities. Currently, approximately 290,000 tons of biodegradable materials are in production for various applications. Additional advantages of bio-based materials include their recyclability, renewability, biodegradability, zero carbon footprint, and non-hazardous properties. They also allow for quick installation, prefabrication, and moisture regulation while exhibiting desirable physical and mechanical properties. Moreover, their inherent breathability improves indoor environmental quality (see Figure 5.2).

### 5.1.3 Applications of Bio-based Materials

New bio-based products are emerging, and their applications are beginning to be explored. They have the potential to take market share from traditional goods made from minerals and petroleum. Biomass conversion, biosynthesis, and bioprocessing cycles provide opportunities for sustainable material creation, further enhanced by recycling organic waste from food, forestry, and agriculture. Many natural processes remain unexplored, presenting significant research and industrial prospects. In the future, factories may resemble greenhouses, where microbes produce materials from recycled biological waste. Nanotechnology and genetic engineering are advancing bio-based materials by modifying feedstocks to optimize their properties. In construction, bio-based materials serve various functions; for example, the fiber content in concrete influences its properties—higher plant particle content lowers density and strength but improves porosity. In 2015, Europe produced 58 million tons of the 322 million tons of global plastic, with bio-based plastics constituting only 0.5–1% of EU usage. Market growth is being driven by the development of new packaging with enhanced qualities and reduced costs, spurred by increasing environmental

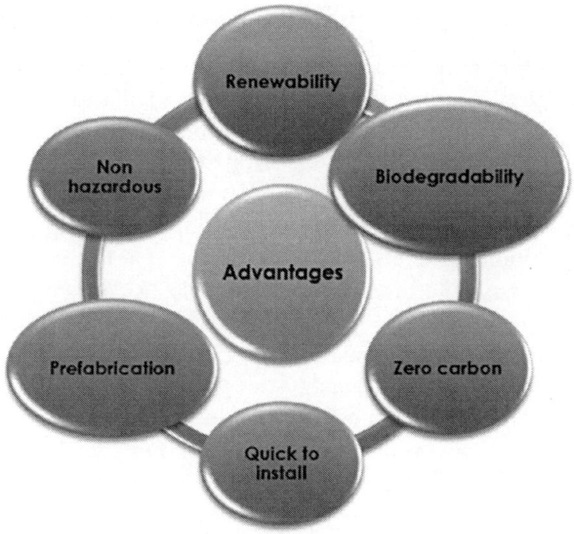


**FIGURE 5.2** Advantages of bio-based materials.

**FIGURE 5.3** Industrial applications of bio-based materials.

awareness. Bio-based products, including biodegradable films and containers made from renewable resources like cellulose and sugarcane, reduce waste and carbon emissions compared to conventional plastics. Their biocompatibility and biodegradability make them suitable for applications in tissue engineering, drug delivery, and wound care, utilizing materials such as chitosan and collagen (Sharma et al., 2019).

## 5.2 REGULATORY FRAMEWORKS FOR BIO-BASED MATERIALS

The goals of zero pollution, biodiversity improvement, and the EU's transition to a circular economy have prompted a re-evaluation of plastic production and disposal. In 2020, only 14% of plastic waste in Europe was locally recycled, while the remainder was exported, landfilled, or incinerated. Enhancing plastic sustainability is crucial due to the prevailing linear production model and the projected tripling of plastic output over the next 20 years. Addressing plastic waste, litter, greenhouse gas emissions, and pollution presents complex challenges (Guo et al., 2023).

Bio-based, biodegradable, and compostable plastics are increasingly replacing conventional plastics. Nearly half of the demand comes from the packaging sector, while the rest includes consumer products, textiles, construction, transportation, and agriculture. These plastics account for over 2 million metric tons annually, representing approximately 1% of global plastic production. By 2030, the EU's zero-pollution action plan aims to reduce marine plastic debris by 50% and microplastics by 30%. The EU's soil strategy emphasizes preventing contamination by promoting the reduction, reuse, and recycling of plastics to achieve a toxic-free environment. This framework for bio-based plastics outlines their benefits and drawbacks, establishes requirements for a positive environmental impact, addresses policy gaps, guides future EU legislation, and fosters a cohesive single market across member states.

### 5.2.1 Regulations and Standards for Bio-based Materials

The EU has enacted regulations to promote bio-based materials for sustainability. The November 2022 Policy Framework on Bio-based, Biodegradable, and Compostable Plastics, part of the Circular Economy Action Plan, emphasizes clear labeling of bio-based content and prioritizes reliable assessment methods over questionable techniques. The Single-Use Plastics Directive bans certain single-use plastics and promotes bio-based alternatives while setting standards for compostable and biodegradable materials. Additionally, the Packaging Waste Directive mandates the use of bio-based materials in packaging to enhance recyclability.

The EU Taxonomy Regulation classifies sustainable economic activities, establishing criteria for environmentally sustainable bio-based materials. Several EU countries have also implemented national laws requiring single-use bags to be biodegradable. For example, Italy mandated that single-use bags contain at least 60% plant-derived carbon by 2021, while France requires a minimum of 30% bio-based material, increasing to 60% by 2025. In the US, both state and federal agencies influence the adoption of bio-based materials. The USDA BioPreferred Program certifies bio-based goods, promoting their use to reduce fossil fuel reliance. The Federal Trade Commission (FTC) establishes guidelines for environmental marketing claims to ensure truthful advertising. Various states, such as California, are advancing legislation to require biodegradable food service items. Asian regulations vary by country. India enforces biodegradable plastics for waste management, Japan has established standards for biodegradable polymers, and China restricts single-use plastics while investing in the development of bio-based materials.

### 5.2.2 International Standards

#### 5.2.2.1 ISO Standards

ISO 16620–2 outlines methodologies for determining a material's bio-based carbon content and provides a framework for certification. ISO 17088 establishes parameters for compostable plastics, ensuring they meet specific criteria for biodegradability under composting conditions. ISO 14021 defines standards for environmental claims, including the appropriate use of terms such as "biodegradable" and "compostable."

#### 5.2.2.2 CEN Standard

EN/TS 16640, which aligns with ISO standards, offers guidance on assessing the bio-based content of plastic materials. EN 16785–1 accurately evaluates bio-based content through a combination of elemental and radiocarbon analysis. The regulatory landscape for bio-based products is evolving globally. The EU, the US, and several Asian nations are making significant efforts to promote sustainability and reduce dependence on fossil fuels. Despite advancements, challenges persist, including the need to harmonize standards and ensure that bio-based materials provide genuine environmental benefits while enhancing consumer understanding (Singh et al., 2024b, 2024f; Verma et al., 2024d).

## 5.3 CERTIFICATION AND LABELING SCHEMES

Certification schemes serve as a means of demonstrating the sustainable production and processing of bio-based products. Bio-based goods are already subject to various certification programs and instruments covering a range of topics, including sustainable feedstock, bio-based content, and end-of-life possibilities (Singh et al., 2024c, 2024e, Verma et al., 2024f). Many sustainable feedstock certification programs, such as RSB, FSC, and ISCC, which were initially developed for biofuels, have been adapted to encompass the material use of biomass.

The standard method for determining a product's bio-based composition is to measure its carbon (C) content using ASTM D6866–10. Labels—particularly ecolabels—are essential tools for informing consumers about the benefits of bio-based products, especially when such items meet predetermined sustainability standards and have been certified (Singh et al., 2024d; Srivastava et al., 2024; Singh & Verma, 2024). These labels provide information on product attributes that are difficult to assess, particularly regarding the environmental impacts of production practices. The EU Ecolabel ("the Flower") and other national ecolabels, such as the German and Nordic Ecolabels, currently promote only a limited number of bio-based products (Verma et al., 2022, 2024e).

Several certification programs exist for bio-based materials to enhance consumer confidence, sustainability, and transparency. The DIN-Geprüft Bio-based Certification in Germany verifies the bio-based content of products made from renewable resources, distinguishing among Bio-based 21–50%, Bio-based 51–86%, and Bio-based >86%. The certification process includes testing for bio-based carbon content, 14C measurement, and elemental analysis (Verma et al., 2023). The certificate can be renewed through monitoring and further assessments after its six-year expiration and

**FIGURE 5.4** Certification procedure.

goes through a series of procedures (see Figure 5.4). SGS Bio-based Materials Certification: SGS certifies under European standard EN 16785–1, aligning with NEN scheme NCS 16785. Transparency: SGS compares the product's total weight to its bio-based carbon content to accurately represent bio-based content.

The Roundtable on Sustainable Biomaterials (RSB) provides a global certification standard for bio-based and sustainable biomaterials. Certified products, per EN 16640, ASTM D6866, or ISO 16620–1, must contain a minimum of 25% bio-based carbon. The RSB standard ensures sustainability across the entire supply chain for advanced and bio-based products (Singh et al., 2023). Green Circle Certified: Green Circle verifies environmental statements, including bio-based content, through unbiased third-party validation. Claims about renewable content, product, raw materials, and production methods are assessed according to ASTM D6866 testing guidelines. This certification ensures the credibility of environmental claims for various bio-based products (Shrivastava et al., 2023). The Blue Angel Ecolabel (Germany) certifies environmentally friendly products made from bio-based materials. Applicants must provide test reports from recognized laboratories, often using ASTM standards, to ensure their products meet bio-based content requirements. These certifications enhance the credibility of claims regarding bio-based materials and will evolve as the sustainable product market grows, establishing clearer regulations and boosting consumer confidence in bio-based products (Singh et al., 2024a).

## 5.4 STANDARDS FOR BIO-BASED CONTENT AND COMPOSITION

The standards and certification schemes for bio-based content play a critical role in promoting the use of renewable resources and ensuring transparency in the marketplace. By adhering to these standards, manufacturers can demonstrate their commitment to sustainability and help consumers make informed choices about the products they purchase.

### 5.4.1 Test Methods for Determining Bio-based Content

Common test procedures for bio-based content of gaseous, liquid, and solid samples are as follows. Method A measures bio-based material using accelerator mass spectrometry (AMS) and isotope ratio mass spectrometry (IRMS). The bio-based content may have an error of 0.1–0.5% (1 rsd), with total uncertainty reaching ±3% (absolute) due to various error sources. To prepare a sample, $CO_2$ is produced in a vacuum manifold, transported to a quartz tube, distilled, measured in calibrated volume, and torch sealed. Information is in Sections 8.6–8.10. The AMS facility then processes the $CO_2$. Method B transforms sample carbon to benzene, using liquid scintillation counting (LSC) to quantify bio-based content, also with a maximum error of 3%. These procedures differentiate between fossil fuel-based and modern carbon inputs by measuring the $^{14}C/^{12}C$ ratio against the carbon-based radiocarbon Standard Reference Material (SRM) for oxalic acid. It is expressed as *fM*, the fraction of modern carbon, related to the original oxalic SRM composition (Zhu et al., 2022).

### 5.4.2 Requirements for Minimum Bio-based Content

Minimum bio-based content limits are outlined by various standards, primarily in the US and Europe, ensuring products meet certification criteria for federal procurement. Under the USDA BioPreferred Program, federal agencies must prioritize bio-based products, with required content ranging from 6% to 95% based on category. A minimum of 25% bio-based content is required for USDA certification unless otherwise authorized. European regulations lack a unified criterion, but some agreements, like those in Japan, set a 25% renewable material threshold for biomass in specific products (see Table 5.1).

## 5.5 STANDARDS FOR BIODEGRADABILITY AND COMPOSTABILITY

Biodegradability and compostability criteria are essential for ensuring that bio-based materials break down under various conditions and support sustainable waste management. The primary guidelines for bio-based, biodegradable, and compostable

**TABLE 5.1**
**Minimum Requirements.**

| Program | Minimum bio-based content |
|---|---|
| **BioPreferred USDA (designated categories)** | 6% to 95% |
| **BioPreferred USDA (non-designated)** | 24% |
| **Japan Industry Commitment** | 25% |
| **European Standards** | No common |

materials include the ASTM D6400 standard, which specifies requirements for polymers intended for commercial and municipal aerobic composting. This standard ensures that the biodegradation rate aligns with that of known compostable materials. The substance must fully break into the soil without leaving visible residue, and the finished compost must pass through a 2 mm sieve (Moretti et al., 2022). Additionally, heavy metals must not be present in the decomposed material, and the finished compost must support plant growth. Biodegradation should occur within 180 days, ideally matching the decomposition rates of organic materials such as paper and leaves. ASTM D6400 testing lasts up to 180 days, with a minimum duration of 90 days, during which samples are exposed to a municipal waste inoculum that simulates typical composting conditions. Another relevant standard, ASTM D6868, specifies labeling requirements for products containing plastics and polymers combined with paper or other substrates intended for aerobic composting. These guidelines complement ASTM D6400.

Additionally, ASTM D5338 outlines testing procedures for the aerobic biodegradation of plastics under controlled composting conditions, which is essential for evaluating compliance with ASTM D6400 and D6868. The European standard EN 13432 establishes specifications for compostable packaging, covering aspects such as compost quality, disintegration, and biodegradation—closely aligning with ASTM D6400. Similarly, ISO 14855 describes the methodology for determining the maximum aerobic biodegradability of plastic materials under controlled composting conditions, serving as a global benchmark for biodegradability assessment.

The effectiveness of these criteria can be influenced by both material properties and environmental conditions. Home composting facilities, for instance, often fail to reach the thermophilic temperatures required for effective biodegradation, raising concerns about the relevance of laboratory test conditions to real-world composting scenarios (Verma et al., 2024b). This disparity may create confusion regarding manufacturers' claims about the compostability of their products. Table 5.2 shows the standards for biodegradability and compostability.

**TABLE 5.2**
**Standards for Biodegradability and Compostability.**

| S.No. | Standard | Focus area | Requirements |
|---|---|---|---|
| 1 | ASTM D6400 | Biodegradable polymers in industrial settings | No heavy metals, 180 days biodegradation |
| 2 | ASTM D6868 | Coatings and additives in biodegradable substrates | Akin to D6400, with an emphasis on composite materials |
| 3 | ASTM D5338 | Testing for aerobic biodegradation | Regulated criteria for composting |
| 4 | EN 13432 | Reusable packaging that can be composted | Criteria for biodegradation and disintegration |
| 5 | ISO 14855 | Aerobic biodegradability is maximum | International biodegradability testing standard |

### 5.5.1 Test Methods for Assessing Biodegradability and Compostability

Evaluating the biodegradability and compostability of bio-based products is essential for ensuring their environmental sustainability. A variety of standardized test procedures are employed, each with specific guidelines and specifications (Singh et al., 2024a). The key test methods used to assess these properties in bio-based materials are outlined below.

The internationally recognized standard ISO 14855 measures the maximum aerobic biodegradability of plastics in controlled composting environments by tracking carbon dioxide ($CO_2$) emissions. The test typically lasts 45 days but can extend to six months at 58°C, using an inoculum derived from mature compost. $CO_2$ evolution is regularly measured to determine the percentage of biodegradation based on the organic carbon content of the test material. ASTM D5338, an American standard, also evaluates plastic degradation under aerobic composting conditions over a period ranging from 90 days to 6 months. This method assesses biodegradation effectiveness by measuring $CO_2$ production.

ISO 14851 determines aerobic biodegradability in aquatic environments by measuring oxygen demand, making it particularly useful for materials that are not suitable for composting. The test tracks oxygen consumption over time in controlled water conditions and applies to substances that are fully or partially soluble. ISO 20200 assesses the extent of plastic breakdown under laboratory conditions that simulate composting environments. This test evaluates material decomposition over time and correlates with other biodegradability assessments, using weight loss and visual inspection to measure disintegration.

The European standard EN 13432 specifies requirements for compostable and biodegradable packaging, considering factors such as ecotoxicity, biodegradation, and disintegration. Under composting conditions, the material must break down within 12 weeks, with at least 90% decomposing within six months. The resulting compost should not be harmful to plant growth or soil organisms. ISO 14855–2, a variation of ISO 14855, focuses on laboratory-scale testing and provides gravimetric measurement of $CO_2$ evolution during biodegradation (Walker & Rothman, 2020) (Table 5.3).

These uniform testing procedures offer a thorough framework for assessing the compostability and biodegradability of bio-based materials, guaranteeing that they satisfy standards for environmental sustainability (Verma & Singh, 2024).

### 5.5.2 Requirements for Biodegradability and Compostability in Different Environments

Bio-based materials decompose under specific conditions for biodegradability. This summary outlines the standard for various settings. Polymers for outdoor use may not break down in uncontrolled environments. Key factors in biodegradation include temperature, humidity, and oxygen levels. Biodegradable plastics decompose quickly in warm, humid climates than in cold, dry ones. Microbial presence is vital; specific microbes and diverse communities in soil, freshwater, and marine environments affect rates. Material properties, shaped by chemical composition, influence

**TABLE 5.3**
**Test Methods.**

| S.No. | Test method | Focus area | Duration | Measurement method |
|---|---|---|---|---|
| 1 | ASTM D5338 | Composting involves aerobic biodegradation | 90 days to 6 months | $CO_2$ evolution |
| 2 | ISO 14851 | Biodegradability of aerobic organisms in water | Variable | Oxygen demand |
| 3 | ISO 20200 | Level of disintegration throughout the composting process | Variable | Weight loss and visual |
| 4 | ISO 14855 | Maximum aerobic biodegradability | 45 days to 6 months | $CO_2$ evolution |
| 5 | ISO 14855–2 | Biodegradability at the laboratory scale | 45 days to 6 months | Gravimetric $CO_2$ evolution |
| 6 | EN 13432 | Compostable packaging that can be recovered | 12 weeks | Ecotoxicity and $CO_2$ evolution |

microbial sensitivity. Some materials only decompose in commercial composting, not outdoors (Atiwesh et al., 2021).

Testing requirements for soil biodegradability include ISO 17556 evaluates how plastics biodegrade in soil, requiring at least 90% degradation within 24 months. ASTM D5988 assesses the pace and degree of biodegradation similarly. NF U52–001 and UNI 11462 may offer additional biodegradation guidelines alongside safety regulations. Compostable materials must meet specific criteria. ASTM D6400 outlines requirements for compostable plastics in municipal facilities. EN 13432 states materials must disintegrate within 12 weeks and achieve 90% biodegradation in six months under industrial composting. In aquatic environments, biodegradability is crucial for products in freshwater or oceans. ISO 14851 uses oxygen demand to evaluate biodegradability in water. Challenges include varying temperature and salinity, complicating assessments of plastic's environmental impact (Abral et al., 2020). Factors like microorganisms, environmental conditions, and material properties influence the need for biodegradability and compostability. Standards such as ASTM D6400 and EN 13432 offer frameworks for evaluating these materials in composting, while ISO standards address biodegradability in soil and water. To reduce environmental impact and ensure effective integration into waste management, biodegradable materials must meet these criteria.

## 5.6 STANDARDS FOR ENVIRONMENTAL IMPACT AND LIFE CYCLE

### 5.6.1 Life Cycle Assessment Standards for Bio-based Materials

Life cycle assessment (LCA) evaluates a product, process, or service's environmental impact by analyzing each stage from extraction and production to consumption and disposal (Curran, 2023). LCA evaluates a process or product's environmental impact, aiding the industrial sector in reorganizing technologies and procedures to mitigate these impacts (Brusseau, 2019). Unlike environmental impact assessment (EIA), which considers only the effects of proposed developments, LCA broadens the environmental assessment scope and standardizes evaluation areas (Židoniene & Kruopiene, 2015). The four methodological steps are as follows: (a) goal and scope definition, (b) inventory analysis, (c) impact assessment, and (d) interpretation (ISO Standard 14040; Brusseau, 2019). The LCA method is adaptable to its goals, and its findings will be specifically developed and implemented. For authentic LCA, key integration requirements must be considered and explicitly stated (Muthu, 2022; ISO Standard 14040; Jolliet et al., 2021). LCA Standards are ISO 14067, which addresses the carbon footprint of goods and how to measure and report it, considering land use changes and biogenic carbon emissions. ISO 14044 outlines the prerequisites and recommendations for LCA, including objective and scope definition, inventory analysis, impact assessment, and result interpretation. Regular and transparent LCAs are essential. ISO 14040, in turn, provides a structured method to assess the environmental effects of products by detailing the fundamentals and structure of LCA (Ita-Nagy et al., 2021). Powder rheology of additive manufacturing powders was also tested by different researchers to investigate the angle of repose, flowability, and electrostatic charge (Neveu et al., 2020; Tripathi et al., 2019a, 2019b, 2019c, 2019d, 2020; Lumay et al., 2019a; Kalman et al., 2019a, 2019b, 2019c).

### 5.6.2 Environmental Impact Standards and Requirements

Fossil-based materials are being replaced by bio-based ones from renewable resources, perceived as more sustainable. However, their environmental impact must be carefully evaluated to ensure they support sustainability goals (Bushi et al., 2022). This involves following certification procedures and standards, assessing materials' environmental performance and bio-based composition. ISO 16620–1 applies to synthetic fibers and bio-based polymers, providing standards for quantifying bio-based content to ensure consistency across products. ASTM D6866 and EN 16785–1 Standards determine bio-based composition using radiocarbon analysis to differentiate between fossil fuel and renewable carbon. Products are categorized based on bio-based carbon content and certification requirements. The regulatory environment in the EU is shifting to favor bio-based enterprises, influenced by programs like the European Green Deal (Atiwesh et al., 2021). Strict feedstock evaluation rules, REACH compliance, and Waste Framework Directive duties mandate eco-friendly waste management techniques.

## 5.7 STANDARDS FOR FOOD CONTACT MATERIALS

Food contact materials (FCMs) are essential for ensuring food quality and safety. They include items like equipment, utensils, and packaging that touch food. Various laws regulate their safety, composition, and labeling. Bio-based FCMs follow the same rigorous regulations as conventional plastics. The main regulation is the EU Framework Regulation (EC) No. 1935/2004, which sets safety standards for all

FCMs. It prohibits materials from altering food composition, transferring harmful substances, or degrading food's organoleptic qualities (Verma et al., 2024c). Plastics Regulation (EU) No. 10/2011 covers plastic articles and materials in contact with food. It lists approved materials for plastic production and provides guidelines for migration testing to assess safety and determine migration limits based on EFSA toxicity data. Good Manufacturing Practice Regulation (EC) No. 2023/2006 requires adherence in food-contact material production, including bio-based alternatives, ensuring low contamination and quality. The Regulation on Recycled Plastic Materials (EC) 2022/1616 aims to speed up the development and approval of innovative recycling solutions for food-contact materials, outlining conditions for registering advanced technologies in a new EU registry.

### 5.7.1 Migration Testing and Compliance Requirements

To ensure consumer safety, bio-based FCMs must meet strict criteria and undergo rigorous migration testing. Research shows minimal migration of nanomaterials from packaging into food, while regulations are still developing. Specific migration testing assesses how much of a substance migrates. EFSA reviews toxicological data to establish limits. Analytical methods like GC-MS and HPLC evaluate the total chemical migration from materials into food. The EU Plastics Regulation (EU) specifies an overall migration limit of 60 mg/kg food using relevant simulants. Compliance requires approved compound manufacturers to use only substances listed in the EU's Plastics Regulation. Good Manufacturing Practices Regulation (EC) No. 2023/2006 mandates the hygienic production of bio-based FCMs to prevent contamination. Documentation for compliance is necessary, and materials must be traceable throughout the supply chain. Labels must accurately represent the types of materials used. Existing techniques for fossil-based plastics are expected to be applicable to bio-based FCMs, although method development may be needed (Dhiman et al., 2024). Numerous migration studies have shown that a wide variety of bio-based materials, such as starch, cellulose, PLA, cassava, and bagasse, can be processed utilizing traditional methods that employ food simulants.

## 5.8 STANDARDS FOR PACKAGING APPLICATIONS

### 5.8.1 Specific Standards for Bio-based Packaging Materials

Bio-based packaging materials, derived from renewable resources, must meet specific requirements for performance, safety, and environmental impact. Key guidelines include the ASTM standard for assessing bio-based content using radiocarbon dating to differentiate between fossil fuel and renewable carbon. This standard applies to various materials and includes Method B, which employs isotope ratio mass spectrometry (IRMS) and accelerator mass spectrometry (AMS) with a total uncertainty of ±3%, and Method C, using LSC with a maximum error of ±3%. European standards (EN) detail radiocarbon analysis for verifying bio-based content, relevant for European manufacturers. ISO standards provide transparency on renewable resources in plastics, enhancing marketability. The ISO/TS standards outline food safety criteria for bio-based packaging that comes into contact with food, ensuring hygienic production and compliance with Good Manufacturing Practices (GMP). Adhering to these regulations ensures product performance, sustainability, and safety, while aiding manufacturers in certifying bio-based content and meeting consumer expectations (Piccinno et al., 2019).

### 5.8.2 Performance Requirements and Testing Methods

Packaging materials derived from biotechnology are engineered to fulfil precise performance standards, guaranteeing sufficient food product safety and environmental sustainability. The key performance parameters and the associated testing procedures that are utilized to assess these materials are covered in this thorough overview. Seal strength is crucial for preventing leaks and maintaining freshness; seals must be durable, especially with flexible materials. Tensile strength measures a material's ability to withstand stress without cracking, essential for enduring storage, transportation, and filling. Tear resistance assesses a material's capacity to resist tearing, vital for package integrity. A low oxygen transmission rate (OTR) is needed to prevent oxidation that causes spoilage. Light-sensitive foods require packaging with effective light barriers to maintain quality. Water vapor transmission rate (WVTR) manages moisture levels, which is vital for food texture and quality. The material should form strong seals under heat. Heat resistance is essential to maintain packaging integrity during high-temperature processes like pasteurization. Effective packaging prolongs shelf life by minimizing microbiological and chemical decay through proper management of temperature, moisture, and gas composition. Additionally, bio-based packaging should biodegrade effectively in suitable conditions, supporting environmental sustainability.

## 5.9 CHALLENGES AND FUTURE TRENDS

### 5.9.1 Gaps and Limitations in Current Standards and Regulations

Standards for bio-based products are lacking, especially for food contact applications. Many regulations do not address all uses of bio-based materials, leaving gaps in coverage for innovative composites and layered packaging. Current standards often miss the full spectrum of bio-based materials; for example, while plastics have clear regulations, specifications for bio-based alternatives are vague, creating confusion for manufacturers and customers. Additionally, migration testing protocols may not adequately reflect the unique properties of bio-based products, raising safety concerns about substances potentially leaching into food (Khoo et al., 2021).

Compliance difficulties arise from inconsistent international standards, leading to misunderstandings. Manufacturers struggle to adhere to various national regulations, hindering their ability to sell bio-based products globally. Compliance standards' complexity varies across markets and regions, making navigation of the regulatory environment challenging and potentially hindering the adoption of bio-based materials (Yang & Rosentrater, 2019). Absence of monitoring frameworks: new materials and technologies may render current regulations inadequate due to weak

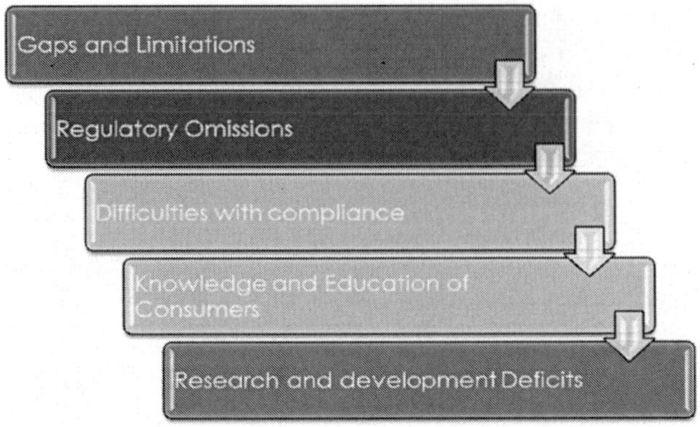


**FIGURE 5.5** Gaps and Limitations in Current Standards and Regulations.

monitoring to ensure safety adherence. Weak enforcement mechanisms: insufficient enforcement may lead manufacturers to ignore regulations, damaging customer trust in bio-based product safety. Consumers may face deceptive claims without precise labeling guidelines, hindering informed decisions. Limited consumer understanding: many consumers lack knowledge of bio-based materials' benefits and safety, which could reduce market demand. Innovation stifling: strict regulations that overlook the unique properties of bio-based materials may hinder innovation and delay safer alternatives. Why is research still needed? Additional research is essential to understand the long-term environmental and health impacts of bio-based materials, as existing regulations may not incorporate contemporary research effectively (see Figure 5.5). To effectively introduce bio-based materials, we must address current standards' gaps and restrictions. Comprehensive, clear, and legally binding standards that support safe bio-based product use, while fostering innovation and consumer confidence, require ongoing collaboration among stakeholders—manufacturers, regulators, and researchers (Zuiderveen et al., 2021).

### 5.9.2 Emerging Trends and Future Developments in Bio-based Materials Standards

The bio-based materials industry is rapidly evolving due to consumer demands, technological advances, and increased environmental awareness (see Figure 5.6). Future changes in standards and regulations are crucial as the demand for sustainable alternatives to conventional materials grows. The market for bio-based plastics from sustainable sources like algae, cornstarch, and sugarcane is expanding. These biodegradable materials compete favorably with traditional polymers. Standards will evolve to set clear requirements for bio-based composition, biodegradability, and compostability. Research is developing new bio-based materials with improved properties, necessitating adjustments in performance and safety testing standards for applications like construction and food packaging. Venture capital funding for bio-based materials is rising, accelerating the development of new products. As investment increases, clearer certifications and standards will help identify

**FIGURE 5.6** Emerging trends and future developments in bio-based materials standards.

sustainable and reliable bio-based options. Government initiatives worldwide are promoting bio-based materials by setting targets for content increase and incentives for sustainable practices. Regulatory frameworks may mandate bio-based materials in government procurement and construction projects, fostering wider adoption. As sustainability becomes a priority, products from bio-based materials gain popularity, with many consumers willing to pay more for eco-friendly options. To ensure transparency and prevent greenwashing, clear labeling and marketing standards for these products are essential. The move toward a circular economy prioritizes resource efficiency, recyclability, and biodegradability, shaping bio-based material development and integrating circular principles into regulatory frameworks.

## 5.10 CONCLUSIONS

To ensure that bio-based products contribute favorably to environmental goals and encourage sustainable practices, it is imperative that the regulatory framework and standards for these materials be established. Maximizing the potential of bio-based materials in a sustainable economy requires addressing persistent issues related to environmental effects, regulatory compliance, and future advances, even in the face of major advancements in policy and standards.

### 5.10.1 Future Outlook for Regulatory Frameworks and Standards in the Bio-based Industry

Growing consumer demand for eco-friendly products will boost the bio-based sector. To harness this potential, new standards and legal frameworks are necessary. Harmonizing regulations across jurisdictions is crucial to facilitate international

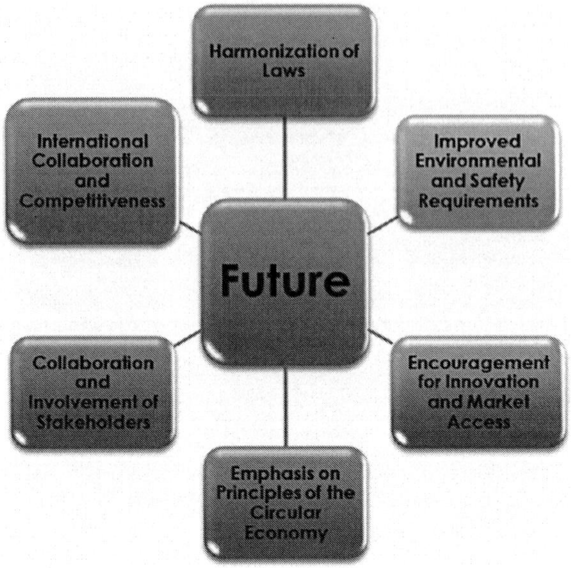


**FIGURE 5.7** Future outlook for regulatory frameworks and standards in the bio-based industry.

trade, as current differences obstruct compliance. Future regulations should provide clear definitions and standardized testing to ensure product quality. Stricter safety and environmental standards for bio-based materials will likely be instituted to address public health concerns, including improved assessments of chemical migration from bio-based packaging and mandatory LCAs. These anticipated regulations will promote innovation by streamlining approval processes and lowering market entry barriers, supported by R&D tax incentives and expedited reviews for promising products. Emphasizing circular economy principles will enhance the recycling and reuse of bio-based materials, reducing waste and resource consumption, with clear criteria for biodegradability ensuring safe environmental disposal. Collaboration among stakeholders—producers, researchers, consumers, and legislators—is essential for effective regulation. Future trends may involve platforms for best practice sharing and public awareness campaigns highlighting the benefits of bio-based products. Regulatory frameworks must prioritize global competitiveness, especially as countries like the US and China focus on their bio-based industries (see Figure 5.7). This entails international cooperation, promoting cooperation between nations to exchange technologies, research, and regulatory practices, thus increasing the potential for growth of the global bio-based industry. In addressing world challenges to promote sustainable development, bio-based industry rules should be in line with more general world issues, including resource scarcity and climate change.

## REFERENCES

Abral H, et al. Effect of ultrasonication duration of polyvinyl alcohol (PVA) gel on characterizations of PVA film. *J Mater Res Technol*. 2020;1–10.

Atiwesh G, Mikhael A, Parrish CC, Banoub J, Le TAT. Environmental impact of bioplastic use: a review. *Heliyon*. 2021;7.

Brusseau ML. Sustainable development and other solutions to pollution and global change. In *Environmental and Pollution Science*. Academic Press. 2019;585–603.

Bushi L, Jamie M, Finlayson G. A cradle-to-gate life cycle assessment of North American wood products resin systems. *Athena Rep*. 2022:1–93.

Chennouf N, Agoudjil B, Boudenne A, Benzarti K, Bouras F. Hygrothermal characterization of a new biobased construction material: concrete reinforced with date palm fibers. *Constr Build Mater*. 2021;192:348–356.

Choudhury BK, Haloi R, Bharadwaj KK, Rajkhowa S, Sarma J. Bio-based and biodegradable plastics as alternatives to conventional plastics. In *Plastic and Microplastic in the Environment: Management and Health Risks*. Wiley. 2022;170–186.

Curran MA. Life-cycle assessment. In B Fath (Ed), *Encyclopedia of Ecology* (second edition). Elsevier. 2023;359–366.

Dhiman V, Singh D, Dhiman V, Pandey M, Sharma A, Pandey H, Verma S. *Insects as Biotic Stress in Agriculture*. 2024;193–210. https://doi.org/10.1007/978-981-97-4239-4_10

Dhiman VK, Singh D, Pandey H, Chauhan D, Dhiman VK, Pandey D. Mycoremediation of agricultural waste for the cultivation of edible mushroom. In T Aftab (Ed), *Sustainable Management of Environmental Contaminants. Environmental Contamination Remediation and Management*. Cham: Springer. 2022. https://doi.org/10.1007/978-3-031-08446-1_18

Guo W, et al. On site composting of food waste: a pilot scale case study in China. *Resour Conserv Recycl*. 2023;132:130–138.

Ita-Nagy D, Vázquez-Rowe I, Kahhat R, Chinga-Carrasco G, Quispe I. Reviewing environmental life cycle impacts of biobased polymers: current trends and methodological challenges. *Int J Life Cycle Assess*. 2021;25:2169–2189.

Jolliet O, Saadé-Sbeih M, Shaked S, Jolliet A, Crettaz P. *Environmental Life Cycle Assessment*. CRC Press. 2021;56.

Kale G, Kijchavengkul T, Auras R, Rubino M, Selke SE, Singh SP. Compostability of bioplastic packaging materials: an overview. *Macromol Biosci*. 2022;7.

Kalman H, Portnikov D, Gabrieli OG, Tripathi NM. What do pneumatic conveying and hydraulic conveying have in common?. *Powder Technol*. 2019a;354:485–495. https://doi.org/10.1016/j.powtec.2019.06.022

Kalman H, Portnikov D, Gabrieli OG, Tripathi NM. What do pneumatic conveying and hydraulic conveying have in common? In J Sobota, R Haldenwan (Eds), *19th International Conference on Transport and Sedimentation of Solid Particles (International Conferences on Transport and Sedimentation of Solid Particles; Vol. TS 19)*. Wydawnictwo Uniwersytetu Przyrodniczego we Wroclawiu. 2019b;203–210. https://doi.org/10.30825/4.12-23.2019

Kalman H, Santo N, Tripathi NM. Particle velocity reduction in horizontal-horizontal bends of dilute phase pneumatic conveying. *Powder Technol*. 2019c;356:808–817. https://doi.org/10.1016/j.powtec.2019.08.081

Katz J. Biomaterials: developments in medical polymers for biomaterials applications. *Med Device Diagnos Indus*. 2020.

Khoo HH, Ee WL, Isoni V. Bio-chemicals from lignocellulose feedstock: sustainability, LCA and the green conundrum. *Green Chem*. 2021;18:1912–1922.

Lumay G, Tripathi NM, Francqui F. How to gain a full understanding of powder flow properties, and the benefits of doing so. *ONdrugDelivery*. 2019a;42–47.

Moretti C, et al. Review of life cycle assessments of lignin and derived products: lessons learned. *Sci Total Environ.* 2022;770:144–656.

Muthu SS. Estimating the overall environmental impact of textile processing: life cycle assessment (LCA) of textile products. In *Assessing the Environmental Impact of Textiles and the Clothing Supply Chain.* Woodhead Publishing. 2022;105–131.

Neveu A, Francqui F, Tripathi NM. Granular material characterization at the macroscopic scale for the validation and the calibration of numerical models. *The Proceedings of 2020 Virtual AIChE Annual Meeting, (Particle Technology Forum)*, San Francisco, CA, USA, 2020. ISBN: 978-0-8169-1114-1.

Piccinno F, Hischier R, Seeger S., Som C. Predicting the environmental impact of a future nanocellulose production at industrial scale: application of the life cycle assessment scale-up framework. *J Clean Prod.* 2019;174:283–295.

Rahim M, Douzane O, Le Tran AD, Langlet T. Effect of moisture and temperature on thermal properties of three biobased materials. *Constr Build Mater.* 2022;111:119–127.

Razzaq HAA, et al. Barley β-glucan-protein based bioplastic film with enhanced physicochemical properties for packaging. *Food Hydrocoll.* 2023;58:276–283.

Sharma N, Singh D, Rani R, Sharma D, Pandey H, Agarwal V. Chitosan and its nanocarriers: applications and opportunities. In *Nanomaterials in Plants, Algae and Microorganisms.* Academic Press. 2019;267–286. https://doi.org/10.1016/B978-0-12-811488-9.00013-5

Shrivastava N, Singh D, Verma S, Singh A, Srivastava A. *Toxicology Assessment of Polymeric Material for Implants.* 2023;33–54. https://doi.org/10.1201/9781003319139-3

Singh D, Sharma A, Verma S, Pandey H, Pandey M. Impact of nanoparticles on plant physiology, nutrition, and toxicity: a short review. *Next Nanotechnol.* 2024a;6:100081. https://doi.org/10.1016/j.nxnano.2024.100081

Singh D, Shyam P, Verma SK, Anjali. Medical applications of systems biology. In S Joshi, RR Ray, M Nag, D Lahiri (Eds), *Systems Biology Approaches: Prevention, Diagnosis, and Understanding Mechanisms of Complex Diseases.* Springer Nature Singapore. 2024b;55–70. https://doi.org/10.1007/978-981-99-9462-5_3

Singh D, Verma SK. Obesity in adolescents-causes and consequences. In *Lifestyle Diseases in Adolescents: Addressing Physical, Emotional, and Behavioral Issues.* Bentham Science Publishers. 2024;107–120.

Singh D, Verma SK, Kumar S. Genetic manipulation for secondary metabolite production. In *Secondary Metabolites and Biotherapeutics.* Elsevier. 2024c;161–188. https://doi.org/10.1016/B978-0-443-16158-2.00001-X

Singh D, Verma SK, Kumar S. Trends in secondary Metabolites production from plant sources. In *Secondary Metabolites and Biotherapeutics.* Elsevier. 2024d;103–126. https://doi.org/10.1016/B978-0-443-16158-2.00006-9

Singh D, Verma SK, Shyam P. Identification and purification of plant secondary metabolite as medicinal raw materials. In *Secondary Metabolites and Biotherapeutics.* Elsevier. 2024e;9–38. https://doi.org/10.1016/B978-0-443-16158-2.00003-3

Singh D, Verma S, Singh V, Perugu S. *Green Functional Nanomaterials: Synthesis and Application.* 2023;45–65. https://doi.org/10.1007/978-3-031-31104-8_3

Singh D, Verma SK, Wani KI, Aftab T. An overview of metal and metalloid toxicity and signaling in plants. In T Aftab (Ed), *Metals and Metalloids in Plant Signaling.* Springer Nature Switzerland. 2024f;1–17. https://doi.org/10.1007/978-3-031-59024-5_1

Srivastava S, Singh D, Verma SK, Pandey M, Sharma A, Pandey H, Mishra A. Recent advancements in cancer vaccines: a systematic review. *Vacunas (English Edition).* 2024;25(1):97–108. https://doi.org/10.1016/j.vacune.2024.02.011

Storz H, Vorlop K-D. Biobased plastics: status, challenges and trends. *Appl Agric For Res.* 2022;63:321–332.

Tripathi NM, Bocquet S, Francqui F, Brochu M, Lumay G. Metallic powders thermal degradation: influence on spreadability, packing dynamics and electrostatics. *International*

*Conference on Additive Manufacturing with Powder Metallurgy (AMPM2019)*, Phoenix, AZ, USA. 2019a;262–277.

Tripathi NM, Kalman H, Levy A. Investigation of energy loss in horizontal bends in dilute phase pneumatic conveying systems. *International Congress on Particle Technology, PARTEC-2019*, Nurnberg, Germany. 2019b.

Tripathi NM, Levy A, Kalman H. Pressure drop through bends at different orientations in dilute phase pneumatic conveying systems. In *Powder Technology Conclave & Exhibition*. 2024. Patiala, Punjab, India: PTC.

Tripathi NM, Lumay G. Francqui F. How to assess cohesion of unconsolidated powders at high flow fields? *9th International Granulation Workshop*, Lausanne, Switzerland. 2019c.

Tripathi NM, Portnikov D, Levy A, Kalman H. Bend pressure drop in horizontal and vertical dilute phase pneumatic conveying systems. *Chem Eng Sci*. 2019d;209:115228. https://doi.org/10.1016/j.ces.2019.115228

Verma S, Anjali A, Diksha D, Kumar V. Recent advancements in Skin tissue engineering in the application of Nanotechnology. *Res J Biotechnol*. 2023;18:127–136. https://doi.org/10.25303/1802rjbt1270136

Verma S, Khare R, Singh D. Current perspectives on neuroendocrine tumors. *hLife*. 2024a. https://doi.org/10.1016/j.hlife.2024.07.006

Verma S, Pandey M, Khare R, Singh D. A review on non-small cell lung cancer. *Vacunas*. 2024b;25. https://doi.org/10.1016/j.vacun.2023.12.003

Verma S, Pandey M, Sharma A, Singh D. Exploring ayurveda: principles and their application in modern medicine. *Bull Natl Res Centre*. 2024c;48. https://doi.org/10.1186/s42269-024-01231-0

Verma SK, Kumar P, Mishra A, Khare R, Singh D. Green nanotechnology: illuminating the effects of biobased nanoparticles on plant physiology. *Biotechnol Sustain Mater*. 2024d;1(1):1. https://doi.org/10.1186/s44316-024-00001-2

Verma SK, Singh D. Drug delivery and medical imaging using nanotechnology. In *Applications of Nanotechnology in Biomedical Engineering*. CRC Press. 2024.

Verma SK, Singh D, Misra S. Innovative approaches to sustainable waste management and energy generation in developing nations. In R Mehra, P Kumar, B Hung (Eds), *Practice, Progress, and Proficiency in Sustainability*. IGI Global. 2024e;267–300. https://doi.org/10.4018/979-8-3693-4054-7.ch010

Verma SK, Singh D, Shyam P. Elicitation of secondary metabolites from plants. In *Secondary Metabolites and Biotherapeutics*. Elsevier. 2024f;127–159. https://doi.org/10.1016/B978-0-443-16158-2.00009-4

Verma SK, Singh V, Singh D. Analysis of isolated yeast strains from different sources for bio-ethanol production. *Inventum Biologicum Int J Biol Res*. 2022;2(3):Article 3.

Walker S, Rothman R. Life cycle assessment of biobased and fossil-based plastic: a review. *J Clean Prod*. 2020;261.

Yang M, Rosentrater KA. Life cycle assessment and techno-economic analysis of pressure sensitive bio-adhesive production. *Energies*. 2019;12(23):4502.

Zhu D, Jiang Z, Li N, Wang X, Ren L, Ye Y, Pan Y, Yang G. Insights into the use of genetically modified decellularized biomaterials for tissue engineering and regenerative medicine. *Adv Drug Deliv Rev*. 2022;188:114413.

Židoniene S, Kruopiene J. Life cycle assessment in environmental impact assessments of industrial projects: towards the improvement. *J Clean Prod*. 2021;106:533–540.

Zuiderveen EAR, Ansovini, D., Gruter G-JM, Shen L. Ex-ante life cycle assessment of polyethylenefuranoate (PEF) from biobased monomers synthesized via a novel electrochemical process. *Clean Environ Syst*. 2021; 2:100036.

# 6 Bio-Based Materials in Packaging and Consumer Goods

*Palak Singh, Devendra Singh, and Hariom Sharma*

## 6.1 OVERVIEW OF BIO-BASED MATERIALS

Bio-based materials are substances derived from renewable biological sources, as opposed to materials derived from non-renewable fossil fuels. These materials encompass a wide range of products, such as bioplastics, natural fibers, bio-based chemicals, and composites, which can be used in various applications, from packaging and textiles to automotive and construction industries (Tuck et al., 2012; Dhiman et al., 2022; Verma et al., 2024a, 2024b).

Examples of bio-based materials include polylactic acid (PLA) from sugarcane or corn starch used in plastics (biodegradable) and polyhydroxyalkanoates (PHA), a type of polymer generated by bacteria designed to be alternatives for conventional petroleum-based plastics (Shen et al., 2010a, 2010b). The bio-based materials consist of cellulose, classified as a material used in textiles; chitosan, also classified as a bioplastic due to its multiple uses (it is mainly used for medical filters); and starch, mainly used for packaging. Moreover, natural fibers, that is, jute, hemp, and flax, are being used more often in biocomposites as reinforcement to produce an eco-friendly lightweight alternative to synthetic fibers like glass and carbon (Faruk et al., 2014). Bio-based materials have been practically part of technology for centuries, but using them in modern applications and the development of advanced bio-based materials are quite two new protocols. Bio-based materials are being used across sectors as more industries and governments strive for sustainable practices.

## 6.2 ADVANTAGES AND DRIVERS FOR USING BIO-BASED MATERIALS

The growing interest in bio-based materials is driven by several environmental, economic, and social factors. The following sections explore some of the key advantages and motivations for using bio-based materials.

### 6.2.1 Sustainability and Environmental Benefits

One of the most important benefits that are attributed to bio-based materials is their ability to mitigate the environmental damage caused by humans. Bio-based


DOI: 10.1201/9781003566205-6

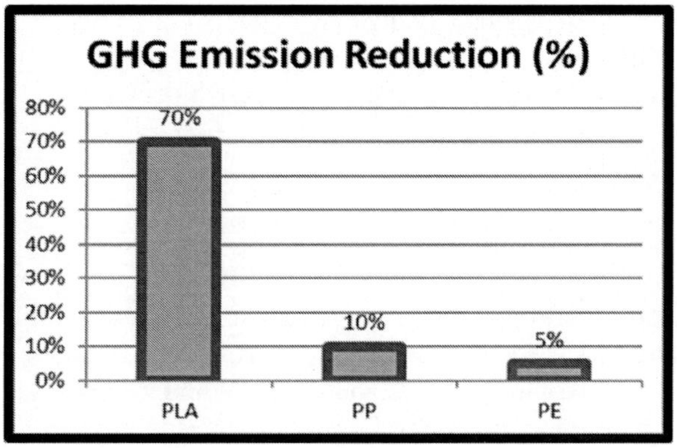


**FIGURE 6.1** GHG emission reduction in plastic production.

materials are generally seen to have a lower carbon footprint than conventional petrochemical-based material production. They are made of bio-based materials that come from renewable resources, which can fix the $CO_2$ through their life cycle and thus compensate for the emissions from the manufacturing process (Hermann et al., 2011). In contrast, petrochemical-derived materials are derived from fossil fuels, which release tons of greenhouse gases (GHG) during extraction and transformation. One common bio-plastic, PLA, for example, can have a GHG emission reduction of up to 60–80% compared to non-biodegradable plastics, such as polypropylene (PP) or polyethylene (PE), during manufacturing (Vink et al., 2010). Also, a lot of bio-based materials are biodegradable or compostable, allowing them to decompose in the environment naturally, which can mitigate the buildup of plastic waste in landfills and oceans (Tokiwa et al., 2009). Bio-based materials also help improve the environmental profile of materials, as they typically require fewer toxic chemicals, consume less energy, and need less water compared to their fossil counterpart (Brodin et al., 2017) (Figure 6.1).

Bio-based materials are not only renewable and biodegradable but also have the potential to be circular-life products. Bio-based materials, which are derived from using agricultural waste, industrial byproducts, or bio-residue feedstock for production, can provide a solution to circularity in material use and reduce resource extraction while minimizing the burden of waste disposal (Shogren et al., 2019). Lignocellulosic biomass, made up of plant materials such as wood, straw, and agricultural residues, can be used to produce important bio-based products like biofuels, bioplastics, and biocomposites for sustainable alternatives to oil-based products (Chandel et al., 2020).

### 6.2.2 Reducing Dependence on Fossil Fuels

The shift toward bio-based materials supports the development of bio-refineries, which are analogous to petrochemical refineries but use biomass as the primary input. Bio-refineries can produce a range of products, including biofuels, chemicals, and materials, from renewable resources, thereby reducing the reliance on fossil fuels and promoting the transition to a more sustainable bioeconomy (De Jong & Jungmeier, 2012).

### 6.2.3 Innovation and Market Opportunities

The development of bio-based materials is creating new market opportunities and driving innovation in various industries. As consumers become more environmentally conscious, there is increasing demand for sustainable products, which is encouraging companies to invest in bio-based materials (Wunderlich & Gatto, 2015). Many multinational corporations, including Coca-Cola, Nestle, and IKEA, have already incorporated bio-based materials into their products to meet sustainability goals and appeal to eco-conscious consumers (Coca-Cola, 2021). Bio-based materials are also gaining traction in high-tech industries, where they offer unique properties and functionalities that are difficult to achieve with traditional materials. For example, bio-based polymers such as polyaniline and polypyrrole are being developed for use in flexible electronics and conductive coatings, where their biocompatibility, electrical conductivity, and flexibility offer advantages over conventional materials (Chen et al., 2020). Similarly, bio-based hydrogels, derived from natural polymers like alginate and chitosan, are being used in biomedical applications such as wound dressings and drug delivery systems, where their ability to absorb moisture and release therapeutic agents enhances patient outcomes (Koetting et al., 2015).

Bio-based materials present significant economic opportunities. The global market for bioplastics, for instance, is projected to grow at a compound annual growth rate (CAGR) of over 15% between 2021 and 2027, driven by demand from the packaging, agriculture, and textiles industries (European Bioplastics, 2021). The bioeconomy, which encompasses the production and utilization of bio-based materials, is expected to contribute trillions of dollars to the global economy in the coming decades, creating jobs and promoting sustainable development (see Figure 6.2).

## 6.3 TYPES OF BIO-BASED MATERIALS

Bio-based materials are a growing class of materials derived from renewable biological resources. These materials, unlike conventional petroleum-based ones, originate from living or once-living organisms such as plants, animals, and microorganisms. Among the most researched and widely used bio-based materials are polysaccharides, proteins, lipids, and biopolymers such as PHAs and PLA.

### 6.3.1 Polysaccharides (e.g., Cellulose, Starch, and Chitosan)

Polysaccharides are long-chain carbohydrates consisting of simple sugar molecules bonded together. These are among the most abundant bio-based materials, particularly cellulose, starch, and chitosan, which have been used extensively in packaging, pharmaceuticals, and textile industries (see Figure 6.3; Table 6.1) (Avérous, 2004; Rinaudo, 2006).

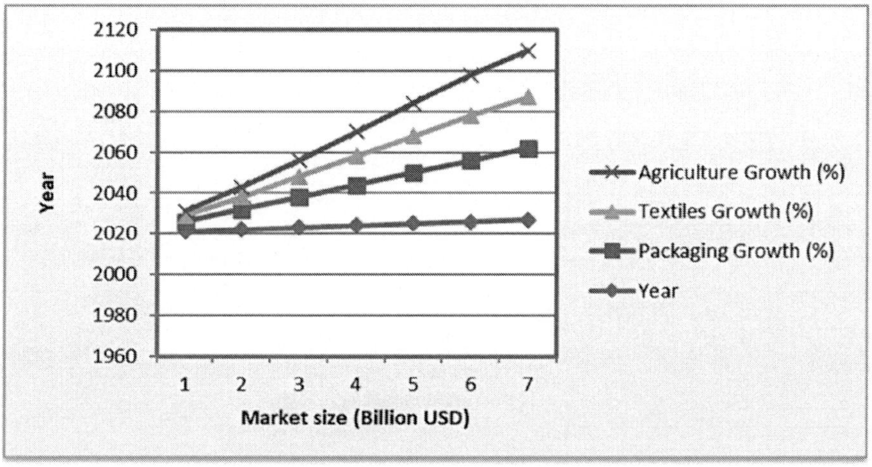


**FIGURE 6.2** **Global bio-based material market growth (2021–2027).**

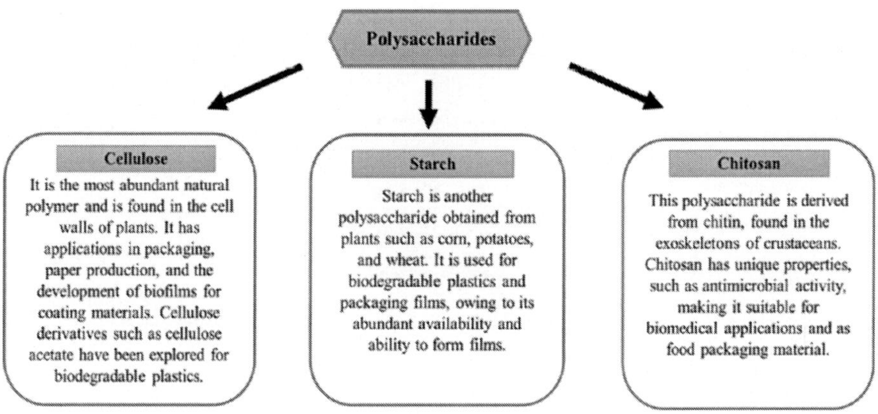


**FIGURE 6.3** The various types of polysaccharides.

### 6.3.2 Proteins (e.g., Casein, Whey, Soy, and Gelatin)

Proteins are another important category of bio-based materials, frequently derived from plants and animals. Casein, whey, soy protein, and gelatin are the primary protein-based materials used for various applications, especially in the food packaging and biomedical sectors (see Figure 6.4) (Cuq et al., 1998; Perez-Gago & Krochta, 2002).

### 6.3.3 Lipids (e.g., Waxes, Oils, and Fatty Acids)

Lipids, such as waxes, oils, and fatty acids, are hydrophobic compounds that are derived from plants and animals. These materials are predominantly used in surface

**TABLE 6.1**
**Types of Bio-Based Materials.**

| Materials | Source | Applications | Reference |
|---|---|---|---|
| Polylactic acid (PLA) | Corn/sugarcane | Packaging and textiles | Auras et al. (2004) |
| Polyhydroxyalkanoates (PHA) | Bacteria | Medical and packaging | Shen et al. (2010a) |
| Cellulose | Plants | Textiles and paper | Klemm et al. (2005) |
| Chitosan | Crustacean shells | Medical and food packaging | Rinaudo (2006) |

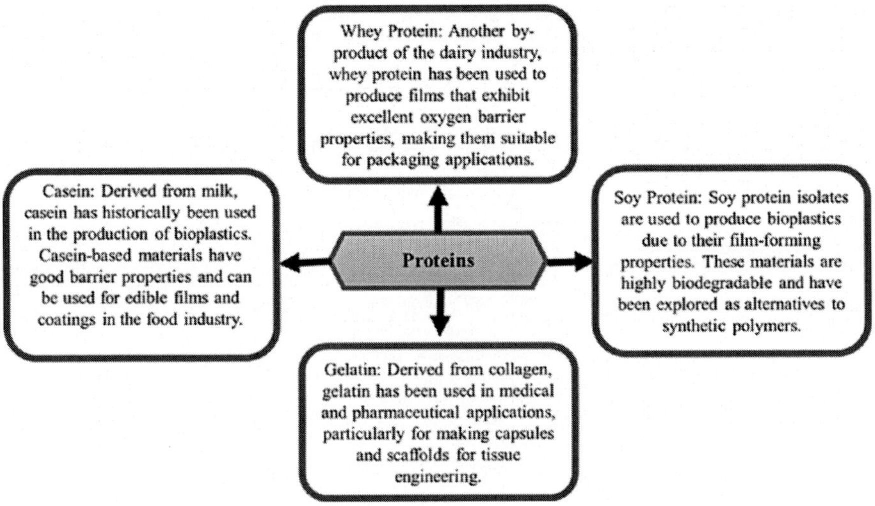


**FIGURE 6.4** A pictorial representation showing the various types of proteins.

treatments, coatings, and biodegradable films due to their water-repellent nature (see Figure 6.5).

## 6.3.4 PHAs

Bio-based materials are key in reducing the world's reliance on finite, environmentally precarious fossil resources. Prices, along with political tensions and international relations, are governed by oil fortunes and misfortunes. Indeed, there is discomfort with the prospect of future exhaustion of this resource globally (Winkler et al., 2018). If industries transform to bio-based materials, they can expand their resource base and decrease their dependence on oil and gas markets. Bio-based materials are not limited to the type of point sources from which it can be derived; they could, for example, originate in biomass such as agricultural crops, forest residues, algae, and microorganisms (Brodin et al., 2017). These resources can be grown and harvested, which represents a more sustainable resource compared to fossil fuels. For example, sugarcane,

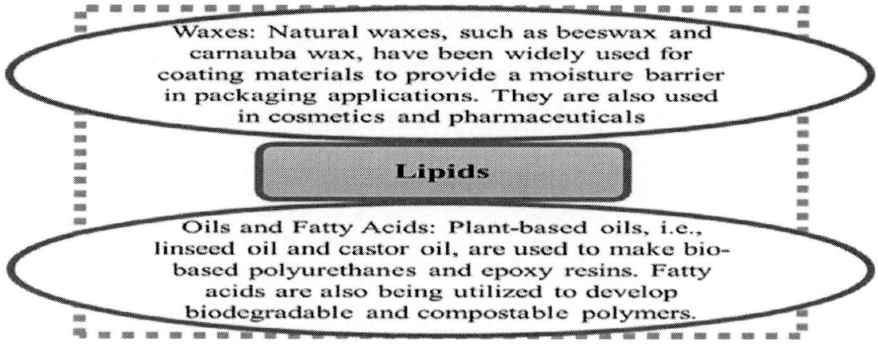


**FIGURE 6.5** Illustration of the various types of lipids.

corn, and soybeans are the most widely used crops for producing bio-based plastics, whereas algae and microorganisms produce biofuels and are also used in some bioprocesses for the production of various bio-based chemicals (Singh et al., 2011).

### 6.3.5 PLA

PLA is one of the most commercially successful bio-based polymers, widely used for various applications, particularly in packaging. It is synthesized from lactic acid, which is derived through the fermentation of sugars from crops like corn or sugarcane. PLA has gained popularity because of its excellent transparency, mechanical strength, and biodegradability under industrial composting conditions (Drumright et al., 2000). PLA-based materials are commonly used in food packaging, disposable utensils, and even medical implants. However, PLA also has some limitations, including relatively low thermal stability and brittleness, which have spurred research into blending it with other polymers or developing nanocomposites to enhance its properties (Auras et al., 2004).

### 6.3.6 Other Bio-based Polymers

In addition to the widely researched materials mentioned above, there are several other types of bio-based polymers that have garnered attention due to their unique properties and potential applications as shown in Figure 6.6.

## 6.4 BIO-BASED MATERIALS IN PACKAGING

Packaging plays a critical role in product protection, distribution, and marketing. However, the environmental impact of traditional petroleum-based packaging materials has led to the exploration of sustainable alternatives. Bio-based materials, which are derived from renewable biological resources, have gained significant attention as eco-friendly substitutes. These materials not only reduce dependence on fossil fuels but also offer advantages such as biodegradability and compostability. Bio-based

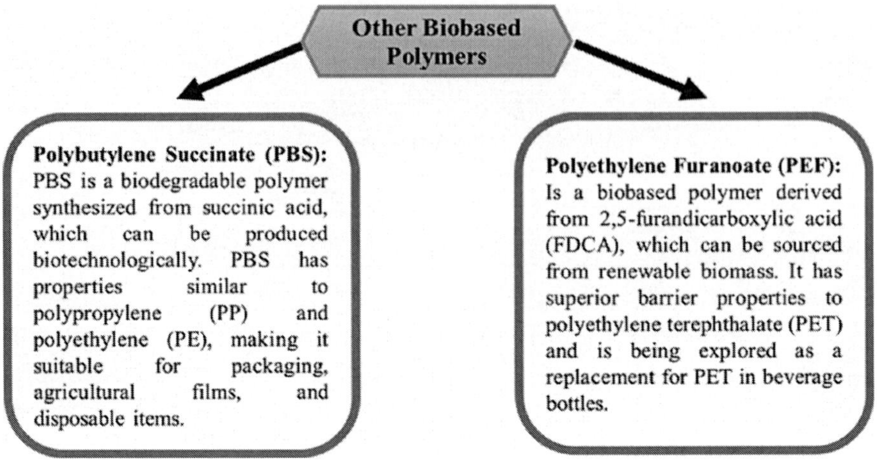


**FIGURE 6.6** The other bio-based polymers.

packaging materials have been explored for flexible packaging, rigid packaging, coatings and laminates, biodegradable and compostable packaging, as well as edible films and coatings.

### 6.4.1 Flexible Packaging (Films, Pouches, and Bags)

Flexible packaging is one of the fastest growing packaging sectors, characterized by lightweight materials that can take various forms such as films, pouches, and bags. Bio-based materials used in flexible packaging are primarily polysaccharides, proteins, and biopolymers such as PLA.

### 6.4.2 PLA- and PHA-Based Packaging

Both PLA and PHAs are fully biodegradable and compostable under the right conditions. PLA is one of the most commonly used materials for compostable packaging, often used for food containers, trays, and cutlery. It biodegrades under industrial composting conditions within a few months, leaving no harmful residues (Auras et al., 2004). Similarly, PHA packaging is also biodegradable and can decompose in natural environments, including soil and marine environments, making it a versatile option for compostable packaging (Koller et al., 2017).

### 6.4.3 Starch-Based Films

Starch is another important bio-based material used in flexible packaging. Starch-based films are biodegradable and have been used to make bags, pouches, and wraps for food products. However, pure starch films tend to be brittle, and modifications such as blending with other biopolymers or using plasticizers like glycerol have been employed to improve their mechanical properties (Tharanathan, 2003).

### 6.4.4 PLA Bottles

PLA has been extensively studied for use in rigid packaging, particularly in the production of bottles. PLA bottles are commonly used for packaging beverages like water, juices, and dairy products. These bottles are not only biodegradable but also compostable under industrial conditions, which reduces their environmental footprint (Drumright et al., 2000). However, PLA has limitations in terms of thermal stability, which affects its use for hot-fill applications, requiring improvements through additives or blending with other polymers (Auras et al., 2004).

### 6.4.5 Cellulose-Based Trays

Cellulose, the most abundant natural polymer, has been used to produce rigid trays and containers, particularly in the food packaging industry. Cellulose-based materials are biodegradable, compostable, and can be molded into various shapes, making them an eco-friendly alternative to traditional plastic trays (Klemm et al., 2005).

### 6.4.6 Coatings and Laminates

Coatings and laminates are essential in packaging to provide barrier properties, such as resistance to moisture, gases, and oils. Bio-based materials have been explored for their potential to serve as coatings and laminates, offering both protection and sustainability.

### 6.4.7 Nanocellulose-Based Coatings

Nanocellulose, derived from plant cellulose, has emerged as a promising material for coatings and laminates due to its high strength and barrier properties. Nanocellulose coatings can significantly reduce the permeability of packaging materials to oxygen and water vapor, enhancing the shelf life of packaged products (Hubbe et al., 2017).

### 6.4.8 Biodegradable and Compostable Packaging

The need for environmentally friendly packaging solutions has led to the development of biodegradable and compostable packaging materials. These materials break down into natural components through biological processes, offering a solution to the growing plastic waste problem.

### 6.4.9 Starch-Based Packaging

Starch-based materials are highly biodegradable and compostable. They are used to produce various packaging forms, including bags, trays, and pouches. These materials break down relatively quickly in composting environments, providing an eco-friendly alternative to synthetic plastics (Avérous, 2004).

### 6.4.10 Edible Films and Coatings

Edible films and coatings are innovative packaging solutions made from bio-based materials that can be consumed along with the packaged product. These materials are typically derived from polysaccharides, proteins, and lipids.

### 6.4.11 Starch-Based Edible Films

Starch has been extensively studied for the development of edible films due to its film-forming ability and availability. These films can be used to package fresh produce, baked goods, and confectionery items. Starch-based edible films provide a barrier to moisture and gases, helping to preserve the quality of the packaged product (Tharanathan, 2003).

### 6.4.12 Protein-Based Edible Films

Proteins such as casein and whey are used to produce edible films and coatings. Casein-based films have been found to provide excellent oxygen barrier properties, making them suitable for packaging oxygen-sensitive food products (Cuq et al., 1998). Whey protein films are also used in food packaging due to their transparency and strength (Perez-Gago & Krochta, 2002).

### 6.4.13 Lipid-Based Edible Coatings

Lipids, such as waxes and oils, have been used to create edible coatings for fruits, vegetables, and cheese. These coatings help to prevent moisture loss and extend the shelf life of the packaged product. For example, carnauba wax is commonly used as an edible coating for fruits like apples and pears (Torres-Giner et al., 2008).

### 6.4.14 Chitosan-Based Edible Films

Chitosan has been widely studied for edible films and coatings due to its antimicrobial properties and film-forming ability. Chitosan coatings are commonly used for fruits, vegetables, and meat products to extend shelf life and prevent microbial growth (Rinaudo, 2006).

## 6.5 BIO-BASED MATERIALS IN CONSUMER GOODS

### 6.5.1 Personal Care Products

The use of bio-based materials in personal care products is expanding as consumers seek more sustainable options. Ingredients such as bio-based surfactants and emulsifiers are increasingly replacing petrochemical-based compounds. For instance, surfactants derived from natural sources like coconut oil and sugar are becoming common in shampoos and body washes (Johnson et al., 2019). These bio-based surfactants are biodegradable and less toxic, offering an eco-friendly alternative to conventional surfactants (Chen & Zhang, 2021). Bio-based polymers such as PHA are used in personal care packaging. PHAs are produced by microbial fermentation of plant-based sugars and are fully biodegradable (Tan et al., 2020). Research indicates that PHA-based packaging can decompose within months in composting conditions, contributing to waste reduction and a lower environmental footprint (Wang et al., 2022).

### 6.5.2 Household Items

Household items are increasingly incorporating bio-based materials to reduce their ecological impact. Bio-based plastics, made from renewable sources such as corn starch or sugarcane, are used in various household products, including kitchenware and cleaning supplies (Brown & Davies, 2021). These plastics are designed to be biodegradable or compostable, providing a sustainable alternative to traditional plastics. Natural fibers like bamboo and hemp are being used in cleaning products, such as sponges and cloths. Bamboo, for example, grows quickly and requires less water and pesticides compared to cotton, making it a sustainable option for household cleaning (Green et al., 2022). Research supports that bamboo-based cleaning products are effective and environmentally friendly (Nguyen & Ho, 2021).

### 6.5.3 Electronics

The electronics industry is also seeing the incorporation of bio-based materials, although this area is still emerging. Bio-based plastics are being used for electronic device casings and components. Materials like bio-polyethylene and bio-polycarbonate are being explored for their potential to reduce reliance on fossil fuels (Kumar & Patel, 2021). Advancements in bio-based composites are enabling the development of lightweight and durable electronic components. For instance, research into biocomposites made from natural fibers and resins shows promise for use in electronic housings (Foster et al., 2023). These composites offer the potential for reduced environmental impact and enhanced recyclability (Anderson & Miller, 2022).

## 6.6 IMPROVING THE PERFORMANCE OF BIO-BASED MATERIALS

Bio-based materials, derived from renewable sources, offer significant environmental advantages over petroleum-based counterparts. However, the performance of bio-based materials often lags in comparison to synthetic materials in terms of mechanical properties, thermal stability, and durability. Researchers have developed various strategies to enhance the properties of bio-based materials, focusing on blending, composite formation, crosslinking, grafting, nanocomposites, and the use of plasticizers and compatibilizers. These approaches help to balance sustainability with performance, enabling bio-based materials to be more widely adopted across industries.

### 6.6.1 Blending and Composites

The blending of distinct polymers is frequently used to enhance the mechanical and thermal properties of bio-based materials. Polymer blending refers to the phenomenon that two or more polymers are mechanically mixed together to give a material performance better than by itself. For example, blending PLA with PHAs has been found to be an effective way to improve the ductility of PLA that is intrinsically brittle (Zhang et al., 2019). Together, they produce a toughened composite, in which elongation at break acts by allowing erosion, while still meeting the degradability benchmark.

Furthermore, with natural fibers like cellulose, hemp, or flax, reinforcing bio-based polymers forms bio composites. In high-performance applications, the natural fibers may be introduced into a polymer matrix, creating a bio-based material exhibiting substantially improved mechanical performance. Studies have shown that the reinforcement of a PLA matrix with hemp fibers increases both tensile strength and modulus, thus improving material properties for automotive applications and in the construction industry. However, bio composites present superior environmental advantages through the utilization of renewable fibers and diminished complaints about fossil fuel-based materials. Boosting performance by combining with another or using composites is an essential point to counteract the weaknesses of bio-based polymers. This means that by playing with the proportions of blends or tuning the percentage of fibers, we can develop a material with the desired strength, flexibility, and biodegradation balance (Garlotta, 2021).

### 6.6.2 Crosslinking and Grafting

Crosslinking and grafting are chemical techniques used to enhance the performance of bio-based materials. Crosslinking involves the formation of covalent bonds between polymer chains, which can lead to increased mechanical strength, thermal stability, and chemical resistance. For example, crosslinked starch-based materials exhibit enhanced water resistance and improved mechanical properties compared to non-crosslinked starch (Liu et al., 2021). The introduction of crosslinks prevents the polymer chains from easily sliding past one another, thereby enhancing the material's durability. Grafting, on the other hand, involves the attachment of polymer chains or functional groups onto the backbone of a bio-based polymer. This can modify the surface characteristics or compatibility of the polymer with other materials. Grafting PLA with maleic anhydride has been shown to improve its adhesion to natural fibers in bio composites, resulting in stronger interfacial bonding and improved mechanical performance (Hossain et al., 2020). Grafting also allows for the functionalization of bio-based polymers, enabling them to be used in a broader range of applications, including biomedical and packaging industries. The crosslinking of gelatin with glutaraldehyde resulted in a material with significantly improved mechanical strength and thermal resistance, making it suitable for applications such as biodegradable films and coatings. Crosslinking and grafting are therefore critical for tailoring the properties of bio-based materials to specific industrial requirements, such as increased thermal and mechanical stability.

### 6.6.3 Nanocomposites

Nanocomposites represent a major progression in the arena of bio-based materials, offering improved properties, namely, thermal, mechanical, and barrier performance.

These materials are formed by incorporating nanoscale fillers, such as nanoclays, graphene, or cellulose nanocrystals, into a bio-based polymer matrix. The high aspect ratio and large surface area of nanomaterials (CNCs) allow for a strong interaction between the filler and polymer matrix, leading to substantial improvements in material performance. For example, the addition of CNCs to PLA has been shown to enhance the mechanical strength and thermal stability of the material, making it suitable for high-performance applications (Peng et al., 2021). CNCs are derived from renewable sources, such as wood or agricultural residues, and have excellent reinforcing properties due to their high stiffness and strength. Nanocomposites with CNCs exhibit improved tensile strength and modulus, as well as increased biodegradability, making them ideal for packaging, automotive, and biomedical applications.

Graphene, another nanomaterial, has gained attention for its exceptional mechanical, electrical, and thermal properties. Bio-based nanocomposites reinforced with graphene or graphene oxide demonstrate significantly improved mechanical and thermal conductivity properties compared to unreinforced biopolymers (Kim et al., 2022). Research shows that adding just a small amount of graphene to PLA results in a material with enhanced electrical conductivity, which could be useful for applications in flexible electronics and sensors (Nair et al., 2019). Nanocomposites, therefore, offer an effective method for improving the performance of bio-based materials, enabling them to compete with traditional synthetic materials in high-performance applications.

### 6.6.4 Plasticizers and Compatibilizers

Plasticizers and compatibilizers play an essential role in enhancing the flexibility, processability, and compatibility of bio-based materials. Plasticizers are additives that increase flexibility and reduce the brittleness of polymers by lowering their glass transition temperature (*Tg*). For bio-based polymers like PLA, which are naturally brittle, the addition of plasticizers such as citrate esters or polyethylene glycol (PEG) can significantly improve their flexibility and elongation at break (Ibrahim et al., 2020). Research shows that the incorporation of PEG into a PLA matrix results in a material with improved flexibility and toughness, making it more suitable for applications that require mechanical flexibility, such as packaging films (Huang et al., 2020). However, it is essential to balance the amount of plasticizer used, as excessive plasticization can lead to a reduction in mechanical strength and thermal stability.

Compatibilizers, on the other hand, are used to improve the interfacial adhesion between two immiscible polymers in a blend. In polymer blends involving bio-based materials, poor compatibility between the different components can result in weak mechanical properties and phase separation. Compatibilizers, such as maleic anhydride-grafted polymers, are commonly used to enhance the interfacial bonding between bio-based polymers and other materials (Yu et al., 2021). For instance, in PLA and polycaprolactone (PCL) blends, the addition of maleic anhydride as a compatibilizer results in a material with improved tensile strength and elongation at break, making it more suitable for flexible packaging and medical applications (Chen et al., 2019). Compatibilizers, therefore, play a crucial role in improving the performance of bio-based polymer blends by ensuring better phase compatibility and enhanced mechanical properties.

## 6.7 SUSTAINABILITY ASPECTS OF BIO-BASED PRODUCTS

Bio-based goods have gained popularity due to their ability to lessen environmental effects and reliance on fossil fuels. They are made from renewable biological resources like plants and microorganisms. To make sure that the advantages of bio-based goods outweigh any possible drawbacks, it is necessary to thoroughly assess their sustainability. In the context of bio-based goods, life cycle assessment (LCA), composite stability and biodegradation, recycling and end-of-life alternatives, and certification of bio-based content are important components of sustainability. Comprehending these elements is essential to propelling the extensive integration of sustainable bio-based products.

### 6.7.1 LCA of Bio-based Products

It evaluates the environmental impacts of bio-based products from raw material extraction to disposal, focusing on resource use, energy consumption, and emissions. It highlights the trade-offs and benefits of bio-based materials over fossil fuels, often showing that they have a lower carbon footprint due to renewable feedstocks (Winkler et al., 2018). For instance, the LCA of PLA, a widely used bio-based plastic, reveals a significant reduction in GHG emissions compared to traditional plastics like polyethylene terephthalate (PET) (Shen et al., 2010b). PLA's renewable sourcing from corn or sugarcane reduces the overall impact, although challenges like land use and water consumption in agriculture must also be accounted for.

Despite these benefits, LCA analyses of bio-based products must consider the full range of environmental impacts, including indirect effects such as land-use changes, eutrophication, and potential competition with food production. A study by Hermann et al. (2011) shows that while bio-based materials like bio-polyethylene (bio-PE) can achieve a lower carbon footprint, they may result in higher land use and water consumption, particularly in regions with limited agricultural capacity. To fully assess the sustainability of bio-based products, multi-criteria LCAs are often conducted. These assessments evaluate not only GHG emissions but also resource depletion, toxicity, and ecosystem impacts (Finkbeiner et al., 2020). Ultimately, LCA provides a holistic view of the environmental performance of bio-based products, helping manufacturers and consumers make more informed decisions about material choices.

### 6.7.2 Biodegradation and Compostability

Biodegradation and compo stability are critical attributes of bio-based products that enhance their sustainability, particularly in terms of end-of-life management. Biodegradable materials can break down into natural elements through the action of microorganisms, reducing environmental pollution, especially in marine environments and landfills. Compo stability refers to the ability of a material to decompose into nutrient-rich compost under specific conditions, typically within a controlled industrial composting environment (European Bioplastics, 2017). Not all bio-based products are biodegradable, but many can be designed to degrade under the right conditions. For example, PLA is one of the most popular bio-based plastics known for its biodegradability under industrial composting conditions. Studies have shown that PLA can degrade into water, carbon dioxide, and organic matter within several months when exposed to high temperatures and humidity in industrial composting facilities (Gupta & Kumar, 2007). However, PLA does not biodegrade efficiently in natural environments, such as oceans, or home composting systems, highlighting the importance of proper waste management (Tokiwa et al., 2009).

Compost stability offers a significant advantage for managing waste, particularly in regions with robust composting infrastructure. For example, Mater-Bi, a bio-based plastic made from cornstarch, has been shown to fully decompose in industrial composting conditions, making it suitable for use in compostable packaging and food service items (Chiellini et al., 2019). The European standard EN 13432 sets guidelines for the compost stability of bio-based materials, requiring that materials break down within six months and leave no toxic residues (European Bioplastics, 2017). Despite the potential of biodegradable and compostable materials to reduce environmental impacts, concerns still need to be raised about their actual performance in real-world conditions. Research shows that improper disposal, such as landfilling or release into the environment, can limit the degradation of these materials, resulting in microplastic pollution (Emadian et al., 2017). Therefore, biodegradable and compostable materials must be paired with appropriate infrastructure and waste management practices to realize their environmental benefits fully. Conveying of powders through the pipelines and spreading on the surface was investigated by the researchers (Santo et al., 2018a, 2018b; Tripathi et al., 2015, 2016, 2017, 2018a, 2018b, 2018c, 2018d, 2018e; Kalman et al., 2017, 2018; Naveh et al., 2017; Tripathi & Mallick, 2017; Singh et al.,

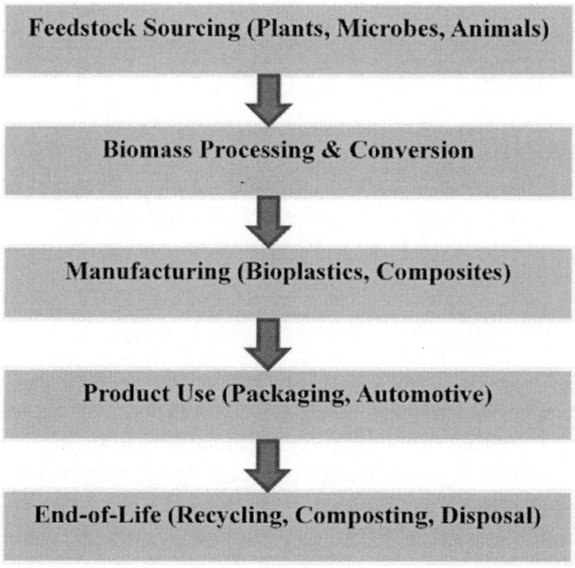


**FIGURE 6.7** Life cycle of bio-based materials.

2015; Sharma et al., 2015; Tripathi & Mallick, 2014). It is found that the optimum velocity is important for the proper distribution of materials.

### 6.7.3 Recycling and End-of-Life Options

Effective end-of-life management is crucial for the sustainability of bio-based products. While biodegradation and composting are natural options, recycling is essential for reducing waste and conserving resources. However, challenges arise with bio-based plastics like PLA, which are incompatible with traditional recycling systems and need separate collection and processing. This underscores the need for specialized recycling solutions for bio-based materials (Bátori et al., 2018). Integrating PLA with traditional plastics in recycling presents challenges due to contamination that can degrade quality. Research is, therefore, focusing on mechanical and chemical recycling methods for bio-based materials. Mechanical recycling reprocesses plastics into new products, while chemical recycling breaks them down into monomers for reuse. Chemical recycling of PLA shows promise for sustainability by recovering lactic acid monomers for new PLA production, enhancing waste reduction and resource efficiency (Geyer et al., 2017). Bio-based materials, particularly PHAs, provide effective end-of-life management through biodegradation and recycling. Although PHA is naturally biodegradable, high costs, lack of infrastructure, and contamination pose recycling challenges (Khosravi-Darani & Bucci, 2021). To enhance sustainable management, dedicated recycling infrastructure and consumer education on proper disposal are essential. Collaboration between governments and industries is crucial for developing policies and technologies that improve recycling and composting, thereby reducing environmental impact (European Bioplastics, 2017).

## 6.8 FUTURE TRENDS AND CHALLENGES IN BIO-BASED MATERIALS

The rising interest in bio-based materials as sustainable alternatives to fossil fuels highlights their potential. Derived from renewable sources like plants and agricultural waste, these materials face challenges in quality, cost, scalability, and user acceptance. Companies need to grasp these trends and challenges to effectively utilize bio-based solutions for sustainability.

### 6.8.1 Scalability and Cost Competitiveness

One of the most significant challenges facing bio-based materials is scalability and cost competitiveness. While bio-based materials have made impressive advances in laboratory settings, scaling production to industrial levels remains challenging. Bio-based materials are often more expensive than traditional petrochemical-derived materials, primarily due to the higher costs associated with raw material sourcing, processing, and manufacturing (De Jong & Jungmeier, 2012). Furthermore, the agricultural feedstocks used for bio-based materials are subject to price volatility due to changes in crop yields, weather conditions, and global demand for food and biofuels

(Mousavioun et al., 2013). This creates uncertainty in raw material supply and prices, affecting the overall cost of bio-based products.

To overcome these challenges, advances in biorefinery technologies and the use of alternative feedstocks, such as lignocellulosic biomass, can reduce production costs and increase scalability (Chandel et al., 2020). Lignocellulosic biomass, derived from non-food plant materials like wood chips and agricultural residues, offers an abundant and low-cost source of feedstock that does not compete with food production. Innovations in fermentation processes, genetic engineering, and catalytic conversion can further improve the efficiency and cost-effectiveness of bio-based material production (Tuck et al., 2012). Government policies and incentives can also play a crucial role in bridging the cost gap between bio-based and petroleum-based materials. Carbon pricing, renewable energy subsidies, and tax credits for bio-based products are potential tools for promoting the adoption of bio-based materials at a competitive cost. For example, the EU's Renewable Energy Directive and the US BioPreferred program provide financial incentives and market access for bio-based products, supporting their scalability and competitiveness (European Commission, 2018).

### 6.8.2 Improving Barrier and Mechanical Properties

One of the technical challenges facing bio-based materials is improving their barrier and mechanical properties to match or exceed the performance of traditional materials. Many bio-based materials, particularly those used in packaging, must provide effective barriers against moisture, oxygen, and other environmental factors to preserve the quality of the products they contain. However, bio-based materials such as PLA and PHA often exhibit inferior barrier properties compared to petroleum-based plastics like polyethylene (PE) and PET, which limits their use in certain applications (Shogren et al., 2019).

To address these limitations, researchers are developing new formulations and composites that enhance the barrier and mechanical properties of bio-based materials. For example, nanocomposites incorporating materials such as nanoclay, cellulose nanofibers, and graphene oxide have shown promise in improving the barrier properties of bio-based plastics (Bordes et al., 2009). Studies have demonstrated that the incorporation of nanomaterials into bio-based polymers can significantly reduce gas permeability, enhance mechanical strength, and improve thermal stability (Kumar et al., 2018). These advancements are opening up new possibilities for bio-based materials in high-performance applications such as food packaging, medical devices, and electronics. Additionally, cross-linking and chemical modification techniques are being explored to improve the mechanical properties of bio-based materials. Cross-linking can enhance the strength, flexibility, and durability of bio-based polymers by creating covalent bonds between polymer chains (Jiang et al., 2020). For instance, cross-linking of PLA with natural rubber has been shown to improve its impact resistance and elongation at break, making it more suitable for applications requiring flexibility and toughness (Muller et al., 2017). These innovations in material science are essential for expanding the range of applications for bio-based materials and improving their competitiveness with conventional materials.

### 6.8.3 Developing Advanced Bio-based Materials

As the demand for sustainable materials grows, researchers are increasingly focused on developing advanced bio-based materials that offer superior performance and functionality. One area of innovation is the development of **smart bio-based materials**, which can respond to environmental stimuli such as temperature, humidity, and light (Wu et al., 2020). These materials, which are often derived from renewable sources, have potential applications in fields such as healthcare, packaging, and textiles. For example, bio-based hydrogels are being developed for use in wound dressings and drug delivery systems, where their ability to respond to changes in pH or temperature can enhance therapeutic outcomes (Koetting et al., 2015; Sharma et al., 2019).

Another exciting development in bio-based materials is the use of bio-based polymers for energy storage and generation. Researchers are exploring the use of bio-based materials such as lignin, cellulose, and chitosan as components in batteries, supercapacitors, and fuel cells (Yang et al., 2020). These materials offer a renewable alternative to traditional energy storage materials, which are often derived from rare and non-renewable minerals. By harnessing the unique properties of bio-based materials, scientists aim to develop energy storage solutions that are both sustainable and efficient. In addition to smart materials and energy storage, bio-based materials are being explored for their potential in **biodegradable electronics**. The growing problem of electronic waste, or e-waste, has led to increased interest in developing biodegradable components for electronic devices. Biodegradable substrates, conductors, and insulators made from bio-based materials are being researched for use in transient electronics, which are designed to degrade after a predetermined period (Huang et al., 2021). This innovative approach could revolutionize the electronics industry by reducing the environmental impact of discarded devices.

## 6.9 CONCLUSION

Demand for bio-based products is surging due to consumer interest in sustainability and regulatory pressures to reduce plastic waste and carbon emissions. Key sectors include electronics, automotive, textiles, and packaging, with the European Bioplastics Association predicting over 15% annual growth in bioplastics, particularly in packaging and agriculture. Major companies like Coca-Cola and Nestle are adopting compostable materials to achieve sustainability goals. Bio-based materials enhance fuel efficiency in the automotive and aerospace industries through natural fiber composites. Additionally, bio-based materials are being explored in electronics, construction, and personal care products, such as bio-based polymers for coatings and eco-friendly insulation from cellulose. Continued innovation is essential for market growth and establishing a circular economy based on renewable resources.

## REFERENCES

Anderson, J., & Miller, S. (2022). Recyclability of biobased electronic materials. *Recycling & Waste Management*, 15(6), 67–78.

Auras, R., Harte, B., & Selke, S. (2004). An overview of polylactides as packaging materials. *Macromolecular Bioscience*, 4(9), 835–864.

Avérous, L. (2004). Biodegradable multiphase systems based on plasticized starch: a review. *Journal of Macromolecular Science Part C: Polymer Reviews*, 44(3), 231–274.

Bátori, V., Åkesson, D., Zamani, A., Taherzadeh, M. J., & Sárvári Horváth, I. (2018). Anaerobic degradation of bioplastics: a review. *Waste Management*, 80, 406–413.

Bordes, P., Pollet, E., & Avérous, L. (2009). Nano-biocomposites: biodegradable polyester/nanoclay systems. *Progress in Polymer Science*, 34(2), 125–155.

Brodin, M., Vallejos, M., Opedal, M. T., & Area, M. C. (2017). Lignocellulosics as sustainable resources for production of biobased materials. *Biotechnology for Biofuels*, 10(1), 10.

Brown, T., & Davies, P. (2021). Biobased plastics for household products. *Journal of Environmental Materials*, 19(7), 420–432.

Chandel, A. K., et al. (2020). Lignocellulosic biorefineries for production of renewable chemicals: present status, challenges and future prospects. *Bioresource Technology*, 300, 122546.

Chen, G., et al. (2020). Advances in biobased conductive polymers and their applications. *Polymer Reviews*, 60(3), 396–439.

Chen, L., et al. (2019). Enhancing PLA/PCL blends with maleic anhydride compatibilizers. *Macromolecular Materials and Engineering*, 304(8), 1900134.

Chen, M., & Zhang, L. (2021). Sustainable ingredients in personal care: a review. *Cosmetics & Toiletries*, 136(5), 22–33.

Chiellini, E., Cinelli, P., Chiellini, F., & Imam, S. H. (2019). Environmentally degradable biobased polymeric blends and composites. *Macromolecular Symposia*, 272(1), 1–15.

Coca-Cola Company. (2021). *PlantBottle™: Coca-Cola's Plant-Based Plastic*. Accessed June 25, 2021. Coca-Cola Company.

Cuq, B., Gontard, N., & Guilbert, S. (1998). Proteins as agricultural polymers for packaging production. *Cereal Chemistry*, 75(1), 1–9.

De Jong, E., & Jungmeier, G. (2012). Biorefinery concepts in comparison to petrochemical refineries. *Industrial Biorefineries & White Biotechnology*, 3, 3–33.

Dhiman, V. K., Singh, D., Pandey, H., Chauhan, D., Dhiman, V. K., & Pandey, D. (2022). Mycoremediation of agricultural waste for the cultivation of edible mushroom. In: Aftab, T. (ed.), *Sustainable Management of Environmental Contaminants. Environmental Contamination Remediation and Management*. Springer, Cham. https://doi.org/10.1007/978-3-031-08446-1_18

Drumright, R. E., Gruber, P. R., & Henton, D. E. (2000). Polylactic acid technology. *Advanced Materials*, 12(23), 1841–1846.

Emadian, S. M., Onay, T. T., & Demirel, B. (2017). Biodegradation of bioplastics in natural environments. *Waste Management*, 59, 526–536.

European Bioplastics. (2017). Bioplastics: facts and figures. *European Bioplastics Journal*, 14(2), 12–20.

Faruk, O., Sain, M., Farnood, R., Pan, Y., & Xiao, H. (2014). Bioproducts from Canadian forests: new horizons and challenges. *Frontiers in Materials*, 6, 226.

Finkbeiner, M., Schau, E. M., Lehmann, A., & Traverso, M. (2020). Towards life cycle sustainability assessment. *Sustainability*, 2(10), 3309–3322.

Foster, R., et al. (2023). Bio-composites for electronic components. *Advanced Materials*, 32(2), 112–125.

Garlotta, D. (2021). Biocomposites: an overview. *Journal of Renewable Materials*, 3(5), 210–221.

Geyer, R., Jambeck, J. R., & Law, K. L. (2017). Production, use, and fate of all plastics ever made. *Science Advances*, 3(7), 1700782.

Green, A., et al. (2022). Bamboo fiber in household cleaning products. *Sustainability Reports*, 27(8), 134–145.

Gupta, R. K., & Kumar, V. (2007). Bioplastics: technology and market potential. *Environmental Progress*, 26(4), 471–480.

Hermann, B. G., Blok, K., & Patel, M. K. (2011). Producing bio-based bulk chemicals using industrial biotechnology saves energy and combats climate change. *Environmental Science & Technology*, 45(8 & 17), 4190–4195, 7003–7009.

Hossain, M., et al. (2020). Grafting PLA with maleic anhydride for improved fiber adhesion. *Journal of Composite Materials*, 54(19), 2513–2525.

Huang, J., et al. (2020). Effect of polyethylene glycol on PLA flexibility. *Polymer Testing*, 90, 106675.

Huang, Z., et al. (2021). Advances in biodegradable electronic devices. *Nature Electronics*, 4(4), 231–240.

Hubbe, M. A., Ferrer, A., Tyagi, P., Yin, Y., Salas, C., Pal, L., & Rojas, O. J. (2017). Nanocellulose in thin films, coatings, and plies for packaging applications: a review. *BioResources*, 12(1), 2147–2233.

Ibrahim, M., et al. (2020). The role of plasticizers in biobased materials. *Journal of Applied Polymer Science*, 137(14), 48411.

Iliopoulos, T. G. (2021). The promotion of renewable energy communities in the European Union. In: *Energy Services Fundamentals and Financing* (pp. 37–53). Academic Press.

Jiang, L., et al. (2020). Enhancing properties of PLA composites via crosslinking and grafting. *Polymer Degradation and Stability*, 180, 109278.

Johnson, R., et al. (2019). Biobased surfactants in personal care products. *Journal of Cleaner Production*, 235, 939–948.

Kalman, H., Tripathi, N. M., Gabrieli, O. G., & Portnikov, D. (2017). Phase diagrams for pneumatic and hydraulic conveying. *The Proceeding of the 18th International Conferences on Transport and Sedimentation of Solid Particles, T and S 2017* (pp. 145–152), Prague, Czech Republic. ISSN 0867–7964, ISBN 978-83-7717-269-8.

Kalman, H., Tripathi, N. M., Gabrieli, O. G., & Portnikov, D. (2018). Effect of fluid on particles conveying in pipes. *The Conference Proceedings of the 8th World Congress on Particle Technology*, Orlando, FL, USA. ISBN: 978-0-8169-1105-9.

Khosravi-Darani, K., & Bucci, D. Z. (2021). Polyhydroxyalkanoate production and application in bioplastics. *Applied Microbiology and Biotechnology*, 94(6), 1467–1487.

Kim, J., et al. (2022). Graphene-reinforced biobased nanocomposites. *Advanced Materials*, 34(2), 2103672.

Klemm, D., Heublein, B., Fink, H. P., & Bohn, A. (2005). Cellulose: fascinating biopolymer and sustainable raw material. *Angewandte Chemie International Edition*, 44(22), 3358–3393.

Koetting, M. C., et al. (2015). Stimuli-responsive hydrogels. *Soft Matter*, 11(1), 27–40.

Koller, M., Salerno, A., Dias, M., Reiterer, A., & Braunegg, G. (2017). Modern biotechnological polymer synthesis: a review. *Food Technology and Biotechnology*, 50(4), 403–419.

Kumar, P., et al. (2018). Nanocomposite coatings for barrier and packaging applications. *Progress in Organic Coatings*, 123, 127–142.

Kumar, S., & Patel, A. (2021). Biobased plastics in electronics: an overview. *Journal of Electronics & Sustainability*, 13(4), 209–220.

Liu, Y., et al. (2021). Crosslinking of starch-based materials for enhanced performance. *Carbohydrate Polymers*, 255, 117492.

Mousavioun, P., et al. (2013). The effect of agricultural waste on the cost competitiveness of biobased products. *Bioresource Technology*, 129, 306–313.

Muller, J., González-Martínez, C., & Chiralt, A. (2017). Cross-linking strategies for improving the properties of biobased packaging materials. *Comprehensive Reviews in Food Science and Food Safety*, 16(6), 1190–1219.

Nair, R., et al. (2019). Enhancing electrical conductivity of PLA using graphene. *Polymer Engineering & Science*, 59(8), 1783–1791.

Naveh, R., Tripathi, N. M., & Kalman, H. (2017). Experimental pressure drop analysis for horizontal dilute phase particle-fluid flows. *Powder Technology*, 321, 355–368. https://doi.org/10.1016/j.powtec.2017.08.029

Nguyen, T., & Ho, M. (2021). Effectiveness of bamboo-based cleaning products. *Journal of Green Chemistry*, 29(5), 512–526.

Peng, X., et al. (2021). Cellulose nanocrystals in PLA nanocomposites. *Journal of Nanocomposites*, 9(1), 74–88.

Perez-Gago, M. B., & Krochta, J. M. (2002). Formation and properties of whey protein films and coatings. *Journal of Dairy Science*, 85(5), 1033–1039.

Rinaudo, M. (2006). Chitin and chitosan: properties and applications. *Progress in Polymer Science*, 31(7), 603–632.

Santo, N., Portnikov, D., Tripathi, N. M., & Kalman, H. (2018a). Experimental study on the particle velocity development profile and acceleration length in horizontal dilute phase pneumatic conveying systems. *Powder Technology*, 339, 368–376. https://doi.org/10.1016/j.powtec.2018.07.074

Santo, N., Tripathi, N. M., & Kalman, H. (2018b). Particle velocity reduction in horizontal-horizontal bends of dilute phase pneumatic conveying. *35th Israeli Conference on Mechanical Engineering (ICME)*, Beer-Sheva, Israel.

Sharma, K., Singh, S., Yadav, M., Yadav, S., & Tripathi, N. M. (2015). A review on the performance of the nanofluid based solar collectors–solar energy. *Asian Review of Mechanical Engineering*, 4(1), 17–26. https://doi.org/10.51983/arme-2015.4.1.2391

Sharma, N., Singh, D., Rani, R., Sharma, D., Pandey, H., & Agarwal, V. (2019). Chitosan and its nanocarriers: applications and opportunities. In: *Nanomaterials in Plants, Algae and Microorganisms* (pp. 267–286). Academic Press. https://doi.org/10.1016/B978-0-12-811488-9.00013-5

Sharma, S., Nath, R., Kityania, S., Nath, D., & Talukdar, A. D. (2025). Market value and demand for bioplastics. In: *Bioplastics* (pp. 247–265). Apple Academic Press.

Shen, L., Haufe, J., & Patel, M. K. (2010a). Product overview and market projection of emerging bio-based plastics. *PRO-BIP 2009*, 243.

Shen, L., Worrell, E., & Patel, M. (2010b). Present and future development in plastics from biomass. *Biofuels, Bioproducts, and Biorefining*, 4(1), 25–40.

Shogren, R. L., et al. (2019). Biodegradable polymers: advances in product design and performance. *Journal of Cleaner Production*, 235, 1157–1169.

Singh, R., Shukla, A., & Tiwari, S. (2011). A review on biopolymer-based films and their applications. *Journal of Polymers and the Environment*, 19(3), 764–778.

Singh, S., Sharma, K., Lal, K., & Tripathi, N. M. (2015). To study the behaviour of nanorefrigerant in vapour compression cycle-a review. *International Journal of Research in Engineering and Technology*, 4(04), 648–652.

Tan, Y., et al. (2020). PHA bioplastics: applications and degradation. *Biodegradable Materials Journal*, 8(3), 299–312.

Tharanathan, R. N. (2003). Biodegradable films and composite coatings: past, present and future. *Trends in Food Science & Technology*, 14(3), 71–78.

Tokiwa, Y., et al. (2009). Biodegradability of plastics. *International Journal of Molecular Sciences*, 10(9), 3722–3742.

Torres-Giner, S., Gimenez, E., & Lagaron, J. M. (2008). Characterization of the morphology and thermal properties of zein prolamine nanostructures obtained by electrospinning. *Food Hydrocolloids*, 22(4), 601–614.

Tripathi, N. M., Kalman, H., & Levy, A. (2017). Analysis of acceleration pressure drop in dilute phase pneumatic conveying. *International Symposium of Reliable Flow of Particulate Solidsv, RELPOWFLO V*, Skien, Norway.

Tripathi, N. M., Levy, A., & Kalman, H. (2016). Initial acceleration pressure drop in dilute phase pneumatic conveying system. *Powder, Granule and Bulk Solids: Innovations and Applications Conference.*

Tripathi, N. M., Levy, A., & Kalman, H. (2018a). Acceleration pressure drop analysis in horizontal dilute phase pneumatic conveying system. *Powder Technology*, 327, 43–56. https://doi.org/10.1016/j.powtec.2017.12.045

Tripathi, N. M., Levy, A., & Kalman, H. (2018b). Bend pressure drop analysis in dilute phase pneumatic conveying. *The Conference Proceeding of Conveying and Handling of Particulate Solids (CHoPS-2018)*, Greenwich, London, United Kingdom.

Tripathi, N. M., & Mallick, S. S. (2014). *An Investigation into Pressure Drop Across Bends for Fluidised Densephase Pneumatic Conveying Systems* (Doctoral dissertation). Thapar University, Patiala, India.

Tripathi, M. N., & Mallick, S. S. (2017). Pneumatic conveying of fly ash: Bend models investigation. *Advanced Materials Proceedings*, VBRI Press, 2(8), 526–531. https://doi.org/10.5185/amp.2017/812

Tripathi, N. M., Santo, N., Kalman, H., & Levy, A. (2018c). Analysis of particle flow in vertical dilute phase pneumatic conveying. *35th Israeli Conference on Mechanical Engineering (ICME)*, Beer-Sheva, Israel.

Tripathi, N. M., Santo, N., Kalman, H., & Levy, A. (2018d). Experimental analysis of particle velocity and acceleration in vertical dilute phase pneumatic conveying. *Powder Technology*, 330, 239–251. https://doi.org/10.1016/j.powtec.2018.02.017

Tripathi, N. M., Santo, N., Levy, A., & Kalman, H. (2018e). Experimental analysis of Bend effects and secondary acceleration in vertical dilute phase pneumatic conveying. *The Conference Proceeding of Conveying and Handling of Particulate Solids (CHoPS-2018)*, Greenwich, London, United Kingdom.

Tripathi, N. M., Sharma, A., Mallick, S. S., & Wypych, P. W. (2015). Energy loss at bends in the pneumatic conveying of fly ash. *Particuology*, 21, 65–73. https://doi.org/10.1016/j.partic.2014.09.003

Tuck, C. O., et al. (2012). Biorefineries: the essentials for success. *Science*, 337(6095), 695–699.

Verma, S. K., et al. (2024a). Green nanotechnology: illuminating the effects of bio-based nanoparticles on plant physiology. *Biotechnology for Sustainable Materials*, 1, 1. https://doi.org/10.1186/s44316-024-00001-2

Verma, S. K., Singh, D., & Misra, S. (2024b). Innovative approaches to sustainable waste management and energy generation in developing nations. In: Mehra, R., Kumar, P., & Hung, B. (eds.), *Municipal Solid Waste Management and Recycling Technologies* (pp. 267–300). IGI Global Scientific Publishing. https://doi.org/10.4018/979-8-3693-4054-7.ch010

Vink, E. T. H., et al. (2010). The eco-profile for current Ingeo polylactide production. *Industrial Biotechnology*, 6(4), 212–224.

Wang, L., et al. (2022). Compostable packaging: a viable alternative. *Sustainable Packaging*, 14(6), 112–126.

Winkler, R., Koellner, T., & Bush, G. (2018). Quantifying the carbon footprint reduction potential of renewable raw materials in consumer goods. *Journal of Cleaner Production*, 196, 470–482.

Wu, J., et al. (2020). Smart biobased materials: stimuli-responsive systems for advanced applications. *Advanced Materials*, 32(28), 2001584.

Wunderlich, S. M., & Gatto, K. A. (2015). Consumer perceptions of sustainable food products and willingness to pay. *Journal of Food Products Marketing*, 21(5), 531–545.

Yang, C., et al. (2020). Renewable biopolymers for energy storage devices. *Green Chemistry*, 22(11), 3613–3629.

Yu, S., et al. (2021). Compatibilizers in biopolymer blends: a review. *Progress in Polymer Science*, 112, 101341.

Zhang, X., et al. (2019). Blending PLA and PHA for enhanced ductility. *Polymer Science Journal*, 12(4), 345–359.

# 7 Lightweight and Sustainable

## *The Evolution of Natural Fiber Reinforced Composites in Automotive and Aerospace Sectors*

*Mufaddal Huzefa Shakir, Akant Kumar Singh, Naveen Mani Tripathi, Rahul Nadda, and Manvendra Yadav*

## 7.1 INTRODUCTION

As the industrial sectors grow, there is a constant demand for improved materials that may be more rigid, light in weight, cost-effective, and sustainable—all of which describe composite materials. In the last few decades, polymers have increasingly penetrated many applications, displacing traditional materials like metals due to their advantages. Polymers offer advantages such as increased ease of manufacturing, decreased weight, improved productivity, and cost savings over conventional materials. However, to meet the stringent strength requirements for many of the applications, the properties of polymers need to be enhanced with fillers and fibers (Lunetto et al., 2023). One particular area of interest has been the development of natural fibers (NFs) obtained from plants and animals. Several benefits of NFs are their low density, low price, vast availability, high biodegradability, and high strength and stiffness (Islam et al., 2024). Reinforcements in NFs can be used in the composites. They are naturally found in different forms, such as bast fibers: flax, ramie, kenaf, sisal, and coir; leaf fibers: abaca, banana, and pineapple; seed fibers: cotton, kapok, and loofah; reed and grass fibers: rice, corn, and wheat. The classification of NFs is shown in Figure 7.1. Several studies have already been done to explore the utility of plant-based NFs as reinforcements in composites. The utilization of NFs in polymer composites results in lower cost, high strength, and eco-friendly materials. Hence, it is observed that NFRCs have great potential for many applications. Each of these fibers has its own advantages, making them suitable for several applications such as composite materials for different load-bearing structures where lightweight

DOI: 10.1201/9781003566205-7

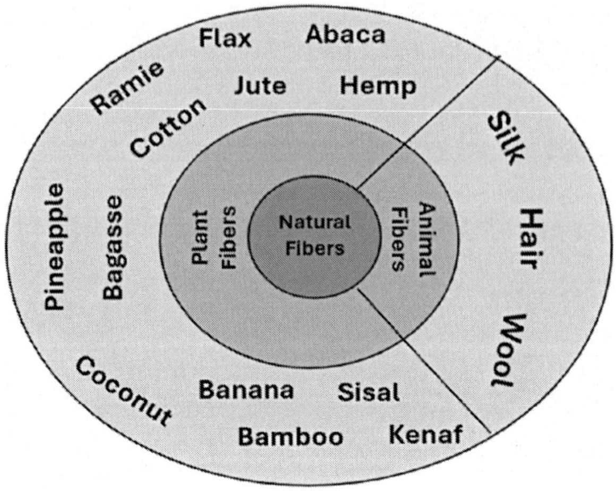


**FIGURE 7.1** Categorization of NFs.

and strength, along with sustainability, are parameters of concern (Dev et al., 2023; Kerni et al., 2020).

NFRCs have gained significant attention as an eco-friendly substitute to synthetic materials, mainly in the automotive and aerospace industries. These sectors are focused on improving fuel efficiency, reducing carbon emissions, and cutting production costs, while maintaining high performance and safety standards. By using NFs such as abaca, ramie, sisal, banana, jute, and cotton as reinforcement in a polymer matrix, NFRCs address these demands effectively (Khan et al., 2024). In the automotive sector, the push for lighter vehicles to enhance fuel efficiency and reduce emissions has driven the adoption of such materials. NFRCs, with their lower density compared to traditional options like fiberglass composites and metals, have proven especially suitable for these applications.

NFRCs are increasingly being used in automotive parts like interior panels, door trims, dashboards, and seat backs, offering a dual advantage of weight reduction and enhanced thermal and acoustic insulation. The use of NFs aligns with the sector's shift toward sustainable manufacturing by reducing energy consumption and environmental impact during production. Beyond their eco-friendly appeal, NFRCs provide practical benefits, such as improved crashworthiness due to the flexibility and impact resistance of NFs, which enhance passenger safety. Its excellent vibration-damping properties also contribute to better ride comfort. Combined with the cost-effectiveness of NFs compared to synthetic alternatives, these features make NFRCs an ideal choice for automotive manufacturers seeking to integrate sustainability with performance and affordability (Patel et al., 2023; Hallal et al., 2013; Rajak et al., 2022).

In the aerospace industry, material selection prioritizes high-performance composites capable of withstanding extreme conditions while delivering superior mechanical properties. While NFRCs may not yet replace advanced materials like carbon fiber in critical structural applications, they are gaining traction for secondary uses where their benefits shine (Feng et al., 2024). NFRCs are well-suited for interior cabin components, luggage bins, seat shells, and non-structural panels, offering advantages like reduced weight, cost-effectiveness, and environmental sustainability. By lowering the overall weight of components, NFRCs contribute to fuel economy and decreased greenhouse gas emissions, key priorities in aviation (Asim et al., 2018). Their adoption also supports the industry's push toward circular economy practices and sustainable development. The growing use of NFRCs in both the aerospace and automotive sectors highlights their potential as lightweight, eco-friendly materials that balance performance, cost savings, and environmental responsibility. As industries strive for greener and more efficient solutions, NFRCs represent a promising path for future material innovation (Mansor et al., 2019).

A brief overview of NFRCs is discussed in this chapter, covering key aspects of their development and application. It begins with a brief discussion on the chemical content and properties of various NFs. The chapter also delves into the fabrication techniques employed in producing NFRCs, highlighting advancements in the field. A significant focus is placed on reviewing studies related to the utilization of NFRCs in the aerospace and automotive sectors. The findings from the literature strongly indicate a growing trend in the adoption of NFRCs in these sectors, driven by their promising balance of performance, sustainability, and cost-effectiveness.

## 7.2 PROPERTIES AND CHEMICAL COMPOSITION OF NFS

### 7.2.1 Chemical Composition

NFs are composed of key organic constituents, including cellulose, hemicellulose, pectin, lignin, and waxes, with their proportions varying depending on the kind of fiber. Factors such as plant location, growth rate, and tissue type influence the cellular composition of fibers (Khalil et al., 2012). Cellulose, a polysaccharide made of long glucose chains, forms the crystalline structure that offers good strength and rigidity to the fiber. Hemicellulose, a branched carbohydrate composed of sugars like xylose, mannose, and glucose, affects the fiber's biodegradability and moisture absorption due to its amorphous nature. Lignin, an aromatic polymer with a water-repellent structure, adds rigidity and strength through its extensive cross-linking. Pectin, a polysaccharide rich in galacturonic acid, plays a critical role in cell adhesion and is primarily located in the middle lamella and primary cell walls. Waxes, mixtures of long-chain alkanes, fatty acids, and alcohols, form a hydrophobic coating on the fiber surface, offering protection and moisture resistance. The composition and proportions of these constituents vary among different NFs, influencing their suitability for specific applications (Sreenivasan et al., 2011; Chokshi et al., 2022). A detailed breakdown of the cellular makeup of commonly used NFs is provided in Table 7.1.

### 7.2.2 Mechanical and Physical Properties

NFs possess unique characteristics shaped by environmental factors such as soil, air, sunlight, and water, which vary across regions and seasons. These variations

**TABLE 7.1**
**Chemical Composition of Different NFs.**

| Natural fiber | Pectin | Waxes | Lignin | Hemicellulose | Cellulose | Reference |
|---|---|---|---|---|---|---|
| Hemp | 0.8 | 0.7 | 4.76 | 11.7 | 75.3 | (Kostic et al., 2008) |
| Kenaf | 2.5–4.8 | – | 7.0–18.0 | 20.2–23.5 | 29–68 | (Guo et al., 2021) |
| Sisal | 0.7–1.2 | 0.25–2.0 | 3.0–18 | 9.0–14 | 43–85 | (Chen et al., 2020) |
| Banana | 3.5 | – | 4.5 | 18.0 | 61–66 | (Benítez-Guerrero et al., 2017) |
| Flax | 0.7 | 1.7 | 2.4 | 12.3 | 82.5 | (Ramesh, 2018) |
| Jute | ~0.1 | ~0.4 | 12–13 | 13.6–20.6 | 61–71 | (Shakir et al., 2024) |
| Bamboo | – | – | 4.0–32 | 35.0 | 24–63 | (Yuan et al., 2021) |
| Pineapple | 3.0 | – | 3.17 | 19.5 | 67.2 | (Daud et al., 2014) |
| Cotton | 0–1.3 | 0.7 | <2 | 5.9 | 80.8–88.0 | (Dittenber & GangaRao, 2012) |
| Ramie | 3.5–6 | 0.8–2.3 | 0.8–1.5 | 13.5–17 | 68–75 | (Li et al., 2016) |
| Bagasse | 9.0 | – | 19–25.3 | 15.7 | 32–55.3 | (Kamarudin et al., 2022) |
| Abaca | <1 | <1 | 5–16 | 19–20 | 63–68 | (Mamun et al., 2015) |

influence plant growth and, subsequently, the properties of NFs. Fibers harvested from younger plants are typically softer and differ in attributes compared to those from mature plants. Proper storage of NFs is essential to preserve their quality, as poor conditions can lead to mold, discoloration, and degradation (Komuraiah et al., 2014). Fibers with higher cellulose content and well-aligned cellulose microfibrils generally exhibit superior properties. Among these, tensile strength stands out as a critical factor, making NFs well-suited for composite materials. Other important properties to consider for applications include density, elastic modulus, and breaking elongation (Bakar et al., 2020). A detailed overview of the physical and mechanical properties of various NFs is presented in Table 7.2.

## 7.3 FABRICATION OF NF-REINFORCED COMPOSITES

The fabrication of NFRCs involves two key steps: creating the mold and reinforcing it with a polymer resin. A variety of processing techniques are available for producing these composites, each tailored to specific application requirements. These methods are generally categorized based on how the reinforcement and polymer matrix are applied to the mold, as well as the curing processes employed. Broadly, composite fabrication techniques are divided into open molding and closed molding methods, determined by the conditions under which the polymer matrix and reinforcement are applied. The kind of fabrication method is affected by many factors, such as the materials used, the accessibility of required tools, and the desired characteristics of the final composite product (Syduzzaman et al., 2020; Marques, 2011). Figure 7.2 shows different fabrication techniques to manufacture polymer composites.

**TABLE 7.2**
**Physio-Mechanical Properties of Different NFs.**

| Fiber | Tensile strength (MPa) | Elastic modulus (GPa) | Breaking elongation (%) | Moisture regain (%) | Density (g/cm³) | Fiber diameter (microns) | Reference |
|---|---|---|---|---|---|---|---|
| Kenaf | 370–920 | 15–45 | 6.0–11.0 | 1.2–3.3 | 1.45–1.5 | 18–240 | (Malik et al., 2021) |
| Sisal | 300–800 | 6.0–25.0 | 3.0–15.7 | 12.0 | 1.2–1.5 | 40–310 | (Iniya & Nirmalkumar, 2021) |
| Hemp | 305–740 | 25–57 | 1.3–4.5 | 13.0 | 1.47 | 16–22 | (Shahzad, 2012) |
| Flax | 235–770 | 21–55 | 1.1–3.2 | 12–14 | 1.4–1.5 | 130–330 | (Malik et al., 2022) |
| Banana | 50–750 | 17.85 | 10.35 | – | 1.3 | 70–260 | (Venkateshwaran & Elayaperumal, 2010) |
| Pineapple | 423–870 | 30–75 | 15.8 | – | 1.5 | – | (Asim et al., 2015) |
| Bamboo | 230–800 | 12–37 | 2.3–3.9 | – | 1.1 | 20–45 | (Liu et al., 2012) |
| Jute | 320–850 | 12–40 | 1.5–2.0 | 12.65 | 1.47 | 50–120 | (Biswas et al., 2013) |
| Abaca | 400–970 | 9–28 | 3–10.0 | 5.81 | 1.5 | 150–260 | (Shakir & Singh, 2023) |
| Ramie | 200–940 | 44–128 | 2.0–3.8 | 12–17 | 1.5 | 50.0 | (Lu et al., 2006) |
| Bagasse | 220–310 | 15–29 | 3.0–6.0 | – | 1.2 | 12–35 | (Xiong, 2018) |
| Cotton | 287–597 | 5.5–13 | 7–8.0 | 8–25 | 1.6 | – | (Ahmad et al., 2015) |

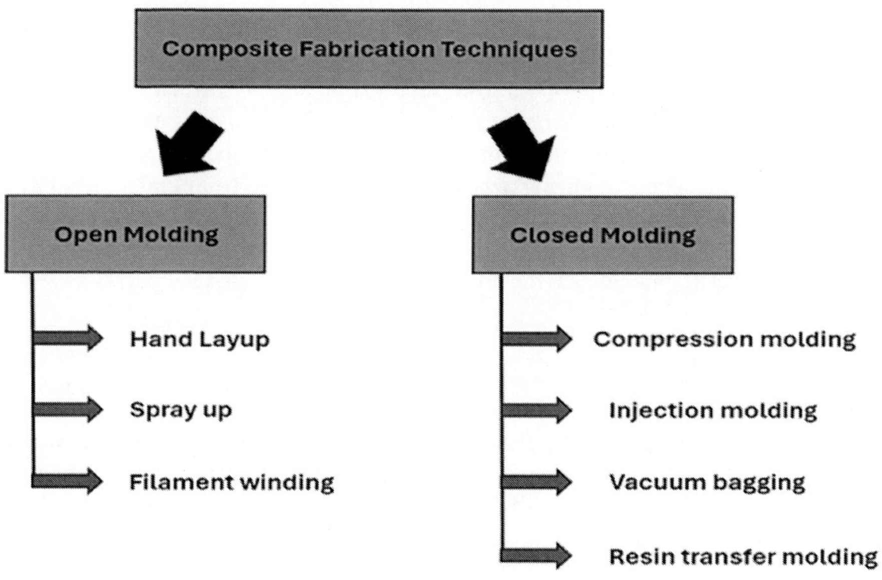


**FIGURE 7.2** Techniques to fabricate NFRCs.

### 7.3.1 Hand Layup

The hand layup process is a manual method for fabricating NF composites, appreciated for its simplicity and adaptability, especially in small-scale production. The process begins with selecting NFs such as flax, hemp, silk, kenaf, or cotton, chosen for their properties and suitability for the intended application. The fibers are cleaned and aligned to ensure consistent quality and strength. A mold is then prepared according to the desired dimensions of the composite part. During the layup process, the NFs are manually saturated with a polymer resin, commonly epoxy or polyester resin, ensuring even distribution and optimal bonding to enhance the composite's performance. The resin-soaked fibers are carefully layered onto the mold in specific orientations to achieve the required strength and stiffness. Attention to detail during fiber alignment and resin application minimizes defects and ensures uniformity. Once the desired thickness and fiber arrangement are achieved, the composite is cured under controlled conditions, such as specific temperatures and pressures, to facilitate proper bonding and polymerization. After curing, the part is extracted from the mold. The result is a lightweight, durable, and eco-friendly material ideal for applications such as automotive components and consumer goods (Syduzzaman et al., 2020; Elkington et al., 2015).

### 7.3.2 Compression Molding

Compression molding is a technique utilized for manufacturing NFRCs to create high-quality composites with excellent mechanical properties. The process commences with the preparation of NFs, which are cleaned, aligned, and sometimes treated with additives or fillers to enhance their performance. These fibers are then placed strategically in a preformed mold cavity, often oriented to optimize the composite's strength and rigidity. Once the fibers are arranged, the mold is closed, and pressure and heat are applied simultaneously. This step ensures that the NFs are compacted and impregnated with a polymer resin, promoting strong bonding and uniformity throughout the composite. The heat and pressure facilitate the resin's flow, ensuring even distribution within the fiber matrix while reducing voids and achieving the desired density and thickness. After a specified curing period, during which the resin undergoes polymerization, the mold is opened, and the completed part is extracted (Jaafar et al., 2019; Rajak et al., 2019). This method produces durable and uniform NFRC components suitable for various applications, including automotive and industrial parts.

### 7.3.3 Vacuum Bagging

Vacuum bagging is a highly effective method for fabricating NF composites, offering superior control over resin infusion and enabling the production of lightweight, high-quality components. The process begins with preparing NFs, which are cleaned, aligned, and sometimes shaped into mats or preforms for uniformity. These fiber preforms are placed on a mold surface, typically coated with a release agent to prevent sticking. A flexible vacuum bag, made of plastic or silicone, is placed over the preform and sealed around the mold edges to create an airtight enclosure. A vacuum pump removes air from the space between the bag and mold, creating a vacuum that compresses the bag tightly against the preform. This pressure compacts the fibers, eliminates trapped air, and aids in resin impregnation. A low-viscosity thermosetting resin is then introduced, either manually or via injection ports, ensuring thorough saturation of the fiber preform under vacuum pressure. The vacuum is maintained throughout the process to ensure uniform resin distribution and minimize voids. The composite is then allowed to cure under controlled conditions, often with added heat, to facilitate resin polymerization and bonding. After curing, the vacuum bag is removed, and the finished composite part is demolded, trimmed, and refined as required. This technique produces durable, lightweight components ideal for high-performance applications (Abdurohman et al., 2018; Correia et al., 2005).

### 7.3.4 Resin Transfer Molding (RTM)

RTM is a commonly utilized technique for producing NFRCs, valued for its capability to fabricate complex parts with great precision and excellent surface quality. The process starts by selecting the fibers, which are then cleaned, aligned, and sometimes preformed into mats or shapes for easier handling. These fiber mats are placed into a closed mold with channels designed to guide the resin flow. After positioning the fibers, the mold is sealed and tightly clamped, creating an enclosed space. A thermosetting resin, typically polyester or epoxy, is injected into the mold under pressure, allowing it to flow through the fibers, displacing any air, and ensuring complete coverage. This pressure also ensures strong bonding between the fibers and resin, with an even distribution throughout the composite. The composite is then cured under controlled conditions, usually involving heat and sometimes additional pressure, which enables the resin to polymerize and achieve the desired mechanical properties. Once curing is done, the mold is opened, and the completed part is extracted, followed by trimming and other necessary finishing processes (Hsiao & Heider, 2012).

### 7.3.5 Injection Molding (IM)

IM is a fabrication process employed to produce NFRCs, offering an efficient way to mass-produce complex and high-quality components. The process starts by preparing the NFs, which are cleaned and typically mixed with a thermoplastic or thermosetting resin, often in pellet form. This fiber–resin blend is then fed into an IM machine, where the polymer is melted and injected into a cavity of the mold with high pressure. The mold is carefully designed to match the shape of the desired part, enabling the creation of detailed geometries with tight tolerances. As the molten mixture fills the mold, the fibers are evenly distributed, and the material takes the mold's shape. After the mold cools, solidifying the composite, it is opened, and the completed part is ejected, typically requiring little to no post-processing (Fu et al., 2020).

## 7.4 STUDIES ON NF-REINFORCED COMPOSITES

Various NFs have been used to reinforce thermosetting polymers, like vinyl ester, polyester, and epoxy, as well as thermoplastic polymers like high-density polyethylene (HDPE), polystyrene, and polypropylene (PP), to create NF-reinforced composites (NFRCs). In one study, abaca and polyethylene composites were made using a hot-press technique, with polyethylene in powder form. Among different fiber contents, the composite with 55 wt.% fibers displayed the highest mechanical properties in terms of tensile modulus and strength, along with the greatest hardness value (Rajakumar et al., 2022). In another study, glass and flax fiber-reinforced PLA-PBS hybrid composite specimens were produced using IM. The combination of flax and glass fibers greatly improved the tensile properties, with the elastic modulus increasing from 3.8 to 4.3 GPa (Ketata et al., 2024). A separate study focused on jute, glass fiber, and polyester hybrid composites, also made utilizing the hot press technique. The mechanical and thermomechanical behavior of the composites was studied, and the composite's properties were further enhanced by applying a 1 kGy dose of gamma radiation, improving the bending and tensile strengths (Das et al., 2021). In another experiment, banana and epoxy composites were created, and their mechanical behavior was analyzed, focusing on the repercussion of fiber length and content on bending and tensile properties. The optimal fiber content and length for banana and epoxy composites were found to be 16 wt.% and 15 mm, respectively. SEM micrographs showed that higher fiber content (over 16 wt.%) led to weaker fiber–matrix interaction, reducing the composite's performance (Venkateshwaran et al., 2011).

The damping and mechanical properties of hemp and epoxy, as well as jute and epoxy, composites were studied using the manual layup method. Results showed that hemp and epoxy composite samples outperformed jute and epoxy composite samples in terms of shear, tensile, and bending strength. However, the damping factor decreased across various frequency ranges from low to high (Singh et al., 2023). The moisture absorption and mechanical properties of pineapple and sisal fiber-reinforced polyester hybrid composites were also investigated. The fabricated composite demonstrated the highest impact strength of 46.7 N/mm². Increasing the sisal fiber content to 50 wt.% improved the mechanical behavior of the pineapple and polyester composites while reducing moisture absorption (Praveena et al., 2021). Another study examined ramie and sisal fiber-reinforced PP hybrid composites, fabricated utilizing the IM process. The hybrid composites reflected a 50% rise in the flexural strength and a 13% rise in tensile strength (Mahesh et al., 2022). Similarly, oil palm and kenaf fiber-reinforced epoxy hybrid composites were created using manual layup. It was noticed that increasing the proportion of kenaf fibers in the composites greatly improved the bending and tensile properties, although the impact properties were higher in pure oil palm composites than in the hybrids (Hanan et al., 2020). A summary of the mechanical properties of different NFRCs is displayed in Table 7.3.

## 7.5 APPLICATIONS OF NF-REINFORCED COMPOSITES

### 7.5.1 Automotive Applications

Composites are becoming increasingly important in the automotive sector owing to their strength, fatigue resistance, corrosion resistance, and lightweight properties. Initially, composites were primarily used in high-end sports and luxury vehicles, where their high cost was justified by the need for materials that could perform well in demanding environments. However, stricter fuel economy and emission regulations worldwide are now driving the use of composites in passenger cars. To

**TABLE 7.3**
**Summary of the Mechanical Properties of NFs.**

| Composite | Fiber content | Flexural strength (MPa) | Tensile strength (MPa) | Flexural modulus (MPa) | Tensile modulus (MPa) | Impact strength | Reference |
|---|---|---|---|---|---|---|---|
| Abaca/polyethylene | 55 wt.% | – | 61.0 | – | 5.0 | 10 J | (Rajakumar et al., 2022) |
| *Calotropis giganteal*/ Palmyra/phenolic resin | 40 wt.% | 55.0 | 45.0 | – | – | 1.25 $KJ/m^2$ | (Sekar et al., 2022) |
| Flax/banana/polyester | 40 wt.% | 167.8 | 113.4 | – | – | – | (Zelalem Biresaw et al., 2022) |
| Flax/glass fiber/PLA-PBS | 10 wt.% glass and flax fibers | – | 53.0 | – | 5.4 | – | (Ketata et al., 2024) |
| Ramie/phenolic resin | – | – | 82.0 | – | 11.5 | – | (Wang et al., 2023) |
| Banana/jute/vinyl ester | 45 wt.% | 43.0 | 21.0 | – | – | 11.0 J | (Ravindran et al., 2021) |
| Jute/glass fiber/polyester | 55 wt.% | 240.0 | 110.0 | 9.5 | 4.0 | 43 $KJ/m^2$ | (Das et al., 2021) |
| Banana/epoxy | 12 wt.% | 37.0 | 17.0 | – | 0.65 | 2.25 J/m | (Venkateshwaran et al., 2011) |
| Jute/epoxy | 5 layers of jute | 76.52 | 53.69 | 3.20 | 1.40 | 97 J | (Singh et al., 2023) |
| Sisal/pineapple/polyester | 20 wt.% | 45.0 | 51.0 | – | – | – | (Praveena et al., 2021) |
| Sisal/ramie/PP | 15 wt.% sisal and ramie fibers | 60.5 | 40.7 | – | – | 39.48 J/m | (Mahesh et al., 2022) |
| Sugar palm/PET fibers/PP | 20 wt.% sugar palm and PET fibers | – | 33.0 | – | – | 13.5 $KJ/ m^2$ | (Vardai et al., 2021) |
| Basalt/coconut/ bio-polyethylene | 12 wt.% basalt and coconut | 45.4 | 31.5 | 0.27 | 0.34 | 12.2 $KJ/m^2$ | (Bazan et al., 2020) |
| Coir/sugarcane/polyester | 30 wt.% | 28.41 | 16.88 | 3.16 | 1.20 | 0.95 J/m | (Arul et al., 2020) |
| Oil palm/kenaf/epoxy | 50 wt.% | 112.0 | 62.0 | 7.9 | 3.6 | 1.2 J | (Hanan et al., 2020) |

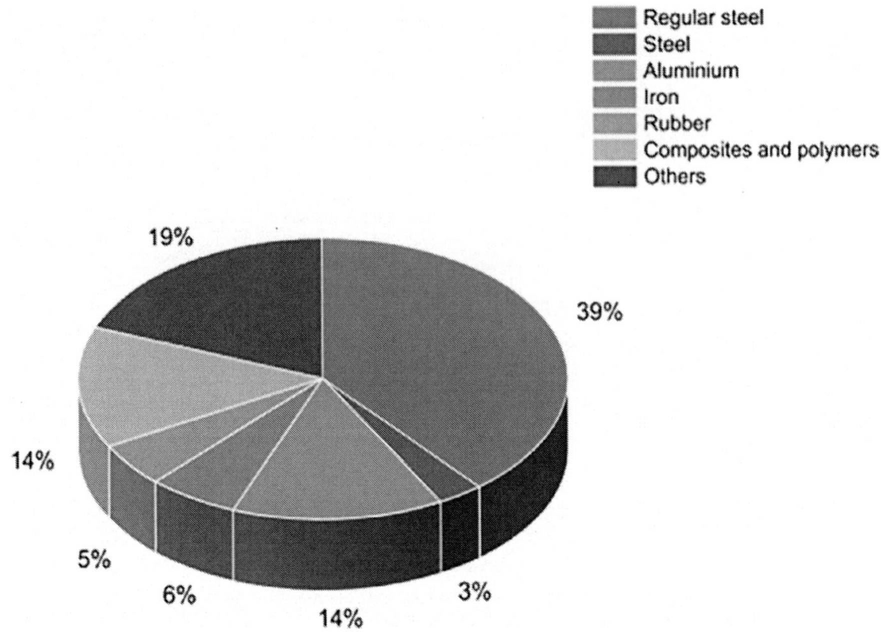


**FIGURE 7.3** Various materials are used in an automobile.

address environmental concerns, automakers are exploring eco-friendly solutions, focusing on green technologies that make vehicles lighter, more cost-effective, and fuel-efficient. NFs are particularly appealing because they are less expensive to produce and have a smaller environmental footprint compared to traditional materials. Shifting from conventional metal components to NFs is a logical step toward sustainability (Njuguna, 2016). Figure 7.3 illustrates the range of materials currently used in vehicles.

#### 7.5.1.1 Brake Pads

Brake pads are critical safety components in vehicles, responsible for slowing or stopping motion by converting kinetic energy into thermal energy. This heat is primarily absorbed by the brake pad, with some dissipated into the environment. Brake pads are made from friction materials that include four key components: reinforcement fibers, binders, friction modifiers, and fillers. Traditionally, asbestos-based fibers were widely used owing to their excellent thermal and mechanical properties. However, the health hazards and environmental concerns associated with asbestos have led to the search for safer alternatives (Borawski, 2020; Blau, 2001). Recent studies have explored NFRCs as eco-friendly options for brake pad applications. For instance, nettle, sisal, and *Grewia optiva* fibers reinforced with a PLA matrix were fabricated using compression molding. These composites demonstrated improved wear resistance and friction performance. Another study explored the mechanical behavior of brake pad composites created with clamshell powder, alumina, and basalt grains, showing hardness comparable to commercially available asbestos when tested using Vickers microindenter under ASTM guidelines (Bajpai et al., 2013).

Some other researchers also aimed at the cutting-edge and sustainable development of the bio-based additive manufacturing processes (Agrawal et al., 2024; Atal et al., 2023; Bhadouria et al., 2022; Ranjan et al., 2024a, 2024b; Ranjan & Tripathi, 2024; Tripathi & Sharma, 2024; Satishkumar et al., 2024a, 2024b; Jafar et al., 2024; Singh et al., 2024; Sharma et al., 2024; Neveu et al., 2020a, 2020b, 2020c).

Further research focused on developing green brake pad materials to replace harmful substances like antimony trisulfide and copper (Gautier di Confiengo & Faga, 2022). These brake pads incorporated by the company, named JMM NFs and synthetic aramid fibers. Performance tests using a full-scale brake dynamometer revealed effective braking, low wear rates, excellent fade resistance, and minimal speed sensitivity (Yun et al., 2010). Banana fibers were also studied for their influence on the properties of brake pads. The experimental outcomes indicated that increasing the banana fiber content raised wear, porosity, water absorption, and compressibility while reducing density and hardness. An optimal fiber content of 10 wt.% or less significantly improved friction and brake fade performance. These advancements underscore the potential of NFs as sustainable and effective alternatives in automotive braking systems (Singh et al., 2019).

#### 7.5.1.2 Bumper and Antiroll Bar

Automobile bumpers are designed to absorb energy during low-speed impacts, protecting both passengers and the vehicle. They must balance flexibility to absorb energy with the strength and rigidity needed to minimize damage to surrounding parts. With increasing safety standards, the development of more adaptive bumper beams has become a priority. Material selection plays a key role in achieving these goals, and the use of natural or synthetic fibers in hybrid designs can enhance safety while reducing vehicle weight (Belingardi et al., 2015; Kulkarni et al., 2017).

One study investigated the crashworthiness of a lightweight hybrid bumper beam made of carbon fiber, replacing the traditional steel beam. Using LS-DYNA simulations, the impact performance of both beams was compared. The hybrid beam matched the steel beam in impact performance while reducing weight by 46%, compared to the 50% reduction achieved by a pure steel beam. The research also noted that increasing the wall thickness of the hybrid beam did not linearly improve crash performance, highlighting the importance of optimizing material usage (Li et al., 2015).

Another study evaluated the mechanical performance of bumper beams made from glass and jute fiber composites. Using a manual layup process, three types of composites were created: one with 40% jute fiber, one with 40% glass fiber, and a hybrid composite combining 30% jute and 10% glass fiber. To enhance bonding, maleated PP was added as a compatibilizer. The hybrid composite outperformed the others in terms of impact resistance and hardness, demonstrating its potential as a strong and lightweight alternative to conventional bumper materials.

These advancements highlight the role of innovative materials and designs in improving automotive safety and efficiency (Olorunnishola & Adubi, 2018). A study examined the tensile, impact, and bending properties of kenaf, glass, and epoxy composites fabricated using the sheet molding compound (SMC) process. Since kenaf fibers are naturally twisted, a fixture was used to align them properly. When compared to glass mat thermoplastic (GMT) materials, the kenaf and glass hybrid composite demonstrated superior flexural and tensile strength, although its impact strength was lower (Davoodi et al., 2010). Another study focused on designing and manufacturing woven glass and carbon hybrid mat fibers as a replacement for traditional glass mat fibers in reinforced automobile bumper beams. Using numerical analysis and classical lamination theory, researchers evaluated the mechanical properties of the hybrid fibers. Impact simulations were conducted to optimize the bumper beam design and a microgenetic algorithm. A multi-objective optimization tool was employed to balance weight, carbon content, and impact performance. The final hybrid bumper beam was 33% lighter than its glass fiber-reinforced counterpart. The study also confirmed that classical laminate plate theory is an effective tool for predicting the mechanical behavior of such composites, with experimental results closely matching the predictions. These outcomes underscore the potential of hybrid composites for creating lightweight, high-performance materials in automotive applications (Kim et al., 2015).

An anti-roll bar is an essential part of a vehicle's suspension system, connecting the suspension on the left and right sides to improve stability. Typically made of U-shaped metal, the anti-roll bar increases suspension stiffness, reducing the risk of rollovers (Gustafsson & Agrawal, 2017). In a study focused on environmental sustainability, researchers explored replacing traditional metal anti-roll bars with hybrid composites made from sugar palm and kenaf fibers. A life cycle assessment conducted using SolidWorks evaluated the environmental impact of these materials. The results showed that sugar palm fibers, due to their lightweight nature, produced 10% lower carbon emissions and required less energy during production compared to kenaf fibers. This makes sugar palm fibers a promising, eco-friendly alternative for manufacturing anti-roll bars (Mastura et al., 2017).

#### 7.5.1.3 Exterior and Interior Components

The use of NFs in automotive manufacturing began in 1996 when Mercedes-Benz introduced jute and epoxy composites for the E-Class models. Audi followed in 2000 with sisal and flax hybrid composites for the door panels of the A2 midrange vehicle (Koronis et al., 2013). Toyota also embraced this innovation by using kenaf and PLA composites for outer wheel covers and combining kenaf and PLA, and ramie biocomposites in translucent vehicle roofs. BMW incorporated NF mats in a thermoplastic matrix for door panels in the 7 Series sedan, and the 2010 Ford Flex featured a storage bin reinforced with 20% wheat straw (Kaur et al., 2023). In further exploration of NFs, a study developed biodegradable cellulosic barkcloth fiber reinforced with epoxy for use in vehicle instrument panels. The fabric, treated with 5 wt.% alkali solution, showed improved mechanical properties. Using the manual layup method, the alkali-treated composites achieved an average tensile strength of 33 MPa, exceeding the industry benchmark of 25 MPa by 32%. These findings highlight barkcloth fabrics as a promising, sustainable material with strong structural integrity for automotive applications (Rwawiire et al., 2015).

Selecting materials for exterior automotive components is a complex challenge due to the need for durability in harsh environmental conditions, including moisture exposure. Unlike interior components, these parts must withstand constant environmental stress. One of the earliest applications of NFs in exterior automotive components was in the Mercedes-Benz Travego travel coach, which featured a transmission and engine cover made from flax and polyester composite (Suddell & Evans, 2005). Daimler Chrysler also innovated by using banana fiber-reinforced PP to produce the spare wheel cover for the Mercedes A-Class. These developments showcase the growing potential of NFRCs in exterior automotive design (Elseify et al., 2021). In another study, researchers explored the use of glass and carbon woven fibers embedded in an epoxy matrix for structural applications. By maintaining a constant weight percentage of 45% across five composite laminates, they tested different stacking sequences. A configuration with carbon fibers on the exterior and glass fibers internally demonstrated the highest bending strength, whereas an alternating carbon and glass arrangement offered superior compressive strength. These findings highlight the critical influence of fiber arrangement on composite performance (Zhang et al., 2012). Additionally, a study developed a bus body component, specifically the section between the headlights, using hemp fiber reinforced with PTP vegetable resin through an SMC process. A comparative model was fabricated using glass fiber and unsaturated polyester. Both natural and synthetic fiber composites underwent life cycle assessments and mechanical tests, including evaluations of water absorption, combustion, tensile strength, flexural strength, and impact resistance. While the hemp and PTP resin composite had better environmental performance, the mechanical properties of the NFRC were generally inferior to those of the glass fiber composite. These studies illustrate the evolving role of NFRCs in exterior automotive applications, balancing sustainability with mechanical performance (Müssig et al., 2006). Table 7.4 summarizes the use of NFRCs across various car models and components.

**TABLE 7.4**
**NFRCS for Various Automotive Components (Hill et al., 2012).**

| Composite | Model | Applications |
|---|---|---|
| Flax/PP | Chevrolet Impala | Trim and rear shelf |
| Sisal/acrylic | BMC 7-Series | Interior door panel |
| Wood/PP | Cadillac Deville | Seatbacks |
| Corn/Sorona | Nissan Leaf | Floor mats |
| Corn/PLA | Mazda 5 Hydrogen | Seat fabric, console |
| Soy/polyurethane | Ford Fusion | Headrests |
| Bamboo/corn/polyethylene | Lexus CT200h | Floor mats |
| Kenaf/polyurethane | Ford Fiesta | Interior door panel |
| Coconut/PP | Ford Focus BEV | Baggage-compartment speakers |
| Soy/polyurethane | Lincoln MKZ | Console door |
| Wheat straw/PP | Ford Flex | Interior storage bins |
| Starch/kenaf/Sorona | Toyota Raum | Spare wheel covers, floor mats |
| Kenaf/polyester | GMC Terrain | Acoustic insulator, ceiling liner, trim, rear shelf |
| Flax/polyethylene | Mercedes-Benz C-class | Transmission and engine cover, underbody floor panels |
| Corn/Sorona | Toyota Pruis | Instrument panels, AC vents |
| Abaca/polyethylene | Mercedes-Benz A-class | Underbody panels |

### 7.5.2 Aerospace Applications

Lightweight materials are crucial in the aerospace industry, where high strength, impact resistance, and excellent thermal and vibration stability are essential. Aircraft manufacturers continually strive to improve fuel efficiency by optimizing engine performance and reducing the weight of structural components and systems. This transition began in 1987 with the introduction of composites in aircraft manufacturing, leading to notable weight reductions. The Boeing 787 exemplifies this innovation, with composites comprising up to 50% of its main structure—the highest proportion in any commercial aircraft. These materials not only enhance impact resistance but also achieve weight reductions of up to 20% compared to traditional aluminum designs. The advent of polymer matrix composites has further transformed aircraft design, enabling their extensive application in wings, fuselage sections, and interiors, including cabin dividers, floorboards, bulkheads, lavatories, wall panels, stowage compartments, and galleys. Collaboration between researchers and manufacturers continues to focus on integrating sustainable and renewable materials into composite manufacturing. This effort underscores the industry's dedication to innovation and environmental responsibility (Hill et al., 2012; Mouritz, 2012).

Research has highlighted the potential of NFRCs in aerospace applications. One study focused on using ramie fiber composites to optimize the weight of aircraft wing box components. Theoretical analysis revealed that replacing 7000 series aluminum alloy with ramie fiber composites could reduce weight by 12%–14%, showcasing their potential for lighter aerospace designs (Boegler et al., 2015). Another study explored the use of NFRCs for airplane interiors, specifically cotton and polyurethane (PU) composites. This research stood out for its use of a PU matrix derived from castor bean oil, ensuring that the composite was entirely sourced from natural materials. The tensile properties of cotton and PU composites were compared with those of cotton and epoxy, fiberglass and PU, and fiberglass and epoxy composites. While the tensile properties of cotton and PU composites were slightly lower than traditional materials, the difference, particularly compared to cotton and epoxy, was minimal. These findings underscore the feasibility of NFRCs as sustainable alternatives to synthetic materials in aviation interiors, paving the way for innovative and eco-friendly advancements in aerospace design (Eloy et al., 2015).

Another study focused on the fabrication of NFRCs using flax, hemp, kraft pulp, and rutabaga fibers. These fibers underwent surface treatments and advanced mechanical processing, producing nanocellulosic composites with exceptional strength and lightweight properties. The study emphasized their suitability for weight-sensitive, high-performance structural components in the aerospace industry (Bhatnagar & Sain, 2005).

Another investigation examined low-weight cork as a sustainable alternative to synthetic sandwich core materials for aerospace structural parts. Cork-based sandwich composites reinforced with carbon fiber laminates were assessed for bending stiffness, acoustic performance, damping characteristics, and resistance to impact damage. Compared to synthetic foam structures like Rohacell 110 IG, cork composites showed slightly lower bending stiffness but excelled in acoustic insulation, damping, and impact resistance. These experimental outcomes display the potential

**TABLE 7.5**
**Utilization of Composites in Different Models of Airbus Aircraft (Bold, 2007).**

| Time period | Aircraft model (Airbus) | Composites applications |
|---|---|---|
| 1970–1980 | A300 | Fairings, radome |
| 1980–1990 | A310 | Rudder, elevators, air brakes, spoilers |
| | A320 | Flaps, engine cowlings, LG door |
| 1990–2000 | A330/A340 | Wet HTP box, Ailerons |
| 2000–2013 | A340–600 | Keel beam, J-nose |
| | A350 | Wing box, fuselage |
| | A380 | Cross beams, wing ribs fuselage |
| | A400 | Upper wing |

of expanded cork composites and NFRCs for aerospace applications, particularly in areas requiring sound insulation and durability, marking a significant step toward eco-friendly material innovations in the sector (Walsh et al., 2017). Table 7.5 outlines the use of composites in various models of Airbus aircraft.

## 7.6 CONCLUSION

This chapter has explored the properties and chemical composition of different NFs, the fabrication of NFRCs, and their utilization in the automotive and aerospace sectors. NFRCs have proven to be a compelling choice over traditional materials because of their lightweight, fuel efficiency, and reduced environmental impact. These advantages have spurred significant research and development in recent years. Despite their potential, several challenges like moisture absorption, dimensional instability, and limited fire resistance still need to be explored. Moreover, advancements in the processing of fiber, resin choice, and hybrid composite design are steadily tackling these obstacles. In summary, the employment of NFRCs in the automotive and aerospace industries marks a crucial step toward sustainability. With ongoing research and innovation, the adoption of these materials is expected to expand further, enabling their utilization in a wider range of structural and non-structural applications.

## REFERENCES

Abdurohman, K., Satrio, T., Muzayadah, N. L., & Teten. (2018, November). A comparison process between hand lay-up, vacuum infusion and vacuum bagging method toward e-glass EW 185/lycal composites. In *Journal of Physics: Conference Series* (Vol. 1130, p. 012018). IOP Publishing. https://doi.org/10.1088/1742-6596/1130/1/012018

Agrawal, A. P., Ali, S., Tripathi, N. M., Kumar, R., & Srivastava, P. (2024). Optimizing structural steel spur gear design for reduced stress concentrations in Industry 4.0 using finite element analysis. *Recent Patents on Mechanical Engineering*, *17*(1), e050724231667. https://doi.org/10.2174/0122127976309163240620080241

Ahmad, F., Choi, H. S., & Park, M. K. (2015). A review: natural fiber composites selection in view of mechanical, light weight, and economic properties. *Macromolecular Materials and Engineering*, *300*(1), 10–24. https://doi.org/10.1002/mame.201400089

Arul, M., Sasikumar, K. S. K., Sambathkumar, M., Gukendran, R., & Saravanan, N. (2020). Mechanical and fracture study of hybrid natural fiber reinforced composite–Coir and sugarcane leaf sheath. *Materials Today: Proceedings*, *33*, 2795–2797. https://doi.org/10.1016/j.matpr.2020.02.677

Asim, M., Abdan, K., Jawaid, M., Nasir, M., Dashtizadeh, Z., Ishak, M. R., & Hoque, M. E. (2015). A review on pineapple leaves fibre and its composites. *International Journal of Polymer Science*, *2015*(1), 950567. https://doi.org/10.1155/2015/950567

Asim, M., Saba, N., Jawaid, M., & Nasir, M. (2018). Potential of natural fiber/biomass filler-reinforced polymer composites in aerospace applications. In *Sustainable composites for Aerospace Applications* (pp. 253–268). Woodhead Publishing. https://doi.org/10.1016/B978-0-08-102131-6.00012-8

Atal, S., Dwivedi, D., & Tripathi, N. M. (2023). Corrosion under insulation (CUI) in oil and gas industries. In *Functional Materials for the Oil and Gas Industry* (pp. 127–138). CRC Press. https://doi.org/10.1201/9781003242550-8

Bajpai, P. K., Singh, I., & Madaan, J. (2013). Tribological behavior of natural fiber reinforced PLA composites. *Wear*, *297*(1–2), 829–840. https://doi.org/10.1016/j.wear.2012.10.019

Bakar, N., Chin, S. C., Siregar, J. P., & Ngien, S. K. (2020). A review on physical, mechanical, thermal properties and chemical composition of plant fibers. In *IOP Conference Series: Materials Science and Engineering* (Vol. 736, No. 5, p. 052017). IOP Publishing. https://doi.org/10.1088/1757-899X/736/5/052017

Bazan, P., Nosal, P., Kozub, B., & Kuciel, S. (2020). Biobased polyethylene hybrid composites with natural fiber: mechanical, thermal properties, and micromechanics. *Materials*, *13*(13), 2967. https://doi.org/10.3390/ma13132967

Belingardi, G., Beyene, A. T., Koricho, E. G., & Martorana, B. (2015). Alternative lightweight materials and component manufacturing technologies for vehicle frontal bumper beam. *Composite Structures*, *120*, 483–495. https://doi.org/10.1016/j.compstruct.2014.10.007

Benítez-Guerrero, M., Pérez-Maqueda, L. A., Artiaga, R., Sánchez-Jiménez, P. E., & Pascual-Cosp, J. (2017). Structural and chemical characteristics of sisal fiber and its components: effect of washing and grinding. *Journal of Natural Fibers*, *14*(1), 26–39. https://doi.org/10.1080/15440478.2015.1137529

Bhadouria, A. S., Kumar, A., Raj, D., Verma, A., Singh, S., Tripathi, P., Kumar, Y., Sinha, A. S. K., Tripathi, N. M., Yeneneh, A. M., & Dwivedi, D. (2022). Corrosion mitigation in oil reservoirs during CO injection. In *Nanotechnology for CO2 Utilization in Oilfield Applications* (pp. 127–145). Elsevier. https://doi.org/10.1016/B978-0-323-90540-4.00014-4

Bhatnagar, A., & Sain, M. (2005). Processing of cellulose nanofiber-reinforced composites. *Journal of Reinforced Plastics and Composites*, *24*(12), 1259–1268. https://doi.org/10.1177/0731684405049864

Biswas, S., Ahsan, Q., Cenna, A., Hasan, M., & Hassan, A. (2013). Physical and mechanical properties of jute, bamboo and coir natural fiber. *Fibers and Polymers*, *14*, 1762–1767. https://doi.org/10.1007/s12221-013-1762-3

Blau, P. J. (2001). *Compositions, Functions, and Testing of Friction Brake Materials and their Additives* (No. ORNL/TM-2001/64). Oak Ridge, TN (United States): Oak Ridge National Lab. (ORNL). https://doi.org/10.2172/788356

Boegler, O., Kling, U., Empl, D., & Isikveren, A. T. (2015). *Potential of Sustainable Materials in Wing Structural Design* (pp. 16–18). Bonn, Germany: Deutsche Gesellschaft für Luft- und Raumfahrt-Lilienthal-Oberth eV.

Bold, J. (2007, October). Airbus composite training. *Proceedings of Visual Product Development (VPD) Conference*, Frankfurt, Germany.

Borawski, A. (2020). Conventional and unconventional materials used in the production of brake pads–review. *Science and Engineering of Composite Materials*, *27*(1), 374–396. https://doi.org/10.1515/secm-2020-0041

Chen, F., Sawada, D., Hummel, M., Sixta, H., & Budtova, T. (2020). Unidirectional all-cellulose composites from flax via controlled impregnation with ionic liquid. *Polymers*, *12*(5), 1010. https://doi.org/10.3390/polym12051010

Chokshi, S., Parmar, V., Gohil, P., & Chaudhary, V. (2022). Chemical composition and mechanical properties of natural fibers. *Journal of Natural Fibers*, *19*(10), 3942–3953. https://doi.org/10.1080/15440478.2020.1848738

Correia, N. C., Robitaille, F., Long, A. C., Rudd, C. D., Šimáček, P., & Advani, S. G. (2005). Analysis of the vacuum infusion moulding process: I. Analytical formulation. *Composites Part A: Applied Science and Manufacturing*, *36*(12), 1645–1656. https://doi.org/10.1016/j.compositesa.2005.03.019

Das, S. C., Paul, D., Grammatikos, S. A., Siddiquee, M. A., Papatzani, S., Koralli, P., . . . & Petousis, M. (2021). Effect of stacking sequence on the performance of hybrid natural/synthetic fiber reinforced polymer composite laminates. *Composite Structures*, *276*, 114525. https://doi.org/10.1016/j.compstruct.2021.114525

Daud, Z., Mohd Hatta, M. Z., Mohd Kassim, A. S., Awang, H., & Mohd Aripin, A. (2014). Exploring of agro waste (pineapple leaf, corn stalk, and napier grass) by chemical composition and morphological study. *BioResources*, *9*(1).

Davoodi, M. M., Sapuan, S. M., Ahmad, D., Ali, A., Khalina, A., & Jonoobi, M. (2010). Mechanical properties of hybrid kenaf/glass reinforced epoxy composite for passenger car bumper beam. *Materials & Design*, *31*(10), 4927–4932. https://doi.org/10.1016/j.matdes.2010.05.021

Dev, B., Rahman, A., Alam, R., Repon, R., & Nawab, Y. (2023). Mapping the progress in natural fiber reinforced composites: preparation, mechanical properties, and applications. *Polymer Composites*, *44*(7), 3748–3788. https://doi.org/10.1002/pc.27376

Dittenber, D. B., & GangaRao, H. V. (2012). Critical review of recent publications on use of natural composites in infrastructure. *Composites Part A: Applied Science and Manufacturing*, *43*(8), 1419–1429. https://doi.org/10.1016/j.compositesa.2011.11.019

Elkington, M., Bloom, D., Ward, C., Chatzimichali, A., & Potter, K. (2015). Hand layup: understanding the manual process. *Advanced Manufacturing: Polymer & Composites Science*, *1*(3), 138–151. https://doi.org/10.1080/20550340.2015.1114801

Eloy, F. D. S., Costa, R. R. C., De Medeiros, R., Ribeiro, M. L., & Tita, V. (2015). Comparison between mechanical properties of bio and synthetic composites for use in aircraft interior structures. *Meeting on Aeronautical Composite Materials and Structures*. São Carlos: University of São Paulo.

Elseify, L. A., Midani, M., El-Badawy, A., & Jawaid, M. (2021). Natural fibers in the automotive industry. In *Manufacturing Automotive Components from Sustainable Natural Fiber Composites* (pp. 1–10). Cham: Springer International Publishing. https://doi.org/10.1007/978-3-030-83025-0_1

Feng, Y., Hao, H., Lu, H., Chow, C. L., & Lau, D. (2024). Exploring the development and applications of sustainable natural fiber composites: a review from a nanoscale perspective. *Composites Part B: Engineering*, 111369. https://doi.org/10.1016/j.compositesb.2024.111369

Fu, H., Xu, H., Liu, Y., Yang, Z., Kormakov, S., Wu, D., & Sun, J. (2020). Overview of injection molding technology for processing polymers and their composites. *ES Materials & Manufacturing*, *8*(20), 3–23. https://doi.org/10.30919/esmm5f713

Gautier di Confiengo, G., & Faga, M. G. (2022). Ecological transition in the field of brake pad manufacturing: an overview of the potential green constituents. *Sustainability*, *14*(5), 2508. https://doi.org/10.3390/su14052508

Guo, A., Sun, Z., & Satyavolu, J. (2021). Experimental and finite element analysis on flexural behavior of mortar beams with chemically modified kenaf fibers. *Construction and Building Materials*, *292*, 123449. https://doi.org/10.1016/j.conbuildmat.2021.123449

Gustafsson, J., & Agrawal, H. (2017). *Investigation of Active Anti-roll Bars and Development of Control Algorithm.* Master's Thesis, Department of Applied Mechanics Chalmers University of Technology Gothenburg, Sweden.

Hallal, A., Elmarakbi, A., Shaito, A., & El-Hage, H. (2013). Overview of composite materials and their automotive applications. In *Advanced Composite Materials for Automotive Applications: Structural Integrity and Crashworthiness* (pp. 1–28). https://doi.org/10.1002/9781118535288.ch1

Hanan, F., Jawaid, M., & Tahir, P. M. (2020). Mechanical performance of oil palm/kenaf fiber-reinforced epoxy-based bilayer hybrid composites. *Journal of Natural Fibers.* https://doi.org/10.1080/15440478.2018.1477083

Hill, K., Swiecki, B., & Cregger, J. (2012). The bio-based materials automotive value chain. *Center for Automotive Research*, *112*, 1–92.

Hsiao, K. T., & Heider, D. (2012). Vacuum assisted resin transfer molding (VARTM) in polymer matrix composites. In *Manufacturing Techniques for Polymer Matrix Composites (PMCs)* (pp. 310–347). Woodhead Publishing. https://doi.org/10.1533/9780857096258.3.310

Iniya, M. P., & Nirmalkumar, K. (2021, February). A review on fiber reinforced concrete using sisal fiber. In *IOP Conference Series: Materials Science and Engineering* (Vol. 1055, No. 1, p. 012027). IOP Publishing. https://doi.org/10.1088/1757-899X/1055/1/012027

Islam, T., Chaion, M. H., Jalil, M. A., Rafi, A. S., Mushtari, F., Dhar, A. K., & Hossain, S. (2024). Advancements and challenges in natural fiber-reinforced hybrid composites: a comprehensive review. *SPE Polymers*. https://doi.org/10.1002/pls2.10145

Jaafar, J., Siregar, J. P., Tezara, C., Hamdan, M. H. M., & Rihayat, T. (2019). A review of important considerations in the compression molding process of short natural fiber composites. *The International Journal of Advanced Manufacturing Technology*, *105*, 3437–3450. https://doi.org/10.1007/s00170-019-04466-8

Jafar, M. R., Tripathi, N. M., Yadav, M., & Nasato, D. S. (2024). Additive manufacturing in the age of Industry 4.0 and beyond. In *Advances in Pre-and Post-Additive Manufacturing Processes* (pp. 213–230). CRC Press. https://doi.org/10.1201/9781003428862-11

Kamarudin, S. H., Mohd Basri, M. S., Rayung, M., Abu, F., Ahmad, S. B., Norizan, M. N., . . . & Abdullah, L. C. (2022). A review on natural fiber reinforced polymer composites (NFRPC) for sustainable industrial applications. *Polymers*, *14*(17), 3698. https://doi.org/10.3390/polym14173698

Kaur, K., Kaur, G., & Singh, A. (2023). Water chestnut starch: extraction, chemical composition, properties, modifications, and application concerns. *Sustainable Food Technology*, *1*(2), 228–262. https://doi.org/10.1039/D2FB00041E

Kerni, L., Singh, S., Patnaik, A., & Kumar, N. (2020). A review on natural fiber reinforced composites. *Materials Today: Proceedings*, *28*, 1616–1621. https://doi.org/10.1016/j.matpr.2020.04.851

Ketata, N., Ejday, M., Grohens, Y., Seantier, B., & Guermazi, N. (2024). Investigation of the hybridization effect on mechanical properties of natural fiber reinforced bio-sourced composites. *Journal of Composite Materials*, 00219983241255751. https://doi.org/10.1177/00219983241255751

Khalil, H. A., Bhat, A. H., & Yusra, A. I. (2012). Green composites from sustainable cellulose nanofibrils: a review. *Carbohydrate Polymers*, *87*(2), 963–979. https://doi.org/10.1016/j.carbpol.2011.08.078

Khan, F., Hossain, N., Mim, J. J., Rahman, S. M., Iqbal, M. J., Billah, M., & Chowdhury, M. A. (2024). Advances of composite materials in automobile applications–a review. *Journal of Engineering Research*. https://doi.org/10.1016/j.jer.2024.02.017

Kim, D. H., Kim, H. G., & Kim, H. S. (2015). Design optimization and manufacture of hybrid glass/carbon fiber reinforced composite bumper beam for automobile vehicle. *Composite Structures*, *131*, 742–752. https://doi.org/10.1016/j.compstruct.2015.06.028

Komuraiah, A., Kumar, N. S., & Prasad, B. D. (2014). Chemical composition of natural fibers and its influence on their mechanical properties. *Mechanics of Composite Materials*, *50*, 359–376. https://doi.org/10.1007/s11029-014-9422-2

Koronis, G., Silva, A., & Fontul, M. (2013). Green composites: a review of adequate materials for automotive applications. *Composites Part B: Engineering*, *44*(1), 120–127. https://doi.org/10.1016/j.compositesb.2012.07.004

Kostic, M., Pejic, B., & Skundric, P. (2008). Quality of chemically modified hemp fibers. *Bioresource Technology*, *99*(1), 94–99. https://doi.org/10.1016/j.biortech.2006.11.050

Kulkarni, A., Vora, R., & Ravi, K. (2017, November). Study design and analysis of automobile bumper for pedestrian safety. In *IOP Conference Series: Materials Science and Engineering* (Vol. 263, No. 6, p. 062060). IOP Publishing. https://doi.org/10.1088/1757-899X/263/6/062060

Liu, D., Song, J., Anderson, D. P., Chang, P. R., & Hua, Y. (2012). Bamboo fiber and its reinforced composites: structure and properties. *Cellulose*, *19*, 1449–1480. https://doi.org/10.1007/s10570-012-9741-1

Li, G., Liu, J., Jiang, G., & Liu, H. (2015). Numerical simulation of temperature field and thermal stress field in the new type of ladle with the nanometer adiabatic material. *Advances in Mechanical Engineering*, *7*(4), 1687814015575988. https://doi.org/10.1177/1687814015575988

Lu, Y., Weng, L., & Cao, X. (2006). Morphological, thermal and mechanical properties of ramie crystallites—reinforced plasticized starch biocomposites. *Carbohydrate Polymers*, *63*(2), 198–204. https://doi.org/10.1016/j.carbpol.2005.08.027

Li, Z., Li, Z., Ding, R., & Yu, C. (2016). Composition of ramie hemicelluloses and effect of polysaccharides on fiber properties. *Textile Research Journal*, *86*(5), 451–460. https://doi.org/10.1177/0040517515592811

Lunetto, V., Galati, M., Settineri, L., & Iuliano, L. (2023). Sustainability in the manufacturing of composite materials: a literature review and directions for future research. *Journal of Manufacturing Processes*, *85*, 858–874. https://doi.org/10.1016/j.jmapro.2022.12.020

Mahesh, A., Rudresh, B. M., & Reddappa, H. N. (2022). Potential of natural fibers in the modification of mechanical behavior of polypropylene hybrid composites. *Materials Today: Proceedings*, *54*, 131–136. https://doi.org/10.1016/j.matpr.2021.08.195

Malik, K., Ahmad, F., & Gunister, E. (2021). A review on the kenaf fiber reinforced thermoset composites. *Applied Composite Materials*, *28*(2), 491–528. https://doi.org/10.3151/jact.12.167

Malik, K., Ahmad, F., Gunister, E., Nakato, T., Mouri, E., Muhammad, M. B., & Ali, S. (2022). A review of flax fiber reinforced thermoset polymer composites: structure and mechanical performance. *Journal of Natural Fibers*, *19*(14), 9656–9680. https://doi.org/10.1080/15440478.2021.1990182

Mamun, A. A., Heim, H. P., Faruk, O., & Bledzki, A. K. (2015). The use of banana and abaca fibres as reinforcements in composites. In *Biofiber Reinforcements in Composite Materials* (pp. 236–272). Woodhead Publishing. https://doi.org/10.1533/9781782421276.2.236

Mansor, M. R., Nurfaizey, A. H., Tamaldin, N., & Nordin, M. N. A. (2019). Natural fiber polymer composites: utilization in aerospace engineering. In *Biomass, Biopolymer-Based Materials, and Bioenergy* (pp. 203–224). Woodhead Publishing. https://doi.org/10.1016/B978-0-08-102426-3.00011-4

Marques, A. T. (2011). Fibrous materials reinforced composites production techniques. In *Fibrous and Composite Materials for Civil Engineering Applications* (pp. 191–215). Woodhead Publishing. https://doi.org/10.1533/9780857095583.3.191

Mastura, M. T., Sapuan, S. M., Mansor, M. R., & Nuraini, A. A. (2017). Environmentally conscious hybrid bio-composite material selection for automotive anti-roll bar. *The International Journal of Advanced Manufacturing Technology*, *89*, 2203–2219. https://doi.org/10.1007/s00170-016-9217-9

Mouritz, A. P. (2012). Materials selection for aerospace. In *Introduction to Aerospace Materials* (pp. 569–600). Woodhead Publishing.

Müssig, J., Schmehl, M., Von Buttlar, H. B., Schönfeld, U., & Arndt, K. (2006). Exterior components based on renewable resources produced with SMC technology—considering a bus component as example. *Industrial Crops and Products*, *24*(2), 132–145. https://doi.org/10.1016/j.indcrop.2006.03.006

Neveu, A., Lumay, G., Pillitteri, S., Monsuur, F., Pauly, T., Ribeyre, Q., Francqui, F., Vandewalle, N., & Tripathi, N. M. (2020a). Physical characterization of blends containing mesoporous particles with a focus on electrostatic properties. In *2020 AIChE Virtual Spring Meeting and 16th Global Congress on Process Safety (2020 AIChE Virtual Spring Meeting and 16th Global Congress on Process Safety)*. AIChE. ISBN: 978-0-8169-1113-4.

Neveu, A., Tripathi, N. M., Francqui, F., Lumay, G., & Rigo, O. (2020b). Characterization of powders spreadability in the framework of additive manufacturing processes. *The Proceedings of 2020 Virtual AIChE Annual Meeting, (Particle Technology Forum)*, San Francisco, CA, USA. ISBN: 978-0-8169-1114-1.

Neveu, A., Tripathi, N. M., Rigo, O., Francqui, F., & Lumay, G. (2020c). Experimental investigation of spreadability of metal powders in recoating process. In *Proceedings—Euro PM2020 Congress and Exhibition (Proceedings—Euro PM2020 Congress and Exhibition)*. European Powder Metallurgy Association (EPMA). ISBN: 978-1-8990-7251-4.

Njuguna, J. (Ed.). (2016). *Lightweight Composite Structures in Transport: Design, Manufacturing, Analysis and Performance*. Woodhead Publishing.

Olorunnishola, A. A. G., & Adubi, E. G. (2018). A comparative analysis of a blend of natural jute and glass fibers with synthetic glass fibers composites as car bumper materials. *IOSR Journal of Mechanical and Civil Engineering*, *15*, 67–71. https://doi.org/10.9790/1684-1503016771

Patel, R. V., Yadav, A., & Winczek, J. (2023). Physical, mechanical, and thermal properties of natural fiber-reinforced epoxy composites for construction and automotive applications. *Applied Sciences*, *13*(8), 5126. https://doi.org/10.3390/app13085126

Praveena, B. A., Kumar, S. V., Manjunath, H. N., Sachin, B., Yadav, S. P. S., Lochan, B. R., . . . & Reddy, J. S. (2021). Investigation of moisture absorption and mechanical properties of natural fibre reinforced polymer hybrid composite. *Materials Today: Proceedings*, *45*, 8219–8223. https://doi.org/10.1016/j.matpr.2021.04.254

Rajak, D. K., Pagar, D. D., Behera, A., & Menezes, P. L. (2022). Role of composite materials in automotive sector: potential applications. *Advances in Engine Tribology*, 193–217. https://doi.org/10.1007/978-981-16-8337-4_10

Rajak, D. K., Pagar, D. D., Menezes, P. L., & Linul, E. (2019). Fiber-reinforced polymer composites: manufacturing, properties, and applications. *Polymers*, *11*(10), 1667. https://doi.org/10.3390/polym11101667

Rajakumar, I. P. T., Raguraman, D., Isaac, J. S., Suthan, R., Bhattacharya, S., Seikh, A. H., . . . & Raghavan, I. K. (2022). Mechanical properties of polymer composites reinforced with alkaline-treated natural fibre. *Advances in Polymer Technology*, *2022*(1), 1458547. https://doi.org/10.1155/2022/1458547

Ramesh, M. (2018). Hemp, jute, banana, kenaf, ramie, sisal fibers. In *Handbook of Properties of Textile and Technical Fibres* (pp. 301–325). Woodhead Publishing. https://doi.org/10.1016/B978-0-08-101272-7.00009-2

Ranjan, R., Singh, D., Pandey, V., Sharma, A., Sinha, A. S. K., & Tripathi, N. M. (2024a). A comprehensive review on the machining process: unconventional as an alternative to conventional machining. In *Advances in Pre-and Post-Additive Manufacturing Processes* (pp. 163–195). CRC Press. https://doi.org/10.1201/9781003428862-9

Ranjan, R., & Tripathi, N. M. (2024). Optimizing solar power: engineering efficiency in CdTe/MAsnI3 double absorber solar cell. In *2nd International Conference on Emerging Materials, Smart Manufacturing & Computational Intelligence (ICEMSMCI-2024)*. Punjab: Chitkara University.

Ranjan, R., Verma, S., & Tripathi, N. M. (2024b). The organic HTL in highly efficient germanium-alloyed CsSnI3 perovskite solar cell: a modelling and computational analysis. *Physica Scripta*, *99*(6), 066010. https://doi.org/10.1088/1402-4896/ad4e15

Ravindran, S., Sozhamannan, G. G., Saravanan, L., & Venkatachalapathy, V. S. K. (2021). Study on mechanical behaviour of natural fiber reinforced vinylester hybrid composites. *Materials Today: Proceedings*, *45*, 4526–4530. https://doi.org/10.1016/j.matpr.2020.12.1077

Rwawiire, S., Tomkova, B., Militky, J., Jabbar, A., & Kale, B. M. (2015). Development of a biocomposite based on green epoxy polymer and natural cellulose fabric (bark cloth) for automotive instrument panel applications. *Composites Part B: Engineering*, *81*, 149–157. https://doi.org/10.1016/j.compositesb.2015.06.021

Satishkumar, P., Haldar, B., Tripathi, N. M., Sharma, A., Desai, D. J., Sharma, V. K., Seenivasan., S., & Babbar, A. (2024a). Effect of interlayer temperatures and heat inputs on porosity and hydrogen solubility in wire arc additive manufactured AA2618 aluminium: a comparative study between pulsed-MIG and CMT methods. In *Advances in Pre-and Post-Additive Manufacturing Processes* (pp. 137–162). CRC Press. https://doi.org/10.1201/9781003428862-8

Satishkumar, P., Kumar, R., Tripathi, N. M., Sharma, A., Desai, D. J., Verma, A., Natarajan., N., & Babbar, A. (2024b). Integrating titanium coating with 3D-Printed GFRP panels: an innovative approach to harnessing composite strengths in engineering applications. In *Advances in Pre-and Post-Additive Manufacturing Processes* (pp. 18–38). CRC Press. https://doi.org/10.1201/9781003428862-2

Sekar, S., Suresh Kumar, S., Vigneshwaran, S., & Velmurugan, G. (2022). Evaluation of mechanical and water absorption behavior of natural fiber-reinforced hybrid biocomposites. *Journal of Natural Fibers*, *19*(5), 1772–1782. https://doi.org/10.1080/15440478.2020.1788487

Shahzad, A. (2012). Hemp fiber and its composites–a review. *Journal of Composite Materials*, *46*(8), 973–986. https://doi.org/10.1177/0021998311413623

Shakir, M. H., & Singh, A. K. (2023). An investigation on the mechanical properties, surface treatments and applications of abaca fibre-reinforced composites. *Advances in Materials and Processing Technologies*, 1–26. https://doi.org/10.1080/2374068X.2023.2247701

Shakir, M. H., & Singh, A. K. (2024). Mechanical properties, surface treatments, and applications of jute fiber reinforced composites—a review. *Polymers for Advanced Technologies*, *35*(1), e6268. https://doi.org/10.1002/pat.6268

Sharma, A., Babbar, A., Singh, K., Singh, A. K., Tripathi, N. M., & Desai, D. J. K. (2024). Effect of minimum quantity lubrication (MQL) method on machining characteristics for ductile substrates: a future direction. In *Modern Hybrid Machining and Super Finishing Processes* (pp. 35–59). CRC Press. https://doi.org/10.1201/9781003327905-3

Singh, A. K., Shakir, M. H., Yadav, S., Yadav, M., Tripathi, N. M., & Fidan, I. (2024). Investigations on the mechanical characterization of additively manufactured polymer composite materials fabricated via MEX. In *Advances in Pre-and Post-Additive Manufacturing Processes* (pp. 64–77). CRC Press. https://doi.org/10.1201/9781003428862-4

Singh, G., Sidhu, S. S., Bains, P. S., & Bhui, A. S. (2019). Improving microhardness and wear resistance of 316L by TiO2 powder mixed electro-discharge treatment. *Materials Research Express*, *6*(8), 086501. https://doi.org/10.1088/2053-1591/ab1bab

Singh, S. P., Dutt, A., & Hirwani, C. K. (2023). Mechanical, modal and harmonic behavior analysis of jute and hemp fiber reinforced polymer composite. *Journal of Natural Fibers*, *20*(1), 2140328. https://doi.org/10.1080/15440478.2022.2140328

Sreenivasan, V. S., Somasundaram, S., Ravindran, D., Manikandan, V., & Narayanasamy, R. (2011). Microstructural, physico-chemical and mechanical characterisation of

Sansevieria cylindrica fibres–an exploratory investigation. *Materials & Design*, *32*(1), 453–461. https://doi.org/10.1016/j.matdes.2010.06.004

Suddell, B. C., & Evans, W. J. (2005). Natural fiber composites in automotive applications. In *Natural Fibers, Biopolymers, and Biocomposites* (pp. 253–282). CRC Press. https://doi.org/10.1201/9780203508206

Syduzzaman, M., Al Faruque, M. A., Bilisik, K., & Naebe, M. (2020). Plant-based natural fibre reinforced composites: a review on fabrication, properties and applications. *Coatings*, *10*(10), 973. https://doi.org/10.3390/coatings10100973

Tripathi, N. M., & Sharma, A. (Eds.). (2024). *Advances in Pre-and Post-Additive Manufacturing Processes: Innovations andApplications*. CRC Press. https://doi.org/10.1201/9781003428862

Vardai, R., Lummerstorfer, T., Pretschuh, C., Jerabek, M., Gahleitner, M., Bartos, A., . . . & Pukanszky, B. (2021). Improvement of the impact resistance of natural fiber–reinforced polypropylene composites through hybridization. *Polymers for Advanced Technologies*, *32*(6), 2499–2507. https://doi.org/10.1002/pat.5280

Venkateshwaran, N., & Elayaperumal, A. (2010). Banana fiber reinforced polymer composites-a review. *Journal of Reinforced Plastics and Composites*, *29*(15), 2387–2396. https://doi.org/10.1177/0731684409360578

Venkateshwaran, N., ElayaPerumal, A., & Jagatheeshwaran, M. S. (2011). Effect of fiber length and fiber content on mechanical properties of banana fiber/epoxy composite. *Journal of Reinforced Plastics and Composites*, *30*(19), 1621–1627. https://doi.org/10.1177/0731684411426810

Walsh, J., Kim, H. I., & Suhr, J. (2017). Low velocity impact resistance and energy absorption of environmentally friendly expanded cork core-carbon fiber sandwich composites. *Composites Part A: Applied Science and Manufacturing*, *101*, 290–296. https://doi.org/10.1016/j.compositesa.2017.05.026

Wang, A., Yin, P., Liu, X., & Xian, G. (2023). Hydrothermal effect on ramie-fiber-reinforced polymer composite plates: water uptake and mechanical properties. *Polymers*, *15*(14), 3066. https://doi.org/10.3390/polym15143066

Xiong, W. (2018). Bagasse composites: a review of material preparation, attributes, and affecting factors. *Journal of Thermoplastic Composite Materials*, *31*(8), 1112–1146. https://doi.org/10.1177/0892705717734596

Yuan, J., Chen, Q., Fang, C., Zhang, S., Liu, X., & Fei, B. (2021). Effect of chemical composition of bamboo fibers on water sorption. *Cellulose*, *28*(11), 7273–7282. https://doi.org/10.1007/s10570-021-03988-3

Yun, R., Filip, P., & Lu, Y. (2010). Performance and evaluation of eco-friendly brake friction materials. *Tribology International*, *43*(11). https://doi.org/10.1016/j.triboint.2010.05.001

Zelalem Biresaw, A., & Sirahbizu Yigezu, B. (2022). Investigation on the mechanical properties of flax/false banana hybrid fiber-reinforced polymer composite. *Advances in Materials Science and Engineering*, *2022*(1), 5696758. https://doi.org/10.1155/2022/5696758

Zhang, J., Chaisombat, K., He, S., & Wang, C. H. (2012). Glass/carbon fibre hybrid composite laminates for structural applications in automotive vehicles. In *Sustainable Automotive Technologies 2012: Proceedings of the 4th International 4Conference* (pp. 69–74). Springer Berlin Heidelberg. https://doi.org/10.1007/978-3-642-24145-1_10

# 8 Bio-Based Materials in Biomedical Applications

*Praneta Yadav, Devendra Singh, Manju Kumari, Vivek Kumar Dhiman, and Hariom Sharma*

## 8.1 INTRODUCTION

Renewable biological resources, such as plants, animals, and microbes, are the source of bio-based products. They are becoming increasingly acknowledged as sustainable substitutes for traditional fossil-based materials, resolving issues with resource depletion and environmental effects. Materials derived from living things or their byproducts are purposefully used to create bio-based materials. These materials can be made using various techniques, such as physical, chemical, and biological ones. Bio-based materials are becoming more important in biomedical applications because of their unique qualities, sustainability, and capacity to improve patient outcomes. These materials offer several benefits over conventional synthetic equivalents since they are derived from renewable biological resources. Bio-based materials highly compatible with human tissues, like collagen, chitosan, and alginate, lower the possibility of adverse reactions. This characteristic is essential for applications where contact with biological systems is unavoidable, such as medication delivery systems and implants. After their purpose is served, the body can safely absorb or digest many bio-based materials, reducing the need for surgical removal. This feature benefits short-term uses like medication delivery systems and wound dressings. By mimicking the extracellular matrix (ECM), these materials can encourage cellular adhesion, growth, and differentiation. Regenerative medicine and tissue engineering depend on this mimicking ability. These days, innovative biomaterials exhibit improved characteristics, including cell adhesion and proliferation, and are made to resemble the characteristics and behaviors of tissue. Thus, a variety of polymeric materials, including silicone, polyvinylidene fluoride (PVDF), poly (methyl methacrylate) (PMMA), and others, can be utilized as coatings or as simple medical devices (Teo et al., 2020). Because of their adjustable characteristics and behaviors, an increasing number of biocompatible PUs have also been found to be useful in the biomedical field in recent decades. Grades for medical-use PU-based polymers are marketed under several brands, including DSM's Carbosil® and Bionate™ (Holland) and Lubrizol's Carbothane™, Pellethane™, and Tecoflex™ (US). Since they have earned the confidence of the medical and surgical sectors, these various fossil-based PUs represent this thriving and prosperous business. They are primarily thermoplastic based, occasionally combined with other polymers, and can be molded and altered to suit the intended use (Ren et al., 2020; Wang et al., 2021). Materials are increasingly being

DOI: 10.1201/9781003566205-8

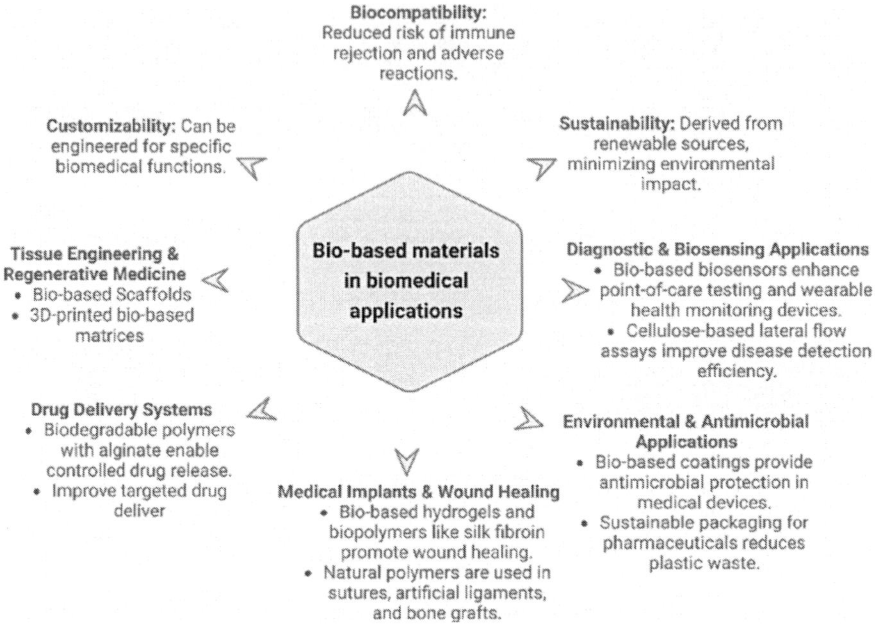


**FIGURE 8.1** An overview of bio-based materials in biomedical applications, highlighting their key advantages—biocompatibility, sustainability, and customizability—as well as their diverse applications.

used to address the growing demand for innovative, long-lasting, or sustainable solutions, which are connected to several variables, including the global trend toward bio-based and green solutions (Arntzen et al., 2021). Competitive properties have mainly been attained using various novel bio-based macromolecular designs in response to these pressing demands and trends. However, with just around 0.1% of the 20 million tons of PU produced annually (Akindoyo et al., 2020), renewable PU remains a niche market today (Ramakrishna et al., 2020). Of these 20 million tons, less than 3% are manufactured for biomedical devices and applications. Like all materials in this field, PUs must go through a lengthy procedure and meet stringent requirements regarding cytotoxicity, acute and sub-chronic toxicity, hemocompatibility, and carcinogenicity to be approved and validated for use in biomedical applications (Figure 8.1).

## 8.2 TYPES OF BIO-BASED MATERIAL

Based on their composition, characteristics, and uses, bio-based materials—made from renewable biological resources—can be divided into several groups.

### 8.2.1 Polymers

Biopolymers are environmentally benign and biodegradable. Agricultural resources, byproducts, chemical processing, and microbiological activity are the origins of biopolymers (Satyanarayana et al., 2020). Another way to make them is to combine two

**TABLE 8.1**
**Common Applications and Features of Bio-based Materials.**

| Material | Source | Key properties | Common applications | Features |
|---|---|---|---|---|
| Polylactic acid (PLA) | Corn starch, sugarcane | Biodegradable, biocompatible, thermoplastic | Implants, sutures, tissue engineering scaffolds | Widely used in 3D printing; may be brittle in some formulations |
| Polyhydroxyalkanoates (PHAs) | Bacterial fermentation of sugars | Biocompatible, biodegradable, tunable mechanical properties | Drug delivery, tissue scaffolds, sutures | PHB is a common type; production costs can be high |
| Chitosan | Crustacean shells, fungi | Antimicrobial, hemostatic, biocompatible | Wound dressings, drug delivery systems, tissue regeneration | Excellent for wound care; solubility can vary with pH |
| Collagen | Animal tissues (bovine, porcine, marine) | Mimics natural extracellular matrix, low antigenicity | Tissue engineering scaffolds, wound dressings, cosmetic applications | High bioactivity; potential for immune reactions if not processed properly |
| Gelatin | Derived from collagen | Biocompatible, biodegradable, thermoreversible gelation | Drug delivery, wound dressings, scaffolds | Easier to process than collagen; lower mechanical strength |
| Cellulose/ nanocellulose | Plants, bacteria | High mechanical strength, hydrophilic, easily modifiable | Wound dressings, scaffolds, drug delivery | Nanocellulose enhances cell adhesion and mechanical support |
| Silk fibroin | Silkworm silk | Biocompatible, strong, controllable degradation rate | Tissue engineering, sutures, drug delivery | Superior mechanical properties; processing can be complex |
| Alginate | Brown algae (seaweed) | Biocompatible, mild gelation via ionic crosslinking | Wound dressings, drug delivery, cell encapsulation | Suitable for 3D printing and injectable gels; limited strength |
| Fibrin | Blood plasma | Natural clot formation, supports cell adhesion, biodegradable | Tissue adhesives, wound healing, scaffolds | Simulates the natural healing mechanisms, although the rapid degradation may present limitations. |
| Hyaluronic acid | Animal tissue, bacterial fermentation | Hydrophilic, biocompatible, lubricating properties | Dermal fillers, joint lubrication, wound healing | Naturally present in the human body; widely used in cosmetics |
| Keratin | Wool, human hair, feathers | Biocompatible, supports cell adhesion, biodegradable | Tissue repair, healing of wounds, and drug delivery | Potential for hair and skin regeneration applications |

*(Continued)*

**TABLE 8.1** *(Continued)*
**Common Applications and Features of Bio-based Materials.**

| Material | Source | Key properties | Common applications | Features |
|---|---|---|---|---|
| Pectin | Fruits (apples, citrus) | Biocompatible, gel-forming, hydrophilic | Dressings for wounds, drug delivery, and tissue frameworks | Useful for hydrogels and controlled drug release |
| Lignin | Plant biomass | Antioxidant, biocompatible, high mechanical strength | Bone scaffolds, drug carriers, wound healing | Underexplored in biomedical applications, but promising |
| Starch-based polymers | Corn, potato, cassava | Biodegradable, non-toxic, hydrophilic | Drug delivery, wound dressings, tissue scaffolds | Poor mechanical strength, often blended with other biopolymers |
| Soy PROTEIN ISOLATE | Soybeans | Biodegradable, supports cell growth, water soluble | Tissue engineering, wound healing, drug delivery | Requires crosslinking for stability |
| Marine spong-es-derived collagen | Marine sponges | High bioactivity, supports tissue regeneration | Bone and cartilage tissue engineering, wound healing | Alternative to mammalian collagen with reduced immune response |
| Decellularized extracellular matrix (dECM) | Mammalian tissues | Biomimetic, supports cell growth, biodegradable | Organ scaffolds, regenerative medicine, tissue engineering | Preserves native tissue structure; immunogenicity varies |
| Mycelium-based biopolymers | Fungal biomass | Biodegradable, customizable mechanical properties | Wound dressings, scaffolds, drug delivery | Emerging in biomedical research; sustainable alternative |
| Shellac | Lac insects | Hydrophobic, biocompatible, bioresorbable | Drug coatings, wound dressings | Used as a natural bioadhesive and protective layer |

biopolymers. The biodegradability of a polymer is determined by its structure, place of origin, and degradation circumstances (Mohanty et al., 2022). During breakdown, the polymer is attacked by microbes, which break it down into smaller molecules and release $CO_2$. Natural fibers enhance the mechanical performance of biopolymers, which otherwise have poor mechanical properties (Van de Velde and Kiekens, 2022). Conventional polymers, including epoxy, polyester, polypropylene, and polyethylene, have been around for a while and have undergone several phases of research and commercialization (Eagan et al., 2021; Alavudeen et al., 2022; Carrillo-Escalante et al., 2021). Creating these polymers has taken a great deal of time and money, leading to their better mechanical performance. However, recycling issues and underlying environmental concerns about their use have shifted the focus to biopolymers (Isikgor and Becer, 2022). Over three-fourths of the 7 billion metric tons of plastic produced wind up in landfills [Zwawi, 2021].

#### 8.2.1.1 Polylactic Acid (PLA)

PLA is appropriately referred to as the "polymer of the 21st century" and is one of the most promising polymers utilized in these applications (Maharana et al., 2019). It is the only one that is simultaneously biocompatible, biodegradable, and bio-based and is made on a larger scale (Pretula et al., 2020). Lactic acid primarily comes from various plant and animal sources, including corn, fish waste, cellulose, starch, and kitchen scraps (Jem and Tan, 2020), and is the source of PLA, an aliphatic bio-based polyester. The first PLA with a low molecular weight was created by Carothers in 1932, while a product with a higher molecular weight was patented by DuPont in 1954. PLA is an excellent substitute for conventional petroleum-based polymers due to its high compostability, biocompatibility, biodegradability, and superior mechanical strength. It is also anticipated to be extensively utilized in various industries, including the food packaging sector, the automobile sector, and the medical sector. With increasing applications and a need for increased production, it appears to be the most environmentally friendly polymer available today. PLA is a traditional bioplastic derived from biomass and can be produced in three steps: fermentation, separation, and polymerization. PLA is a biodegradable polyester polymer using monomers made from sustainable sugarcane, cornstarch, and other plentiful natural resources (Hamad et al., 2020). NatureWorks Co. (Blair, NE, USA) has adopted a large-scale, intensive method from recombinant corn that reduces production costs (DeStefano et al., 2020; Luo et al., 2020). Temperature, consternation, catalyst type, and residence time can all affect the molecular weights of the polymer produced by ring-opening polymerization. It would also be possible to regulate the final polymer's L- and D-lactic acid unit ratio and sequence (Zaaba and Jaafar, 2020). PLA and its degradation products, mainly $H_2O$ and $CO_2$, are non-toxic and non-carcinogenic to humans, making them perfect for biomedical uses, including drug delivery systems, clips, and sutures. Additionally, PLA's superior thermal processability compared to other biomaterials allows it to be manufactured through various techniques such as fiber spinning, blow molding, film casting, and melt extrusion (Ren, 2019).

#### 8.2.1.2 Polyhydroxyalkanoates (PHAs)

Polyesters known as PHAs have a distinctive link of esters that accumulate as energy and carbon stores, as well as a limited supply of nitrogen, helping to provide energy. Monomers of 3-hydroxy fatty acids make up the structure of PHAs. They are thought to be naturally biocompatible and biodegradable. PHAs are classified based on the number of carbon atoms in their monomer units. Short-chain-length PHAs (scl-PHAs) consist of three to five carbon atoms, medium-chain-length PHAs (mcl-PHAs) range from six to fourteen carbon atoms, and long-chain-length PHAs (lcl-PHAs) contain fifteen or more carbon atoms. PHAs have properties with thermoplastic substances because of their molecular weight. PHAs and traditional thermoplastics are comparable in flexibility, stability, and durability. Mechanical and physical characteristics of scl-PHAs are similar to those of traditional polymers. They are often hard, rigid structures with a crystalline character. mcl-PHAs are amorphous thermoplastics with variable levels of crystallinity that possess elastomeric and adhesive qualities (Licciardello et al., 2019). The composition of the monomers has a significant impact on the chemical and physical characteristics of PHAs. Water does not dissolve PHAs. Hydrophobicity, melting point, glass transition temperature, and degree of crystallinity are typically used to describe polymers. From hard crystalline to elastic, the PHA family has a broad range of mechanical characteristics (Cavaliere et al., 2021). One of PHA's most unique qualities is its biodegradability. It comes naturally from microorganisms and, if buried in soil, can decompose in seven months into $CO_2$ and water (Paduani, 2020). Various natural bacteria release enzymes that break down PHA in the environment, converting it into nutrients that dissolve in water and then use the produce as food. Equivocal esterases and their subclasses work to break down PHA. In the rat gastrointestinal tract, close to the PHA insert, we observed the activity of the enzyme lipases, a subclass of esterases, and thought that lipases were complicit in the PHA chemical reactions occurring in the body. The degradation of PHA embedded in the target host organism's specialized cells increases the possibility that this event may be paired with the release of bioactive substances like antifungal, antibacterial, or anticancer medications. For example, if a drug is saturated in a cell, it will eventually deteriorate and release the material, acting as a self-activating operative. PHB is a common material for implants. PHB-made implants are significant because the body does not see them as foreign objects, which prevents rejection or reaction from the immune system (Liechty et al., 2019). PHB's remarkable qualities make it the perfect material for antibacterial chemical delivery and nano-entrapment.

### 8.2.2 Proteins

Protein-based bio-based materials are becoming increasingly popular as environmentally friendly substitutes for various applications, especially in the textile, biomedical, and packaging industries. These substances take advantage of the special qualities of proteins, which are polymers composed of amino acids.

#### 8.2.2.1 Collagen

Collagen is a versatile substance used in pharmaceuticals, dentistry, and medicine. It can be manufactured into gels or solids and has served as a suture material for centuries, competing with catgut in surgery. Its applications include anti-thrombogenic surfaces, bone-filling materials, wound dressings, and as a drug delivery vehicle. Recent research focuses on collagen-based systems like hydrogels, injectable microspheres, and membranes for treating oral wounds and promoting tissue regeneration. The collagen molecule consists of three polypeptide chains with a distinctive twist and is characterized by high glycine content. Key amino acids, including proline, influence its structure, while hydrogen bonds stabilize the helix. Abnormalities in collagen can lead to disorders such as Ehlers–Danlos syndrome and osteoporosis. Collagen serves as an effective biomaterial for delivering proteins, DNA, and medications, supporting advancements in tissue engineering and drug delivery. Its potential for biomedical applications is promising for future developments.

#### 8.2.2.2 Gelatine

On the other hand, gelatine is a good barrier substance that shields against moisture, oxygen, and other impurities, making it a good option for food packaging (Flórez et al., 2022). Additionally, they have shown promise in various medical applications in the healthcare industry, such as tissue engineering, medication delivery systems, and wound dressings. In tissue engineering, gelatin has been used as a support framework to encourage cell division (Bello et al., 2020). Gelatin helps reduce damage to non-target cells and prevents drug degradation. Its properties benefit cancer treatment, drug delivery, and diagnostics. Hydrolytic breakdown of collagen produces gelatin, whose unique amino acids provide therapeutic benefits. As collagen breaks down in tissues, it forms a gel that cracks when dried (Alipal et al., 2021). Either acid or alkaline hydrolysis can produce gelatin once collagen has been isolated (Bello et al., 2020). Collagen's unique triple-helix structure provides stability and tensile strength. Its isoelectric point typically ranges from pH 5–6, depending on type and source. Collagen is used in pharmaceuticals, cosmetics, and culinary arts. Its three-dimensional structure is supported by hydrogen bonding and covalent cross-links, enhancing mechanical strength. A sizable worldwide business has emerged due to the rising demand for collagen and collagen-derived products in nutraceutical, cosmetic, and medicinal applications (Vieira et al., 2023). Gelatine comes in tablet, granule, or powder form and often needs water to dissolve. Its low toxicity, biodegradability, and biocompatibility promote cell adhesion, differentiation, and proliferation. The body breaks it down via natural metalloproteinases and enzymes without causing an immune response. Due to these qualities, gelatine is used in food, cosmetics, pharmaceuticals, and photography. It acts as a gelling, texturizing, stabilizing, and emulsifying agent in dairy, bread, beverages, and confectionery. Gelatin, high in protein, can replace fats and carbohydrates in nutritious foods. Collagen, the main protein in humans and animals, supplies protein for gelatin, primarily found in the skin, tendons, ligaments, and bones (Alipal et al., 2021). Gelatin is a low-energy choice because it contains no fat or cholesterol. It also has defensive colloids that may benefit your health (Alipal et al., 2021).

### 8.2.3 Polysaccharides

Because of their special qualities, polysaccharides—a noteworthy class of bio-based materials originating from natural sources—show various applications. Because these polymeric carbohydrates are non-toxic, biodegradable, and derived from renewable resources, they can be used in various settings, including food packaging and biomedical applications.

#### 8.2.3.1 Chitosan

Chitosan is a polymer mostly made up of D-glucosamine and, to a lesser extent, randomly β-(1–4)-linked N-acetyl-D-glucosamine units. Chitin is the second most prevalent polysaccharide, and can be made by deacetylating chitin. Recent advancements now allow the extraction of chitin and chitosan from diverse sources, including insects and microorganisms, although crustacean shells remain the main source of chitin (Peniche et al., 2018; Zargar et al., 2019; Sharma et al., 2019). Fungal sources, particularly, have drawn more attention because of possible benefits, such as high solubility, deacetylation, and a homogenous polymer length (Ghormade et al., 2019). Generally speaking, chitosan can be obtained through two different methods, namely, chemical and biological. Strong acid and alkaline treatments are used in the chemical approach, which is the most widely used industrial process (El-Knidri et al., 2018). Enzymes and microorganisms aid biological processes; however, chitin's high crystallinity hinders scalable enzymatic deacetylation. The source and extraction method of chitosan are vital depending on its intended use, influencing product qualities for biomedical applications. These qualities shape chitosan's mechanical and biological traits (Kim, 2018). Chitosan's adaptability allows its use in various forms like membranes, scaffolds, and gels, making it suitable for biomedical applications such as tissue engineering and drug delivery. However, it has low solubility at pH > 7.0 and poor mechanics in moist conditions, prompting efforts to modify its structure or combine it with other materials. Chitosan, a polycationic polymer formed by protonating D-glucosamine, has a positive charge at physiological pH. Despite its versatility, cationic polymers can be neurotoxic, causing CNS damage via inflammasome activation linked to particle size. There is limited information on chitosan's neurotoxicity, but its derivatives also show neuroprotective effects against neurological disorders like stroke and Alzheimer's disease.

#### 8.2.3.2 Alginates

Alginates are polysaccharides obtained from brown algae's kelp/*Sargassum* algae (Zhang et al., 2020). Alginates' physical and chemical characteristics have been the subject of much research since their discovery in the late 19th century (Zhong et al., 2020). Alginates are readily available, abundant in sources, and non-toxic. In the human body, it has been demonstrated to be both biocompatible and biodegradable (Hernandez-Gonzalez et al., 2020). Alginates are widely used in food and as vital biomaterials in pharmaceuticals. Their properties, such as high absorption, non-toxicity, and biocompatibility, make them suitable for wound healing. They are used in various forms, including hydrogels and topical treatments. Alginate dressings, including bandages, can absorb excess fluid, maintain a moist environment, reduce infection, and enhance healing compared to gauze. To improve mechanical properties, alginates are often combined with synthetic polymers (Uyen et al., 2020). The amount of synthetic polymers, the kind and strength of cross-linking, and the addition of nanoparticles and antimicrobial agents all impact the therapeutic effectiveness of composite wound dressings.

## 8.3 PROPERTIES OF BIO-BASED MATERIALS

### 8.3.1 Biocompatibility

The ability of a substance to interact with biological systems without causing negative reactions makes it a crucial characteristic for materials utilized in biomedical applications. Compared to conventional synthetic materials, bio-based materials—made from sustainable biological resources—frequently show improved biocompatibility. The main features of biocompatibility in bio-based materials are covered in this overview, along with their benefits, mechanisms, and implications for biomedical applications. The capacity of a material to carry out its intended function in a biological setting without producing toxicity or an unfavorable immunological response is known as biocompatibility. Materials utilized in scaffolds for tissue engineering, medication delivery systems, and implants must have this quality. Numerous bio-based products come from natural sources such as bacteria, plants, and algae. Polysaccharides such as hyaluronic acid, chitosan, and alginate are examples. For these materials to successfully integrate into biological tissues, they usually include inherent biological qualities that support cell adhesion, proliferation, and differentiation.

### 8.3.2 Biodegradability

A key characteristic of bio based materials is their biodegradability, which dictates both their applicability for different uses and their influence on the environment. Bio-based materials come from renewable biological resources, and their chemical makeup, the environment, and the presence of microbes all affect their biodegradability. The main cause of biodegradation is the activity of microorganisms like fungi and bacteria. Enzymes found in these organisms have the ability to degrade complex biopolymers into simpler ones, which can then be converted into biomass, carbon dioxide, and water. For instance, certain microbial strains that use polysaccharides as energy sources can easily break down polysaccharides like cellulose and starch (Verma et al., 2024b).

### 8.3.3 Mechanical Properties

Bio-based materials are increasingly utilized in various applications, particularly in the biomedical industry, due to their mechanical properties. Key properties include tensile strength, flexural strength, impact resistance, and hardness. Tensile strength measures a material's resistance to being torn apart and varies with composition and processing. For instance, PLA has a tensile strength of 50–70 MPa for moderate-strength applications. Chitosan, modified or blended, offers high tensile strength, beneficial in biomedical fields. Flexural strength indicates a material's ability to withstand bending forces. Natural fiber composites can achieve flexural strengths over 90 MPa depending on fiber type and matrix. Sandwich composites with lightweight cores reach flexural strengths around 93.5 MPa. Impact resistance is crucial for materials facing sudden forces; bio-based sandwich composites have shown remarkable resistance, with peak forces up to 2700 N in drop testing. Modified hydrogels also absorb energy well due to their viscoelastic properties. Hardness indicates resistance to deformation, and biopolymers like PLA can have their hardness enhanced with other polymers. Adding natural fibers significantly increases the hardness of bio-based composites, with flax fiber composites exhibiting notable hardness due to their reinforcement capability.

### 8.3.4 Chemical Properties

Bio-based materials possess unique chemical properties that influence their functionality and suitability in various sectors, particularly in biomedical applications. Made from renewable biological resources, their key characteristics include biodegradability, chemical composition, reactivity, and response to environmental stimuli. The chemical structure significantly affects these properties. Organic polymers, composed of proteins (like gelatin) or polysaccharides (such as cellulose and starch), have functional groups affecting their mechanical and thermal qualities. Synthetic bio-based polymers, like polycaprolactone, are created from bio-derived monomers and can be copolymerized for enhanced versatility. The applicability of bio-based materials across different fields is largely dictated by their chemical features, with biodegradability and unique compositions promoting sustainability. Ongoing research aims to improve these attributes through innovative formulations and processing techniques, expanding bio-based materials' potential in various industries.

### 8.3.5 Thermal Properties

The thermal properties of bio-based materials are vital for industries such as biomedical devices, textiles, and construction. Thermal conductivity measures a material's heat conduction ability, and bio-based materials may struggle with this compared to traditional synthetics, although advances in engineering are improving performance. Phase change materials (PCMs) based on biotechnology absorb and release large amounts of latent heat during transitions. Additives like multiwall carbon nanotubes (MWCNTs) enhance the low conductivity of bio-based PCMs from fatty acids, improving their thermal management suitability. Natural fiber composites improve thermal conductivity, with materials like hemp or flax fibers enhancing performance due to larger contact areas with polymers. Heat capacity, the energy required to change a material's temperature, is generally favorable in bio-based materials. Bio-based PCMs exhibit significant latent heat storage; for instance, soybean oil-based PCMs have around a 29.38°C melting point and a latent heat capacity of 149.2 J, valuable for energy storage. Specific heat capacity varies with chemical structure; PLA provides effective thermal management due to its favorable properties. The phase shift ability is key for bio-based materials as thermal regulators. Bio-based PCMs excel in applications requiring temperature stability, changing states while absorbing or releasing heat. Their phase change temperatures can be tailored by modifying chemical compositions or adding conductive fillers. The thermal stability of bio-based materials depends on their chemistry and processing. PLA maintains integrity in thermal applications, with a glass transition temperature ($Tg$) around 60°C and a melting temperature ($Tm$) of 150°C, showcasing significant thermal stability. Many bio-based polymers demonstrate good heat stability, like PLA, which shows noticeable weight loss above 300°C, making it suitable for moderate-temperature applications. Additives like stabilizers can enhance thermal stability, with inorganic fillers improving resistance and reducing flammability compared to pure polymers.

## 8.4 MECHANISMS OF BIOCOMPATIBILITY AND BIODEGRADABILITY

The creation and use of biomaterials in biomedical domains depend heavily on the processes of biocompatibility and biodegradability. Designing materials that can safely decompose within the body and interact favorably with biological systems is made easier with an understanding of these mechanisms.

### 8.4.1 Cellular Response to Bio-based Materials

The cellular response to bio-based materials is a complex interplay of biochemical and biophysical interactions that significantly influence tissue engineering and regenerative medicine. Understanding these responses is crucial for developing effective biomaterials that promote healing and integration within the body. This detailed exploration encompasses various aspects, including the mechanisms of interaction, the role of material properties, and the implications for tissue regeneration.

#### 8.4.1.1 Cell Adhesion and Proliferation

The biological response to bio-based materials starts with cell adhesion, predominantly influenced by surface characteristics such as topography, chemistry, and roughness. Surface chemistry regulates protein adsorption, affecting cell adhesion through integrins. Changes in surface chemistry can enhance or hinder cell adhesion, influencing subsequent processes like migration and proliferation. Topographical features at the micro- and nanoscale influence cell orientation and activity. Research shows that mimicking the natural ECM can promote fibroblast differentiation and proliferation. Mechanical properties strongly affect cellular responses. Stiffness determines stem cell fate; that is, stiffer substrates may lead to osteogenic differentiation, while softer ones support neural differentiation, crucial for scaffold design. Dynamic mechanical stimulation via platforms like the Universal Biomaterial-on-Chip (UBoC) improves cell viability and proliferation in flow conditions, simulating physiological settings. Bioactive substances, like growth factors (e.g., BMPs), can modulate cellular responses to promote tissue regeneration. Ions from bio-based material degradation can influence cellular activity; for instance, calcium ions enhance osteoblast activity. Cells sense and respond to mechanical stimuli through calcium signaling pathways, affecting proliferation and differentiation. The mechanical environment and material characteristics influence focal adhesion development, vital for tissue regeneration and cell migration. Understanding these cellular reactions informs tissue engineering significantly. Scaffold design can be optimized for specific cellular activities, like bone or cartilage regeneration, by adjusting physical and chemical scaffold properties. Biocompatibility assessment involves analyzing cellular responses to determine the suitability of bio-based materials for biomedical applications, favoring those that enhance cell adherence and proliferation. Integrating bioactive substances with bio-based materials aligns regenerative strategies with advancements in tissue engineering and personalized therapy.

### 8.4.2 Enzymatic Degradation

The primary process in the breakdown of many polymers, including bio-based materials and synthetic plastics, is enzymatic degradation. Enzymes catalyze the hydrolysis of polymer chains into smaller, easier-to-manage molecules in this process. Developing efficient waste management and material recycling techniques requires an understanding of the mechanisms, types of enzymes involved, and factors influencing enzymatic degradation (Dhiman et al., 2022; Verma et al., 2024a). Enzymes initially adhere to the polymer's surface. Because it affects the effectiveness of subsequent hydrolysis, this step is crucial. Surface hydrophobicity, surface roughness, and the presence of functional groups on the polymer surface are some of the variables that might affect the adsorption process (Abrusci et al., 2019). Enzymes catalyze the hydrolysis of polymer chains after they are linked. Oligomers, dimers, and monomers are created as a result of this reaction, which breaks down ester or amide linkages in polyesters or polyamides. Cutinize has been demonstrated to efficiently hydrolyze ester linkages in poly (butylene succinate) (PBS) and other aliphatic polyesters. In the bio-fragmentation phase, the polymer chains are further broken up into tiny pieces that microbes can absorb. Enzymes such as oxidoreductases and hydrolases, which alter the structure of polymers to improve their biodegradability, facilitate bio-fragmentation (Yao et al., 2022). Microbial cells absorb hydrolysis products for biomass or energy metabolism, essential for biodegradation and carbon return to ecosystems. This process ends when microbial metabolism converts breakdown products into carbon dioxide and water. Enzymes in polymer breakdown have specific functions; for example, hydrolases that hydrolyze ester bonds in biodegradable polyesters such as PBS and PCL. Oxidoreductases, such as laccases and peroxidases, support the breakdown of resistant plastics like polystyrene and polyethylene. Carboxylesterases target ester linkages in aliphatic polyesters, facilitating further microbial breakdown. Degradation effectiveness is influenced by factors such as polymer structure; higher molecular weights and crystalline structures resist degradation, while amorphous regions are more vulnerable. Environmental conditions affect enzyme activity; temperature, pH, and moisture optimize catalytic effectiveness. Increased enzyme quantities enhance hydrolysis until substrate saturation is reached. Certain surfactants improve enzyme adsorption and degradation efficiency. Microbial communities boost enzymatic degradation by producing complementary enzymes for complex polymer breakdown.

### 8.4.3 Environmental Factors Affecting Biodegradability

Environmental factors influence the biodegradation of materials like plastics and organic waste. Understanding these factors is essential for developing efficient biodegradable materials and improving waste management. Temperature affects enzymatic and microbial activities, with most microbes thriving between 20°C and 60°C.

Higher temperatures can boost enzyme activity and degradation speed, as seen in industrial composting at around 58°C. High temperatures increase polymer chain mobility, making them susceptible to microbial attack, while lower temperatures slow metabolic processes and hinder biodegradation. Moisture is essential for microorganisms; sufficient water is required for enzyme reactions and growth. In dry conditions, microorganisms may become dormant or die, reducing biodegradation rates. Water is crucial for hydrolytic degradation, aiding polymer chain breakdown and enhancing microbial–substrate interaction (Millet et al., 2019). pH levels affect microbial metabolism. Most bacteria prefer neutral-to-slightly acidic (pH 6–7) conditions, but various microbes have unique pH preferences. Extreme pH can hinder biodegradation by slowing enzyme and microbial activity. The stability and activity of bacterial enzymes depend on pH, with some enzymes being more effective at specific pH levels, impacting the overall breakdown process. Biodegradation can be aerobic or anaerobic, relying on oxygen availability. Aerobic conditions—where microbial diversity is crucial—enable microorganisms to decompose organic waste into carbon dioxide and water, a faster process than anaerobic breakdown, which produces methane, a potent greenhouse gas (Chamas et al., 2020).

## 8.5 BIOMEDICAL APPLICATIONS OF BIO-BASED MATERIALS

Because of their nontoxicity, biodegradability, and biocompatibility, bio-based materials are becoming more and more popular in biomedical applications. These materials, which come from natural sources including bacteria and plants, are being thoroughly researched for a number of medicinal applications, such as wound healing, medication administration, and tissue engineering.

### 8.5.1 Drug Delivery Systems

Targeted delivery using bio-based materials enhances chemotherapy localization at tumors, reducing systemic toxicity. Liposomal formulations, compared to conventional methods, show greater efficacy and fewer side effects. Drug delivery techniques may boost anti-cancer drug effectiveness and lower toxicity. Long-circulating carriers like liposomes exploit the "enhanced permeability and retention" effect to preferentially exit tumor blood vessels. New liposome structures are being developed for various drugs. Future delivery systems will feature true molecular targeting, with constructs like immunoliposomes combining tumor-recognizing biological elements with delivery technologies (Krishna and Pandit, 2020). Liposomal drug delivery systems that are currently licensed offer stable formulation, enhanced pharmacokinetics, and some "passive" or "physiological" targeting to tumor tissue (Bhat et al., 2020). These carriers do not specifically target tumor cells. Design changes that protect liposomes from unwanted interactions with membranes and proteins also hinder them from interacting with tumor cells. Instead, liposomes remain in the tumor stroma as drug-loaded deposits after extravasating into the tissue. Medication is eventually released from liposomes through enzymatic breakdown or phagocytosis, allowing diffusion to tumor cells. Direct targeting of cancer cells via antibody or ligand interactions is the hallmark of the next generation of drug carriers. Gene therapy also aligns with the concept of using biomaterials for targeted drug delivery (Li et al., 2021; Quadros et al., 2021). To treat genetic abnormalities or promote tissue repair, biomaterials can deliver therapeutic genes to specific cell types. Gene therapies precisely target sites where needed, enhancing therapeutic potential and minimizing off-target effects.

### 8.5.2 Tissue Engineering and Regenerative Medicine

Bio-based materials play a crucial role in regenerative medicine by supporting tissue repair and healing. In bone regeneration, scaffolds made from collagen and hydroxyapatite create a conducive environment for osteoblasts, promoting bone growth and structural integrity. For cartilage repair, hydrogels and bio-based scaffolds replicate the mechanical properties of cartilage, aiding in joint recovery and enhancing durability. In wound healing, bio-based dressings derived from gelatin or alginate help retain moisture, serve as protective barriers against infections, and accelerate tissue repair. Additionally, in nerve regeneration, bio-based scaffolds provide structural support while releasing neurotrophic factors that promote nerve cell survival, growth, and guided regeneration.

### 8.5.3 Wound Healing

Because of its advantageous qualities, such as biocompatibility, biodegradability, and the capacity to replicate the natural ECM, bio-based materials are being used increasingly in wound healing. By creating an environment that is favorable for cell migration, proliferation, and tissue regeneration, these materials improve the healing process.

In order to encourage cell migration and inhibit the formation of scabs, bio-based products aid in preserving the ideal moisture content at the wound site. During the healing process, the danger of infection is decreased by the natural antibacterial qualities of many bio-based materials. These substances promote a number of cellular functions that are essential for efficient wound healing, including fibroplasia (development of fibrous tissue), angiogenesis (new blood vessels), and epithelialization (regeneration of skin). Bio-based dressings are used to treat traumatic injuries or surgical incisions, offering a barrier of protection and accelerating the healing process. Bio-based materials help increase the healing rates of conditions, including diabetic ulcers and pressure sores, by improving cellular activity and moisture retention. For burn wounds, bio-based hydrogels and dressings are very helpful because they help control exudate, relieve pain, and promote healing.

### 8.5.4 Medical Implants and Devices

Due to their advantageous qualities, such as biocompatibility, biodegradability, and the capacity to replicate the natural ECM, bio-based materials are increasingly being utilized in the creation of medical implants and devices. These materials offer several benefits over conventional synthetic materials, particularly regarding safety and environmental impact, and are derived from renewable biological sources. Bio-based materials decrease the likelihood of rejection or adverse reactions as they interact effectively with biological tissues. Many bio-based materials are designed to degrade at a rate that aligns with tissue regeneration, thereby eliminating the need for additional procedures to remove implants. Certain bio-based materials can release growth factors and other bioactive compounds that facilitate healing and tissue integration. Uses for medical implants include screws, plates, and scaffolds for bone restoration made from bio-based materials such as PLA and hydroxyapatite. As new bone forms, the body can absorb them due to their biodegradable nature. In dental implants, coatings of collagen and hydroxyapatite enhance osseointegration and accelerate the healing process. Furthermore, bioactive ceramics are employed to promote bone formation around dental implants. Bioresorbable stents made from bio-based polymers like PLA offer a temporary solution for maintaining artery patency while encouraging natural healing, without the long-term complications associated with permanent metallic stents. Natural polymers such as gelatin and alginate are used in soft tissue restoration procedures, including dermal fillers and breast implants, due to their compatibility with human tissues. Alginate or gelatin-based bio-based hydrogels can serve as dressings to cover wounds and create a moist environment that promotes healing.

## 8.6 CONCLUSION

In many industries, especially biomedical applications, the field of bio-based materials is leading the way in a revolutionary transition toward sustainability. As we have seen, these materials have many benefits over conventional synthetic substitutes, such as biodegradability, biocompatibility, and the possibility of less environmental impact. Bio-based materials are advancing significantly in fields like tissue engineering, wound healing, medication delivery systems, and surgical applications. Their efficacy and safety in medical contexts are increased by their capacity to replicate natural biological structures. Innovative studies are constantly expanding the potential of bio-based materials. Advances in 3D printing, smart materials, and innovative composites are increasing their usefulness and relevance in a number of industries. Even though bio-based materials have a lot of promise, there are still a number of obstacles to overcome, such as finding sustainable raw materials, being profitable, getting past performance constraints, navigating regulatory environments, and raising customer awareness. It will be essential to address these issues if bio-based treatments are to be widely used. Trends for the bio-based materials have a promising future, thanks to trends showing rising investment, advances in technology, and a growing focus on the ideas of the circular economy. The market for bio-based materials is anticipated to grow substantially as governments enact supportive legislation and customers seek out more environmentally friendly goods. In summary, one of the most important steps toward a more sustainable future is the incorporation of bio-based materials into biomedical applications and beyond. We can create solutions that not only satisfy present demands but also protect the environment for coming generations by utilizing the potential of renewable resources and cutting-edge technologies. Overcoming current obstacles and achieving the full potential of bio-based materials in a variety of industries will require a sustained dedication to research and interdisciplinary cooperation.

## REFERENCES


Abrusci, C., Pablos, J.L., Marín, I., Espí, E., Corrales, T., & Catalina, F. Comparative effect of metal stearates as pro-oxidant additives on bacterial biodegradation of thermal- and photo-degraded low density polyethylene mulching films. *Int. Biodeterior. Biodegrad.* 2019;83:25–32.

Akindoyo, J.O., Beg, M.D.H., Ghazali, S., Islam, M.R., Jeyaratnam, N., & Yuvaraj, A.R. Polyurethane types, synthesis and applications—a review. *RSC Adv.* 2020;6(115): 114453–114482.

Alavudeen, A., Rajini, N., Karthikeyan, S., Thiruchitrambalam, M., & Venkateshwaren, N. Mechanical properties of Banana/Kenaf fiber-reinforced hybrid polyester composites: effect of woven fabric and random orientation. *Mater. Des.* 2022;66:246–257.

Alipal, J., Nayan, N., Sahari, N., et al. A review of gelatin: properties, sources, process, applications, and commercialisation. *Mater. Today Proc.* 2021;42:240–250.

Arntzen, C., Dale, B., Beachy, R., Burgess, R., Gallagher, P., Hardy, R., Johnson, D., Kirk, T., Kishore, G., & Klibanov, A. *Bio-Based Industrial Products: Priorities for Research and Commercialization.* Iowa State University, Department of Economics, Staff General Research Papers, 2021.

Bello, A.B., Park, H., Lee, S.H., et al. Engineering and functionalization of gelatin biomaterials: from cell culture to medical applications. *Tissue Eng. Part B Rev.* 2020;26(2):164–180.

Bhat, M., Shenoy, D.S., Udupa, N., & Srinivas, C.R. Optimization of delivery of betamethasone—dipropionate from skin preparation. *Indian Drugs.* 2020;32:211–214.

Carrillo-Escalante, H.J., Alvarez-Castillo, A., Valadez-Gonzalez, A., & Herrera-Franco, P.J. Effect of fiber-matrix adhesion on the fracture behavior of a Carbon fiber reinforced thermoplastic-modified epoxy matrix. *Carbon Lett.* 2021;19:47–56.

Cavaliere, C., Capriotti, A.L., Cerrato, A., Lorini, L., Montone, C.M., Valentino, F., Laganà, A., & Majone, M. Identification and quantification of polycyclic aromatic hydrocarbons in polyhydroxyalkanoates produced from mixed microbial cultures and municipal organic wastes at pilot scale. *Molecules.* 2021;26:539.

Chamas, A., Moon, H., Zheng, J., Qiu, Y., Tabassum, T., Jang, J.H., & Suh, S. Degradation rates of plastics in the environment. *ACS Sustain. Chem. Eng.* 2020;8:3494–3511.

DeStefano, V., Khan, S., & Tabada, A. Applications of PLA in modern medicine. *Eng. Regen.* 2020;1:76–87.

Dhiman, V.K., Singh, D., Pandey, H., Chauhan, D., Dhiman, V.K., & Pandey, D. Mycoremediation of agricultural waste for the cultivation of edible mushroom. In T. Aftab (Ed.), *Sustainable Management of Environmental Contaminants. Environmental Contamination Remediation and Management.* Springer, Cham, 2022. https://doi.org/10.1007/978-3-031-08446-1_18

Eagan, J.M., Xu, J., Di Girolamo, R., Thurber, C.M., Macosko, C.W., LaPointe, A.M., Bates, F.S., Coates, G.W. Combining polyethylene and polypropylene: enhanced performance with PE/i PP multiblock polymers. *Science.* 2021;355:814–816.

El-Knidri, H., Belaabed, R., Addaou, A., Laajeb, A., & Lahsini, A. Extraction, chemical modification and characterization of chitin and chitosan. *Int. J. Biol. Macromol.* 2018;120(Pt A):1181–1189.

Flórez, M., Guerra-Rodríguez, E., Cazón, P., & Vázquez, M. Chitosan for food packaging: recent advances in active and intelligent films. *Food Hydrocoll.* 2022;124:107328.

Ghormade, V., Pathan, E.K., & Deshpande, M.V. Can fungi compete with marine sources for chitosan production? *Int. J. Biol. Macromol.* 2019;104(Pt B):1415–1421.

Hamad, K., Kaseem, M., Yang, H.W., Deri, F., & Ko, Y.G. Properties and medical applications of polylactic acid: a review. *Express Polym. Lett.* 2020;9:435–455.

Hernandez-Gonzalez, A.C., Tellez-Jurado, L., & Rodriguez-Lorenzo, L.M. Alginate hydrogels for bone tissue engineering, from injectables to bioprinting: a review. *Carbohydr. Polym.* 2020;229:115514.

Isikgor, F.H., & Becer, C.R. Lignocellulosic biomass: a sustainable platform for the production of bio-based chemicals and polymers. *Polym. Chem.* 2022;6:4497–4559.

Jem, K.J., & Tan, B. The development and challenges of poly (lactic acid) and poly (glycolic acid). *Adv. Ind. Eng. Polym. Res.* 2020;3:60–70.

Kim, S. Competitive biological activities of Chitosan and its derivatives: antimicrobial, antioxidant, anticancer, and anti-inflammatory activities. *Int. J. Polym. Sci.* 2018; 170–172.

Krishna, R, & Pandit, J.K. Carboxymethylcellulose-sodium based transdermal drug delivery system for propranolol. *J. Pharm. Pharmacol.* 2020;48:367–70.

Li, M., Sun, J., Zhang, W., Zhao, Y., Zhang, S., & Zhang, S. Drug delivery systems based on CD44-targeted glycosaminoglycans for cancer therapy. *Carbohydr. Polym.* 2021;251:117103.

Licciardello, G., Catara, A.F., & Catara, V. Production of polyhydroxyalkanoates and extracellular products using Pseudomonas Corrugata and P. Mediterranea: a review. *Bioengineering.* 2019;146:105.

Liechty, W.B., Kryscio, D.R., Slaughter, B.V., & Peppas, N.A. Polymers for drug delivery systems. *Annu. Rev. Chem. Biomol. Eng.* 2019;1:149–173.

Luo, F., Fortenberry, A., Ren, J., & Qiang, Z. Recent progress in enhancing poly(lactic acid) stereocomplex formation for material property improvement. *Front. Chem.* 2020;8:688.

Maharana, T., Mohanty, B., & Negi, Y.S. Melt-solid polycondensation of lactic acid and its biodegradability. *Prog. Polym. Sci.* 2019;34:99–124.

Millet, H., Vangheluwe, P., Block, C., Sevenster, A., Garcia, L., & Antonopoulos, R. The nature of plastics and their societal usage. In R.E. Hester (Ed.), *Plastic and the Environment.* Royal Society of Chemistry's, London, UK, 2019, pp. 1–20.

Mohanty, A.K., Misra, M., & Drzal, L.T. Sustainable bio-composites from renewable resources: opportunities and challenges in the green materials world. *J. Polym. Environ.* 2022;10:19–26.

Paduani, M. Microplastics as novel sedimentary particles in coastal wetlands: a review. *Mar. Pollut. Bull.* 2020;161:111739.

Peniche, C., Argüelles-Monal, W., & Goycoolea, F.M. Chitin and Chitosan: major sources, properties and applications. In M.N. Belgacem & A. Gandini (Eds.), *Monomers, Polymers and Composites from Renewable Resources.* Elsevier, Amsterdam, 2018, pp. 517–542.

Pretula, J., Slomkowski, S., & Penczek, S. Polylactides—methods of synthesis and characterization. *Adv. Drug Deliv. Rev.* 2020;107:3–16.

Quadros, M., Momin, M., & Verma, G. Design strategies and evolving role of biomaterial assisted treatment of osteosarcoma. *Mater. Sci. Eng. C.* 2021;121:111875.

Ramakrishna, S., Tian, L., Wang, C., Liao, S., & Teo, W.E. (Eds.). *Medical Devices.* Woodhead Publishing, 2020, pp. 137–153.

Ren, H., Bull, J.L., & Meyerhoff, M.E. Transport of nitric oxide (NO) in various biomedical grade polyurethanes: measurements and modelling impact on NO release properties of medical devices. *ACS Biomatter. Sci. Eng.* 2020;2(9):1483–1492.

Ren, J. *Biodegradable Poly(Lactic Acid): Synthesis, Modification, Processing.* Springer, Heidelberg, Germany: Dordrecht, The Netherlands: London, UK: New York, NY, USA: Beijing, China, 2019.

Satyanarayana, K.G., Arizaga, G.G.C., & Wypych, F. Biodegradable composites based on lignocellulosic fibers—an overview. *Prog. Polym. Sci.* 2020;34:982–1021.

Sharma, N., Singh, D., Rani, R., Sharma, D., Pandey, H., & Agarwal, V. Chitosan and its nanocarriers: applications and opportunities. In *Nanomaterials in Plants, Algae and Microorganisms* (pp. 267–286). Academic Press, 2019. https://doi.org/10.1016/B978-0-12-811488-9.00013-5

Teo, A.J.T., Mishra, A., Park, I., Kim, Y.-J., Park, W.-T., & Yoon, Y.-J. Polymeric biomaterials for medical implants and devices. *ACS Biomatter. Sci. Eng.* 2020;2(4):454–472.

Uyen, N.T.T., Hamid, Z.A.A., Tram, N.X.T., & Ahmad, N. Fabrication of alginate microspheres for drug delivery: a review. *Int. J. Biol. Macromol.* 2020;153:1035–1046.

Van de Velde, K., Kiekens, P. Biopolymers: overview of several properties and consequences on their applications. *Polym. Test.* 2022;21:433–442.

Verma, S.K., Kumar, P., Mishra, A., et al. Green nanotechnology: illuminating the effects of bio-based nanoparticles on plant physiology. *Biotechnol. Sustain. Mater.* 2024a;1:1. https://doi.org/10.1186/s44316-024-00001-2

Verma, S.K., Singh, D., & Misra, S. Innovative approaches to sustainable waste management and energy generation in developing nations. In R. Mehra, P. Kumar, & B. Hung (Eds.), *Municipal Solid Waste Management and Recycling Technologies* (pp. 267–300). IGI Global Scientific Publishing, 2024b. https://doi.org/10.4018/979-8-3693-4054-7.ch010

Vieira, H., Cabral, A.E., Botelho, A., et al. Current and expected trends for the marine chitin/chitosan and collagen value chains. *Mar. Drugs.* 2023;12:605.

Wang, C., Zolotarskaya, O., Ashraf, K.M., Wen, X., Ohman, D.E., & Wynne, K.J. Surface characterization, antimicrobial effectiveness, and human cell response for a biomedical grade polyurethane blended with a mixed soft block PTMO-quat/PEG copolyoxetane polyurethane. *ACS Appl. Mater. Interfaces.* 2021;11(23):20699–20714.

Yao, Z., Seong, H.J., & Jang, Y.-S. Environmental toxicity and decomposition of polyethylene. *Ecotoxicol. Environ. Saf.* 2022;242.

Zaaba, N.F., & Jaafar, M. A review on degradation mechanisms of polylactic acid: hydrolytic, photodegradative, microbial, and enzymatic degradation. *Polym. Eng. Sci.* 2020;60:2061–2075.

Zargar, V., Asghari, M., & Dashti, A. A review on chitin and chitosan polymers: structure, chemistry, solubility, derivatives, and applications. *ChemBioEng. Rev.* 2019;2:204–226.

Zhang, C., Wang, W., Zhao, X., Wang, H., & Yin, H. Preparation of alginate oligosaccharides and their biological activities in plants: a review. *Carbohydr. Res.* 2020;494:108056.

Zhong, H., Gao, X., Cheng, C., Liu, C., Wang, Q., & Han, X. The structural characteristics of seaweed polysaccharides and their application in gel drug delivery systems. *Mar. Drugs.* 2020;18:658.

Zwawi, M. A review on natural fiber bio-composites, surface modifications and applications. *Molecules.* 2021;26:404.

# 9 Bio-Based Materials for Sustainable Energy Applications

*Palak Singh, Devendra Singh, Gaurav Yadav, Naveen Mani Tripathi, Sunil Kumar Verma, and Dmitry Portnikov*

## 9.1 INTRODUCTION

Bio-based materials (BBMs) are renewable, environmentally friendly, and have the potential to alleviate the environmental issues associated with fossil fuel-based systems; interest in them has increased in response to the growing demand for sustainable energy solutions worldwide. A strategy to lower carbon emissions and promote circular economies is through the use of bio-based products, which are made from biological resources (renewable), that is, plant biomass, agricultural waste, and microorganisms (Li et al., 2021). These materials have special qualities that make them appropriate for a range of energy-related applications, for example, insulation for buildings, energy conversion, and energy storage. Due to the growing need for the use of non-renewable resources, eco-friendly products, recycling, and reuse are widely acknowledged by the industrial as well as scientific communities. According to this perspective, biocomposites made from biofiber and biopolymers are very appealing since they may provide the necessary qualities and functions at an affordable cost (Satyanarayana et al., 2009).

Energy storage devices like batteries and supercapacitors are witnessing the incorporation of bio-based polymers such as lignin, cellulose, and chitosan for their role as functional electrolytes and separators (Sharma et al., 2019; Kondawar et al., 2022). The inherent biodegradability and tunable properties of these materials position them as promising alternatives to petroleum-based counterparts. Similarly, bio-based carbon materials derived from biomass precursors exhibit excellent electrochemical properties, making them suitable for use as electrodes in energy storage systems (Zhou et al., 2023). In addition to energy storage, BBMs are gaining attention in energy conversion systems, particularly in photovoltaic and thermoelectric technologies. The use of bio-based polymers as light-harvesting layers in organic solar cells (OSCs) demonstrates their potential to achieve efficient energy conversion with reduced environmental impact (Kim et al., 2020). Moreover, advances in the synthesis of bio-based thermoelectric materials have opened up possibilities for waste heat recovery and conversion into usable energy (Gupta et al., 2021).

DOI: 10.1201/9781003566205-9

For energy generation, biofuels from feedstocks, for example, algae and agricultural residues, are emerging as viable alternatives to traditional fuels (Dhiman et al., 2022; Verma et al., 2024a, 2024b). These biofuels are not only renewable but also exhibit a lower environmental footprint compared to fossil fuels (Zhang et al., 2022). Still, bio-based hydrogen production technologies, including fermentation and photocatalysis, are being explored as sustainable pathways to meet the rising demand for clean energy (Liang et al., 2021).

## 9.2 BBMS FOR ENERGY STORAGE

The application of BBMs in energy storage has gained noteworthy traction due to their environmental benefits, cost-effectiveness, and potential for high-performance systems. From batteries to supercapacitors and hydrogen storage, BBMs are becoming integral components in advanced energy technologies.

### 9.2.1 Batteries

#### 9.2.1.1 Anode Materials

Bio-derived carbon materials have arisen as promising contenders for anodes due to their higher surface area, conductivity, and tunable porosity. For instance, carbon materials synthesized from biomass like coconut shells and lignin have demonstrated excellent lithium-ion storage capabilities (Zhou et al., 2023). These anodes not only reduce the dependence on graphite but also offer better sustainability profiles. Additionally, lignocellulosic biomass precursors are being explored for their ability to form hierarchical porous structures, enhancing ion transport during battery cycling (Li et al., 2021). This approach aligns with the growing demand for renewable alternatives in the production of high-performance battery components.

#### 9.2.1.2 Cathode Materials

BBMs also show promise for cathodes, particularly in sodium-ion and potassium-ion batteries. For example, polydopamine derived from natural amino acids has been used as a cathode coating material, improving electrochemical stability and capacity retention (Kim et al., 2020). Furthermore, hybrid materials combining bio-polymers and inorganic elements offer a green route to achieving high energy density and long cycle life. Biopolymer-derived cathodes, such as those from chitosan, exhibit enhanced charge transfer capabilities, making them suitable for next-generation batteries (Gupta et al., 2021).

#### 9.2.1.3 Electrolytes and Separators

Bio-based polymers (chitosan and cellulose) have found applications in solid-state electrolytes and separators due to their ion-conducting and biodegradable properties (Kondawar et al., 2022). These materials are functionalized with ionic liquids or salts to achieve improved ionic conductivity, which is crucial for high-performance batteries. Nanocellulose-based separators, derived from plant sources, are particularly attractive due to their thermal stability, mechanical strength, and compatibility with lithium-ion chemistries. The integration of such BBMs helps mitigate the environmental challenges posed by conventional polymer-based separators (Zhang et al., 2022).

#### 9.2.1.4 Supercapacitors

Supercapacitors, acknowledged for its fast-charging and high-power density abilities, are increasingly adopting BBMs for electrodes and electrolytes. Biomass-derived carbon materials, such as those obtained from bamboo or coconut shells, exhibit high specific surface areas, making them ideal for use in supercapacitors (Zhou et al., 2023). Biopolymer hydrogels, including alginate and carrageenan, are used as flexible electrolytes in supercapacitors, offering excellent ionic conductivity and eco-friendliness (Gupta et al., 2021). The incorporation of these materials provides a pathway to develop sustainable and high-performance energy storage devices.

#### 9.2.1.5 Hydrogen Storage

The use of materials (bio-based) in hydrogen storage is being explored due to their lightweight, high specific surface area, and ability to form stable complexes with hydrogen molecules. Activated carbons derived from biomass, such as wood and agricultural residues, have shown high hydrogen adsorption capacities under cryogenic conditions (Liang et al., 2021). Furthermore, biopolymers like polylactic acid (PLA) are being functionalized with metal–organic frameworks (MOFs) to create hybrid materials for enhanced hydrogen storage (Zhang et al., 2022). These bio-based hydrogen storage solutions represent a green alternative to traditional systems, contributing to the development of sustainable hydrogen energy technologies.

## 9.3 BBMS FOR ENERGY CONVERSION

BBMs are revolutionizing energy conversion technologies by offering sustainable, eco-friendly, and efficient alternatives. These materials play critical roles in biofuels, fuel cells, and solar cells, enhancing performance while reducing environmental impact.

### 9.3.1 Biofuels

#### 9.3.1.1 First-Generation Biofuels

Production of first-generation biofuel has led to sustainable feedstock technologies that address food supply and environmental issues, replacing traditional sources like sugarcane, corn, and vegetable oils to improve biofuel sustainability and minimize the impact of using food crops. These materials are processed into ethanol and biodiesel through fermentation and transesterification, respectively. While they provide immediate renewable energy solutions, their reliance on food crops has led to worries about the safety of food and land use changes. Efforts to improve the efficiency of first-generation biofuels include optimizing feedstock cultivation and using advanced fermentation techniques to maximize ethanol yield.

#### 9.3.1.2 Second-Generation Biofuels

Biofuels of the second generation are produced from non-food biomass such as agro-residues, lignocellulosic biomass, and forest waste. These biofuels address the limitations of first-generation alternatives by utilizing waste materials and reducing competition with food production. Lignocellulosic biomass is pretreated and hydrolyzed to release fermentable components (carbohydrates, cellulose, and hemicelluloses), which are subsequently converted into ethanol or biodiesel. The use of enzymes and microbial consortia for efficient bioconversion has been a significant area of research (Singh et al., 2022a, 2022b). In the last two decades, there has been a significant increase in interest from governments, large corporations, and academic researchers in second-generation bioethanol because it offers a compelling renewable substitute for depleting fossil fuels. Because it will help in slowing down the climatic changes and is understood as non-competitive with the feed and food industry, its production is also generally welcomed by the public. However, due to the higher research costs and manufacturing costs, commercial cellulosic ethanol production is still in its infancy, and significant work is needed to make it more widely available and lucrative (Aditiya et al., 2016; Tan et al., 2008).

#### 9.3.1.3 Third-Generation Biofuels

Biofuels of the third generation are produced from algae, offering superior yields compared to traditional crops. Algal biofuels can be produced in various forms, including biodiesel, bioethanol, and biogas. Additionally, algae capture $CO_2$ during growth, contributing to carbon neutrality (Liang et al., 2021). Advances in genetic engineering and bioreactor designs have enhanced algal biomass production and lipid extraction, making third-generation biofuels more economically viable (Zhang et al., 2022).

Intense research is being done on substitutes, that is, second- and third-generation biofuel (bioethanol)-making technologies, because rising levels of bioethanol production and population growth raise questions about the long-term sustainability of first-generation bioethanol. These concerns include the threat to the security of global feed and food, the demand for water and land resources, and the possibility of soil contamination from distillation residues, as shown in Table 9.1 (Sharma et al., 2020; Bhatia et al., 2022; Bertrand et al., 2016).

### 9.3.2 Fuel Cells

#### 9.3.2.1 Proton Exchange Membrane Fuel Cells (PEMFCs)

PEMFCs utilize bio-based proton-conducting membranes, such as those derived from chitosan and cellulose, to achieve high efficiency in hydrogen-based energy

**TABLE 9.1**
**Comparison of Biofuel Generations.**

| Aspect | First-generation biofuels | Second-generation biofuels | Third-generation biofuels |
|---|---|---|---|
| Feedstock | Edible crops | Non-edible biomass | Algae |
| Production process | Fermentation, transesterification | Enzymatic, thermochemical | Cultivation, extraction |
| Advantages | Established technology | Reduces food competition | High yield, $CO_2$ capture |
| Challenges | Food vs. fuel conflict | High processing costs | Scaling algae cultivation |

conversion. These membranes exhibit excellent mechanical properties and ionic conductivity, ensuring durable and eco-friendly alternatives to synthetic membranes (Kim et al., 2020). Efforts are ongoing to functionalize these bio-based membranes with nanoparticles to improve proton transport and reduce fuel crossover issues (Gupta et al., 2021).

#### 9.3.2.2 Direct Methanol Fuel Cells (DMFCs)

In DMFCs, BBMs are utilized as catalysts and membranes to convert methanol into electricity. Lignin and polydopamine-derived carbon materials have been explored as cost-effective and sustainable alternatives to platinum-based catalysts (Li et al., 2021). Bio-derived membranes from plant polymers such as cellulose are also gaining attention for their methanol barrier properties and biodegradability, addressing critical challenges in DMFC technology.

#### 9.3.2.3 Microbial Fuel Cells (MFCs)

MFCs leverage the metabolic activity of microorganisms to convert organic waste into electricity. Bio-based electrodes made from biomass-derived carbon and conducting polymers enhance the efficiency of electron transfer between microbes and the anode (Liang et al., 2021). Efforts to integrate BBMs as biofilm supports have shown promise in improving the long-term performance and scalability of MFC systems (Singh et al., 2022b).

### 9.3.3 Solar Cells

#### 9.3.3.1 Dye-Sensitized Solar Cells (DSSCs)

Bio-based dyes derived from plant pigments, such as anthocyanins and chlorophyll, are increasingly used in DSSCs for their light-harvesting properties. These natural dyes are non-toxic, abundant, and capable of achieving comparable efficiencies with synthetic alternatives under optimal conditions (Kim et al., 2020). Additionally, cellulose-based substrates are being developed as flexible and sustainable alternatives to traditional glass or plastic substrates in DSSCs (Gupta et al., 2021).

#### 9.3.3.2 OSCs

Bio-based polymers and small molecules have revolutionized OSCs by offering lightweight and flexible alternatives to silicon-based solar cells. Polymers like polythiophene derived from renewable monomers exhibit excellent light absorption and charge transport properties, making them more suitable for OSC applications. Research is continuing to enhance the stability and efficacy of OSCs by incorporating bio-based additives and hybridizing with inorganic materials, as shown in Table 9.2 (Liang et al., 2021).

### 9.3.4 Biofiber

Since 2800 BC, biofibers have been cultivated for a variety of domestic and household purposes with little utilization, but their use in industrial sectors has increased due to growing awareness of global climate change and the decline in petroleum

**TABLE 9.2**
**Bio-Based Contributions to Fuel Cells.**

| Fuel type | Cell bio-based component | Function | Advantages |
|---|---|---|---|
| PEMFCs | Chitosan-based membranes | Proton conduction | Biodegradable, cost-effective |
| DMFCs | Lignin-derived catalysts | Methanol oxidation | Reduced reliance on platinum |
| MFCs | Biomass-derived electrodes | Electron transfer | Enhances conductivity |

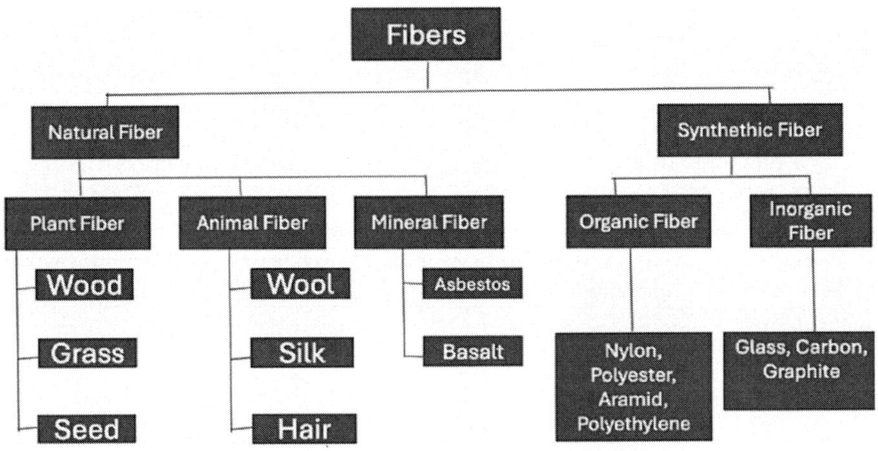


**FIGURE 9.1** Classification of fibers.

products. In particular, biofiber-based materials were used in airplanes as early as the 1940s. Biofibers come from a variety of biological sources, including minerals, plants, and animals (Satyanarayana et al., 2009). They are used in composite materials as reinforcements. Biofibers, especially jute, flax, hemp, and kenaf, have drawn a lot of interest for structural applications due to their remarkably specialized mechanical qualities-to-cost ratio (see Figure 9.1) (Vinod et al., 2020).

## 9.4 BBMS FOR ENERGY EFFICIENCY

BBMs play a pivotal role in enhancing energy efficiency in buildings and infrastructure. By utilizing natural, renewable resources, these materials contribute to lower energy consumption and carbon footprints in construction and insulation applications.

### 9.4.1 Insulation Materials

Insulation materials derived from bio-based sources not only improve thermal performance but also offer sustainability advantages. Their ability to reduce energy loss and biodegradability makes them suitable alternatives to synthetic insulation materials.

### 9.4.2 Cellulose Insulation

Cellulose insulation, made from recycled paper or plant fibers, is one of the most extensively used bio-based insulating materials. It has a high thermal resistance (*R*-value) and provides excellent soundproofing properties. Studies show that cellulose insulation can reduce heating as well as cooling energy needs by up to 20% in buildings (Zhang et al., 2023). The material is treated with borate compounds to enhance fire resistance and pest control. Its low embodied energy makes it a preferred choice for green buildings.

### 9.4.3 Hemp Insulation

The stems of the hemp plant are used to make insulation. Its natural fibers promote indoor air quality by regulating moisture and offering strong temperature resistance. According to research, hemp insulation has a thermal conductivity of approximately 0.04 W/m·K, which is on par with synthetic insulations like mineral wool (Li et al., 2021). Hemp insulation is also carbon-negative, as the plant absorbs $CO_2$ during growth. Its durability and resistance to mold make it suitable for a variety of climates. Since the net greenhouse gas emissions from its production and installation fall within the negative range, hemp concrete is regarded as carbon-negative. Carbon sequestration—the material capability to store carbon (C) in a stable form within itself—is a beneficial mechanism that benefits hemp concrete (Jami et al., 2019). Industrial processes or workshops transform hemp-based goods, which can be used for new construction projects or renovations (Dlimi et al., 2019).

### 9.4.4 Wood Fiber Insulation

High insulation performance and sustainability are combined in wood fiber insulation, which is made from sawmill waste. It comes in hard boards, loose-fill forms, and flexible mats. This material offers extra advantages, including breathability and vapor permeability, along with thermal insulation values comparable to those of traditional choices. It is particularly advantageous for timber-framed constructions due to its compatibility with natural building systems. Research indicates that wood fiber insulation can reduce energy demands in residential buildings by up to 15% annually. The characteristics related to moisture transmission and wood's thermal behavior are assessed on a macroscopic level. The type of wood employed, the direction of measurement (parallel or perpendicular to the fibers), moisture content, and density are some of the variables that affect wood's thermal conductivity (Abahri et al., 2015).

## 9.5 BUILDING MATERIALS

BBMs are also used to create high-performance building materials, promoting energy efficiency and sustainability. BBMs for building are acquired from renewable biological sources and are gaining attention for their sustainability, environmental benefits, and unique properties.

### 9.5.1 Bio-Based Composites

Bio-based composites consist of natural fibers, for example, jute, flax, or hemp, combined with bio-resins. These materials are lightweight, strong, and provide thermal insulation, making them ideal for use in panels, roofing, and flooring (Kim et al., 2020). For example, hemp–lime composites, commonly referred to as "hempcrete," are used for insulating walls. These composites offer excellent thermal inertia and reduce the need for active heating and cooling systems. The use of bio-based composites can decrease building operational energy by approximately 30% over their life cycle.

### 9.5.2 Bio-Based Adhesives

Bio-based adhesives acquired from soy protein, lignin, or starch are eco-friendly alternatives to synthetic adhesives. They are used in laminates, plywood, and engineered wood products. Research shows that lignin-based adhesives demonstrate high bonding strength and resistance to humidity, making them more suitable for structural applications. These adhesives also exhibit lower VOC emissions, enhancing indoor air quality. Innovations in nanotechnology have enabled the development of bio-based adhesives with enhanced thermal stability and durability (see Table 9.3).

## 9.6 CHALLENGES AND FUTURE VIEWPOINTS

Despite its bright future, the field of BBMs for energy applications faces several obstacles. To ensure sustainable energy solutions and achieve broad acceptance, these obstacles must be removed.

### 9.6.1 Technical Challenges

#### 9.6.1.1 Performance and Durability

BBMs often face issues of mechanical and thermal stability when exposed to varying operational conditions. For example, bio-based proton-conducting membranes

**TABLE 9.3**
**Comparison of Bio-Based Insulation Materials.**

| Property | Cellulose insulation | Hemp insulation | Wood FIBER INSULATION |
|---|---|---|---|
| Thermal conductivity | 0.032–0.040 W/m·K | 0.04 W/m·K | 0.038–0.045 W/m·K |
| Density | 30–65 kg/m³ | 30–50 kg/m³ | 50–270 kg/m³ |
| Fire resistance | Moderate (with borate) | High | High |
| Moisture regulation | Moderate | Excellent | Excellent |
| Carbon impact | Low | Carbon-negative | Carbon-negative |

in fuel cells degrade faster than their synthetic counterparts (Smith et al., 2021). Enhancing the thermal resistance of these materials through crosslinking or blending with nanomaterials can address these shortcomings.

#### 9.6.1.2 Cost-Effectiveness

The production of BBMs, especially those derived from advanced feedstocks, often incurs higher costs compared to conventional materials. For instance, algae-derived biofuels are promising but remain expensive due to cultivation and extraction challenges (Zhao et al., 2020). Research on more efficient processing technologies could reduce these costs.

#### 9.6.1.3 Scalability

Scaling up BBM production from laboratory to industrial levels is a significant challenge. Limited availability of feedstock, high resource requirements, and the lack of efficient production infrastructure hinder large-scale adoption (Brown et al., 2022).

## 9.7 ENVIRONMENTAL IMPACT

### 9.7.1 Life Cycle Assessment (LCA)

LCA is a critical tool for assessing the environmental impact of BBMs through their total life cycle, starting from the extraction of raw material to recycling or disposal. LCA allows researchers and companies to quantify the environmental consequences of bio-based products and compare them to conventional materials. For instance, while biofuels are frequently hyped as a cleaner substitute to fossil fuels, LCA studies reveal that their production can result in unintended environmental costs, such as increased water use and land competition, especially in the case of first-generation biofuels (Zhou et al., 2021). Additionally, the cultivation of feedstocks like corn or soy for biofuel production may lead to soil depletion and nutrient runoff, which contributes to environmental degradation (Müller et al., 2022). In contrast, biofuels of the second generation made from waste biomass or non-food crops show improved environmental outcomes as they do not compete with food production and tend to require less land and water. However, even second-generation biofuels require careful LCA studies to identify potential environmental trade-offs, such as the energy intensity of the conversion processes (Hansen et al., 2023). For BBMs used in packaging, textiles, and construction, LCA can provide insights into energy consumption, carbon emissions, and waste generation, helping industries optimize their production and improve sustainability (Kramer et al., 2020). Comprehensive LCA studies are essential to evaluate the true environmental benefits of BBMs. For example, while biofuels reduce greenhouse gas emissions, their manufacture can lead to water use conflicts and soil degradation if not managed sustainably (Chen et al., 2021).

### 9.7.2 Sustainability Metrics

Sustainability metrics are essential for evaluating the overall sustainability of BBMs. These metrics go beyond environmental impacts to include social, economic, and technical dimensions. For BBMs, the primary environmental metrics often include carbon footprint, energy use, and recyclability, but other factors like water usage, biodiversity impact, and land use efficiency also play significant roles (Karnan et al., 2021). The challenge is to create a comprehensive set of metrics that can be applied to a wide range of BBMs, each of which may have different impacts depending on the production process and the nature of the raw materials. For instance, carbon footprint analysis of bio-based polymers (e.g., PLA) can vary significantly based on feedstock (corn vs. sugarcane) and the method of processing (fermentation vs. chemical synthesis) (Singh et al., 2021). Standardized sustainability metrics for BBMs are lacking. Metrics that assess carbon footprint, energy consumption, and recyclability would help industries make informed decisions about material choices (Patel et al., 2020).

## 9.8 FUTURE RESEARCH DIRECTIONS

### 9.8.1 Advanced Materials and Processes

Developing BBMs with superior properties, such as high conductivity or thermal resistance, is critical. For instance, combining bio-based polymers with graphene or carbon nanotubes has shown potential for creating high-performance electrodes in supercapacitors (Singh et al., 2022b). Bio-based composites incorporated with CNTs have been shown to enhance the mechanical properties and electrical conductivity of bioplastics, making them more viable for energy-related applications such as electronic components or storage systems (Wang et al., 2023). These advancements not only improve the performance of the BBMs but also open the door for more sustainable alternatives to current energy storage technologies. Another innovative approach involves the use of bio-based carbon materials derived from agricultural waste in the development of high-performance electrodes for batteries and capacitors. This not only addresses the issue of energy storage but also provides a valuable use for waste biomass, promoting circular economy principles (Jones et al., 2021). Research is also ongoing to develop bio-based thermoplastic elastomers that can withstand extreme temperatures, further broadening the scope of applications for BBMs in high-performance energy systems (Martin et al., 2020).

### 9.8.2 Integration with Renewable Energy Systems

Integrating BBMs into renewable energy systems like solar or wind power storage systems can enhance their sustainability. For example, bio-based adhesives and composites can be used in wind turbine manufacturing, reducing dependency on fossil-derived components (Taylor et al., 2023). Energy storage solutions may also be made easier by combining BBMs with renewable energy technology. For example, bio-based batteries and capacitors are being developed for large-scale storage applications using materials such as cellulose-based electrolytes and carbon produced from lignin. These bio-based substitutes hold the potential to lessen dependency on traditional battery components like cobalt and lithium, whose extraction and disposal methods can have negative environmental effects. Bio-based energy storage

**TABLE 9.4**
**Challenges and Solutions for Bio-Based Materials in Energy Applications.**

| Challenge | Description | Proposed solution | Reference |
|---|---|---|---|
| Performance and durability | Limited mechanical and thermal stability | Crosslinking and nanomaterial blending | Smith et al., 2021 |
| Cost-effectiveness | High production costs | Process optimization and feedstock diversification | Zhao et al., 2020 |
| Scalability | Infrastructure and resource limitations | Development of efficient industrial-scale technologies | Brown et al., 2022 |
| Environmental impact | Water and soil resource conflicts | Sustainable resource management practices | Chen et al., 2021 |
| LCA and sustainability | Lack of standardized metrics | Creation of global sustainability standards | Patel et al., 2020 |

systems can, therefore, contribute to the development of cleaner, more sustainable energy storage technologies, promoting the transition to renewable energy sources (see Table 9.4).

## 9.9 CONCLUSION

BBMs stand at the forefront of sustainable innovation, offering eco-friendly alternatives across diverse energy applications. Their deployment in energy storage systems, such as batteries and supercapacitors, has shown promise in improving efficiency while minimizing reliance on environmentally harmful materials. Similarly, advancements in energy conversion technologies, including biofuels and solar cells, underscore the capacity of bio-based components to deliver high performance while reducing carbon footprints. In the realm of energy efficiency, bio-based insulation and construction materials are enabling greener buildings, driving energy savings, and enhancing thermal comfort. Despite their immense potential, challenges remain. Issues of durability, cost, and scalability hinder widespread adoption, and these must be tackled through strategic innovations such as nanomaterial integration, process optimization, and feedstock diversification. Moreover, comprehensive life cycle assessments are vital to ensure that these materials genuinely contribute to environmental sustainability without unintended consequences like resource depletion or habitat disruption. Collaboration among academia, industry, and policymakers will be instrumental in overcoming these barriers. Investments in research and development, coupled with supportive policy frameworks, can accelerate the transition from laboratory-scale solutions to commercial applications. Education and awareness initiatives are equally important to promote acceptance of BBMs among industries and consumers alike.

Looking ahead, the synergy between BBMs and renewable energy systems offers an exciting frontier. For instance, integrating BBMs into wind turbines, solar panels, and energy storage devices can further align energy technologies with sustainability goals. Additionally, advancements in synthetic biology and material science hold the potential to engineer BBMs with superior properties tailored for specific energy applications. In conclusion, BBMs embody the vision of a sustainable future where renewable resources drive innovation and mitigate the environmental impacts of energy consumption. By addressing the current challenges and capitalizing on emerging opportunities, these materials can become a cornerstone of global efforts to achieve energy security, environmental protection, and economic sustainability. Their role in shaping a greener, more resilient energy landscape is not just promising but imperative.

## REFERENCES

Abahri, K., Belarbi, R., & El Hachem, C. (2015, May). *Macro-Hydric Characterization of Bio-Based Materials*. Bayonne, France: University Meetings on Civil Engineering.

Aditiya, H.B., Mahlia, T.M.I., Chong, W.T., Nur, H., & Sebayang, A.H. (2016). Second generation bioethanol production: a critical review. *Renewable and Sustainable Energy Reviews*, 66, 631–653.

Bertrand, E., Vandenberghe, L.P.S., Soccol, C.R., Sigoillot, J.-C., & Faulds, C. (2016). First generation bioethanol. In *Green Fuels Technology: Biofuels; Green Energy and Technology*, eds. Soccol, C.R., Brar, S.K., Faulds, C., & Ramos, L.P. Cham, Switzerland: Springer International Publishing, pp. 175–212. ISBN 978-3-319-30205-8.

Bhatia, L., Bachheti, R.K., Garlapati, V.K., & Chandel, A.K. (2022). Third-generation biorefineries: a sustainable platform for food, clean energy, and nutraceuticals production. *Biomass Conversion & Biorefinery*, 12, 4215–4230.

Brown, E., Thomas, M., & Lee, Y. (2022). Scalability of bio-based material production: a review of industrial challenges. *Industrial Biotechnology Insights*, 9(1), 89–103.

Chen, H., Lin, Y., & Robertson, A. (2021). Environmental impacts of bio-based fuel production: a life cycle assessment perspective. *Sustainability Science and Technology*, 14(6), 512–525.

Dhiman, V.K., Singh, D., Pandey, H., Chauhan, D., Dhiman, V.K., & Pandey, D. (2022). Mycoremediation of agricultural waste for the cultivation of edible mushroom. In *Sustainable Management of Environmental Contaminants. Environmental Contamination Remediation and Management*, ed. Aftab, T. Cham: Springer. https://doi.org/10.1007/978-3-031-08446-1_18

Dlimi, M., et al. (2019). Dynamic assessment of the thermal performance of hemp wool insulated external building walls according to the Moroccan climatic zoning. *Journal of Energy Storage*, 26, 101007.

Gupta, R., Singh, A., & Mehra, N. (2021). Advances in biobased thermoelectric materials for sustainable energy. *Journal of Sustainable Materials Science*, 9(3), 78–90.

Hansen, T., Zhao, L., & Kim, H. (2023). Second-generation biofuels: a review of life cycle impacts and potential improvements. *Renewable and Sustainable Energy Reviews*, 15(2), 321–333.

Jami, T., Karade, S.R., & Singh, L.P. (2019). A review of the properties of hemp concrete for green building applications. *Journal of Cleaner Production*, 239, 117852.

Jones, M., Li, J., & Wang, H. (2021). Bio-derived carbon materials for high-performance energy storage devices. *Journal of Renewable Energy Materials*, 11(4), 512–528.

Karnan, R., Perumal, S., & Raj, M. (2021). A holistic approach to sustainability metrics for bio-based materials. *Sustainable Materials and Technologies*, 29(3), 154–165.

Kim, J., Lee, H., & Park, S. (2020). Biobased polymers in organic photovoltaics: progress and perspectives. *Solar Energy Materials & Solar Cells*, 210, 1109–1124.
Kondawar, S., Patil, A., & Sharma, R. (2022). Application of biopolymers in energy storage systems: a review. *Polymer Science Journal*, 18(2), 124–139.
Kramer, G., Hernandez, S., & Lin, Y. (2020). Life cycle assessment of bio-based packaging materials. *Journal of Bio-Based Materials*, 17(1), 66–80.
Li, X., Chen, Y., & Zhang, T. (2021). Renewable resources and biobased materials for energy storage applications. *Energy Storage Materials*, 25, 56–72.
Liang, H., Zhao, F., & Jiang, D. (2021). Biobased hydrogen production technologies: progress and challenges. *Hydrogen Energy Journal*, 46(5), 2200–2215.
Martin, T., Park, J., & Kim, S. (2020). Thermoplastic elastomers from bio-based sources for high-temperature energy applications. *Energy Materials and Systems*, 10(3), 181–189.
Müller, J., Schmidt, A., & Zhang, Y. (2022). Soil degradation and water scarcity impacts in biofuel production. *Agricultural Sustainability Journal*, 10(4), 110–124.
Patel, D., Kumar, N., & Sharma, R. (2020). Developing sustainability metrics for bio-based materials: a framework proposal. *Materials Science and Sustainability*, 7(5), 300–315.
Satyanarayana, K.G., Arizaga, G.G.C., & Wypych, F. (2009). Biodegradable composites based on lignocellulosic fibers—an overview. *Progress in Polymer Science*, 34(9), 982–1021.
Sharma, B., Larroche, C., & Dussap, C.-G. (2020). Comprehensive assessment of 2G bioethanol production. *Bioresource Technology*, 313, 123630.
Sharma, N., Singh, D., Rani, R., Sharma, D., Pandey, H., & Agarwal, V. (2019). Chitosan and its nanocarriers: applications and opportunities. In *Nanomaterials in Plants, Algae and Microorganisms*. Academic Press, pp. 267–286. https://doi.org/10.1016/B978-0-12-811488-9.00013-5
Singh, P., Kumar, R., & Sharma, V. (2022a). Microbial fuel cells and bioenergy conversion. *BioEnergy Research*, 14(2), 89–101.
Singh, R., Gupta, P., & Patel, M. (2021). Carbon footprint of bio-based polymers: comparative analysis of PLA and PHA. *Polymer Degradation and Stability*, 95(4), 839–846.
Singh, V., Gupta, A., & Mehta, S. (2022b). Bio-based hybrid nanocomposites for high-performance supercapacitors. *Nano Energy Systems*, 15(2), 233–247.
Smith, J., Wang, R., & Davis, T. (2021). Improving the durability of bio-based proton conducting membranes for fuel cell applications. *Journal of Renewable Energy Materials*, 12(3), 245–260.
Tan, K.T., Lee, K.T., & Mohamed, A.R. (2008). Role of energy policy in renewable energy accomplishment: the case of second-generation bioethanol. *Energy Policy*, 36, 3360–3365.
Taylor, P., Morgan, D., & Ahmed, Z. (2023). Bio-based composites in renewable energy systems: applications and opportunities. *Renewable Energy Materials*, 10(1), 120–135.
Verma, S.K., Singh, D., & Misra, S. (2024a). Innovative approaches to sustainable waste management and energy generation in developing nations. In *Municipal Solid Waste Management and Recycling Technologies*, eds. Mehra, R., Kumar, P., & Hung, B. IGI Global Scientific Publishing, pp. 267–300. https://doi.org/10.4018/979-8-3693-4054-7.ch010
Verma, S.K., et al. (2024b). Green nanotechnology: illuminating the effects of bio-based nanoparticles on plant physiology. *Biotechnology for Sustainable Materials*, 1, 1. https://doi.org/10.1186/s44316-024-00001-2
Vinod, A., et al. (2020). Renewable and sustainable biobased materials: an assessment on biofibers, biofilms, biopolymers and biocomposites. *Journal of Cleaner Production*, 258, 120978.
Wang, J., He, L., & Liu, Z. (2023). Carbon nanotube reinforced bio-based polymers for supercapacitor applications. *Materials Chemistry and Physics*, 59(7), 522–530.
Zhang, Y., Wu, H., & Liu, Q. (2022). Biofuels and their role in sustainable energy. *Renewable Energy Reports*, 35, 1–15.

Zhang, Y., Wu, H., & Liu, Q. (2023). Advances in bio-based insulation materials for sustainable construction. *Renewable Materials Journal*, 10(2), 101–115.

Zhao, L., Kaur, P., & Cheng, X. (2020). Economic viability of algae-derived biofuels: current challenges and future directions. *Biofuels and Bioproducts Research*, 8(4), 198–210.

Zhou, L., Wang, Z., & Li, J. (2023). Carbon materials from biomass for supercapacitors. *Electrochemical Energy Reviews*, 6(1), 45–62.

Zhou, Y., Lee, S., & Wang, L. (2021). Environmental trade-offs in the production of first-generation biofuels: a life cycle assessment. *Environmental Impact Review*, 48(3), 202–210.

# 10 Bio-Based Materials for Water Treatment and Purification

*Vinay Kumar Dhiman, Vivek Kumar Dhiman, Ankush Chauhan, Simran Sharma, Devendra Singh, and Pooja Adtani*

## 10.1 INTRODUCTION

The rising global population, coupled with increasing water stress and climate change, is expected to drive further growth in water consumption worldwide. Access to sanitation and potable water is a fundamental human right and is essential for maintaining healthy livelihoods. A significant portion of the pollutants targeted for removal originates from industrial wastewater, as industries are major water consumers. The rapid expansion of global enterprises has intensified water pollution, increasing wastewater treatment as a critical global concern. The term "wastewater" refers to water contaminated and discharged in substantial quantities, requiring purification to meet societal needs and mitigate water scarcity. Regulatory authorities are increasingly emphasizing the treatment and recycling of wastewater to conserve water and achieve zero industrial effluent discharge. Therefore, a secure, creative, and cost-effective technology is required for easy adoption in current industrial systems. Bio-based materials are increasingly accepted as viable options for water treatment owing to their sustainability, biodegradability, and cost-effectiveness. This reassures industry professionals of the economic feasibility of these solutions (Wu et al., 2022). Biological resources are renewable and derivatives of bio-based materials. Bio-based materials are derived from resources of biological origin, such as agri-waste, including agricultural leftovers (e.g., residues from sugarcane, wheat, and maize cobs), animals, and microorganisms from algae, bacteria, and fungi. The primary constituents of biomaterials are the natural biopolymer cellulose, hemicellulose in abundance, and rigid molecules such as lignin and chitin, along with other naturally occurring compounds such as starch and protein. Biomaterials have attracted substantial research attention in different industries, from food packs to textiles to the chemical industry and new fields such as tissue engineering, owing to their exceptional biological compatibility, biological degradation variety, renewability, plenty, cost-effectiveness, environmental sustainability, and numerous reactive sites. Thus, they are expected to function as sustainable substitutes for petroleum-derived materials.

DOI: 10.1201/9781003566205-10

A typical wastewater treatment sequence may involve a combination of physical (e.g., membrane technologies, adsorption, and ion exchange), chemical (e.g., coagulation, chemical oxidation, and electrochemical treatments), and biological (e.g., biofilters and sequential batch reactors) treatment processes. Although these methods are effective at removing contaminants, they also have certain limitations (Pandey et al., 2022; Singh et al., 2022; Thakur et al., 2021; Tyagi et al., 2024). Furthermore, the chemicals used in these processes can sometimes react with other substances in wastewater, forming new compounds with unknown health effects. Conversely, adsorption using activated carbon is eco-friendly, efficient, easy to operate, and resilient to harmful compounds. Despite these advantages, its widespread application is restricted by its high cost, challenges related to carbon regeneration after saturation, and decreased adsorption efficiency over time. Biomaterials have emerged as a compelling option for several investigations because of their cost-effectiveness, efficiency, and biowaste separation processes. Bioadsorbents or biosorbents are biomaterials of biological origin that may exhibit adsorption capabilities comparable to those of chemical adsorbents. The principal biomacromolecules—such as cellulose and hemicellulose (HC) as structural backbones, along with lignin and other plant constituents composed of proteins—are the primary components of biomaterials. The presence of functional groups formed by carbon, oxygen, nitrogen, and hydrogen allows bioadsorbents to establish necessary bonds with contaminants, ensuring the effectiveness of these solutions (Lustenberger & Castro-Muñoz, 2022). Biosorption binds pollutants without sludge and offers a cost-effective and eco-friendly alternative to traditional methods. Biomaterials sourced from natural or synthetic materials interact with biological systems. Bioadsorbents extract heavy metals and pesticides owing to their cost-effectiveness, availability, diverse functional groups, varying pore sizes, and low operational costs. Research on biomaterials is growing in the fields of medicine, biosensors, and wastewater treatment. Kinetic biomaterials are vital for wastewater purification and the identification of harmful contaminants. Recent advancements have made biomaterials effective adsorbents, potentially supporting costly synthetic options. The various biomaterials used in biosorption may effectively reduce pollutants and bridge the gap between laboratory and industrial applications. Recent biomaterial advancements include multiple capabilities, including the capacity of biosorbents to remove organic–inorganic contaminants from wastewater. This chapter discusses the different types of biomaterials and their applications in wastewater treatment.

## 10.2 TYPES OF BIO-BASED MATERIALS AND THEIR APPLICATIONS IN WATER TREATMENT AND PURIFICATION

Biomaterials derived from nature or synthesized can be used for various applications. Various approaches have been used for these materials (see Figure 10.1).

### 10.2.1 Naturally Occurring Biomaterials

Natural biomaterials derived from plants, animals, and microorganisms that produce bacteria and fungi are widely utilized in wastewater treatment. These materials

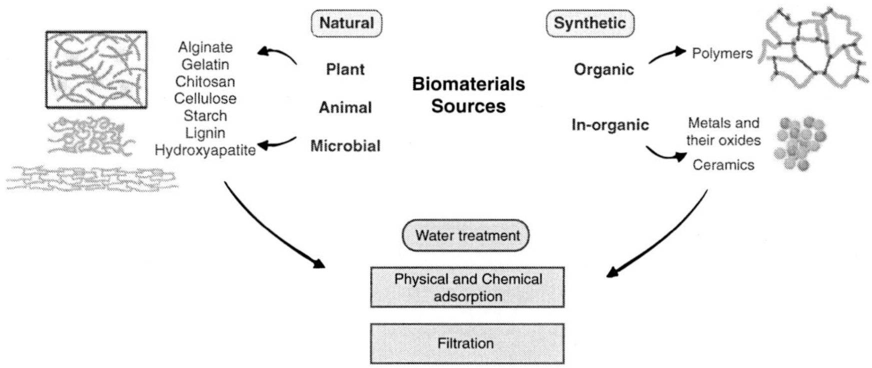


**FIGURE 10.1** Overview of bio-based material sources, types, and their utility in water purification.

originate from diverse biological sources, including sewage sludge, byproducts of the sugar industry, rice husks, orange peels, and eggshells (see Table 10.1). They offer several advantages over synthetic alternatives, including accessibility, cost-effectiveness, and environmentally friendly disposal.

#### 10.2.1.1 Categories of Naturally Occurring Biomaterials

i. **Chitin and Chitosan**: These are composed of natural amino polymers and are particularly effective for capturing pollutants from wastewater. They are biodegradable, biocompatible, environmentally sustainable, and economically viable. In 1811, French natural history professor Henri Braconnot discovered chitin in mushrooms, and Odier discovered it in insects in 1823. Professor C. Rouget synthesized chitosan in 1859, which spurred early 20th-century research. The alkaline N-deacetylation of chitin yields chitosan, a complex polysaccharide biopolymer. Chitosan is a biomaterial with several scientific uses, owing to its green process for compatibility, degradability, and non-toxicity. Chitin and chitosan are employed in the treatment of drinking water and typically for wastewater, where nanomaterials and nanocomposites used for nano-adsorption can purify water (Sarode et al., 2019). Chitosan has unusual viscosity, solubility, polyelectrolyte behavior, film formation, metal chelation, and suitable optical and structural characteristics. Owing to their coagulation and flocculation, and adsorption capabilities for organic and inorganic compounds, several studies have suggested that chitin and chitosan can remove toxins from contaminated water. Chitosan beads and their composites, including polyaniline, glutaraldehyde, and alginate, have demonstrated varying adsorption capacities for $Cu^{2+}$. The honeycomb-structured crosslinked glucan and chitosan composite showed significant adsorption within 40 min and saturation at 60 min. Endothermic heavy metal adsorption was optimal at pH 1–7 and increased with temperature. Glucan and chitosan hydrogels show adsorption capacities of 342, 232, 184, 395, and 269 mg/g (Jiang et al., 2019).

**TABLE 10.1**
**Natural Biomaterials for Water Treatment.**

| Biomaterials | Source | Applications in water treatment | Mechanism of action | Advantages | Limitations |
|---|---|---|---|---|---|
| Chitosan | Derived from chitin (exoskeletons of crustaceans, insects) | Removal of heavy metals, dyes, oils, and turbidity | Chelation with heavy metals, electrostatic interactions with dyes and oils, coagulation, and flocculation | Biodegradable, non-toxic, abundant, and versatile in removing multiple contaminants | Low solubility in water requires chemical modification for enhanced efficiency |
| Cellulose | Plant based (wood, cotton, and agricultural residues) | Adsorption-based heavy metals, organic pollutants, and oil | Hydrogen bonding and surface interactions with pollutants | Renewable, biodegradable, and chemically modifiable for enhanced functionality | Native cellulose has low adsorption capacity and requires functionalization |
| Starch | Plant based (corn, potato, rice, and wheat) | Removal of dyes, nitrate ions, and heavy metals | Adsorption through hydrogen bonding, electrostatic interactions, and chemical modification | Abundant, biodegradable, and chemically modifiable for targeted pollutant removal | Native starch has poor adsorption and requires modification for high efficiency |
| Lignin | Plant biomass (wood and agricultural waste) | Adsorption-based heavy metals, dyes, and organic pollutants | Adsorption through aromatic and polar functional groups | Abundant, cost-effective, and suitable for porous carbon preparation | Adsorption efficiency depends on source and extraction method; complex modification processes |
| Algae (e.g., Spirulina and Chlorella) | Aquatic biomass (marine and freshwater algae) | Adsorption-based heavy metals, dyes, and nutrients (e.g., nitrates and phosphates) | Biosorption through functional groups like carboxyl, hydroxyl, and amine | Renewable, easily cultivable, and effective in nutrient removal | Requires harvesting and processing; lower adsorption in untreated forms |
| Activated carbon | Biomass (coconut shells, wood, and lignin) | Adsorption-based organic pollutants, pesticides, and heavy metals | High surface area and porous structure facilitate pollutant adsorption | High-efficiency and versatile applications | High cost and limited reusability; non-renewable raw materials for some sources |

| | | | | | |
|---|---|---|---|---|---|
| *Moringa oleifera* Seeds | Seeds of the Moringa tree | Coagulation and removal of turbidity, bacteria, and organic matter | Proteins in seeds act as natural coagulants, binding and precipitating contaminants | Affordable, biodegradable, and effective in reducing turbidity and microbial load | Limited application to specific types of contaminants; requires frequent replenishment |
| Peat Moss | Sphagnum mosses | Removal of heavy metals, oil, and organic pollutants | High cation exchange capacity and adsorption on humic substances | Low-cost, renewable, and effective for oil and heavy metal adsorption | Decomposition over time and low mechanical stability |
| Cactus mucilage | Extracted from cactus (e.g., *Opuntia* spp.) | Removal of turbidity, heavy metals, and pathogens | Natural flocculant that binds suspended particles and pathogens | Non-toxic, biodegradable, and abundant in arid regions | Effectiveness depends on species and mucilage preparation methods |
| Guar gum | Seeds of the guar plant (*Cyamopsis tetragonoloba*) | Removal of suspended solids, heavy metals, and dyes | Acts as a flocculant to aggregate contaminants | Renewable, non-toxic, and effective in enhancing sedimentation | Requires chemical modification for high efficiency in water treatment |
| Bacterial cellulose | Produced by bacteria (e.g., *Acetobacter xylinum*) | Adsorption-based heavy metals and dyes | Porous structure allows efficient trapping and adsorption of pollutants | Renewable, biodegradable, and tunable porosity | High production cost and limited scalability |
| Natural clays (e.g., bentonite) | Naturally occurring minerals | Removal of turbidity, heavy metals, and organic contaminants | Adsorption via cation exchange and surface interactions | Abundant, cost-effective, and reusable | Limited effectiveness for specific organic pollutants |
| Eggshell membrane | Waste from eggshells | Removal of heavy metals and organic pollutants | Adsorption via protein-rich structure with reactive sites | Waste valorization, cost-effective, and biodegradable | Limited adsorption capacity compared to modified materials |
| Neem leaves/seeds | Neem tree (*Azadirachta indica*) | Removal of heavy metals, dyes, and pathogens | Bioadsorption through functional groups and natural antibacterial properties | Abundant, biodegradable, and eco-friendly | Limited adsorption capacity for specific pollutants |

*(Continued)*

**TABLE 10.1 (Continued)**
**Natural Biomaterials for Water Treatment.**

| Biomaterials | Source | Applications in water treatment | Mechanism of action | Advantages | Limitations |
|---|---|---|---|---|---|
| Coconut husk fibers | Coconut tree | Adsorption-based oil, organic pollutants, and suspended solids | Physical entrapment and surface adsorption through lignocellulosic structure | Renewable, biodegradable, and low-cost | Limited chemical reactivity; effective primarily for oil-based pollutants |
| Rice husk ash | Agricultural byproduct | Adsorption-based heavy metals and organic pollutants | Silica-based surface provides high adsorption capacity | Abundant, low-cost, and effective for industrial wastewater treatment | Requires preprocessing for optimal adsorption efficiency |
| Banana peels | Fruit waste | Removal of heavy metals like lead, copper, and chromium | Adsorption through natural functional groups (carboxyl, hydroxyl) | Biodegradable, inexpensive, and readily available | Low adsorption capacity; requires modification for enhanced performance |
| Corncob | Agricultural residue | Adsorption-based dyes, heavy metals, and oils | Adsorption via lignocellulosic structure | Abundant, renewable, and low-cost | Low efficiency in untreated form; requires chemical or physical modification. |
| Fish scales | Fish processing waste | Removal of heavy metals such as lead and chromium | Adsorption via bioactive compounds like collagen and hydroxyapatite | Waste valorization, cost-effective, and eco-friendly | Limited scalability and adsorbent regeneration issues |
| Seaweed (e.g., kelp, Ulva) | Marine biomass | Adsorption of heavy metals, dyes, and nutrients | Biosorption through functional groups like hydroxyl, sulfate, and carboxyl | Renewable, abundant, and capable of nutrient recovery | Requires processing; lower efficiency for organic pollutants |
| Sawdust | Wood processing waste | Adsorption-based heavy metals, dyes, and oils | Adsorption via lignocellulosic matrix | Abundant, low-cost, and sustainable | Requires chemical modification for specific pollutants |

| | | | | | |
|---|---|---|---|---|---|
| Orange peels | Citrus fruit waste | Removal of heavy metals and dyes | Functional groups like carboxyl and hydroxyl facilitate adsorption | Biodegradable, abundant, and waste valorization | The native form has low adsorption efficiency and requires chemical modification. |
| Peanut shells | Agricultural waste | Adsorption-based heavy metals, dyes, and oils | Adsorption via lignocellulosic structure | Inexpensive, biodegradable, and sustainable | Low adsorption efficiency; may require modification |
| Cuttlefish bone | Marine waste | Adsorption-based heavy metals, such as lead and cadmium | Adsorption through calcium carbonate structure and porosity | Waste valorization, abundant, and eco-friendly | Limited to specific heavy metals |
| *Aloe vera* gel | *Aloe vera* plant | Removal of dyes, heavy metals, and pathogens | Natural flocculant and adsorption through polysaccharides | Non-toxic, biodegradable, and readily available | Requires processing and limited adsorption capacity for non-polar pollutants |
| Bagasse (sugarcane residue) | Sugarcane processing waste | Removal of dyes, heavy metals, and organic pollutants | Adsorption via lignocellulosic fibers | Renewable, abundant, and cost-effective | Requires chemical treatment for high adsorption efficiency |
| Cereal bran | Wheat, rice, or oat bran | Adsorption-based heavy metals, dyes removal | Adsorption via protein and fiber content | Biodegradable, inexpensive, and sustainable | Low adsorption efficiency without modification |

ii. **Cellulose**: Cellulose is gaining attention because of its microporous membranes, which excel in precise filtration and significantly reduce pollutants. The most prevalent carbohydrate polymer found in nature, cellulose, has shown significant promise for use in industrial water treatment applications. Biosorbents derived from various plant sources, including vegetable extracts, seeds, stems, leaves, aloe vera, tamarind, date palm, and lime tree, have been studied for their effectiveness in wastewater treatment (Hokkanen et al., 2016). Cellulose modification procedures are categorized into direct modification, which bonds functional groups to hydroxyl groups and chains monomers in varying solutions. Cellulose nanocrystals (CNCs) derived from natural sources are ideal for environmental engineering because of their low density, high mechanical properties, and hydrophilic nature. Nanoscale extraction enhances the antifouling properties of CNC membranes by eliminating defects. HC, comprising 20–35% of lignocellulosic biomass, is the second most common polymer. Its branched structure and active hydroxyl groups facilitate chemical modification, expanding its potential as a biosorbent material (Khadir et al., 2021). HC-based biosorbents represent an intriguing possibility for water treatment because they have the potential to reduce environmental stress and, in turn, lower C emissions. However, the industrial use of HC-based adsorbents has not yet been realized, and more research is required to accomplish both the industrial manufacturing and practical application of these adsorbents (Khadir et al., 2021).

iii. **Lignin** is a complex biopolymer composed of three distinct monomeric phenolic substructures. These substructures originated from three separate aromatic alcohol precursors. Lignin-based biosorbents have been created using a variety of modification procedures to serve as viable alternatives to costly commercial sorbents for wastewater treatment. Owing to their high selectivity, stability, and regeneration capabilities, magnetic adsorbents based on lignin have the potential for use in water purification applications that are both cost-effective and environmentally friendly (Fouda-Mbanga et al., 2021). Lignin is potentially an effective adsorbent for removing pollutants from wastewater due to its physical porosity, which facilitates the adsorption of contaminants. Its molecular structure contains alcohol hydroxyl, carbonyl, and methoxyl groups, which can interact with specific heavy metal ions through ion exchange and chelation processes.

iv. **Starch:** A versatile and abundant natural polymer composed of amylose and amylopectin has garnered attention in the chemical, food, and pharmaceutical industries. However, native starch has a limited to low adsorption capacity due less functional groups. Chemical modification techniques have enhanced the adsorption capability of starch and broadened its application. For example, crosslinkers such as epichlorohydrin (ECH) and N, N-methylene bis-acrylamide are used to produce starch-based hydrogels with cationic and anionic groups that are capable of adsorbing both types of dyes, thus benefiting wastewater treatment (Dong et al., 2024). Cationic starch, synthesized by reacting starch with 3-chloro-2-hydroxypropyl trimethyl ammonium chloride, was intercalated into montmorillonite to create a nanocomposite that was effective in adsorbing basic dyes.

v. **Alginate:** Alginate, a polysaccharide from brown seaweeds and its derivatives, is synthesized through functionalization, enhancing the mechanical strength and stability for environmental applications via methods such as esterification and oxidation. Antibiotic contamination of wastewater in hospitals poses significant environmental challenges. Traditional sodium alginate (SA) hydrogels have limited mechanical strength and adsorption efficiency, prompting the development of modified matrices for improved wastewater treatment. Recent advancements include composite hydrogels with improved adsorption, such as those combining graphene oxide and polyacrylamide, which maintain their efficiency at high temperatures. Enzyme immobilization techniques have been studied; however, enzyme activity during recycling remains a challenge. The modified SA matrices were evaluated for the removal of heavy metals, antibiotics, and toxins from wastewater. The hydroxyl and carboxyl groups in crosslinked SA gels aid heavy metal chelation. However, the mechanical strength of SA is a concern, especially under acidic conditions. Recent studies have focused on enhancing the structural integrity and recycling by functionalizing SA with attapulgite foam, melamine, and magnetic substances (Zhang et al., 2022). Biomass immobilization, particularly that using bacteria and algae, is a viable method for heavy metal adsorption and offers renewable biosorbents. However, challenges include finding low-cost modified SA materials that provide high biological adsorption capacity and mechanical stability, as well as efficient separation of biosorbents from treated aqueous solutions. Additionally, modified alginate composites show promise for dye removal from wastewater, offering an alternative to traditional methods, such as photocatalysis and flocculation and coagulation. These composites can adsorb various dyes, including methylene blue, crystal violet, and malachite green, with minimal secondary pollution, thus making them a sustainable option for dye-containing wastewater treatment (Nasrollahzadeh et al., 2021).

vi. **Gelatin:** Another polymer that shows promise for use in water purification is gelatin, which has a high adsorption capacity and the ability to generate a hydration layer that is exceptionally stable (Thakur et al., 2017). Alginate expands and dissolves in alkaline and water-based solutions; however, this can be addressed by converting it into a hydrogel. Gelatin, a water-soluble biopolymer created from partially hydrolyzed collagen, contains functional groups that can bind to contaminants in wastewater systems. However, it has a low mechanical strength and high degradation under moist conditions, limiting its practical use. Nanoparticles can enhance mechanical stability and reduce deterioration rates. Modifying gelation molecular chains with functional groups can produce materials that can effectively remove pollutants from wastewater.

vii. **Hydroxyapatite:** Hydroxyapatite, a member of the phosphate family, is notably hard and found in animal bones and teeth. It has attracted interest for environmental pollution management because of its C and P sites, which are essential for pollutant removal. Hydroxyapatite can be synthesized from sources such as seashells, corals, eggshells, and bones from various animals. It effectively treats water by removing contaminants such as heavy metals, organic pollutants, phosphorus, and radionuclides (Ibrahim et al., 2020). Its high surface area and crystalline structure make it effective for adsorbing metal ions, such as lead, cadmium, copper, chromium, and arsenic, in neutral to slightly alkaline water. The material uses mechanisms such as adsorption, ion exchange, and precipitation to efficiently remove pollutants from industrial wastewater and for drinking water purification. Additionally, hydroxyapatite removes excess phosphates, preventing eutrophication in ecosystems affected by nutrient pollution, especially agricultural runoff. One of the unique applications of hydroxyapatite is the removal of radionuclides. This property is particularly useful in nuclear waste management, where hydroxyapatite can adsorb and immobilize radioactive materials, thereby reducing their environmental impact (Zhang & Liu, 2020). The material is also used in desalination processes to reduce salinity levels in water, although its efficiency in this regard is moderate compared to that of more established desalination methods.

To understand the use of natural biomaterials for water treatment, several examples are discussed below.

#### 10.2.1.2 Plant-Based Sources

i. **Cactus-based biomaterials** have been explored as natural substitutes for synthetic components to reduce turbidity, chemical oxygen demand (COD), and heavy metals in wastewater. Early studies have shown that cactus effectively removes turbidity from synthetic water, and further research has confirmed that cactus powder can serve as a coagulant in industrial wastewater treatment. In addition, cactus juice is a viable natural flocculant when combined with alum to successfully remove organic pollutants and heavy metals. Notably, cactus from *Opuntia ficus-indica* effectively removes chromium from wastewater in the paint manufacturing industry (Alzharani, 2024). Cactus leaves serve as biosorbents for effluent treatment, effectively removing metals from wastewater, but with lower efficiency than other low-cost adsorbents, such as natural wool. Species such as *Opuntia streptacantha* show a high efficiency in metal and dye removal (Nouri et al., 2021). Cactus fibers, which are rich in nutrients, may also serve as organic fertilizers and offer an alternative to chemical fertilizers. Cactus has been used as a packing medium to treat municipal wastewater, with dried pieces of *Opuntia imbricata* showing effective COD removal (Sosa et al., 2006). Cactus biopolymers have been explored for sludge conditioning, delivering satisfactory results compared to commercial polyelectrolytes and inorganic conditioners like $FeCl_3$ and $Al_2(SO_4)$. Additionally, cactus cell cultures and intact plants have shown the potential for transforming toxic textile dyes into less harmful metabolites, contributing to the field of phytoremediation and its associated biochemical processes.

ii. **Luffa plant:** Belonging to the *Cucurbitaceae* family, typical of tropical and subtropical climates. The main species, *Luffa cylindrica* and *Luffa* *acutangula*, are found in many countries, including China, Iran, Korea, Japan, India, Brazil, Australia, Africa, and America. China is a major producer, especially in six provinces. Historically, Japan used luffa as a filter around 1890–1895. It is primarily composed of cellulose, HC, and lignin, although these percentages can vary slightly depending on environmental conditions (Santos et al., 2024). Luffas consist of a fibrous matrix that provides structural strength and are non-toxic and widely used for bio-based materials. They are common in products such as sponges, insulation, pillow filling, potholders, and gloves. Additionally, luffa has been used in water treatment for years to effectively remove pollutants. Its porous structure adsorbs pollutants, particularly positively charged molecules, owing to its negatively charged cellulose structure. Luffas have been explored for the removal of heavy metals and dyes from water (Bhagwat & Jaspal, 2022). Untreated luffa removed 6.711 mg/g of Cd (II) at pH 7 using the Langmuir model. Another study showed 98% efficiency and 47.39 mg/g capacity for Cu removal (Khadir et al., 2021).

iii. **Sisal:** *Agave sisalana*, a plant native to Mexico and belonging to the Asparagaceae family, is prized for its strength, making it ideal for rope making. Lightweight, durable, low cost, and naturally polymeric, sisal has been widely incorporated into composite materials. Its applications span from automotive and furniture manufacturing to the construction, plastics, and paper industries. Brazil is the world's largest producer and exporter of sisal fiber, with extensive sisal farming that significantly supports its economy. Sisal fibers have a density of 1.45 g/cm$^3$, cellulose content of 85–88%, and lignin content of 4–5% (Saxena et al., 2011). Researchers have explored the potential of sisal for water and wastewater treatment. Recent studies have modified sisal with metals such as Fe, Co, Zr, Al, and Cr to enhance its adsorptive properties. Fe is used because of its low toxicity and insolubility in Fe-modified sisal, which is effective for fluoride removal through ligand exchange mechanisms (Khadir et al., 2021).

### 10.2.2 Microbial-Based Biomaterials

Microorganisms are essential for water treatment because they facilitate pollutant removal via biological mechanisms. Microorganisms, including bacteria, fungi, algae, and protozoa, can degrade organic pollutants, remove heavy metals, and aid in the overall purification of water. Microbial use in water treatment is fundamental to both natural and constructed systems, as it provides an eco-friendly and economical approach to wastewater management (Rangabhashiyam et al., 2014). Biomass can adsorb specific compounds from water through biosorption, which differs from bioaccumulation that relies on active transport in living organisms. Biosorption with dead biomass is ongoing and is driven by the affinity between the biosorbent and sorbate. Microbial treatment in biosorption improves water treatment efficiency. Bioaccumulation involves living cells absorbing contaminants, while biosorption is the continuous uptake of contaminants by biological materials (Rangabhashiyam et al.,

2014). Biosorbents used in their natural or enhanced states effectively remove pollutants from wastewater. Chemically modified biosorbents improve removal efficiency, whereas immobilizing non-living biomass enhances biosorption capacity, increases mechanical strength, and simplifies biosorbent separation post-removal.

i. **Algae:** Algae represent a diverse group of microorganisms found in both aquatic and terrestrial ecosystems. Known for their effective pollutant-binding properties, algae possess cell walls rich in functional groups such as amino, hydroxyl, carboxyl, and sulfate (Yaashikaa et al., 2021). These characteristics render them highly effective biosorbents for industrial wastewater treatment. Algal biomass, especially when dead, serves as an efficient biosorbent owing to its rapid growth, substantial yield, enhanced efficiency, and large surface area. To optimize the biosorption of heavy metals, algae must have high biosorption capacity, selectivity, and mechanical strength. Algal strains such as *Chaetophora elegans* and *Cladophora* sp. have demonstrated notable effectiveness in removing heavy metals, along with frequently available algae such as *Chlorella* sp. and *Spirulina* sp. from aqueous solutions. *Oedogonium hatei* is used as a biosorbent for extracting Ni(II) ions from aqueous solutions. The highest monolayer adsorption capacities, as determined by the Langmuir adsorption isotherm, were 40.9 mg $g^{-1}$ for untreated algae and 44.2 mg $g^{-1}$ for acid-treated algae (Gupta et al., 2010). *Ulva lactuca* is a green marine algae that provides a natural biosorbent for the extraction of acidic ions (II) of Cu, Zn, Cd, and Pb. The biosorption capacity increased with increasing pH, with the pseudo-first-order kinetic model best fitting the kinetic data (Areco et al., 2012). Research on algae as biosorbents has focused primarily on the removal of metals such as copper, zinc, and nickel. The experimental data align well with models such as the Freundlich, Toth, Khan, and Radke–Prausnitz models, as well as the pseudo-second-order kinetic model. For instance, immobilized freshwater algae were studied using Ca-alginate beads as a control mechanism for biosorption of copper, zinc, and nickel, yielding maximum biosorption capacities of 75.6, 55.2, and 30.4 mg $g^{-1}$, respectively (Bayramoğlu & Arıca, 2009).
ii. **Fungi:** Fungal biomass, which is common in the fermentation industry, has been widely researched as a biosorbent for removing harmful contaminants. Fungal cell walls contain various functional groups that are essential for heavy metal chelation. Activated carbon is used in almost every water treatment material because the adsorption capability of fine C particles facilitates large-area adsorption for dye removal. However, the disadvantage of using activated carbon for dye removal is its disposal; mycelia have been used for dye biosorption (Dhiman et al., 2022). Fungal biosorbents absorb dyes based on their characteristics and interactions during binding, the type of fungus, the preparation of the biosorbent, and the circumstances of biosorption. Species such as *Rhizopus arrhizus*, *Mucor* sp., *Saccharomyces* sp., and *Aspergillus niger* are commercially exploited, while newer species, including *Phanerochaete chrysosporium* and *Schizophyllum commune*, are being extensively investigated as efficient and cost-effective biosorbents for metal ion removal due to their availability and affordability.

**TABLE 10.2**
**Microbials in Water Treatment.**

| Microorganism | Contaminants removed | Key functional groups | Biosorption mechanism | Treatment efficiency | Advantage |
|---|---|---|---|---|---|
| **Algae** | Heavy metals commonly found (e.g., Cu, Pb, Cd, Ni), dyes (e.g., acid black 1, methylene blue) | Amino, hydroxyl, carboxyl, sulfate | Adsorption on cell surface and cell wall | High biosorption capacity, especially in modified forms | Efficacy in acidic and neutral pH; can be chemically modified for enhanced performance |
| **Fungi** | Heavy metals commonly found (e.g., Cu, Cd, Hg, Zn), dyes (e.g., reactive black 5, methylene blue), phenolic compounds | Amine, hydroxyl, carbonyl, amide groups | Ion exchange, surface adsorption, complexation | High capacity for removing a range of contaminants | Immobilization improves mechanical strength and reusability |
| **Bacteria** | Metal ions interaction (e.g., Cu, Zn, Cr), organic pollutants (e.g., dyes, phenols) | Carboxyl, amine, phosphonate, hydroxyl | Adsorption, ion exchange, chemical reduction | Versatile; can treat diverse organic and inorganic contaminants | Widely abundant and easy to cultivate; efficient in various pH ranges |

iii. **Bacterial:** The bacterial cell wall is primarily in contact with contaminants in the environment, and metal ions or dyes are adsorbed on the surface or within the cell wall. Functional groups within the cell wall, including carboxyl, amine, phosphonate, and hydroxyl groups, play crucial roles in the biosorption process. Bacterial biosorption has attracted significant interest from researchers because of its abundant availability, versatility, and widespread presence in terrestrial biomass (Gupta & Diwan, 2017). The ability of bacteria to biosorb a wide range of organic and inorganic contaminants in wastewater is well documented (see Table 10.2).

## 10.3 SYNTHETICALLY OCCURRING BIOMATERIALS

Synthetic biomaterials are primarily either organic or inorganic. Organic biomaterials mostly include polymers that are adaptable and can be customized for many medicinal uses. Inorganic biomaterials include metals, their oxides, and ceramics, which are often used for their strength, durability, and biocompatibility. These materials are essential for applications such as implants, prosthetics, and tissue engineering (see Table 10.3).

**TABLE 10.3**
**Synthetic Biomaterials for Water Treatment.**

| Synthetic biomaterial | Composition/source | Applications in water treatment | Mechanism of action | Advantages | Limitations |
|---|---|---|---|---|---|
| Polyacrylamide (PAM) | Synthetic polymer from acrylamide | Flocculation of suspended solids, removal of turbidity | Forms bridges between particles to enhance floc formation | High efficiency in flocculation, customizable molecular weight | Potential toxicity and limited biodegradability |
| Polyaniline (PANI) | Conductive polymer | Adsorption of heavy metals, dyes, and organic pollutants | Adsorption via π-π interactions, hydrogen bonding, and ion exchange | High chemical stability, reusability, and conductivity | Synthesis complexity and potential environmental impact |
| Polyvinyl alcohol (PVA) | Synthetic polymer | Adsorption of dyes, oils, and heavy metals | Crosslinking creates a porous structure for trapping pollutants | Biocompatible, water soluble, and easy to modify | Requires crosslinking agents for effective use |
| Polyurethane foams | Polyurethane-based material | Removal of oils, organic solvents, and dyes | Adsorption via porous and hydrophobic structures | Lightweight, highly absorbent, and reusable | Limited adsorption for polar contaminants |
| Polyethylene glycol (PEG) | Synthetic polyether polymer | Adsorption of heavy metals and dyes | Functionalized PEG interacts with contaminants | Biocompatible, hydrophilic, and low toxicity | Requires modification for specific applications |
| Synthetic zeolites | Crystalline aluminosilicates | Adsorption of ammonia, heavy metals, and organic pollutants | Ion exchange and molecular sieving | High chemical stability and reusability | Limited adsorption capacity for non-ionic pollutants |
| Polysulfone membranes | Sulfonated polymer | Filtration of particulates, heavy metals, and pathogens | Physical sieving and surface adsorption | High chemical and thermal stability | Fouling and low selectivity for dissolved contaminants |
| Polypropylene microfibers | Synthetic polymer fibers | Removal of oil and hydrophobic organic pollutants | Adsorption via hydrophobic interactions | Durable, reusable, and cost-effective | Limited to hydrophobic pollutants |

| | | | | | |
|---|---|---|---|---|---|
| Synthetic hydrogels | Crosslinked polymer networks | Adsorption of dyes, heavy metals, and oils | Hydrophilic and porous network traps pollutants | High water absorption, customizable properties | Mechanical instability in some applications |
| Nanocellulose-based composites | Modified cellulose fibers | Removal of heavy metals, dyes, and oils | Surface functionalization enhances adsorption properties | Renewable, biodegradable, and tunable functionality | Production and scaling challenges |
| Synthetic clay composites | Engineered layered silicates | Removal of dyes, heavy metals, and organic pollutants | Adsorption via cation exchange and surface interactions | High surface area, reusability, and customizable properties | Costlier than natural clays; synthesis requires precise control |
| Polystyrene microspheres | Synthetic polymer microspheres | Removal of heavy metals, organic pollutants, and pathogens | Adsorption through surface functionalization | Easy to functionalize and effective for targeted pollutant removal | Non-biodegradable and potential microplastic pollution |
| Synthetic biopolymers (e.g., PLA and PCL) | Polylactic acid (PLA), polycaprolactone (PCL) | Adsorption of dyes and heavy metals | Surface functionalization allows pollutant trapping | Biodegradable, tunable properties, and eco-friendly | Lower adsorption capacity compared to advanced synthetic adsorbents |

### 10.3.1 Organic Biomaterials

Primarily composed of carbon-based molecules. Polylactic acid (PLA) and polyglycolic acid (PGA) are synthetic polymers. They provide flexibility, biocompatibility, and the ability to be tailored to specific applications.

i. **PLA:** Biopolymer, known as PLA, is generated from starch feedstock obtained from renewable resources. Thus, they have the potential to be economically useful. As a result, the emergence of such biopolymers, with compatibility, degradability, superior mechanical strength, and processability, has made them most considerable for water treatment rapidly (Milovanovic et al., 2021). The fabrication of porous PLA materials involves the integrated development of technologies such as electrospinning, foaming, freeze-drying, and thermally induced phase separation, as well as 3D printing. PLA materials are frequently used to treat oily or colored wastewater products.

ii. **Polyhydroxyalkanoates (PHAs)**: PHAs are promising new biopolymers used as an alternative to traditional petroleum-based chemical plastics. This is owing to the fact that PHAs are compatible with old plastic products and have good biodegradability (Hiraishi & Khan, 2003). Short-chain PHAs and medium-length-chain PHAs are two categories that may be readily differentiated from one another. The properties of these PHAs can be modified by varying the strain, feedstock, and fermentation process used. This structural diversity provides PHAs with a variety of qualities that are not present in biomaterials, such as PLA. As a result, PHAs can be used in a variety of applications. Samorì et al. (2015) coupled thermochemical and biotransformation processes in order to valorize the carbon content and chemical energy of anaerobic digested sludge.

iii. **PGA:** PGA is a biodegradable polymer with significant potential for wastewater treatment, owing to its eco-friendly nature and rapid hydrolytic degradation. Upon degradation, PGA produces glycolic acid, a non-toxic byproduct that can be metabolized by microorganisms, making it an ideal material for sustainable water remediation. In biological wastewater treatment processes, PGA is often used as a bio-carrier or scaffold for microbial biofilms to support the growth of microorganisms that efficiently degrade organic pollutants (Wang et al., 2024). Functionalized PGA adsorbs heavy metals and organic contaminants with modifications that enhance its affinity. It is utilized in controlled-release systems as a remediation agent to ensure effective treatment. PGA-derived membranes have been explored for microfiltration and ultrafiltration, providing eco-friendly alternatives to synthetic membranes. Despite its biodegradability and sustainability, PGA faces challenges such as high production costs and limited stability under prolonged hydrolysis, hindering its large-scale use.

### 10.3.2 Inorganic Biomaterials

i. **Zeolites:** Zeolites are 3D aluminosilicate minerals containing cations ($Na^+$, $K^+$, $Ca^{2+}$, and $Mg^{2+}$) and water to balance their charge. Adsorption is an effective wastewater treatment technology, and experimental conditions are crucial for pollutant removal. Regenerating zeolites is vital for waste reduction. Natural zeolites have been studied for the removal of metal ions from wastewater, whereas clinoptilolite has been evaluated as an adsorbent for dyes and phenolic compounds. Clinoptilolite effectively removes metal cations from copper mine effluents, including $Al^{3+}$, $Cd^{2+}$, $Cu^{2+}$, $Fe^{3+}$, $Ni^{2+}$, $Pb^{2+}$, and $Zn^{2+}$. Clinoptilolite has a strong selectivity for heavy metal ions in the following order: $Pb^{2+}$, $Cd^{2+}$, $Cu^{2+}$, $Co^{2+}$, $Cr^{3+}$, $Zn^{2+}$, $Ni^{2+}$, and $Hg^{2+}$. Stylianou et al. (2007) examined the absorption of $Cu^{2+}$ by clinoptilolite in aqueous solution at various temperatures (25–60°C), pH values (1–4), and agitation speeds (0–400 rpm). Zeolite was the most effective at removing copper ions at a pH of 4. Research has indicated that agitation speed affects the level of elimination. The most efficient elimination (37.3%) occurred at 400 rpm and 25°C.

ii. **Metals and their oxides:** Metal and metal oxide nanoparticles have garnered significant attention in chemistry, material science, and physics, owing to their unique properties and versatile applications (Naseem & Durrani, 2021). These offer promising solutions for wastewater treatment and can be attributed to their unique properties and remarkable pollutant removal efficiency. This category includes nanoparticles, nanocomposites, and other metal-based nanomaterials, characterized by their high surface area and reactivity (Rana et al., 2023). These characteristics make them highly effective as adsorbents and catalysts, capable of efficiently removing contaminants from wastewater without the need for harmful chemicals or high-energy input. Their small size and high surface-area-to-volume ratio further enhance their suitability for advanced water treatment applications.

    **Iron oxide nanoparticles** are extensively used in wastewater treatment to remove heavy metals owing to their high surface area, magnetic properties, and ease of separation. Variants such as magnetic magnetite ($Fe_3O_4$), maghemite ($\gamma$-$Fe_2O_4$), and nonmagnetic hematite ($\alpha$-$Fe_2O_3$) are frequently employed as nanosorbents (Jabbar et al., 2022). These nanoparticles exhibit a high adsorption capacity for hazardous metals such as Pb, Zn, Hg, Ni, Cd, and Cr, which pose risks even at low concentrations. Ligands such as EDTA, glutathione, and copolymers enhance their adsorption efficiency.

    **Silver nanoparticles (AgNPs)** are known for their potent antibacterial properties and are widely used for water disinfection. AgNPs show strong antibacterial properties, even at low concentrations, and are effective against several wastewater pathogens (Zahoor et al., 2021). Hybrid formulations, such as polymer–silver composites, further enhance AgNPs' capacity for heavy metal and organic pollutant removal.

**$TiO_2$ nanoparticles** are among the most promising photocatalysts for water purification because of their high photoactivity, stability, and low cost. $TiO_2$ efficiently degrades organic pollutants, heavy metals, and harmful pathogens through UV-induced photocatalytic reactions. Adding elements such as Zn, Cr, and Ag reduces the bandgap energy and enhances the photocatalytic performance under visible light. $TiO_2$ nanoparticles are used in anti-biofouling coatings and membrane technologies for wastewater treatment. Their durability and eco-friendliness make $TiO_2$ nanoparticles a sustainable option for long-term water treatment applications (Mathur et al., 2015).

**ZnO nanoparticles** exhibit exceptional photocatalytic and antibacterial properties, making them suitable for pollutant removal and pathogen control. ZnO nanoparticles generate ROS, which causes oxidative stress and disrupts bacterial membranes. These are effective in breaking down organic pollutants in wastewater (Deak et al., 2019). ZnO combined with other materials, such as chitosan or bentonite, enhances antibacterial activity and pollutant adsorption.

**Copper nanoparticles** are effective against a broad spectrum of bacteria and are being increasingly used in wastewater treatment. Cu-based nanoparticles, such as $CuFe_2SO_4$, exhibit excellent sorption capacities for organic and inorganic pollutants. Although slightly less effective than silver nanoparticles, CuNPs are cost-efficient and versatile, making them a valuable addition to water treatment technologies (Morsi et al., 2017).

Metal and metal oxide nanoparticles offer transformative solutions for wastewater treatment by efficiently removing pollutants and pathogens. Their unique properties, such as their high surface area, magnetic activity, and photocatalytic potential, enable diverse applications.

iv. **Ceramics:** Ceramic membranes have emerged as a promising technology for water treatment because of their exceptional durability, robustness, and efficiency in handling harsh applications, such as oil–water separation, saline water treatment, and the removal of refractory organic and emerging contaminants (Dong et al., 2022). These membranes are valued for their high mechanical, thermal, and chemical stabilities, as well as their resistance to fouling, which contributes to a longer lifespan and reduced maintenance compared to polymeric alternatives. However, the widespread adoption of ceramic membranes is challenging, primarily because of their high capital and processing costs. The use of expensive materials, such as alumina and zirconia, coupled with energy-intensive sintering processes, increases the costs. Additionally, issues such as fouling, limited trade-offs between permeability and selectivity, and potential secondary pollution hinder the operational efficiency (Dong et al., 2022).

## 10.4 CONCLUSION

Bio-based materials offer sustainable and efficient solutions for water treatment, addressing critical challenges, such as water scarcity and pollution. Derived from renewable resources, these materials, such as cellulose, chitosan, and lignin, are cost-effective, biodegradable, and effective in adsorbing contaminants, such as heavy metals, dyes, and pathogens. Natural biomaterials provide eco-friendly alternatives, whereas synthetic options enhance their functionality and scalability. Despite their advantages, challenges such as high cost, scalability, and the need for modification remain. Continued innovation is essential for improving efficiency and bridging the gap between research and industrial application. Owing to their versatility and environmental benefits, bio-based materials hold great promise for revolutionizing water purification technologies. Their adoption can play a vital role in achieving the global water sustainability goals.

## REFERENCES

Alzharani, A. A. (2024). Activated Opuntia ficus-indica as an efficient adsorbent for iodine and heavy metals Cr (III), Cu (II) and Ni (II) ions uptake for wastewater treatment. *Journal of Umm Al-Qura University for Applied Sciences*, 1–19.

Areco, M. M., Hanela, S., Duran, J., & dos Santos Afonso, M. (2012). Biosorption of Cu (II), Zn (II), Cd (II) and Pb (II) by dead biomasses of green alga Ulva lactuca and the development of a sustainable matrix for adsorption implementation. *Journal of Hazardous Materials*, *213*, 123–132.

Bayramoğlu, G., & Arıca, M. Y. (2009). Construction a hybrid biosorbent using Scenedesmus quadricauda and Ca-alginate for biosorption of Cu (II), Zn (II) and Ni (II): kinetics and equilibrium studies. *Bioresource Technology*, *100*(1), 186–193.

Bhagwat, S. B., & Jaspal, D. (2022). Applications of luffa based biomaterials in textile waste water. In *Textile Wastewater Treatment: Sustainable Bio-nano Materials and Macromolecules* (Vol. 2, pp. 71–88). Springer.

Deak, G., Dumitru, F. D., Moncea, M. A., Panait, A. M., Baraitaru, A. G., Olteanu, M. V., Boboc, M. G., & Stanciu, S. (2019). Synthesis of ZNO nanoparticles for water treatment applications. *International Journal of Conservation Science*, *10*(2).

Dhiman, V. K., Singh, D., Pandey, H., Chauhan, D., Dhiman, V. K., & Pandey, D. (2022). Mycoremediation of agricultural waste for the cultivation of edible mushroom. In T. Aftab (Ed.), *Sustainable Management of Environmental Contaminants: Eco-friendly Remediation Approaches* (pp. 471–482). Springer International Publishing. https://doi.org/10.1007/978-3-031-08446-1_18

Dong, Y., Ghasemzadeh, M., Khorsandi, Z., Sheibani, R., & Nasrollahzadeh, M. (2024). Starch-based hydrogels for environmental applications: a review. *International journal of biological macromolecules*, 131956.

Dong, Y., Wu, H., Yang, F., & Gray, S. (2022). Cost and efficiency perspectives of ceramic membranes for water treatment. *Water Research*, *220*, 118629.

Fouda-Mbanga, B., Prabakaran, E., & Pillay, K. (2021). Carbohydrate biopolymers, lignin based adsorbents for removal of heavy metals (Cd2+, Pb2+, Zn2+) from wastewater, regeneration and reuse for spent adsorbents including latent fingerprint detection: a review. *Biotechnology Reports*, *30*, e00609.

Gupta, P., & Diwan, B. (2017). Bacterial exopolysaccharide mediated heavy metal removal: a review on biosynthesis, mechanism and remediation strategies. *Biotechnology Reports*, *13*, 58–71.

Gupta, V. K., Rastogi, A., & Nayak, A. (2010). Biosorption of nickel onto treated alga (Oedogonium hatei): application of isotherm and kinetic models. *Journal of Colloid and Interface Science*, *342*(2), 533–539.

Hiraishi, A., & Khan, S. (2003). Application of polyhydroxyalkanoates for denitrification in water and wastewater treatment. *Applied Microbiology and Biotechnology*, *61*, 103–109.

Hokkanen, S., Bhatnagar, A., & Sillanpää, M. (2016). A review on modification methods to cellulose-based adsorbents to improve adsorption capacity. *Water Research*, *91*, 156–173.

Ibrahim, M., Labaki, M., Giraudon, J.-M., & Lamonier, J.-F. (2020). Hydroxyapatite, a multifunctional material for air, water and soil pollution control: a review. *Journal of Hazardous Materials*, *383*, 121139.

Jabbar, K. Q., Barzinjy, A. A., & Hamad, S. M. (2022). Iron oxide nanoparticles: preparation methods, functions, adsorption and coagulation/flocculation in wastewater treatment. *Environmental Nanotechnology, Monitoring & Management*, *17*, 100661.

Jiang, C., Wang, X., Wang, G., Hao, C., Li, X., & Li, T. (2019). Adsorption performance of a polysaccharide composite hydrogel based on crosslinked glucan/chitosan for heavy metal ions. *Composites Part B: Engineering*, *169*, 45–54. https://doi.org/10.1016/j.compositesb.2019.03.082

Khadir, A., Negarestani, M., Kheradmand, A., Azad, A., & Sillanpää, M. (2021). The utilization of biomaterials for water purification: dyes, heavy metals, and pharmaceuticals. In S. S. Muthu & A. Khadir (Eds.), *Novel Materials for Dye-containing Wastewater Treatment* (pp. 27–58). Springer Singapore. https://doi.org/10.1007/978-981-16-2892-4_2

Lustenberger, S., & Castro-Muñoz, R. (2022). Advanced biomaterials and alternatives tailored as membranes for water treatment and the latest innovative European water remediation projects: a review. *Case Studies in Chemical and Environmental Engineering*, *5*, 100205.

Mathur, A., Kumari, J., Parashar, A., Chandrasekaran, N., & Mukherjee, A. (2015). Decreased phototoxic effects of $TiO_2$ nanoparticles in consortium of bacterial isolates from domestic waste water. *PLoS One*, *10*(10), e0141301.

Milovanovic, S., Markovic, D., Pantic, M., Pavlovic, S. M., Knapczyk-Korczak, J., Stachewicz, U., & Novak, Z. (2021). Development of advanced floating poly (lactic acid)-based materials for colored wastewater treatment. *The Journal of Supercritical Fluids*, *177*, 105328.

Morsi, R. E., Alsabagh, A. M., Nasr, S. A., & Zaki, M. M. (2017). Multifunctional nanocomposites of chitosan, silver nanoparticles, copper nanoparticles and carbon nanotubes for water treatment: antimicrobial characteristics. *International Journal of Biological Macromolecules*, *97*, 264–269.

Naseem, T., & Durrani, T. (2021). The role of some important metal oxide nanoparticles for wastewater and antibacterial applications: a review. *Environmental Chemistry and Ecotoxicology*, *3*, 59–75.

Nasrollahzadeh, M., Sajjadi, M., Iravani, S., & Varma, R. S. (2021). Starch, cellulose, pectin, gum, alginate, chitin and chitosan derived (nano)materials for sustainable water treatment: a review. *Carbohydrate Polymers*, *251*, 116986. https://doi.org/10.1016/j.carbpol.2020.116986

Nouri, H., Abdedayem, A., Hamidi, I., Najjar, S. S., & Ouederni, A. (2021). Biosorption of lead heavy metal on prickly pear cactus biomaterial: kinetic, thermodynamic and regeneration studies. *Cellulose Chemistry and Technology*, *55*, 919–932.

Pandey, H., Singh, D., Dhiman, V. K., Dhiman, V. K., & Pandey, D. (2022). Extremophiles and their application in bioremediation. In A. B. Gunjal, R. Thombre, & J. A. Parray (Eds.), *Physiology, Genomics, and Biotechnological Applications of Extremophiles* (pp. 188–206). IGI Global. https://doi.org/10.4018/978-1-7998-9144-4.ch009

Rana, G., Dhiman, P., Kumar, A., Sharma, G., Verma, Y., & Chauhan, A. (2023). Functionalization of two-dimensional MXene-based nanomaterials for water purifications and energy conversion applications: a review. *Materials Science in Semiconductor Processing*, *165*, 107645. https://doi.org/10.1016/j.mssp.2023.107645

Rangabhashiyam, S., Suganya, E., Selvaraju, N., & Varghese, L. A. (2014). Significance of exploiting non-living biomaterials for the biosorption of wastewater pollutants. *World Journal of Microbiology and Biotechnology*, *30*, 1669–1689.

Samorì, C., Abbondanzi, F., Galletti, P., Giorgini, L., Mazzocchetti, L., Torri, C., & Tagliavini, E. (2015). Extraction of polyhydroxyalkanoates from mixed microbial cultures: impact on polymer quality and recovery. *Bioresource Technology*, *189*, 195–202.

Santos, M. A., De Conto, J. F., Borges, G. R., & Egues, S. M. (2024). Luffa cylindrica as a biosorbent in wastewater treatment applications: a comprehensive review. *Cellulose*, *31*(17), 10115–10142. https://doi.org/10.1007/s10570-024-06206-y

Sarode, S., Upadhyay, P., Khosa, M. A., Mak, T., Shakir, A., Song, S., & Ullah, A. (2019). Overview of wastewater treatment methods with special focus on biopolymer chitin-chitosan. *International Journal of Biological Macromolecules*, *121*, 1086–1100.

Saxena, M., Pappu, A., Haque, R., & Sharma, A. (2011). Sisal fiber based polymer composites and their applications. In *Cellulose Fibers: Bio-and Nano-polymer Composites: Green Chemistry and Technology* (pp. 589–659). Springer.

Singh, D., Pandey, H., Dhiman, V. K., Thakur, K., Sharma, I., Dhiman, V. K., Singh, V., & Pandey, D. (2022). Chapter 10—Assessment of environmental impacts of metal/metalloid pollution on plants. In T. Aftab & K. Hakeem (Eds.), *Metals Metalloids Soil Plant Water Systems* (pp. 217–232). Academic Press. https://doi.org/10.1016/B978-0-323-91675-2.00022-6

Sosa, G., Estrada-Rivera, R., Garza García, Y., & Rodriguez-Martinez, J. (2006). Azo dye decolorization in a down flow aerobic biofilm reactor with Opuntia imbricata as support. *Moscow University Chemistry Bulletin*, *61*, 31–34.

Stylianou, M. A., Hadjiconstantinou, M. P., Inglezakis, V. J., Moustakas, K. G., & Loizidou, M. D. (2007). Use of natural clinoptilolite for the removal of lead, copper and zinc in fixed bed column. *Journal of Hazardous Materials*, *143*(1–2), 575–581.

Thakur, K., Pandey, H., Sharma, I., Dhiman, V., & Pandey, D. (2021). Siderophores and their applications in heavy metal detoxification. In *Plant-Microbial Interactions and Smart Agricultural Biotechnology* (pp. 253–278). CRC Press.

Thakur, S., Govender, P. P., Mamo, M. A., Tamulevicius, S., & Thakur, V. K. (2017). Recent progress in gelatin hydrogel nanocomposites for water purification and beyond. *Vacuum*, *146*, 396–408.

Tyagi, S., Dhiman, V. K., Dhiman, V. K., Pandey, H., Singh, D., Sharma, A., Sharma, P., Kumar, R., Lee, K.-J., & Saharan, B. S. (2024). Plant defense strategies and biomarkers against heavy metal-induced stress: a comprehensive review. *ACS Agricultural Science & Technology*, *4*(2), 129–143. https://doi.org/10.1021/acsagscitech.3c00271

Wang, Z., Li, C., Yang, W., Wei, Y., & Li, W. (2024). Denitrification performance and microbiological mechanisms using polyglycolic acid as a carbon source. *Water*, *16*(9), 1277.

Wu, S., Shi, W., Li, K., Cai, J., & Chen, L. (2022). Recent advances on sustainable bio-based materials for water treatment: fabrication, modification and application. *Journal of Environmental Chemical Engineering*, *10*(6), 108921.

Yaashikaa, P. R., Kumar, P. S., Saravanan, A., & Vo, D.-V. N. (2021). Advances in biosorbents for removal of environmental pollutants: a review on pretreatment, removal mechanism and future outlook. *Journal of hazardous materials*, *420*, 126596.

Zahoor, M., Nazir, N., Iftikhar, M., Naz, S., Zekker, I., Burlakovs, J., Uddin, F., Kamran, A. W., Kallistova, A., & Pimenov, N. (2021). A review on silver nanoparticles: classification, various methods of synthesis, and their potential roles in biomedical applications and water treatment. *Water*, *13*(16), 2216.

Zhang, P., Yuan, L., Zeng, J., Zou, K., Liu, B., Qing, T., & Feng, B. (2022). Alginate production of Pseudomonas strains and its application in preparation of alginate-biomass hydrogel for heavy metal adsorption. *International Journal of Biological Macromolecules*, *222*, 1511–1521.

Zhang, X., & Liu, Y. (2020). Nanomaterials for radioactive wastewater decontamination. *Environmental Science: Nano*, *7*(4), 1008–1040.

# 11 Challenges and Future Directions in Bio-Based Materials

*R.G. Abaszade, E.A. Khanmamadova, and Naveen Mani Tripathi*

## 11.1 INTRODUCTION

Building capacities for the anticipatory governance of emerging technologies, conducting responsible research and innovation, and placing discussions of these innovations within broader discussions about societal futures have all gained prominence in recent years (Grunwald, 2017; Guston, 2014). In relation to menthol production, our research is situated in the framework of both the anticipatory governance of synthetic biology and the broader end-to-end evaluation of biosynthetic menthol along both upstream and downstream dimensions for a range of players and sectors. One method that is pertinent to an end-to-end evaluation is the real-time technological appraisal (Robert et al., 2013). These are very new developments in synthetic biology: menthol is prepared in the laboratory by genetic engineering of bacteria, yielding a product called biosynthetic menthol (Toogood et al., 2015). The next section outlines broader research on the different methodologies for the obtention of menthol, showing how traditional feedstock sources and synthetic menthol move to new biosynthetic routes. Markets are ready for microbial cell factory-derived compounds. Front-runners of this business, such as the USA-based Amyris and Switzerland-based Evolva, have harnessed yeast for orange and woody flavors and fragrances (Schempp et al., 2018). Meanwhile, many other workers in the biotechnology and fermentation field have begun to probe the microbial synthesis of menthol and many other terpenoids (Leffingwell, 2015). The value of such novel biosynthetic products will involve economic and sustainability goals directly meeting significant environmental and societal problems (Wurtzel and Kutchan, 2016). The text will follow multiple dimensions of the sociotechnical transition from conventional to biosynthetic menthol production, highlighting critical attention and expectations inherent in such a turn. This case research on the transition from conventional to biosynthetic menthol underlines how uncertainties are managed within public practices and how these adapt to navigate the challenges posed by novel technologies (Meckin and Balmer, 2018). The discussion resonates with broader themes in climate change mitigation, particularly the need for a paradigm shift in energy production—moving from fossil fuels to renewable sources (REN21, 2015). Projections underline the increasingly rising

DOI: 10.1201/9781003566205-11

importance of bioenergy in the global energy mix: IRENA projects that 108 EJ per year in biomass use would be needed for achieving the SE4All objective of doubling renewable energy's share in the energy mix by year 2030 (Nakada et al., 2014; World Energy Outlook, 2013; GEA, 2012). Bioenergy use shows significant change between 2011 and 2035, which is complemented by global energy trends (Bruckner et al., 2014). According to the Intergovernmental Panel on Climate Change, depending on different climate policies and programs, the use of bioenergy could attain 200 EJ per year toward the end of this century. Most important among these resources are agricultural residues (Chum et al., 2011). The IPCC Special Report on Renewable Energy estimates that their contribution could fall between 15 and 70 EJ/yr in 2050. More recent analyses bring these figures down to 13–30 EJ annually by 2030, then leveling off at 27–30 EJ by mid-century. Residue biomass is thus becoming a crucial vector for driving the global energy landscape toward sustainability and underlines its contribution to a low-carbon transition of energy (Daioglou et al., 2016).

## 11.2 ACQUISITION AND ANALYSIS OF NEW MATERIALS FROM BIOMASS

The research was carried out by synthesizing new bio-based linear aliphatic poly (ester amides) (PEAs) with controlled regioselectivity using a two-step procedure. First, the synthesis of one diol–diamide derivative, followed by a conventional polycondensation between alcohol and acid, was effectively performed. Because diacid and ethanolamine undergo a condensation process with high conversion rates, NMR analyses verified that diol–diamide could be obtained in a single step. Subsequently, a vast collection of novel polymers was created and described. Using sophisticated spectroscopic methods, such as $^{1}H$ and $^{13}C$ NMR and $^{1}H$-$^{13}C$ HSQC NMR spectroscopy, the chemical reactivity of functional groups, including acids, alcohols, and amines, was thoroughly investigated. The intricate processes of diesterification, monoesterification, monoamidation, and diamidation that take place when dicarboxylic acids interact with amines or alcohols have also been extensively documented. All of the synthesized polymers exhibited good stability according to thermal degradation experiments. The regiocontrol during synthesis was the cause of the elevated melting temperatures found in the thermal property study results. Significantly, a greater crystallization rate was discovered for these polymers, which is a crucial property for effective polymer melt processing. Since they behaved very similarly to polyamides in terms of solubility, they could be considered as biodegradable substitutes where the need for eco-friendly materials is significant.

### 11.2.1 New Bio-based Linear PEA Synthesis

The search for sustainable materials and the circular economy has generated much interest in developing bio-based materials and innovative processing techniques, resulting in a wide variety of experimental bio-based foams. Most of these foams show thermal and mechanical properties comparable to expanded polystyrene. On the other hand, many basic knowledge gaps still exist, limiting our understanding of the principles governing such performances. This chapter discusses developments in this fast-changing field regarding material selection, processing methods, performance comparisons, and attempts to model physical properties; the issue of applying theoretical models to actual bio-based systems is also discussed. The trend in this area is to develop completely bio-based and compostable foams, although partially bio-based polyurethane foams have been performing satisfactorily for many years. In this regard, an investigation of the correlation between the structure and properties of foam was conducted using data on thermal conductivity, cell size, and density. Models of heat transfer have served to identify some key variables for which the structural complexity of bio-based foams poses a challenge. Despite these advances, the literature remains sparse in reporting data directly relevant to the insulation capabilities of such materials. While not a robust predictive tool, this approach underscores how existing data can guide the identification of promising materials and methods for future development in the field (Thi Hang et al., 2019).

### 11.2.2 Thermal Insulation Using Bio-based Foams: Selection, Processing, Modeling, and Performance

Most of these foams boast equal thermal and mechanical performances similar to that of expanded polystyrene. However, key knowledge gaps remain in the principles that define such performance. This chapter highlights progress made in this constantly changing field in the areas of material selection, processing methodologies, performance comparisons, and attempts at physical property modeling. Also included are the problems that exist in trying to apply theoretical models to practical bio-based systems. Most attention is given to low thermal conductivity—one of the key attributes sought for temperature-sensitive applications in frozen or refrigerated food, medical supplies, and vaccine packaging, as well as for building insulation. Models of heat transport have been instrumental in identifying key variables where the structural complexity of bio-based foams presents challenges. Despite these advances, literature still needs to be more extensive in reporting data directly relevant to the insulation capabilities of such materials. Dimensional analysis has provided some insight, producing a power-law correlation between thermal conductivity and relative density. Although this approach is not a good predictive tool, it does point out how the available data can be used to guide the search for promising materials and methods for further development in the field (Rebecca et al., 2021).

### 11.2.3 Biomass's Potential Contribution to the World's Energy Grid

Numerous studies have been conducted on the potential contribution of biomass to the global primary energy supply. These evaluations are typically carried out using resource-focused methodologies, which quantify the biomass's physical resources for energy conversion. Subsequent classifications lead to inventories of possible bioenergy resources, which are then evaluated for energy-use feasibility. Actually, the studies account for future demand for both food and non-food crops; on the supply side, they take into account land-use efficiency and the amount of agricultural and forestry land available due to current demand for food and non-food items. As a result, many studies have concentrated more on demand, comparing the competitiveness of power generated from biomass and biofuels or estimating the amount of biomass required to meet predetermined targets for climate-neutral energy. Based on anticipated energy end-use demand and the competitiveness of alternative energy sources, these evaluations might be interpreted as attempts to estimate the market penetration of bioenergy (Verma et al., 2024a, 2024b; Chen et al., 2019).

### 11.2.4 The Amount of Water Needed for the Production of Energy Crops, Biomass-Based Power, and Fuel

The process by which water vapor diffuses from the internal tissues of the leaves through the stomata into the surrounding environment in the opposite direction to that taken by carbon dioxide is known as transpiration. The transpiration pull also draws water from both the soil and the plant surfaces by means of evaporation. These are collectively known as evapotranspiration losses. Water use efficiency (WUE) can be described as the output of photosynthesis, or biological, or economic yield per unit of water used, and is usually expressed in terms of transpiration, evapotranspiration, or applied water per unit of water. Agronomics, engineering, and basin-level planning can use the WUE at a variety of scales, including leaf, plant, field, and ecosystem. WUE is commonly defined in this context as the quantity of dry aboveground biomass generated per unit of evapotranspiration water. Impacts from the emerging bioenergy business go far beyond future water availability and consumption. Large-scale energy crop production may lead to a considerable rise in water appropriation, probably equivalent to the quantity utilized for worldwide cropland evapotranspiration, according to the chapter's conclusion. While such expansion might exacerbate water scarcity in countries already facing stress, others currently not under strain may also experience worsening conditions. However, there are nations where these impacts are unlikely to be as pronounced. Although irrigation for energy crops is a key factor, rain-fed agriculture also plays a substantial role in water-scarce regions. Because water will be an increasingly limiting factor, one of the most critical areas of future research is the determination of bioenergy potential in light of competition for water. It is possible to ascertain basin-scale capacity to sustain large-scale biomass production for energy using tools that forecast future water supply and consumption. Biomass production can have ancillary benefits like erosion control, flood prevention, the creation of carbon sinks, and income from biomass sales when integrated into land-use planning appropriately.

### 11.2.5 Evaluation of Biofibers, Biofilms, Biopolymers, and Biocomposite Materials as Sustainable and Renewable Resources

The current state of the world has influenced the development of new bio-based materials by their crucial advantages in replacing hazardous and synthetic materials. The development of processing methods, characterizations, future applications, and strategies to get around the limitations of biofibers, biopolymers, biofilms, and biocomposites has thus been the main topic of the current perspective section. Researchers and scientists are given vital information about advanced bio-based materials and their composites for possible use in biomedical, commercial, and engineering fields. By using renewable bio-based materials in engineering and construction, waste, landfills, and toxic emissions will be decreased, improving sustainability and creating a cleaner, greener environment (Rebecca et al., 2021; Verma et al., 2024a, 2024b; Chen et al., 2019).

#### 11.2.5.1 Biofibers and Biofibers from Plants

Minerals, plants, and animals are some of the natural sources of fiber. Their origin, geographic location, extraction techniques, and treatment procedures all affect their physical and chemical characteristics. These natural fibers are widely used for structural purposes and as reinforcement in polymer matrices. Typically, bio-fibers are divided into two categories: plant-based and animal-based. The following sections go over these biofibers' intricate structure.

Depending on where they come from, wool, another common textile fiber, also shows physicochemical divergence. As a result, wool fibers such as alpaca, angora, and qiviut have diameters that fall between 12 and 29 μm, 12 and 16 μm, and 15 and 20 μm, respectively. In essence, wool is made up of keratin, a structural protein that is also found in chicken feathers and animal hair (see Figure 11.1).

#### 11.2.5.2 Development of Biorefineries in Cattle Production Systems: Uses, Difficulties, and Prospects for Further Study

The policies of many institutions are already focused on the attainment of sustainable development goals, one of which is to reduce natural and energy resource consumption. Regarding these aspects, livestock farming is among those anthropogenic sources that have become one of the most important sources for the emission of GHG gases and acidifying gases; this, therefore, requires careful assessment for ecological

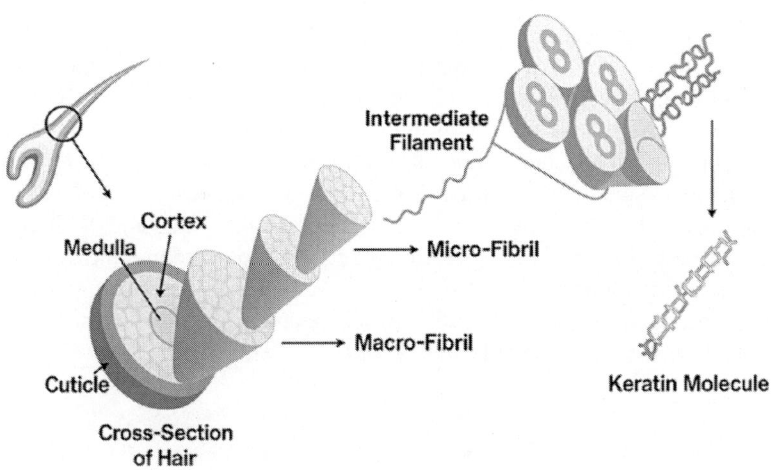


**FIGURE 11.1** The image you provided shows the structure of keratin, specifically in human hair.

impact. Livestock is a resource-intensive production that involves huge water, soil, and energy use; it also includes biowaste management and vaporization. The understanding and optimization of these processes are fundamental for the reduction of their environmental impact. One of the efficient ways is to effectively utilize residual bioresources and organic waste, which would definitely help in converting discarded material into value-added products while consuming a minimum number of natural resources. Biorefinery processes, in which bioresources are transformed into bioenergy or bioproducts, are hence considered very promising. In particular, biorefinery methods applied to animal-derived wastes might disclose new resources while simultaneously reducing resource consumption. A total of 293 scientific publications from the last 11 years on biorefinery applications in livestock biomass are relevant points of interest. The main recent results show a strongly increasing energy production, besides farming of insects in new, alternative ways. To achieve this, efforts must be focused on reducing costs, ensuring economies of scale, managing processes more effectively, and creating synergy among industrial sectors. These are the basic elements that must be in place to allow an effective circular economy system to function. The current review systematically analyzed existing research and presented the driving hypotheses in a wider context of sustainable resource management to provide practical solutions to mitigate environmental impacts from livestock farming (Kluge et al., 2008; Ramamoorthy et al., 2015; Li et al., 2019; Lavery et al., 2019; Bharath et al., 2019; Giovanni et al., 2024; Yeedo et al., 2024; Bentsena and Møller, 2017).

#### 11.2.5.3 Materials Made of Bio-based Ordered Porous Materials at Nano- to Microscales

Bio-based ordered porous materials are another intriguing class of materials with ordered porosity. One of the major obstacles is the fabrication of pores in the intermediate range of 10–1000 nm, which is crucial to bridging the gap between current capabilities and potential applications, even though templating techniques have demonstrated great promise in the creation of bioOPMs (bio-based ordered porous materials) with micro-, meso-, and macro-pores. However, there is a very small selection of precursors that are appropriate for bioOPMs. Unfortunately, carbonization techniques have been used to produce most bioOPMs in microporous and mesoporous structures thus far, compromising the unique properties of the original biopolymer. It is critically necessary to expand the use of polysaccharides and bio-based resins that have been explored thus far for the synthesis of bioOPMs. Novel techniques, such as breath figure methods, selective etching, and both soft- and hard-templating chemistries, are specifically required to overcome the drawbacks of current templating approaches. For bioOPMs to reach their full potential in a range of applications, drying and solidification processes must also keep up with it (Yeedo et al., 2024; Bentsena and Møller, 2017; Abaszade, 2022).

#### 11.2.5.4 Utilizing Biomass-Derived Materials for a Range of İndustrial Applications

Below is a description of how to synthesize it: first, 10 mL of concentrated sulfuric acid at 96% is put into a beaker. Then, 1 g of high-dispersion graphite with 99.9995% purity was mechanically poured into a petri dish together with 0.5 g of sodium hydroxide and placed atop the sulfuric acid. Next, the remaining 13 g of sulfuric acid was added to the system under magnetic stirring. To the system, a total of 3 g of potassium permanganate is added portion-wise over a period of 2 hours at about 20°C using an ice bath. Upon the cessation of stirring, the system undergoes a steam of hot water bath and is kept at about 35°C. The resulting mixture becomes a chestnut-gray paste. In a mixture, add 46 mL of distilled water and place it in the boiling water bath at 109°C. After an hour of boiling, add 250 mL of distilled water to it; mix it very little. Then, add 100 mL of perhydrol; immediately, the color changes, indicating the formation of graphene oxide. After cooling down, the mixture is filtered, and the solid residue on the filter paper is washed with distilled water. The final substance is dried at 50°C in a vacuum dryer for 2 hours until it is completely dry. On complete drying, the substance was obtained in powder form. 1.8 g of material was recovered, which shows a 40% yield loss. The structural quality, purity, and surface morphology of graphene oxide were characterized by SEM, XRD, Raman spectroscopy, IR luminescence, and temperature-dependent resistance measurement techniques. Elemental analysis showed that the C-to-O atomic ratio in the graphene oxide sample is 1.42. It means that oxidation proceeds under the most intense conditions when an extraordinary interlayer spacing increase was observed. Using the microscope, within the magnification range of 1400× to 4000×, thickness and length were measured as 0.31 nm and 14.8 nm, respectively. EDX measurements performed on the synthesized graphene oxide sample resulted in a composition of 70% carbon and 30% oxygen. Accompanying images confirm that the oxidation of graphite is characterized by a decrease in the number of layers, a reduction in crystallinity, and an increase in amorphization.

The structural features of layers that provide information on the mass, thickness, and length of the layer are obtained by analyzing SEM and EDX data to ascertain the elemental composition (see Figure 11.2). The Debye–Scherrer equation was used to calculate the number of layers and the thicknesses of graphite flakes. The results of Raman scattering investigations revealed that the thickness of the graphene oxide affected the flakes. The I2D/IG ratio and the intensities of the D, G, 2D, D+D", 2D/, and ID/IG peaks were measured. By creating a thin layer of graphene oxide flakes,

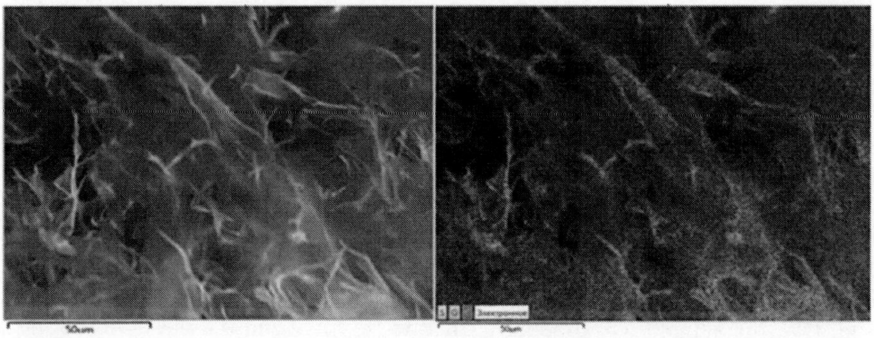

**FIGURE 11.2** Graphene oxide samples as per scanning electron microscope (SEM).

the temperature dependence was examined. The absorption of infrared electromagnetic radiation by graphene oxide atomic groups, which results in the excitation of molecular vibrations by photons, was investigated using infrared spectroscopy. This phenomenon results from a molecule's ability to absorb only quantized amounts of energy that correspond to the vibrational frequencies (both stretching and deformation modes) of its atomic bonds when exposed to infrared radiation (Abaszade, 2022).

#### 11.2.5.5 Using Graphene Oxide/Sulfur Compounds to Model Voltage–Ampere Characteristic Structures

For the oxygen-to-carbon ratio of 3.54 and a gray ratio of 42.54, negative differential resistance on the curve of volt–ampere characteristics was observed. For two samples of structures based on the graphene oxide and sulfur compound, modeling of the current–voltage characteristics, including both positive and negative differential conductivity, and the entire volt–ampere characteristic curve was produced. It is demonstrated that polynomials of the third and fourth orders accurately represent the regions with negative differential conductivity for first-type samples, whereas a polynomial of the fifth order accurately represents the areas with negative differential conductivity for second-type samples (Guliyeva et al., 2023).

The volt–ampere characteristics were detected for both models of the second-type samples: a linear model within the range of 6–17 V for the sample just described, and a second-order polynomial model for the sample in the range of 21–32 V. The curve's greatest failure is estimated to occur at U = 26 V. It is known that the power function and the second-order polynomial both adequately characterize the volt–ampere properties of such systems across the whole stress range. MATLAB R2021a is used to perform simulations. The models produced are quite reliable [31–35]. A novel idea has been developed to produce GO layers that are thinner and more functionalized based on the hammer approach. Weighing was done on plastic sample containers containing 2 g of graphite, 1 g of $NaNO_3$, and 6 g of $KMnO_4$. A graduated cylinder with a capacity of 250 mL was filled with 46 mL of 95–98% sulfuric acid ($H_2SO_4$). The test involved filling the ice bath with ice. This process was carried out twice because the suggested chemistry required two reactions. Following the sample setup, two-necked and three-necked flasks were removed, together with the relevant flask thermometers. The flask was then charged with the ice bath. Graphite was weighed, followed by stepwise addition of $NaNO_3$, and finally, $H_2SO_4$. A thermometer was then introduced into the flask, and initial gas evolution started. After that, it was further chilled to 0°C for 20 minutes in an ice bath. Once it reached 0°C, the ice bath was removed, and the mixture was kept at a higher temperature. To reach 20°C in a shorter amount of time, a hot plate was also used from the beginning of this procedure. When the temperature reached 20°C over the next 2 hours, $KMnO_4$ was gradually added. An ice bath was employed at temperatures between 20 and 25°C, as the addition of $KMnO_4$ raised the temperature. The mixture was stirred at the added 20–25°C for 4 hours. When the temperature reached 35°C, stirring continued for an additional 30 minutes. After that, an ice bath was filled with the combination. Using DW, the sample was gradually filled to 92 mL from the tap. After rising to 90°C, the temperature dropped to 70°C. The temperature was maintained between 70 and 75°C, and the flame was regulated on a hot plate. The contents of the necked

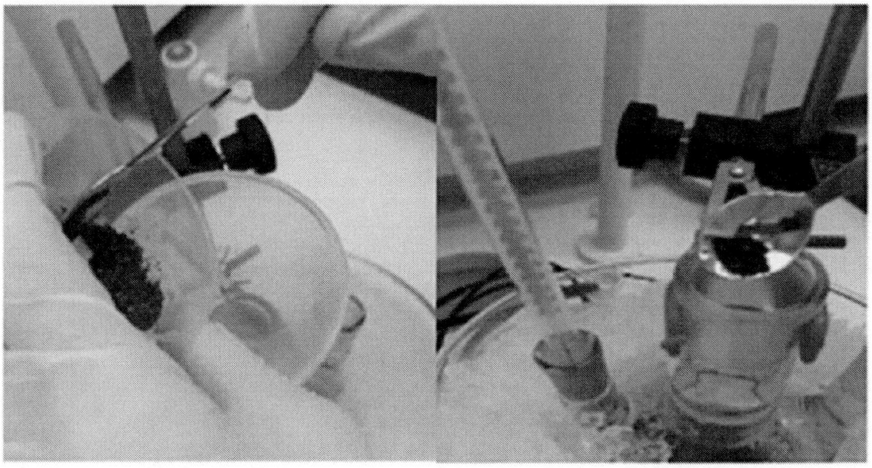

**FIGURE 11.3** Graphene oxide synthesis experiment.

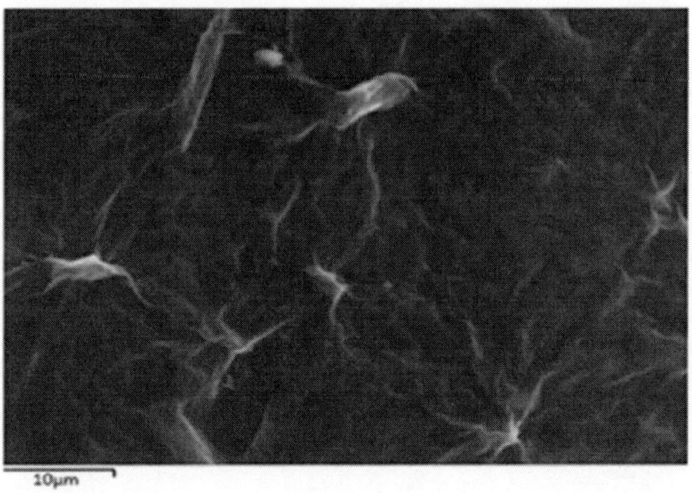


**FIGURE 11.4** SEM image of graphene oxide.

flask were then transferred to a 1-L Erlenmeyer flask after 15 minutes. To facilitate dispersion and reduce product loss, 280 mL of DW was used to rinse the same necked flask. After leaving the combinations for 12 hours, the filtration procedure was initiated. Figure 11.4 shows the results of the obtained layer's SEM investigation. In order to remove sulfuric acid from our products during the filtration process, we employed DW for washing (Abaszade, 2022; Guliyeva and Abaszade, 2023; Moradi et al., 2024; Aliyev et al., 2018; Figarova et al., 2021, 2023; Abaszade et al., 2022, 2023; Sharma et al., 2024; Ivanichok et al., 2024; Boychuk et al., 2022; Khan and Mariatti, 2023) (see Figure 11.3).

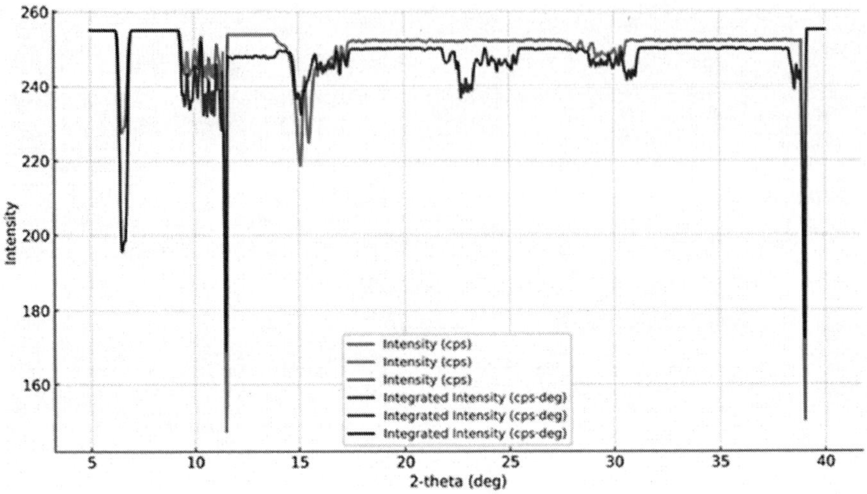


**FIGURE 11.5** X-ray diffraction of graphene oxide.

One of the most popular methods for examining the structural characteristics of graphene oxide samples is XRD scattering. Using a D2 Phaser diffractometer (Bruker, Germany), phase analysis by X-ray diffraction was performed using CuKα radiation ($\lambda$ = 1.5406 Å) throughout a 2θ range of 0.50° to 80°. The structural characteristics of graphene oxide produced by the Hammer method were examined using XRD phase analysis. There are two distinct prominent peaks in the spectrum, which show sharp peaks for carbon. Because there are lower particle sizes present, these are broad.

The X-ray diffraction peak of GO is clearly visible at an angle of 9.8°, as illustrated in Figure 11.5. The interlayer distance, *d*, is determined to be 0.905 nm using Bragg's law. This peak is indexed as 001. Another peak is observed at 27° corresponding to an interlayer distance of 0.334 nm, which is assigned the index 002 (Moradi et al., 2024; Aliyev et al., 2018; Figarova et al., 2021).

## 11.3 USES OF BIOMASS-DERIVED MATERIALS IN A RANGE OF INDUSTRIAL OPERATIONS

### 11.3.1 Biomass's Contribution to the Paris Agreement

According to the most recent global climate change mitigation scenarios, biomass energy will play a significant role in the primary global energy mix in order to keep global temperature below 2°C by 2100. Global energy supply from biomass accounted for 52.3 EJ at the end of 2017, and biomass energy supply worldwide increased on average by 2.23% yearly during the last four years. To be compatible with $CO_2$ concentration routes that are compatible with climate mitigation scenarios, this growth rate must, however, increase (Figarova et al., 2023; Abaszade et al., 2022, 2023; Sharma et al., 2024; Ivanichok et al., 2024; Boychuk et al., 2022). This research suggests a bioenergy deployment approach for light-duty cars that combines the production of electricity and liquid fuel from sugarcane. Steam cogeneration of power is based on commercially viable technologies of first-generation ethanol production from sugarcane waste and from planted forests. When combined with BECCS for $CO_2$ emissions captured during sugar fermentation, this approach results in net negative $CO_2$ emissions. This offers a significant reduction in GHG emissions and will contribute toward the objectives of the Paris Agreement (see Figure 11.6). This approach offers relatively lower consumer costs compared to conventional gasoline vehicles and, in addition, has net GHG emissions from the transport sector that are reduced compared with electric vehicles.

Although many such technological options are available, few of their respective economic and scaling implications have been researched in the literature. For instance, this scenario has not been addressed in any of the annual IEA Energy Technology Perspectives reports to date. This research provides an older technology that,

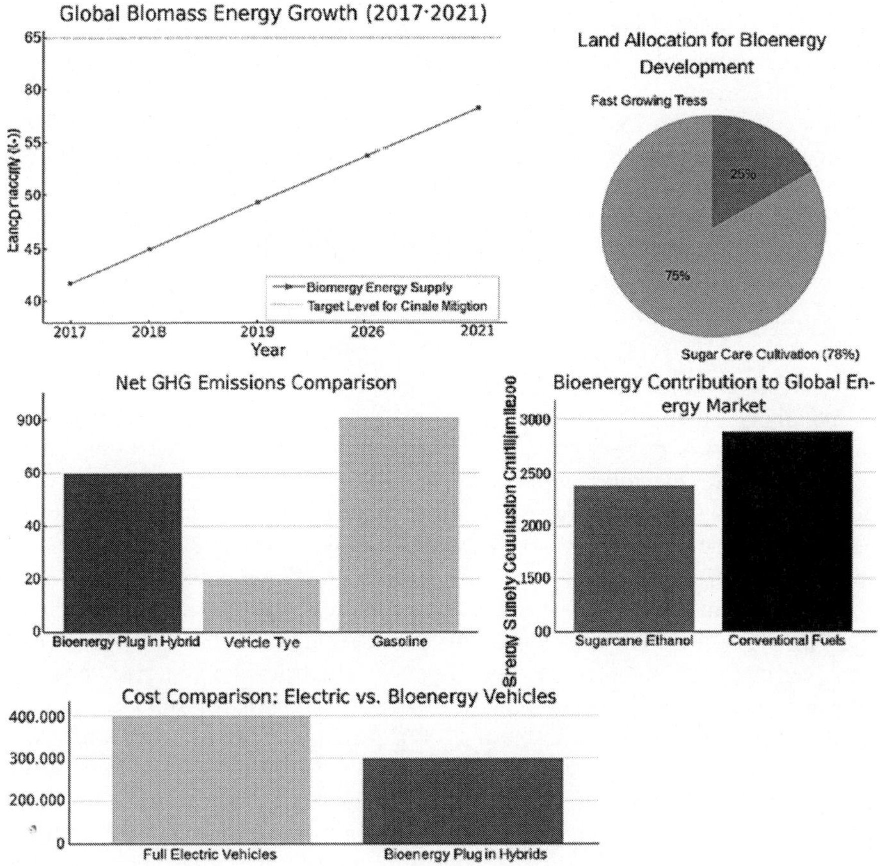


**FIGURE 11.6** Comprehensive visualization of bioenergy deployment and its impact on climate mitigation.

at one time, served as a complement to current electric transport methods, but frames low-cost BECCS' advantages regarding capturing $CO_2$ from ethanol production. We demonstrate herein that relatively small battery electricity storage, combined with ethanol derived from biomass, can meet most greenhouse gas reduction targets for climate change without additional societal costs. The research frequently refers to carbon taxes, which would favor the transition to flex-fuel plug-in hybrid vehicles and increase investors' returns. The global renewable energy market could be supplied by 60 million hectares of sugarcane plantations, producing over 2,000 million barrels of gasoline-equivalent per year, which could fuel around 1 billion vehicles in the world. Although the added value of electricity generated from sugarcane waste and fast-growing trees like eucalyptus is best utilized in tropical countries, temperate countries may find other proven clean energy sources more suitable for producing the additional electricity required for such hybrid vehicles. There is competition from huge battery systems seen in fully electric vehicles. The predicted total GHG benefits of electric cars are, however, smaller than those of the suggested bioenergy solution because of their high cost and the uncertainty surrounding their future pricing. The development of this bioenergy technology also meets the aim of sustainable development. This is commonly shared between both the ethanol and bioelectricity pathway and the BECCS approach (Khan and Mariatti, 2023; Serrano et al., 2014; Qi et al., 2011; Imtenan et al., 2015; Sharon et al., 2013; Zhang and Balasubramanian, 2016; Yasin et al., 2015; Zhu et al., 2010; Alptekin et al., 2015; Anand et al., 2011; Valentino et al., 2012).

#### 11.3.1.1 Developments, Opportunities, and Sustainability of Nature-Derived Polymers and Their Composites for Energy Storage Applications in Batteries and Supercapacitors

Energy storage devices play a major role in energy usage regulation, starting from transportation to electronic devices. In all these materials—batteries, supercapacitors, and others—their efficiency and lifetime depend on the types of materials used. Material selection is quite a critical issue that has to be considered in enhancing the capacity of energy storage systems. Whereas their performance has been satisfactory, processing, synthesis, use, and waste management of these materials pose serious risks to the environment. For example, traditional material waste is an environmental challenge because such waste is hazardous and therefore calls for an effective means of managing it. On the one hand, the interest in research and development of bio-based polymers is relatively recent; their development thus opens an avenue for greener materials and technologies. Biopolymer fabrication is derived either by chemical routes or by fermentation from biological sources or genetically modified microorganisms (Rakopoulos et al., 2006; Atmanli, 2016; Moreira et al., 2019; Ahmed et al., 2023; Anthony et al., 2021; Du et al., 2019; Morales and Ruiz, 2016). The outstanding dielectric properties of biopolymer composites create enormous opportunities for energy storage dielectric layers in high-performance batteries and supercapacitors. Depending on the production method, bio-based polymers could be further categorized into three types: (1) bio-based polymers obtained directly from biological material; (2) bio-based polymers obtained by fermentation or use

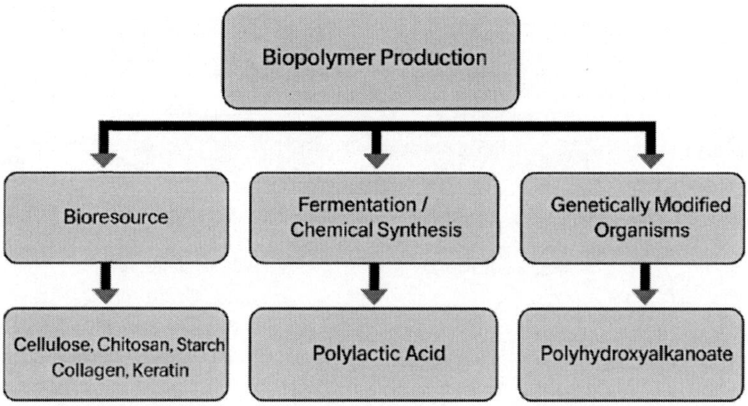


**FIGURE 11.7** Biopolymer classification based on the synthesis method.

of chemical agents with regard to biological material; and (3) bio-based polymers synthesized using microorganisms like bacteria modified by genetic engineering (see Figure 11.7). Throughout the book, critical focus has been given to how, with the processing methodologies employed, the properties of the resultant material are influenced by the chemistry of both monomers and polymers. Each of the chapters concludes with a detailed discussion regarding challenges and a future outlook, emphasizing strategies on how the sustainability of the polymers might be improved in this emerging area (Li et al., 2018). Because polyolefins are volumetric commodities and engineering plastics with excellent mechanical properties, high chemical stability, and good electrical insulation, their applications range across a variety of fields. Commercial polyolefins are fossil resources derived. In the absence of alternative products, conscience dictates that scientists work on bio-based polyolefins from renewable resources. Apart from being carbon-neutral, it is very important for accomplishing a sustainable economy (Rumjit et al., 2022; Rajeh et al., 2019; Leo et al., 2022; Wanga et al., 2020; Zacharopoulou and Lemonidou, 2017; Ammala et al., 2011; Boaen and Hillmyer, 2005; Feldman, 2001; Mathers, 2012; Williams and Hillmyer; 2008; Torres Galvis and De Jong, 2013; Cavani et al., 2017; Jiao et al., 2017; Banks and Kukes, 1985).

### 11.3.2 Bio-based Composites

Biomass materials like starch, sucrose, and cellulose are fermented by certain microbes to produce corresponding bio-based alcohols. Again, rapid thermal cracking of biomass produces syngas (synthesis gas) that could be catalytically transformed into corresponding olefins using the well-known Fischer–Tropsch reaction as shown in Figure 11.8 (Holland et al., 2015; Abaszade et al., 2024; Xiaohan et al., 2024; Jun, 2022; Mensah et al., 2023; Barbara and Philip, 2019; D'Amato et al., 2017; Edurne and Vincent, 2019; Muhammad et al., 2022; Kang et al., 2015; Li et al., 2024; Mousavi-Avval et al., 2023; Foong et al., 2023). Recent works, among others by

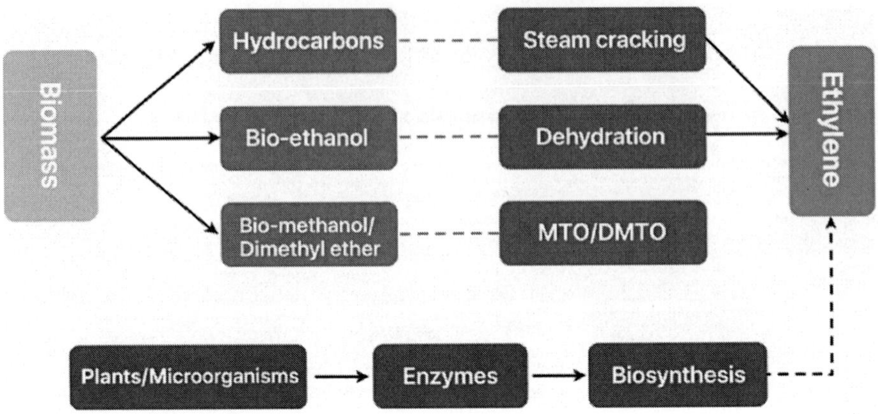


**FIGURE 11.8** Processes of bio-ethylene production.

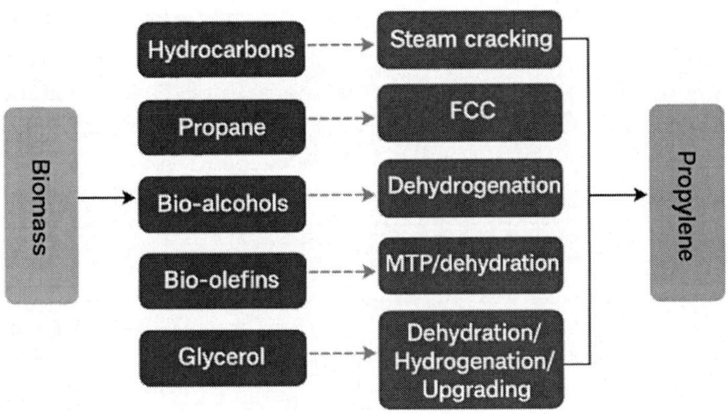


**FIGURE 11.9** Processes of bio-propylene production.

Zhang et al., have shown an improved version of the ZSM-5/SiC foam catalyst for enhanced activity and selectivity toward propylene. Besides methanol, other interesting bioalcohols that are promising feedstocks for propylene synthesis include isopropanol, ethanol, butanol, and glycerin. Bio-based propylene can also be made by metathesis of bio-based ethylene and butylene. The direct conversion of glycerol into propylene in the presence of highly effective catalysts has also substantially improved, offering a more streamlined process with high selectivity (Knifton et al., 2001). Comparatively, polypropylene has greater mechanical strength and thermal stability than polyethylene since its applications can be provided in a wider range of various branches of industry. Similar to biomass conversion to bio-polyethylene, this conversion could occur with the participation of cracking, gasification, fermentation, metathesis, and dehydrogenation, as shown in Figure 11.9. Bio-based polyolefins (polybutylene, polybutadiene, and polyisoprene) are potential derivatives of

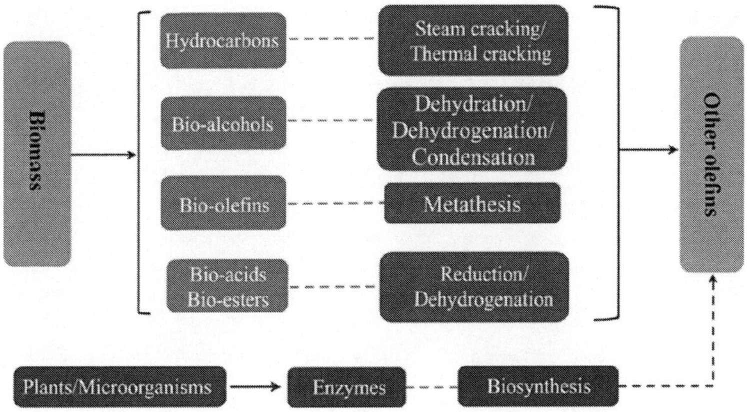


**FIGURE 11.10** Processes of other bio-olefin production.

polyethylene and polypropylene (see Figure 11.10). Other olefins are also formed by biomass steam cracking, but in minor quantities. Long-chain bio-based olefins can be prepared by metathesis reactions of bio-based ethylene and propylene; the dehydration of bioalcohols is often an important intermediate step. For example, the dehydration of butanol gives 1-butene, which is subsequently isomerized to isobutene; hence, these routes offer many possibilities for green chemical synthesis (Kozlowski and Davis, 2013; Blass et al., 2014; Amghizar et al., 2017; Abdelrahman et al., 2017; Ma et al., 2018; Martijn et al., 2021). The bio-based alcohol itaconic acid was an important starting material for synthesis via a series of conversion steps into methyl tetrahydrofuran, and then into isoprene by dehydration. Other current areas of investigation concern materials such as lignin-based polymers, renewable resource-based polyolefins, and long-chain aliphatic polymers within green polymer science, both in academia and industry. All these polymers are of high value for substituting traditional petrochemical-based materials using renewable biomass as their feedstock. Whereas some polymers, such as PA 11, have already found industrial uses, the market share remains extremely low (Table 11.1). It has to answer the challenges at a development pace equal to the rapidity needed for a really sustainable economy.

The properties of bio-based polymers must rival or exceed those of conventional commercial polymers. Additionally, these polymers should demonstrate greater cost-effectiveness. Two promising strategies to achieve this are as follows: (1) synthesizing polymers or their analogs to replace existing commercial counterparts, such as transitioning from petrochemical-derived polyolefins to similar materials derived from renewable biomass; and (2) designing novel polymeric materials with unique architectures, superior performance attributes, and alternative processing capabilities. Biomass precursors like lignin, cellulose, and fatty acids are inexpensive, abundant, and chemically diverse starting materials for these syntheses (Li et al., 2018).

The design of monomers and polymers for new, value-added, or greener commodity products is the perfect example of applying principles of physics and chemistry. One key element of such a design involves leveraging the distinctive chemical

**TABLE 11.1**
**Overview of Recent Advances in Biopolymer-Based Materials and Their Applications.**

| No | Key Points | Details |
|---|---|---|
| 1. | Diverse biopolymers from bioresources | Cellulose, starch, chitosan, etc., are examples of biopolymers sourced from bioresources. The chapter provides detailed information about these bio-polymers and their classification based on the source. |
| 2. | High mechanical stiffness and electrochemical performance of gel membranes | The study showed that gel membranes exhibited high mechanical stiffness and significant electrochemical performance with higher ionic conductivities. |
| 3. | Chitosan biopolymer characteristics | Chitosan, derived from the outer skin of shell-bearing marine animals, consists of N-acetylated-D-gluco-samine and p-(l 4)-D-glucosamine sub-units. It has properties like antimicrobial action, mucoadhesion, and biocompatibility, but poor mechanical |
| 4. | Properties and application of PLA | PLA is biodegradable, thermally processable, and has high mechanical strength and weather resistance. It was used to prepare an electrolyte for Li-ion batteries, which showed excellent cyclic performance and coulombic efficiency. |
| 5. | Applications of biobased polymers | Biobased polymers, due to their biodegradable and biocompatible properties, have potential applications as alternatives to traditional liquid electrolytes in various electrochemical devices, reducing dependency on fossil fuel-based polymers. |
| 6. | Blending nature-based polymers with synthetic polymers | Blending enhances properties such as mechanical strength, crystallinity, and electrical conductivity. For example, the blend of CMC with polyethylene oxide exhibits electrical conductivity of 5.19 x 10- 5 S/cm, making it suitable for solid polymer electrolytes. |
| 7. | Recent developments in biopolymer-based energy storage devices | There have been significant improvements in using biopolymer materials for energy storage devices. However, microframeworks still need improvement to meet the expected energy-power densities. |
| 8. | Advances in biopolymer-based materials for energy storage applications | The review discusses the recent advances in using green biobased polymers for battery and SC energy storage applications, potentially reducing dependency on traditional sources and significantly advancing energy storage technology. |

structures of renewable feedstocks. For example, the rigid aromatic rings in lignin present outstanding opportunities for the development of glassy polymers. Moreover, the intrinsic aromatic components in lignin make the generation of fluorescent materials with very simple processing viable, hence highlighting its versatility in material science (Knifton et al., 2001).

Further basic studies on lignin molecular structure, chemical transformations, design of polymer materials, and processing for the production of high-value-added materials are necessary. Fatty acids, due to their reactive groups such as double bonds, hydroxyls, and carboxyls, are suitable starting materials for making long-chain aliphatic polymers. Importantly, parts of those polymers, similar to linear polyethylene, have long sequences of methylene chains. However, since ester and amide precursors are used in their synthesis, unique properties such as polarity, degradability, and compostability exist that are not commonly associated with polyethylene. These issues are discussed in more detail in the following sections. Sustainable polymers derived from biomass are a very new research area, and its progress attracts great attention all around the world. It covers interdisciplinary domains of material science, chemistry, biology, bionics, and advanced materials processing. Advances in chemical synthesis, technologies of bio-refineries, and analytical methods should enable the discovery of new active species from renewable resources in the future, enriching the repertoire of starting materials. New sustainable polymeric materials, with enhanced and tailored properties, are therefore envisaged to reach the market shortly and provide environmentally friendly and innovative solutions (Holland et al., 2015; Abaszade et al., 2024; Xiaohan et al., 2024; Jun, 2022; Mensah et al., 2023; Barbara and Philip, 2019; D'Amato et al., 2017; Edurne and Vincent, 2019; Muhammad et al., 2022; Kang et al., 2015; Li et al., 2024; Mousavi-Avval et al., 2023).

#### 11.3.2.1 Bio-based Acrylic Pressure-Sensitive Adhesives

In fact, strict regulations involved with adhesives, together with increasing demands for sustainable resources, set pressing requests on research to find environmentally friendly and economic ways of synthesizing PSAs. The aim of this section is to consider the recent strategies and methodologies proposed to achieve acrylic adhesive from renewable raw materials, including vegetable oils, lignin, carbohydrates, and terpenes. The focus of this work is on new challenging approaches for the design of PSA properties to specific application needs. Within the past decade, a tremendous step forward has been taken toward the replacement of PSAs with bio-based alternatives through abundant renewable feedstock and built-up knowledge. However, developing sustainable processes for the separation and synthesis of compatible monomers, and selection of polymerization techniques, remain major challenging issues in achieving the required performance characteristics of PSAs. Emerging information about macromolecular topologies could allow revolutionary answers for several persistent challenges in current PSA development. Their resolution will thus allow overcoming the existing limitations for enabling the progress of this scientific and industrial area toward green and high-performance adhesive technologies (Cavani et al., 2017; Jiao et al., 2017; Banks and Kukes, 1985; Knifton et al., 2001; Kozlowski and Davis, 2013; Blass et al., 2014; Amghizar et al., 2017; Abdelrahman et al., 2017; Ma et al., 2018).

#### 11.3.2.2 Green Process Intensification for the Production of Bio-based Chemicals and Fuels Through the Use of Microreactor Technology

In recent years, microreactor technology has been one of the primary means for process intensification that allows broad prospects toward green and sustainable chemical syntheses. Microreactors have been attracting much attention during the last ten years for improving catalytic biomass conversions (Li et al., 2024). This perspective will be specifically directed at a concise overview of intensified green processes in microreactors with regard to fine chemicals and fuels production from biomass. The discussion will then proceed to describe the advantages that microreactor technology holds in terms of process intensification, showing how the said technology meets the criteria and principles for green chemistry and engineering. Typical syntheses—such as furanic platform compounds and derivatives from monosaccharides and disaccharides, liquid-phase oxidation, hydrogenation of lignocellulosic biomass derivatives, and biodiesel synthesis—are examples of applications of this kind. The next section will discuss the prospects for further research. Solid-handling techniques for catalysts, feeds, and products; process integration in groups or a single microreactor; creating comprehensive databases for biomass transformation; photocatalysis; and the use of novel solvents in micro-flow systems are among the subjects discussed.

### 11.3.3 Challenges and Improvement Strategies

#### 11.3.3.1 Problems in Producing Bio-based Goods and Ethanol Together

One of the viable and sustainable solutions to the current energy crisis is the co-production of ethanol and other bio-based products. The technique efficiently uses a variety of feedstock materials, from industrial byproducts to agricultural leftovers. The whole process meets not only the demand for energy but also assists in waste recycling by transforming discarded materials into useful output. Besides, the integration of co-products, such as bio-based chemicals, animal feed, and bio-plastics, enhances the economic viability of the system through diversification of revenues and reduction in dependence on fossil fuels (Mensah et al., 2023). While these are the advantages, the co-production strategy is not devoid of challenges. These are issues of feedstock variability and availability, scalability of production technologies, and energy-intensive processing in certain stages. Besides, more optimization and innovation in bioprocessing techniques are necessary for high yields with minimal environmental impacts. This calls for addressing all these issues if ethanol co-production is ever to become one of the cornerstones of sustainable energy solutions.

#### 11.3.3.2 Improvement Strategy

For commercial and industrial-scale manufacturing, the co-production gaps and problems outlined above necessitate the enhancement of current technologies and techniques for the optimization of specific co-production process phases (see Figure 11.11).

**FIGURE 11.11** The overview of ethanol and bio-based product co-production.

#### 11.3.3.3 Techno-Economic Evaluations and Environmental Effects of Bio-based Goods

The results indicate that bio-based products are clear environmental winners compared to their fossil-based counterparts due to lower consumption of fossil resources and relatedly lower GHG emissions. On the other hand, bio-based products can give rise to trade-offs in impacts from categories like acidification and ecotoxicity. Regarding energy crops, in contrast to agricultural residues, the production of the crop itself becomes a major source of both environmental and economic burdens, besides significantly contributing to the cost of logistics and feedstock transportation (Li et al., 2024). By enhancing nature with various bio-based products, including $CO_2$ sequestration at times of feedstock growth with presumed carbon storage in soils, the costs of these bio-based products will also depend on feedstock prices, logistics, investment in capital, and scale of production. Considering sustainability, one has to look into not only the environment but also the economic and social aspects. Integrating these factors into ongoing and future research will yield even more holistic and applicable findings that move the use of bio-based products further forward. Innovation might act as the driver of the switch from fossil-based to bio-based products, since the production systems have been optimized with respect to environmental, social, and economic factors. In addition, drivers regarding techno-economic feasibility and environmental impact at a commercial scale have to be identified. This information will be very useful for various stakeholders, such as policymakers, investors, manufacturers, and farmers themselves, in cooperation and further development within this field (Foong et al., 2023).

## 11.4 CONCLUSION

In recent years, growing interest has been observed in developing an anticipatory governance of rising technologies with an emphasis on responsible research and innovation. The biosynthetic synthesis of menthol and the anticipatory governance of synthetic biology in general were covered in this chapter. Applying a cradle-to-gate analysis pertaining to menthol production processes and considering multiple stakeholders, this research has developed a real-time methodology for technology appraisal. This step away from conventional sources of menthol to biosynthetic menthol created with genetically modified bacteria is definitely drastic enough toward more ingenious ways of production. The fact that microbial cell factories can play a role in flavor and scent production, into which menthol falls—as practiced by companies like Amyris and Evolva—proves that already biosynthetic products play their role in hitting economic and sustainability targets. Challenges ranging from environmental sustainability to social justice, infrastructure and economic strategies, and public perception are all woven into the sociotechnical shift to biosynthetic menthol. Insight into the ways public groups negotiate novel technologies' uncertainties might inform the adaptation strategies that could diffuse out of this sociotechnical transition. The world is moving more quickly toward using renewable energy sources, such as biomass, to replace fossil fuels due to the urgency of addressing climate change. Currently, biomass makes up a large proportion of global energy consumption and, in the future, will probably become more significant. One of the largest biomass sources is agricultural wastes, which are likely to provide very large shares of future bioenergy supplies. This analysis underpins some of the complex challenges and opportunities involved in shifting toward biosynthetic production powered by renewable energy. Knowing how to grapple with these dynamics will become more and more important if one is to skillfully navigate the intricate interplay among technology, society, and sustainability.

## REFERENCES

Abaszade, R.G. (2022). Synthesis and analysis of flakes graphene oxide, *Journal of Optoelectronic and Biomedical Materials*, 14(3), 107–114.

Abaszade, R.G., Aliyev, E.M., Mammadov, A.G., Khanmamadova, E.A., Guliyev, A.A., Aliyev, F.G., Zapukhlyak, R.I., Budak, H.F., Kasapoglu, A.E., Margitych, T.O., Singh, A., Arya, S., Gür, E., Stetsenko, M.O. (2024). Investigation of thermal properties of gadolinium doped carbon nanotubes, *Physics and Chemistry of Solid State*, 25(1), 142–147.

Abaszade, R.G., Mammadova, A.G., Khanmammadova, E.A., Bayramov, I.Y., Namazov, R.A., Popal, Kh.M., Melikova, S.Z., Qasımov, R.C., Bayramov, M.A., Babayeva, N.I. (2023). Electron paramagnetic resonance study of gadoliniumum doped graphene oxide, *Journal of Ovonic Research*, 19(2), 259–263.

Abaszade, R.G., Mammadov, A.G., Kotsyubynsky, V.O., Gur, E.Y., Bayramov, I.Y., Khanmamadova, E.A., Kapush, O.A. (2022). Modeling of voltage-ampere characteristic structures on the basis of graphene oxide/sulfur compounds, *Technical and Physical Problems of Engineering*, 14(2), 302–306.

Abdelrahman, O.A., Park, D.S., Vinter, K.P., Spanjers, C.S., Ren, L., Cho, H.J., et al. (2017). Renewable isoprene by sequential hydrogenation of itaconic acid and dehydra-decyclization of 3-Methyl-Tetrahydrofuran, *ACS Catalysis*, 7, 1428–1431.

Ahmed, Sh., Sharma, P., Bairagi, S., Rumjitf, N.P., Garg, Sh., Ali, A., Lai, Ch.W., Mousavi, S.M., Hashemi, S.A., Hussain, Ch.M. (2023). Nature-derived polymers and their composites for energy depository applications in batteries and supercapacitors: advances, prospects and sustainability, *Journal of Energy Storage*, 66, 107391.

Aliyev, E.M., Khan, M.M., Nabiyev, A.M., Alosmanov, R.M., Bunyad-zadeh, I.A., Shishatskiy, S., Filiz, V. (2018). Covalently modified graphene oxide and polymer of intrinsic microporosity (PIM-1) in mixed matrix thin-film composite membranes, *Nanoscale Research Letters*, 13(1), 359.

Alptekin, E., Canakci, M., Ozsezen, A.N., Turkcan, A., Sanli, H. (2015). Using waste animal fat based biodiesels–bioethanol–diesel fuel blends in a DI diesel engine, *Fuel*, 157, 245–254.

Amghizar, I., Vandewalle, L.A., Van Geem, K.M., Marin, G.B. (2017). New trends in olefin production, *Engineering*, 3, 171–178.

Ammala, A., Bateman, S., Dean, K., Petinakis, E., Sangwan, P., Wong, S., et al. (2011). An overview of degradable and biodegradable polyolefins, *Progress in Polymer Science*, 36, 1015–1049.

Anand, K., Sharma, R.P., Mehta, P.S. (2011). Experimental investigations on combustion, performance and emissions characteristics of neat karanji biodiesel and its methanol blend in a diesel engine, *Biomass Bioenergy*, 35, 533–541.

Anthony, L.S., Vasudevan, M., Perumal, V., Ovinis, M., Raja, P.B., Edison, T.N.J.I. (2021). Bioresource-derived polymer composites for energy storage applications: brief review, *Journal of Environmental Chemical Engineering*, 9, 105832.

Atmanli, A. (2016). Comparative analyses of diesel–waste oil biodiesel and propanol, n-butanol or 1-pentanol blends in a diesel engine, *Fuel,* 176, 209–302.

Banks, R.L., Kukes, S.G. (1985). New developments and concepts in enhancing activities of heterogeneous metathesis catalysts, *Journal of Molecular Catalysis*, 28, 117–131.

Barbara, R., Philip, Sh. (2019). Anticipating governance challenges in synthetic biology: insights from biosynthetic menthol, *Technological Forecasting & Social Change*, 139, 311–320.

Bentsena, S.N., Møller, M.I. (2017). Solar energy conserved in biomass: sustainable bioenergy use and reduction of land use change, *Renewable and Sustainable Energy Reviews*, 71, 954–958.

Bharath, K.N., Manjunatha, G.B., Santhosh, K. (2019). Failure analysis and the optimal toughness design of sheep wool reinforced epoxy composites. In *Failure Analysis in Biocomposites, Fibre-Reinforced Composites and Hybrid Composites*. Woodhead Publishing, pp. 97–107.

Blass, S.D., Hermann, R.J., Persson, N.E., Bhan, A., Schmidt, L.D. (2014). Conversion of glycerol to light olefins and gasoline precursors, *Applied Catalysis A*, 475, 10–15.

Boaen, N.K., Hillmyer, M.A. (2005). Post-polymerization functionalization of polyolefins, *Chemical Society Reviews*, 34, 267–275.

Boychuk, V., Zapukhlyak, R., Abaszade, R., Kotsyubynsky, V., Hodlevsky, M., Rachiy, B., Fedorchenko, S. (2022). Solution combustion synthesized $NiFe_2O_4$/reduced graphene oxide composite nanomaterials: morphology and electrical conductivity, *Physics and Chemistry of Solid State*, 23(4), 815–824.

Bruckner, T.A., Bashmakov, I., Mulugetta, Y. (2014). Energy systems. In *IPCC WGIII AR5*. Geneva, CH: IPCC.

Cavani, F., Ballarini, N., Cericola, A. (2017). Oxidative dehydrogenation of ethane and propane: how far from commercial implementation? *Catalysis Today*, 127, 113–131.

Chen, H., Lin, Z., Tan, C. (2019). Classification of different animal fibers by near infrared spectroscopy and chemometric models, *Microchemical Journal*, 144, 489–494.

Chum, H., Faaij, A., Moreira, J., Berndes, G., Dhamija, P., Dong H, et al. (2011). Bioenergy. In Edenhofer, O., Pichs-Madruga, R., Sokona, Y., Seyboth, K., Matschoss, P., Kadner, S., editors. *IPCC Special Report on Renewable Energy Sources and Climate Change Mitigation*. Cambridge, United Kingdom and New York, NY, USA: Cambridge University Press.

D'Amato, D., Droste, N., Allen, B., Kettunen, M., Lahtinen, K., Korhonen, J., Leskinen, P., Matthies, B.D., Toppinen, A. (2017). Green, circular, bio economy: a comparative analysis of sustainability avenues, *Journal of Cleaner Production*, 168, 716–734.

Daioglou, V., Stehfest, E., Wicke, B., Faaij, A., Van Vuuren, D.P. (2016). Projections of the availability and cost of residues from agriculture and forestry, *GCB Bioenergy*, 8, 456–470.

Du, Z., Su, Y., Qu, Y., Zhao, L., Jia, X., Mo, Y., Yu, F., Du, J., Chen, Y. (2019). A mechanically robust, biodegradable and high-performance cellulose gel membrane as gel polymer electrolyte of lithium-ion battery, *Electrochimica Acta,* 299, 19–26.

Edurne, A.I., Vincent, B. (2019). Strengthening the socio-ethical foundations of the circular economy: lessons from responsible research and innovation, *Journal of Cleaner Production*, 233, 280–291.

Feldman, D. (2001). Polymer barrier films, *Journal of Polymers and the Environment*, 9, 49–55.

Figarova, S., Aliyev, E., Abaszade, R., Figarov, V. (2023). Negative thermal expansion of sulphur-doped graphene oxide, *Advanced Materials Research*, 1175, 55–62.

Figarova, S.R., Aliyev, E.M., Abaszade, R.G., Alekberov, R.I., Figarov, V.R. (2021). Negative differential resistance of graphene oxide/sulphur compound, *Journal of Nano Research*, 67, 25–31.

Foong, S.Y., Chan, Y.H., Yiin, Ch.L., Lock, S.S.M., Loy, A.Ch.M., Lim, J.Y., Yek, P.N.Y., Mahari, W.A.W., Liew, R.K., Peng, W., Tabatabaei, M., Aghbashlo, M., Lam, S.Sh. (2023). Sustainable $CO_2$ capture via adsorption by chitosan-based functional biomaterial: a review on recent advances, challenges, and future directions, *Renewable and Sustainable Energy Reviews*, 181, 113342.

GEA. (2012). *Global Energy Assessment—Toward a Sustainable Future*. Cambridge, UK and New York, NY, USA: Cambridge University Press and Laxenburg, Austria: The International Institute for Applied Systems Analysis.

Giovanni, F., Giorgio, P., Stefania, P., Francesco, M., Andrea, P. (2024). Biorefinery development in livestock production systems: applications, challenges, and future research directions, *Journal of Cleaner Production*, 440, 140858.

Grunwald, A. (2017). Assigning meaning to NEST by technology futures: extended responsibility of technology assessment in RRI, *Journal of Responsible Innovation*, 4(2).

Guliyeva, N.A., Abaszade, R.G., Khanmammadova, E.A., Azizov, E.M. (2023). Synthesis and analysis of nanostructured graphene oxide, *Journal of Optoelectronic and Biomedical Materials*, 15(1), 23–30.

Guston, D.H. (2014). Understanding anticipatory governance, *Social Studies of Science*, 44(2), 218–242.

Holland, R.A., Eigenbrod, F., Muggeridge, A., Brown, G., Clarke, D., Taylor, G. (2015). A synthesis of the ecosystem services impact of second-generation bioenergy crop production, *Renewable and Sustainable Energy Reviews*, 46, 30–40.

Imtenan, S., Masjuki, H.H., Varman, M., Rizwanul, F.I.M., Sajjad, H., Arbab, M.I. (2015). Effect of n-butanol and diethyl ether as oxygenated additives on combustion—emission-performance characteristics of a multiple cylinder diesel engine fuelled with diesel–jatropha biodiesel blend, *Energy Convers Manage*, 94, 84–94.
International Energy Agency. (2013). *World Energy Outlook.* Paris, FR: International Energy Agency.
Ivanichok, N., Kolkovskyi, P., Ivanichok, O., Kotsyubynsky, V., Boychuk, V., Rachiy, B., Bembenek, M., Warguła, Ł., Abaszade, R., Ropyak, L. (2024). Effect of thermal activation on the structure and electrochemical properties of carbon material obtained from Walnut Shells, *Materials*, 17(11), 2514.
Jiao, Y., Fan, X., Perdjon, M., Yang, Z., Zhang, J. (2017). Vapor-phase transport (VPT) modification of ZSM-5/SiC foam catalyst using TPAOH vapor to improve the methanol-to-propylene (MTP) reaction, *Applied Catalysis A*, 545, 104–112.
Jun, Y. (2022). Green process intensification using microreactor technology for the synthesis of bio-based chemicals and fuels, *Chemical Engineering and Processing—Process Intensification*, 177, 109002.
Kang, Y., Jeong, S.G., Wi, S., Kim, S. (2015). Energy efficient Bio-based PCM with silica fume composites to apply in concrete for energy saving in buildings, *Solar Energy Materials and Solar Cells*, 143, 430–434.
Khan, J., Mariatti, M. (2023). In-situ graphene oxide reduction via inkjet printing using natural reducing inks, *Flexible and Printed Electronics,* 8, 035009.
Kluge, J.A., Rabotyagova, O., Leisk, G.G., Kaplan, D.L. (2008). Spider silks and their applications, *Trends in Biotechnology*, 26, 244–251.
Knifton, J.F., Sanderson, J.R., Stockton, M.E. (2001). Tert-butanol dehydration to isobutylene via reactive distillation, *Catalysis Letters*, 73, 55–57.
Kozlowski, J.T., Davis, R.J. (2013). Heterogeneous catalysts for the Guerbet coupling of alcohols, *ACS Catalysis*, 3, 1588–600.
Lavery, L.A., Bhavan, K., Wukich, D.K. (2019). Biofilm and diabetic foot ulcer healing: all hat and no cattle, *Annals of Translational Medicine*, 7.
Leffingwell, J.C. (2015). Biotechnology: conquests and challenges in flavors & fragrances, *Leffingwell Reports*, 7(2), 1–11.
Leo, B.F., Manimaran, M., Rumjit, N.P., Lai, C.W. (2022). Safety aspects and environmental impacts of nanomaterials in energy storing devices. In *Encyclopedia of Energy Storage*. Elsevier, pp. 656–666.
Li, L., Hou, L., Cheng, J., Simmons, T., Zhang, F., Zhang, L.T., Linhardt, R.J., Koratkar, N. (2018). A flexible carbon/sulfur-cellulose core-shell structure for advanced lithium–sulfur batteries, *Energy Storage Materials,* 15, 388–395.
Li, Y., Sun, H., Zhang, Y., Wang, X., Gao, M., Sun, X., Wang, Q. (2024). Research progress for co-production ethanol and biobased products, *Industrial Crops & Products*, 212, 118351.
Li, Y.-B., Liu, H.-H., Wang, X.-C., Zhang, X.-X. (2019). Fabrication and performance of wool keratin functionalized graphene oxide composite fibers, *Materials Today Sustainability*, 3, 100006.
Ma, Z., Liu, C., Niu, N., Chen, Z., Li, S., Liu, S., et al. (2018). Seeking brightness from nature: J-Aggregation-Induced emission in cellulolytic enzyme lignin nanoparticles, *ACS Sustainable Chemistry & Engineering*, 6, 3169–3175.
Martijn, A.D., Resat, A., Alexandre, S., José, M.A., Filip, E.D.P. (2021). Biobased acrylic pressure-sensitive adhesives, *Progress in Polymer Science*, 117, 101396.
Mathers, R.T. (2012). How well can renewable resources mimic commodity monomers and polymers? *Journal of Polymer Science Part A: Polymer Chemistry*, 50, 1–15.
Meckin, R., Balmer, A. (2018). Everyday uncertainty work: making sense of biosynthetic menthol, *Engaging Science, Technology, and Society*, 4, 458–484.

Mensah, S.B., Jeanne, L., Johann, F.G. (2023). Biobased acrylic acid production in a sugarcane biorefinery: a techno-economic assessment using lactic acid, 3-hydroxypropionic acid and glycerol as intermediates, *Chemical Engineering Research and Design*, 193, 367–382.

Moradi, R., Pour Khalili, N., Khanmamadova, E., Abaszade, R. (2024). Functionalized carbon nanostructures for flexible electronics. In *Handbook of Functionalized Carbon Nanostructures*. Cham: Springer, pp. 2581–2614.

Morales, M.E., Ruiz, M.A. (2016). Microencapsulation of probiotic cells: applications in nutraceutic and food industry, *Nutraceuticals,* 1, 627–668.

Moreira, J.R., Pacca, S.A., Goldemberg, J. (2019). The role of biomass in meeting the Paris agreement, *IOP Conference Series: Earth and Environmental Science*, 354, 012107.

Mousavi-Avval, S.H., Sahoo, K., Nepal, Pr., Runge, T., Bergman, R. (2023). Environmental impacts and techno-economic assessments of biobased products: a review, *Renewable and Sustainable Energy Reviews*, 180, 113302.

Muhammad, F.J., Ziaur, R., Nauman, I., Miroslav, C., Jakub, C., Ahmed, B.E. (2022). Biobased phase change materials from a perspective of recycling, resources conservation and green buildings, *Energy & Buildings*, 270, 112280.

Nakada, S., Saygin, D., Gielen, D. (2014). *Global Bioenergy: Supply and Demand Projections.* Abu Dahbi, UAE: IRENA.

Qi, D.H., Chen, H., Geng, L.M., Bian, Y.Z. (2011). Effect of diethyl ether and ethanol additives on the combustion and emission characteristics of biodiesel–diesel blended fuel engine, *Renewable Energy*, 36, 1252–1258.

Rajeh, A., Morsi, M.A., Elashmawi, I.S. (2019). Enhancement of spectroscopic, thermal, electrical and morphological properties of polyethylene oxide/carboxymethyl cellulose blends: combined FT-IR/DFT, *Vacuum*, 159, 430–440.

Rakopoulos, C.D., Antonopoulos, K.A., Rakopoulos, D.C., Hountalas, D.T., Giakoumis, E.G. (2006). Comparative performance and emissions study of a direct injection diesel engine using blends of diesel fuel with vegetable oils or bio-diesels of various origins, *Energy Conversion and Management*, 47, 3272–3287.

Ramamoorthy, S.K., Skrifvars, M., Persson, A. (2015). A review of natural fibers used in biocomposites: plant, animal and regenerated cellulose fibers, *Polymer Reviews*, 55, 107–162.

Rebecca, M., Keith, V., Greg, C., Shan, J. (2021). Biobased foams for thermal insulation: material selection, processing, modelling, and performance, *RSC Advances*, 11, 4375–4394.

REN21. (2015). *Renewables 2015 Global Status Report.* Paris, FR: REN21 Secretariat.

Robert, J.S., Miller, C.A., Milleson, V. (2013). Introduction: ethics and anticipatory governance of nano-neurotechnological convergence. In Hays, S.A., Robert, J.S., Miller, C.A., Bennett, I., editors, *Nanotechnology, the Brain, and the Future. Yearbook of Nanotechnology in Society.* Springer, p. 3.

Rumjit, N.P., Thomas, P., Lai, C.W., Wong, Y.H., George, V., Basilraj, P., Bin Johan, M.R. (2022). Recent advancements of supercapacitor electrode materials derived from agriculture waste biomass. In *Encyclopedia of Energy Storage.* Elsevier.

Schempp, F.M., Drummond, L., Buchhaupt, M., Schrader, J. (2018). Microbial cell factories for the production of terpenoid flavor and fragrance compounds, *Journal of Agricultural and Food Chemistry*, 66(10), 2247–2258.

Serrano, M., Oliveros, R., Sánchez, M., Moraschini, A., Martínez, M., Aracil, J. (2014). Influence of blending vegetable oil methyl esters on biodiesel fuel properties: oxidative stability and cold flow properties, *Energy*, 65, 109–115.

Sharma, Bh., Singh, A., Sharma, A., Dubey, A., Gupta, V., Abaszade, R.G., Sundramoorthy, A.K., Sharma, N., Arya, S. (2024). Synthesis and characterization of zinc selenide/graphene oxide (ZnSe/GO) nanocomposites for electrochemical detection of cadmium ions, *Applied Physics A*, 130, 297.

Sharon, H., Jai, Sh.R.P., Jenis, F.K., Murali, S., Muthusamy, R. (2013). Fueling a stationary direct injection diesel engine with diesel-used palm oil–butanol blends—an experimental study, *Energy Conversion and Management*, 73, 95–105.

Thi Hang, N.N., Frédéric, B., Andreas, B., Véronique, B., Philippe, G. (2019). Synthesis of new biobased linear poly(ester amide)s, *European Polymer Journal*, 121, 109314.

Toogood, H.S., Cheallaigh, A.N., Tait, S., Mansell, D.J., Jervis, A., Lygidakis, A. (2015). Enzymatic menthol production: one-pot approach using engineered Escherichia coli, *ACS Synthetic Biology*, 4(10), 1112–1123.

Torres Galvis, H.M., De Jong, K.P. (2013). Catalysts for production of lower olefins from synthesis gas: a review, *ACS Catalysis*, 3, 2130–2149.

Valentino, G., Corcione, F.E., Iannuzzi, S.E., Serra, S. (2012). Experimental study on performance and emissions of a high speed diesel engine fuelled with nbutanol diesel blends under premixed low temperature combustion, *Fuel*, 92, 295–307.

Verma, S.K., Kumar, P., Mishra, A., Khare, R., Singh, D. (2024a). Green nanotechnology: illuminating the effects of bio-based nanoparticles on plant physiology, *Biotechnology for Sustainable Materials*, 1, 1. https://doi.org/10.1186/s44316-024-00001-2

Verma, S.K., Singh, D., Misra, S. (2024b). Innovative approaches to sustainable waste management and energy generation in developing nations. In Mehra, R., Kumar, P., Hung, B., editors. *Municipal Solid Waste Management and Recycling Technologies*. IGI Global Scientific Publishing, pp. 267–300. https://doi.org/10.4018/979-8-3693-4054-7.ch010

Wanga, Zh., Ganewatta, S.M., Chuanbing, T. (2020). Sustainable polymers from biomass: bridging chemistry with materials and processing, *Progress in Polymer Science*, 101, 101197.

Williams, C., Hillmyer, M. (2008). Polymers from renewable resources: a perspective for a special issue of polymer reviews, *Polymer Reviews*, 48, 1–10.

Wurtzel, E.T., Kutchan, T.M. (2016). Plant metabolism, the diverse chemistry set of the future, *Science,* 353, 6305.

Xiaohan, W., Jinwon, J., Yanqun, S., Jingang, L., Hongjie, Z., Zhibin, H., Yonghao, N. (2024). Starting materials, processes and characteristics of bio-based foams: a review, *Journal of Bioresources and Bioproducts*, 9, 160–173.

Yasin, M.H.M., Mamat, R., Yusop, A.F., Aziz, A., Najafi, G. (2015). Comparative study on biodiesel–methanol–diesel low proportion blends operating with a diesel engine, *Energy Procedia*, 75, 10–16.

Yeedo, Ch., Yeling, Zh., Cosima, St., Yi, L., Orlando, Rojas, J. (2024). Biobased ordered porous materials in the nano-to microscales, *Current Opinion in Colloid & Interface Science*, 73, 101822.

Zacharopoulou, V., Lemonidou, A. (2017). Olefins from biomass intermediates: a review, *Catalysts*, 8, 2/1–19.

Zhang, Z., Balasubramanian, R. (2016). Investigation of particulate emission characteristics of a diesel engine fueled with higher alcohols/biodiesel blends, *Applied Energy*, 163, 71–80.

Zhu, L., Cheung, C.S., Zhang, W.G., Huang, Z. (2010). Emissions characteristics of a diesel engine operating on biodiesel and biodiesel blended with ethanol and methanol, *Science of the Total Environment*, 408, 914–921.

# 12 Reinforcing Sustainable Bio-Based Material for Smart Manufacturing

*Ankita Singh, Juveriya Israr,
Sunil Kumar Verma, Swapnil Pandey,
Virendra Nath Pathak, and Sankalp Misra*

## 12.1 INTRODUCTION

### 12.1.1 Sustainable Bio-Based Materials for Smart Manufacturing

"Sustainable manufacturing" means producing and delivering products in an eco-friendly way, using resources responsibly, and reducing waste throughout the entire process—from the factory to the customer. In today's world, manufacturing is linked to almost everything we do. It provides the products and services we rely on for health, safety, and well-being. To fully grasp the idea of "sustainable manufacturing," we need to define its goals, explain how it is put into practice, and outline how progress is measured, as described in previous research (Karuppiah et al., 2024). The "design of manufacturing" strategy is essential for reaching sustainability in manufacturing because of the high energy and resource consumption in this sector. Businesses can achieve their social and environmental objectives while also increasing their bottom line by committing to sustainable practices. Sustainable life cycle design (SLD) has been more important in numerous technical and industrial domains in the last few decades, according to Marín-González et al. (2022). Nevertheless, when it comes to completely incorporating sustainability, there is still a disconnect between theory and practice. The economic, environmental, and social problems require further effort despite the growing emphasis on sustainable development. Rather than relying on the traditional "3Rs" method, sustainable manufacturing follows the "6Rs": reduce, redesign, reuse, recover, remanufacture, and recycle. According to Eslami et al. (2021), the "reduce" step in the 6R process involves minimizing energy consumption and resource utilization during production in order to decrease waste at the usage stage. Reusing items or their components after their initial life cycle is an important part of the "reuse" concept, which aims to decrease resource consumption (Verma & Kumar, 2022). "Recycle" means to turn old, used, or otherwise discarded resources into something fresh (Eslami et al., 2021). In terms of the "Recover" phase, it takes place when the first lifetime components are gathered, dismantled, cleaned, and made ready for the second life cycle (Eslami et al., 2021). In


DOI: 10.1201/9781003566205-12

order to make the product more maintainable, the term "Redesign" refers to the process of using techniques like Design for Environment (DfE) (Munaro et al., 2022). Remanufacturing, according to Munaro et al. (2022), involves taking a product that has been used before and bringing it back to its original condition by recycling as many parts as possible while keeping its functionality. According to Singh et al. (2024e), the total mass of synthetic products is anticipated to surpass that of all living beings on Earth, which is estimated at approximately 1.1 trillion metric tons in the early 2020s. This mass includes concrete buildings, metal automobiles, rubber gloves, and plastic bottles. The exponential growth of synthetic (or conventional and artificial) materials, often known as the anthropogenic mass, marks the beginning of the Anthropocene epoch, when people are considered the primary force shaping the planet. Climate change, ecosystem services, and biodiversity loss are all known to be influenced by a variety of human activities, including farming and city planning.

## 12.2 BIO-BASED MATERIALS: ECOLOGICAL DRIVERS FOR SUSTAINABLE DEVELOPMENT

Pollution and soil erosion, two detrimental environmental impacts, will worsen in tandem with the increasing demand for materials and resources by humans. Therefore, in order to secure humanity's future, it is essential to investigate eco-friendly substitutes for synthetic materials. A number of UN Sustainable Development Goals (SDGs) can be advanced through the production and use of biodegradable, bio-based products like environmentally friendly plastics, which are essential in today's fast-paced world (Moshood et al., 2022).

The most ubiquitous substance in the world, plastic, has changed the way we live our lives and even changed the way medicine is practiced by making disposable medical equipment and supplies more affordable. A significant portion of the more than 8 billion tons of plastic that have been manufactured since the 1950s ends up in the environment, highlighting the negative consequences of our reliance on and use of plastic (Chapman et al., 2024). Poly (ethylene terephthalate) (PET) and other traditional plastics are devastating to land and sea ecosystems because they do not break down in nature. A new study on microplastics (plastic fragments smaller than 5 mm) has provided a comprehensive overview of the ways in which plastic pollution is harming the marine biome. According to the study, zooplankton are eating microplastics instead of food, which reduces their ability to consume and release particle-bound organic carbon into the ocean and disrupts the oxygenation process worldwide (Lear et al., 2021). The widespread use of single-use plastic items, such as masks and gloves, during the coronavirus (COVID-19) pandemic is expected to worsen plastic pollution (Winton et al., 2022).

The zero-waste movement has experienced meteoric growth in the past ten years. Many regions of the world have made plastic-free shopping the standard, and there are many different strategies to make the switch, from using bioplastics to using natural plastics like fiber cloth (Lear et al., 2021). Despite the significant progress made, some obstacles remain for bio-based materials (Marín-González et al., 2022). Misinformation and misunderstanding surrounding bio-based materials, especially bioplastics, threaten to derail an otherwise promising industry's efforts to create sustainable products in the face of rising public awareness of the issue and shifting consumer preferences.

In general, biodegradability is not a property of bio-based materials, according to Damayanti et al. (2021). This includes Bio-Polyethylene (Bio-PE) and Bio-Poly (ethylene Terephthalate) (Bio-PET). The term "bioplastic" is used inaccurately on the labels of many plastic products, contributing to public confusion about bio-based plastics. This, in turn, causes problems with waste disposal, as bioproducts are not properly sorted when they are no longer needed, which has a major impact on the environment. All polymers that are either biodegradable or derived from biological sources are collectively known as bioplastics. In theory, bio-based plastics can mimic the characteristics and longevity of traditional plastics while being composed of sustainable materials like plant biomass (Verma et al., 2024a).

## 12.3 APPROACHES DEMONSTRATING IMPROVED BIO-BASED PACKAGING

Bio-based packaging is an innovative approach to reducing plastic waste and promoting sustainability (Karuppiah et al., 2024). Here are some cutting-edge methods being explored. Traditional packaging materials often lead to challenges with reuse or recycling because separating them for processing takes a lot of time. This makes waste disposal harder and adds to environmental issues. Biodegradable packaging has advantages over synthetic plastics, such as being able to break down naturally and using renewable resources (Karuppiah et al., 2024). Examples include biopolymers and paper-based products. These materials have been shown to reduce moisture loss, prevent leaks, lower oxidation risks, enhance scent, and improve qualities like color retention and resistance to microbial growth.

Materials derived from sustainable sources are known as biopolymers or bio-based polymers. Their primary characteristic is their ease of decomposition, which renders them eco-friendly. The biodegradability of these polymers is due to their composition, which is derived from renewable resources and biological sources. Used in packaging, they decompose in the soil, decreasing pollution and functioning as fertilizers. When compared to recycling, this method uses less energy. It is more challenging to process natural polymers for sustainable development due to their complicated structures. Using methods such as enzymatic procedures and reactive extrusion (REX), bio-based polymers are made more user-friendly (Tarasova et al., 2024). Bio-based materials constitute the whole third generation of food packaging. According to Tarasova et al. (2024), there are three main categories of these materials: those derived from plants or other forms of biomass, those made by chemically processing biomass materials, and those made by either naturally occurring or genetically modified organisms.

## 12.4 USE OF NANOCOMPOSITES IN FOOD PACKAGING

In the case of bio-nanocomposite materials, one of the constituent parts is a continuous phase and the other is a non-continuous one. With the advent of nano-based products, nanotechnology has seen a surge in its application to food packaging in the past few years (Singh et al., 2023). The security and freshness of food packaging can be enhanced using this method. The addition of nanofillers, which are extremely small particles, to food packaging increases the product's shelf life by interacting with it and acting as an antibacterial agent. Packaging materials now make use of inorganic nanostructures rather than manufactured chemicals. Using both continuous and discontinuous components, these nanocomposites are composed of many phases and are manufactured from renewable resources.

## 12.5 INNOVATION FOR STRENGTHENING BIO-BASED MATERIAL APPLICATION

Alternatives to synthetic fibers, such as glass and carbon, are being explored, including natural fibers such as jute, flax, and hemp. The building, packaging, aerospace, and automotive sectors are just a few examples of the many existing uses for natural fiber composites in applications where extreme strength is not critical. By 2020, the United States Department of Agriculture (USDA) and Energy (DOE) had already settled on a target. Ten percent of the basic chemical components used in industries should come from renewable plant resources instead of non-renewable ones like petroleum. By 2050, the aim is to increase the share of these renewable plant resources to 50% as the obstacles to their use are gradually removed. This means more and more industrial materials would be sourced from plants rather than synthetic or fossil-based alternatives. Natural fibers have some appealing qualities compared to traditional materials. They offer good strength relative to their weight, can be easily molded into different shapes, are lightweight, and resist rusting and wear over time (see Figure 12.1).

However, natural fibers are not perfect. They absorb a lot of moisture, which can affect their durability, and they have an uneven structure, making them behave differently in different directions. Several studies have been conducted to test how natural fibers perform when combined with different plastics and resins, like polyester or epoxy (Vishwash et al., 2024). These studies aimed to understand how natural fibers affect the strength and durability of these composites. All these studies have highlighted that the bond between the fiber and the material it is combined with (called the matrix) is crucial. How well the fiber sticks to the matrix affects the strength and performance of the final product. The better the bond, the better the material is able to handle stress.

## 12.6 BIODEGRADABLE POLYMERS: A STEP TOWARD GREEN MANUFACTURING

The environmental impact of conventional petroleum-derived plastics has prompted significant investigation into sustainable substitutes. Biodegradable polymers, originating from renewable resources and amenable to microbial degradation, present a viable solution to the escalating plastic pollution dilemma. This study explored biodegradable polymers' characteristics, classifications, and applications, emphasizing their potential to transform manufacturing processes and foster a circular economy (Srivastava et al., 2024).

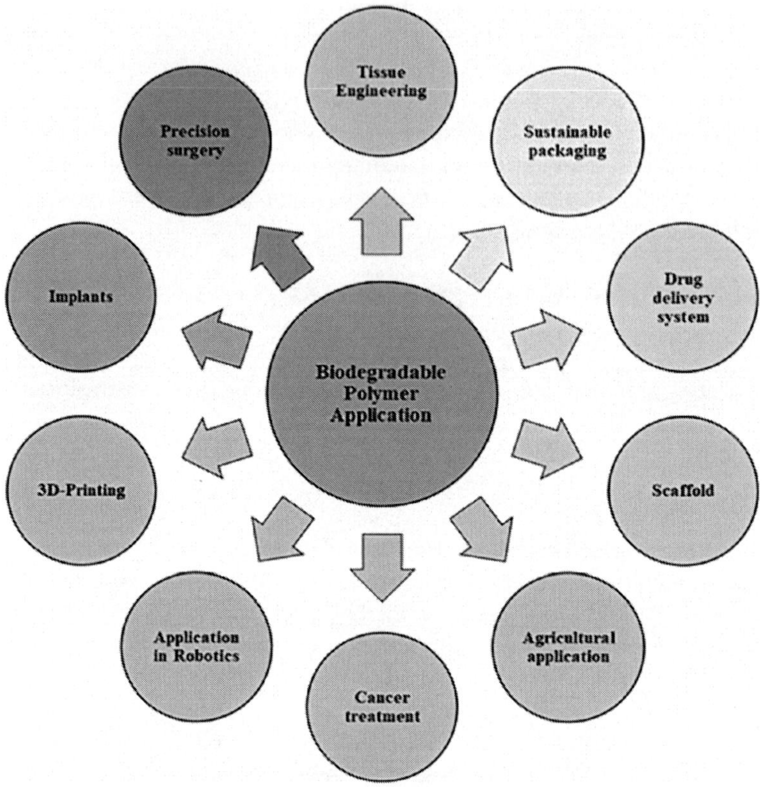


**FIGURE 12.1** Demonstrating varied applications of biodegradable polymer.

## 12.7 PROPERTIES OF BIODEGRADABLE POLYMERS

Biodegradable polymers are characterized by their ability to decompose under specific environmental conditions, primarily through the action of microorganisms such as bacteria and fungi. The degradation process results in the formation of natural byproducts such as carbon dioxide, water, and biomass. Several factors influence the rate of biodegradation, including polymer composition, microstructure, environmental temperature, humidity, and the presence of microorganisms. Table 12.1 enlists various commercially important synthetic fossil-based and bio-based polymers' environmental characteristics and average costs.

## 12.8 TYPES OF BIODEGRADABLE POLYMERS

Biodegradable polymers can be broadly classified into two categories: natural and synthetic.

**TABLE 12.1**
**Comparison of Certain Commercially Important Synthetic Fossil-Based and Bio-Based Polymers' Environmental Characteristics and Average Costs.**

| Polymer | Price (US$ per kg) | AP cradle-to-gate (kg $SO_2$-eq per ton polymer) | GWP cradle-to-gate (ton $CO_2$-eq per ton polymer) | Biodegradation (ocean) | Biodegradation (industrial) |
|---|---|---|---|---|---|
| ***Fossil based and durable*** | | | | | |
| PVC | 1.9 | 3 | 1.5–2.2 | NA | NA |
| HDPE | 1.4–1.6 | 6–22 | 1.8–2.6 | NA | NA |
| PET | 1.2–1.4 | 10–18 | 2.4–5 | NA | NA |
| LDPE | 1.36 | 27 | 1.9–3.1 | NA | NA |
| PS | 0.7–1.5 | NA | 3.2 | NA | NA |
| ***Fossil based and degradable*** | | | | | |
| PCL | NA | NA | NA | 6 weeks | 4–6 weeks |
| PBS | 4.5 | NA | NA | >1 year | 2–5 months |
| PBAT | 4.1 | NA | NA | >1 year | 2–3 months |
| PVA | 2 | NA | NA | 4 months | 1–2 weeks |
| ***Bio-based and durable*** | | | | | |
| bioPE | 1.8–2.4 | 30 | 0.68 | NA | NA |
| bioPET | NA | 13–75 | 2–5.5 | NA | NA |
| PEF | NA | NA | 2.1 | NA | 9 months |
| ***Bio-based and degradable*** | | | | | |
| P4HB | 3–8 | NA | NA | 1–6 months | 4–6 weeks |
| PGA | NA | NA | NA | 1–2 months | 2–3 months |
| bioPBS | NA | 75 | 2.2 | >1 year | >3 months |
| P3HB | 3–8 | 14–25 | −2.3–4 | 1–6 months | 1–4 months |
| PLA | 2–3 | 7–21 | 0.5–2.9 | >1.5 years | 6–9 weeks |

- **Natural biodegradable polymers:** these polymers are derived from renewable resources such as plants, animals, and microorganisms. Examples include:
  - **Polysaccharides:** starch, cellulose, chitin, and alginate—commonly found in plant cell walls and exoskeletons of crustaceans (see Figure 12.2).
  - **Proteins:** collagen, gelatin, and casein—derived from animal sources.
  - **Polyhydroxyalkanoates (PHAs):** produced by microorganisms through fermentation of sugars or lipids (Verma et al., 2024c).
- **Synthetic biodegradable polymers:** these polymers are synthesized by monomers derived from renewable or petroleum-based sources. Examples include:
  - **Polylactic acid (PLA):** produced from the fermentation of corn starch or sugarcane (Singh et al., 2024b).

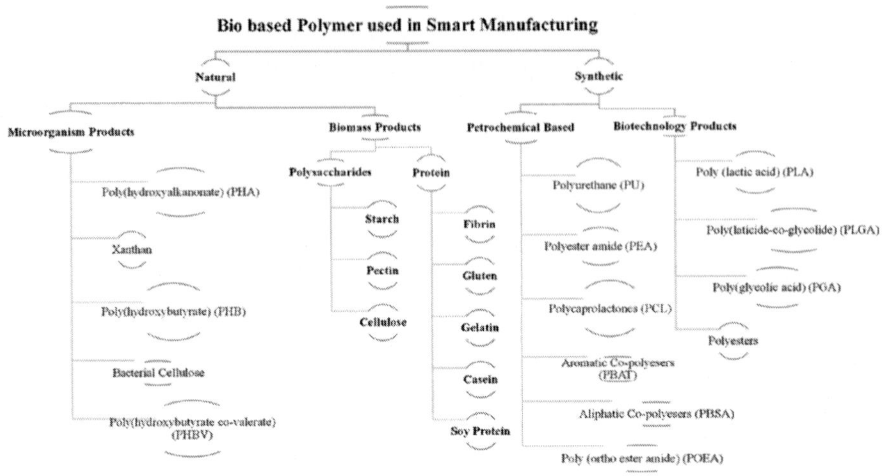


**FIGURE 12.2** Bio-based polymers derived from different sources utilized in smart manufacturing.

- **Polybutylene succinate (PBS):** a petroleum-based polymer with excellent biodegradability.
- **Polycaprolactone (PCL):** a synthetic polyester with a slow degradation rate, often used in long-term applications (Singh et al., 2024c).

## 12.9 APPLICATIONS FOR BIODEGRADABLE POLYMERS

The versatility of biodegradable polymers has led to their adoption in a wide range of applications:

- **Packaging:** biodegradable polymers are increasingly used for food packaging, shopping bags, and disposable containers, reducing the environmental impact of traditional plastic waste.
- **Agriculture:** biodegradable films are used for mulching, controlling weed growth, and releasing fertilizers in a controlled manner.
- **Medicine:** biodegradable polymers find applications in drug delivery systems, tissue engineering scaffolds, and sutures, providing biocompatible and absorbable solutions.
- **Textiles:** biodegradable fibers are used to produce eco-friendly clothing and fabrics with reduced environmental footprint (Dhiman et al., 2022, 2024).
- **3D printing:** biodegradable polymers are gaining popularity in 3D printing for creating prototypes, medical implants, and consumer products (Verma & Hussain, 2020).

## 12.10 ADVANTAGES OF BIODEGRADABLE POLYMERS

- **Environmental sustainability:** biodegradable polymers reduce reliance on fossil fuels, minimize plastic waste, and promote a circular economy (Verma et al., 2024b).
- **Biocompatibility:** many biodegradable polymers are biocompatible, making them suitable for medical and food packaging applications (Singh et al., 2024a).
- **Compostability:** biodegradable polymers can be composted, contributing to soil enrichment and reducing landfill waste.
- **Versatile properties:** biodegradable polymers exhibit a wide range of properties, allowing for customization based on specific applications.

## 12.11 MICROBE-BASED POLYMERS

An extended series of sugars or sugar acids makes up a polysaccharide. The charge, repetition, and structure of these polymers can vary, and they can be either branched or unbranched. They are also classified as homopolymers or heteropolymers. A wide variety of bacteria produce polysaccharides, which they either store in their cells (as glycogen) or release into the environment (as free exopolysaccharides that contribute to the biofilm matrix or as capsular polysaccharides attached to the surface of the cell). Sedimentary bacteria produce a wide variety of exopolysaccharides, such as cellulose, hyaluronate, and alginate, which are used as matrix components (Misra et al., 2017). According to Semwal et al. (2023) and Dixit et al. (2023), alginate provides a protective barrier for cells in the biofilm matrix of some species of *Bacillus* and *Pseudomonas*. Dense hydrogels that can hold a lot of water are made when alginates combine with divalent cations. For enterobacterial infections, cellulose manufacturing is a great boon. Proteinaceous curli fibers are stabilized by mortar-like structures formed by phosphoethanolamine cellulose, which is produced by many soil microorganisms in the rhizosphere (Dixit et al., 2021; Kar et al., 2024). These fibers improve resilience in high-shear situations and allow for robust cell interactions in complicated biofilms.

Furthermore, bacterial polysaccharides exhibit unique material characteristics. Chemical synthesis of polysaccharides requires expertise, it is costly, and usually only works for small molecular weights; only a few types are successfully synthesized. Thus, these polymers cannot be manufactured without microbial cell factories. Polysaccharides are able to bind water well and allow for crosslinking and intermolecular interactions due to the presence of hydrophilic groups like hydroxy and carboxyl groups (Gupta et al., 2023; Majhi et al., 2023).

Bacteria also produce a wide variety of biopolymers, including proteinaceous components and extracellular DNA. Both in bacterial pathogenicity and in the creation of bio-based products, these chemicals play an important role. When cells divide, they release their intracellular DNA into the environment, which is known as extracellular DNA. The stalks mushroom-shaped microcolonies of *P. aeruginosa* are examples of biofilms where extracellular DNA is added to the biofilm structure through the lysis of specific cell populations. Extracellular DNA interacts with positively charged polysaccharides and cations to improve the biofilm matrix's adherence and stability—one of its many functions—due to its strong negative charge. Abiotic stress resistance and nutrition supply while hunger are two additional benefits (Dubey & Misra, 2024).

## 12.12 CHALLENGES AND FUTURE DIRECTIONS

Despite their potential, biodegradable polymers face several challenges:

- **Cost:** biodegradable polymers are often more expensive than traditional plastics, hindering their widespread adoption.
- **Performance:** some biodegradable polymers may have inferior mechanical properties or processing characteristics compared to conventional plastics.
- **End-of-life management:** proper composting infrastructure is crucial to realizing the full environmental benefits of biodegradable polymers.

Future research focuses on developing cost-effective, high-performance biodegradable polymers with improved properties. Biorefinery technologies and genetic engineering advancements are expected to enhance the production efficiency and sustainability of biodegradable polymers (Shrivastava et al., 2023).

## 12.13 CONCLUSION

Biodegradable polymers represent a significant step toward green manufacturing and sustainable development. Their ability to decompose into natural byproducts offers a solution to the plastic pollution crisis and promotes a circular economy. Continued research and development, coupled with supportive policies and infrastructure, will pave the way for the widespread adoption of biodegradable polymers across various industries. We have also highlighted the advancements in comprehending the functions and possible applications of bacterial biopolymers as bio-based materials in this chapter. Furthermore, this chapter has laid the groundwork for drug discovery programs and the development of novel bio-based materials by outlining approaches to overcome the difficulties associated with biopolymers as a cutting-edge bio-based resource.

## ACKNOWLEDGMENTS

The authors acknowledge the financial assistance from the Council of Science & Technology, Uttar Pradesh (UPCST) project (Project ID 2707). The authors also thank the in-house seed grant project (File No. FY- 2023–24/SMG/001) funded by Shri Ramswaroop Memorial University, Lucknow-Deva Road, Barabanki- 225003, U.P., India.

## REFERENCES

Chapman A, Sen KK, Mochida T, Yoshimoto Y, Kishimoto K (2024) Overcoming barriers to proactive plastic recycling toward a sustainable future. *Environ Chall* 19:101040.

Damayanti D, SupriyaDi D, Amelia D, Saputri DR, Devi YL, Auriyani WA, Wu HS (2021) Conversion of Lignocellulose for bioethanol production, applied in bio-polyethylene terephthalate. *Polymers* 13:2886.

Dhiman V, Singh D, Dhiman V, Pandey M, Sharma A, Pandey H, Verma S (2024) Insects as biotic stress in agriculture, 193–210. https://doi.org/10.1007/978-981-97-4239-4_10

Dhiman VK, Singh D, Pandey H, Chauhan D, Dhiman VK, Pandey D (2022). Mycoremediation of agricultural waste for the cultivation of edible mushroom. In Aftab T (ed) *Sustainable Management of Environmental Contaminants. Environmental Contamination Remediation and Management*. Springer, Cham. https://doi.org/10.1007/978-3-031-08446-1_18

Dixit R, BiSht N, Misra S, Gupta SC, Chauhan PS (2023) *Bacillus* consortia modulate transcriptional and metabolic machinery of *Arabidopsis* plants for salt tolerance. *Curr Microbiol* 80:77.

Dixit VK, Misra S, Mishra SK, Joshi N, Chauhan PS (2021) Rhizobacteria-mediated bioremediation: insights and future perspectives. In Parray JA, Mahmoud AHAE, Sayyed R (eds) *Soil Bioremediation: An Approach Towards Sustainable Technology*. John Wiley & Sons, 193–211.

Dubey R, Misra S (2024) Biostimulants: an eco-friendly regulator of plant stress tolerance and sustainable solution to future agriculture. *Proc Indian Natl Sci Acad* 1–8. https://doi.org/10.1007/s43538-024-00328-4

Eslami Y, LezocHe M, Panetto H, Dassisti M (2021) On analysing sustainability assessment in manufacturing organisations: a survey. *Int J Prod Res* 59:4108–4139.

Gupta S, Misra S, Kumar M, Mishra SK, Tiwari S, Narayan S, Agrawal L, Chauhan PS (2023) Enhancement of drought tolerance in transgenic *Arabidopsis thaliana* plants overexpressing Chickpea Ca14–3–3 gene. *J Plant Growth Regul* 42:1544–1557.

Kar S, MiShRa SK, Misra S, Agarwal R, Kumar S, Chauhan PS (2024) Endophytic alkalotolerant plant growth-promoting bacteria render maize (*Zea mays* L.) growth under alkaline stress. *Curr Microbiol* 81:43.

Karuppiah K, Sankaranarayanan B, Lo HW (2024) A systematic literature review on the evolution of sustainable manufacturing practices: key findings and implications. *Clean Eng Technol* 100798.

Lear G, KingSbury JM, Franchini S, Gambarini V, Maday SD, Wallbank JA, Weaver L, Pantos O (2021) Plastics and the microbiome: impacts and solutions. *Environ Microbiome* 16:1–9.

Majhi B, Semwal P, Mishra SK, MiSra S, Chauhan PS (2023) Arsenic stress management through arsenite and arsenate-tolerant growth-promoting bacteria in rice. *Int Microbiol* 18:1–5.

Marín-González F, Moganadas SR, Paredes-Chacín AJ, Yeo SF, Subramaniam S (2022) Sustainable local development: consolidated framework for cross-sectoral cooperation via a systematic approach. *Sustainability* 14:6601.

Misra S, Pandey S, Dixit V, Mishra SK, Khan MH, AgarwaL L, Chauhan PS (2017) Soil microbiome for enhanced crop productivity. In Kalia VC, Shouche Y, Purohit HJ, Rahi P (eds) *Mining of Microbial Wealth and Metagenomics*. Springer Nature Singapore, 227–247.

Moshood TD, Nawanir G, Mahmud F, Mohamad F, Ahmad MH, Abdul Ghani A (2022) Biodegradable plastic applications towards sustainability: a recent innovations in the green product. *Clean Eng Technol* 6:100404.

Munaro MR, Tavares SF, Bragança L (2022) The ecodesign methodologies to achieve buildings' deconstruction: a review and framework. *Sustain Prod Consum* 30:566–583.

Semwal P, Misra S, Misra A, Kar S, Majhi B, Mishra SK, Srivastava S, Chauhan PS (2023) Endophytic *Bacillus* strains enhance biomass and bioactive metabolites of *Gloriosa superba. Ind Crops Prod* 204:117296.

Shrivastava N, Singh D, Verma S, Singh A, Srivastava A (2023) Toxicology assessment of polymeric material for implants, 33–54. https://doi.org/10.1201/9781003319139-3

Singh D, Shyam P, Verma SK, Anjali (2024a) Medical applications of systems biology. In Joshi S, Ray RR, Nag M, Lahiri D (eds) *Systems Biology Approaches: Prevention, Diagnosis, and Understanding Mechanisms of Complex Diseases*. Springer Nature Singapore, 55–70. https://doi.org/10.1007/978-981-99-9462-5_3

Singh D, Verma SK, Kumar S (2024b) Genetic manipulation for secondary metabolite production. In *Secondary Metabolites and Biotherapeutics*. Elsevier, 161–188. https://doi.org/10.1016/B978-0-443-16158-2.00001-X

Singh D, Verma SK, Kumar S (2024c) Trends in secondary metabolites production from plant sources. In *Secondary Metabolites and Biotherapeutics*. Elsevier, 103–126. https://doi.org/10.1016/B978-0-443-16158-2.00006-9

Singh D, Verma SK, Shyam P (2024d) Identification and purification of plant secondary metabolite as medicinal raw materials. In *Secondary Metabolites and Biotherapeutics*. Elsevier, 9–38. https://doi.org/10.1016/B978-0-443-16158-2.00003-3

Singh D, Verma S, Singh V, Perugu S (2023) Green functional nanomaterials: synthesis and application, 45–65. https://doi.org/10.1007/978-3-031-31104-8_3

Singh D, Verma SK, Wani KI, Aftab T (2024e) An overview of metal and metalloid toxicity and signaling in plants. In Aftab T (ed) *Metals and Metalloids in Plant Signaling*. Springer Nature Switzerland, 1–17. https://doi.org/10.1007/978-3-031-59024-5_1

Srivastava S, Singh D, Verma SK, Pandey M, Sharma A, Pandey H, Mishra A (2024) Recent advancements in cancer vaccines: a systematic review. *Vacunas* 25:97–108. https://doi.org/10.1016/j.vacune.2024.02.011

Tarasova E, Krasnou I, Enkhsaikhan G, Abousharabla I, Nunes CC, Karthegesu D, Savale N, Kontturi E, Krumme A (2024) Reactive extrusion of cellulose esters in ionic liquid: exploring properties and performance across different cellulose types and degrees of polymerization. *Cellulose* 7:1–8.

Verma S, Hussain R (2020) Comparative analysis of growth kinetics, concentration of sugar and alcohol tolerance of various yeast strains isolated from five different sources. *Int J Res Rev* 2:2582–5208.

Verma S, Kumar P (2022) Biofuel policies in India: an assessment of policy barriers, 44–64. https://doi.org/10.4018/978-1-6684-5269-1.ch004

Verma SK, Kumar P, Mishra A, Khare R, Singh D (2024a) Green nanotechnology: illuminating the effects of bio-based nanoparticles on plant physiology. *Biotechnol Sustain Mat*. https://doi.org/10.1186/s44316-024-00001-2

Verma SK, Singh D, Misra S (2024b) Innovative approaches to sustainable waste management and energy generation in developing nations. In Mehra R, Kumar P, Hung B (eds) *Practice, Progress, and Proficiency in Sustainability*. IGI Global, 267–300. https://doi.org/10.4018/979-8-3693-4054-7.ch010

Verma SK, Singh D, Shyam P (2024c) Elicitation of secondary metabolites from plants. In *Secondary Metabolites and Biotherapeutics*. Elsevier, 127–159. https://doi.org/10.1016/B978-0-443-16158-2.00009-4

Vishwash B, Sachidananda KB, Shivakumar ND (2024) Examination of the implications of fibers' alignment and volume fraction on the engineering constants of natural fiber epoxy lamina using analytical research. *Ind Crops Prod* 222:119615.

Winton D, Marazzi L, Loiselle S (2022) Drivers of public plastic (mis) use—new insights from changes in single-use plastic usage during the Covid-19 pandemic. *Sci Tot Environ* 849:157672.

# 13 Optimizing Bio-Based Processes for Sustainable Development

*Juveriya Israr, Sunil Kumar Verma, Laxmi Rathor, and Sankalp Misra*

## 13.1 INTRODUCTION

There is an increasing demand for optimal designs and bio-based process operation (BPO) systems globally. Optimizing BPO is a crucial initiative toward sustainable development and a green economy. The estimated annual expenditure for non-conventional energy sources, including bio-based materials, bio-fuels, and ecosystem services, will be around >1 trillion USD in the coming few decades. In addition, the bio-economy has already gained its importance in various enterprises, accumulating >2 trillion Euros in funds around the globe. However, the literature demonstrating the optimization of upstream and downstream operations related to bio-based materials is limited. Therefore, BPO is warranted for the enhancement of small-scale production processes and sensitizing sensor-based technologies. The novel model of green economy utilizes the complete capacity of sustainable and eco-friendly biomass sources, such as domestic, agricultural, and industrial wastes, and converts them into high-value products (Semwal et al., 2023a). Biorefinery technology holds a promising future in utilizing bio-resources (Sharma & Verma, 2014; Singh et al., 2024f; Verma & Hussain, 2020). The value of products is demonstrated to be enhanced under the green economy campaign through animal and plant sources. Remarkably, a huge volume of customer and societal needs is being satiated by bio-based products of added value, including health essentials and fuel products. The novel concept of the green economy is mentioned as the bio-based economy, which is demarcated from the conventional bio-economy comprising agriculture-based wastes, agroforest wastes, etc. Considering the pace of recent advances in the optimization of bio-based technologies, it is evident that scaling up the upstream and downstream processes for harvesting bio-based products is crucial to enhancing economic sustainability. In addition, food-based processing enterprises establish new ventures to gain advantages of resources and bio-based products. This can also release the pressure of food security, which is typically based on limited land area, and most of the land is under environmental challenges. Nevertheless, this will also help to develop policies and regulations for exploiting recent technology, safeguarding waste utilization, and resource planning of bio-based products. A sustainable bio-based economy is not

DOI: 10.1201/9781003566205-13

only achieved by renewability but also comprises initial agricultural, agroforestry, and aquaculture production (Gupta et al., 2023; Dixit et al., 2021, 2023). However, numerous challenges might affect the processing of industrial applications, including waste product management and agricultural processing. Importantly, the green economy is recognized for making substantial contributions to climate change issues by fully utilizing all biological resources. This encompasses the development of new bio-based products and resolutions that can replace non-renewable products, as well as bioactive compounds and non-toxic substitutes to harmful pesticides used in agriculture and aquatic environments. In addition to replacing fossil-based products, there will also be an emphasis on nature-based solutions that can, for instance, minimize or eliminate the need for chemical inputs in agriculture (Verma et al., 2024d). To achieve genuine sustainability in bio-based products and relevant optimization that meets all three sustainability criteria—environmental, economic, and social—it is essential to implement careful regulation and monitoring.

## 13.2 OPTIMIZING BIO-BASED PROCESSES FOR SUSTAINABLE DEVELOPMENT

Human activities are causing serious environmental problems due to the harmful life cycle of synthetic materials like plastics. For instance, less than 2% of plastic waste is recycled globally, and plastic pollution has become a major threat to both society and the planet. Even with higher recycling rates, plastics—whether made from petroleum or plant sources—release harmful chemicals, microplastics, environmental stresses, and nanoplastics into the environment and food chain (Mishra et al., 2023; Misra et al., 2017). In contrast, natural polymers produced from renewable resources like food waste, plants, or ocean-based materials can break down naturally and rejoin the environment without causing harm. These bio-based materials offer a promising alternative to conventional plastics, as they reduce the environmental damage caused by traditional plastic production and disposal. As consumer habits shift and governments introduce policies to support sustainability, bio-based materials are expected to play a significant role in the global economy (Singh et al., 2024b). However, despite their potential benefits, challenges still prevent the large-scale adoption of these materials. These obstacles have slowed the transition to a bio-based economy, leaving much work to be done before these solutions can fully replace traditional plastics. The bioeconomy plays a vital role in the global effort to build resource-efficient and low-carbon economies. According to the European Commission, the bioeconomy involves using renewable biological materials and waste to create valuable products such as food, animal feed, bio-based materials, and energy. In the European Union, there is a strong focus on developing advanced biotechnologies to modernize traditional industries and allow economic growth without causing environmental harm. However, there are concerns about whether switching from fossil-based to bio-based materials can truly address critical issues like food security, resource depletion, and climate change. For example, producing bio-based materials on a large scale from traditional sources like food crops and forest products can lead to problems like overuse of farmland, intensive farming practices, and

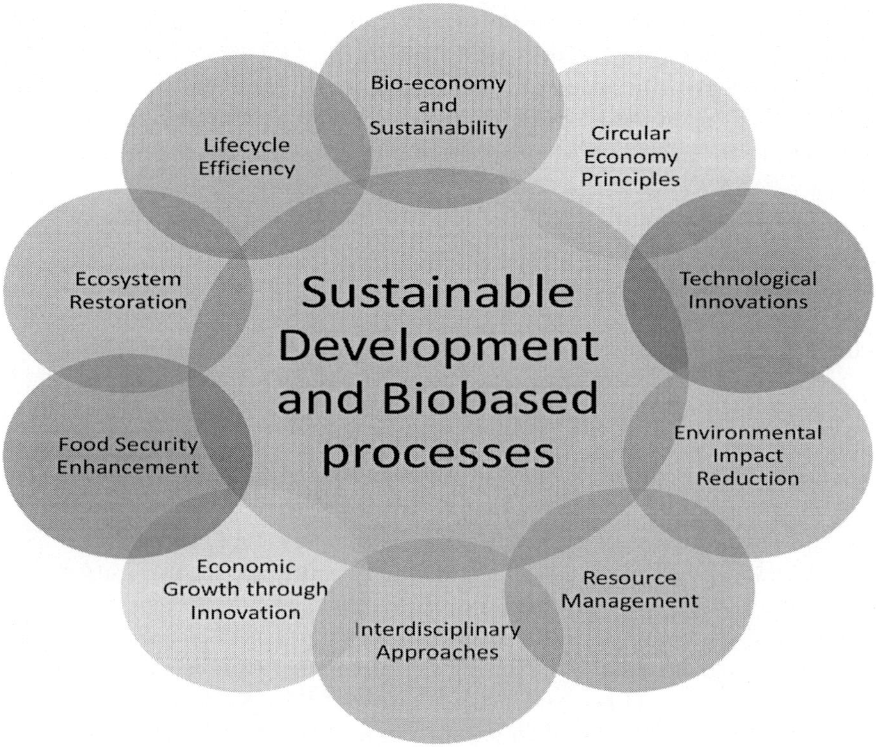


**FIGURE 13.1** Keyword combinations are categorized by identifying the relationship between sustainable development and the optimization of bio-based processes.

changes in land use. Using alternative resources such as algae, perennial crops, or agricultural residues could help, but creating affordable products from these alternatives requires better and more advanced technologies (see Figure 13.1). To make bio-based industries competitive, it is essential to use a variety of raw materials and improve how efficiently these materials are converted into products, which will lower costs and boost their success.

## 13.3 UNIT OPERATIONS AND THEIR OPTIMIZATION

Bio-based automation typically involves interaction between control regulations and real processes, which in turn comprise numerous sensors. Standard single-input comparative internal control systems are typically employed in this context. They operate quickly and reliably, ensuring a functional system that avoids material runoffs and maintains technical parameter levels. However, unit operations generally meet similar kinds of challenges globally. Furthermore, automation can also be challenged by technological shortcomings and the cost-effectiveness of controllers. These challenges can be surpassed by maintaining and optimizing the control systems (Singh et al., 2024e). Operators frequently control the parameters manually to synchronize scheduling decisions with other process constraints. For example, this may involve temporarily easing the workload of a critical process step by sequentially filling upstream buffer tanks. However, due to the properties and variability of raw materials, it may not be straightforward to implement a reliably functioning automatic pH controller.

## 13.4 SUPERVISORY MULTIVARIABLE CONTROL AND BATCH CONTROL

Multiple units' control and operations must be coordinated by the supervisory layer. The identification of process models and, by extension, state estimates, might be hindered by unreliable and/or late observations. When systems are impacted by significant uncertainty, lack reproducibility, or when the underlying processes are unknown, modeling can still be challenging even with adequate data (Dhiman et al., 2024). Optimization and control strategies based on models are made much more difficult by this. Current Process Analytical Technology (PAT) can fix the missing supervisory layer measurements. Those who took part in the business survey were asked to identify the ways in which they utilize chemometric and inferential PAT. No one asked them to divulge the specific technologies used or the locations and frequencies of the procedure's measuring sites because of worries about intellectual property. The Quality by Design framework and the PAT project are efforts by the US Food and Drug Administration to promote a paradigm shift in the historically conservative pharmaceutical industry toward PAT and feedback control. We take a look at the pharmaceutical industry to show how PAT is crucial for feedback regulation. A thorough description of the methods and technologies used is provided. However, process layers and analytical methods are so sidetracked from the main discussion (Singh et al., 2024c). The readers, therefore, need background information to critically assess the company's justifications for control and operations improvements. In terms of material and heat integration, large-scale continuous bioprocesses may be in sync with chemical processes; however, small-scale processes, particularly those in the pharmaceutical industry, are unlikely to employ significant recycling (Verma et al., 2024c). The plant's overall control challenge is affected by the sequentially ordered unit activities in the resulting processes. In most cases, slow responses with limited variable interactivity are the outcome of having enough buffer capacity and a sequential process layout. As a result, manual control becomes a viable option instead of automated methods, although it does so at the expense of capital expenditure. In large-scale continuous chemical processes, which often show a high degree of automation, the difference is clearly demonstrated in numerous papers.

One of the most important aspects of bio-based processing is batch operation. Six out of nine survey takers are involved in batch activities, lending credence to this assertion. According to this view, bio-based processes are similar to specialty and fine chemicals but distinct from the continuous, large-scale activities seen in the bulk chemical industries. While there is a wealth of information on batch control at the unit operating level, the function of automatic and human control within a batch plant as a whole is mostly unexplored (Verma et al., 2024f). The discussion of plant-wide control is complicated by the link between reactive scheduling and process control. The level of automation in continuous operations is usually higher than in batch facilities, according to an industrial survey. Separate sections will be devoted to supervisory control and operation because upstream and downstream processes are commonly divided in academia and business.

## 13.5 CONTROL OF UPSTREAM BIOPROCESSES

Research consistently shows that upstream operations stand to benefit the most from improved control. Reasons for this include possible gains in output and throughput, as well as the rising utility demand for different fermenters. The upstream control problem's multivariate and nonlinear character presents a significant barrier. These findings are relevant to the multiphase solid–liquid–gas system in which several species interact, including substrate, nutrients, intermediate metabolites, and end products. Inhibition, pH-sensitive metabolic and enzymatic kinetics, and heat and mass transfer phenomena are all parts of biological reaction systems that display complicated dynamic interactions. Because of the need to assess crucial performance metrics in labs that rely on biological features, measurement errors, and time delays may be significantly amplified. A rapid change in raw material qualities is a major red flag. When compared to processes mostly impacted by outside disturbances, which are easier to quantify, the complexity introduced by the vast natural variety of bacteria or enzymes is striking. Delays in accessing information and the many potential sources of uncertainty in processes and inputs make feedback control a challenging task. It is not always an easy undertaking to develop observers for biological systems (Singh et al., 2024d). However, PAT has kept the modeling and management of batch and continuous bioreactors at the forefront of industry attention. The steady improvement of models for industrial fermentation is evidence of advancements in this area. First, time-series modeling was used to conduct data-driven analyses of industrial fermentation, after which partial least squares modeling was used. Subsequent research built on this by validating first-principles models experimentally and demonstrating their vast capabilities. Hybrid modeling of bio-based systems is an exciting new area of study because our current understanding of basic processes, including complicated metabolic networks, is very limited.

## 13.6 CONTROL OF DOWNSTREAM PROCESSES

Downstream operating expenditures can exceed 80% of total processing costs in extreme circumstances, according to researchers; this percentage may even be higher for specific recombinant DNA-derived products. A more cautious estimate would place it somewhere in the range of 40–60%. On top of that, utility costs can add up quickly in downstream activities, where process units are common. In the pursuit of operational cost minimization through the application of PSE approaches, these frequently represent suitable concerns for investigation. Concentration evaporators, solvent recovery distillation columns, and membrane-based systems are typical units of interest (Srivastava et al., 2024). There are 45 complex problems that may necessitate advanced control methods, and these devices are typically operated continuously or semi-continuously (Verma & Singh, 2024). Chromatographic separations, on the other hand, could lead to higher operational costs, particularly in the case of non-reusable resins or solvents. This also applies to dead-end cake filtration systems that rely heavily on filter assistance. Primarily used in pharmaceutical procedures, single-use equipment is expensive. The potential reduction in operational expenditure by PSE is highly application specific and so hard to predict. Because operators frequently need to synchronize sequential and parallel batch units, scheduling rather than process control presents a number of downstream issues. This can happen even in a continuous process because of the properties of different unit operations. Scheduling can be complicated by cycle time transitions or batch-continuous interfaces, and biological factors can also impact cycle time uncertainty (Verma et al., 2024a). Complex duties include coordinating regular cleaning-in-place (CIP) or sterilization treatments, starting machines, or replenishing process additives when they run out. When utility systems are spread out over different process cells or manufacturing lines, the problem gets more complex. Since inefficient downtime lowers throughput and significantly raises the proportion of capital expenses in product pricing, effective scheduling is crucial. Equipment failures or faults caused by high-viscosity slurries can occasionally make scheduling difficult, even in reactively managed operations (Singh & Verma, 2024). For the purpose of the industrial survey, we asked each participant to identify the facility's ideal maintenance plan. Because most plant-wide maintenance plans cover a lot of ground, we gave respondents plenty of options on how to respond.

## 13.7 MANUFACTURING EXECUTION SYSTEMS (MSE)

Product specifications, quantities, and applicable process attributes are detailed in production orders distributed by the MSE. In modern production planning, just-in-time (JIT) or pull manufacturing is a common strategy. The continuous flow of materials within a discrete system is fundamental to the lean production concept, which is strongly related to this theory. In most cases, this is accomplished via Kanban (pull) methods. In addition, keeping small quantities of both in-process and completed goods in stock is an important part of lean manufacturing's emphasis on waste reduction (Singh et al., 2024a). Given the potential presence of large numbers of varied semi-finished components prior to final assembly in discrete-parts production, this becomes particularly pertinent. Lean approaches have recently been the subject of a comprehensive study that confirms their significance in the industrial sector. Process industries are considered to be a good fit for lean principles. Reductions in inventory and lot size are the primary targets of cost-saving efforts. Inventory leanness does not, however, ensure improved company performance according to an analysis of enterprises across a number of sectors. The in-process inventory volume in non-discrete process industries is mostly self-regulating because it has to be managed in equipment like buffer tanks and intermediate bulk containers. However, it is also clear that pull production is becoming more popular in this area. However, changeovers cause very large losses, and these activities have very extensive lead times (Verma et al., 2024e). This changes the balance between production costs and supply chain expenses; therefore, evaluating changeover costs thoroughly is required. Future demand projections should form the basis of any thorough production strategy. Therefore, accurate forecasting is essential for well-planned production. In most cases, statistical methods such as smoothing are used to create forecasts from previously collected sales or order data. While certain mechanical parts may show seasonal patterns, others may allow you to manually include canceled orders, and still others may let you manually include customer forecasts or competitor data into your estimates. Because of its versatility, batch plants are often used for more than one product or function. Batch plants do not, however, eliminate the intricate interplay among scheduling, control, and planning. The conditions are difficult for ongoing operations that create small quantities in brief campaigns (Mani et al., 2017). According to the industrial survey, there are instances where production campaigns are quite brief, which emphasizes the need for production planning that aligns with the capabilities of the facility and staff. When compared to fully automated large-scale continuous chemical plants, small-scale facilities show poorer homogeneity regardless of the operating mode. The latter's operational frameworks are well-suited to the automation pyramid model, which allows for precise task allocation by expert staff. The industrial review found that smaller facilities had more ambiguity and that instead of using control experts, all of the facilities that were examined used lean production experts to optimize their plants or sites thoroughly.

## 13.8 ENTERPRISE RESOURCE PLANNING

Considering the constraints mentioned earlier, the primary objective of resource planning is to establish guidelines that permit the cost-effective management of manufacturing facilities. Finding out what kinds of tasks employees perform and what kind of education they have is part of this process, as is determining what amount of automation is economically appropriate. Along with campaign lengths and market features, the ideal inventory quantity should be determined and used; occasionally, a discussion of batch versus continuous manufacturing might suffice as well. These evaluations are very similar to those needed in chemical processing; however, keep in mind the automation and operational constraints that were mentioned before (Verma et al., 2022). Managing technological innovation outside of one's expertise has been identified as a major issue for managers in the modern day, according to a previous assessment. Consistent with previous research, this one makes the case against putting money into technology just because it is good. This exemplifies how the growth in computer-integrated manufacturing has often fallen short of projections. Due to their application-specific nature and intrinsic multidisciplinarity, many bio-based enterprises necessitate extensive knowledge and expertise beyond what is typically taught in traditional education programs in order for managers to effectively oversee their operations. Along with this, both tactical initiatives and strategies, which are decisions made in the long run, are typically impacted by a number of qualitative elements. Consequently, using quantitative multiobjective optimization or other objective methodologies to establish judgments is not always viable (Verma et al., 2023). Public law, especially that which addresses product safety and environmental issues, will have an impact on the operations of food and biopharmaceutical industries, and these companies must be prepared to deal with these repercussions (see Table 13.1).

**TABLE 13.1**
**Comparison Between the Characteristics of Bio-Based Materials from Different Sources.**

| Characteristic | Petroleum-derived materials (plastics) | Bio-based materials |
|---|---|---|
| Origin | Fossil resources | Biological resources |
| Decomposition time | 400–1000 years for plastics | Months to a few years |
| Biodegradability | Non-biodegradable | biode3gradable |
| Sustainability | Low | High |
| Greenhouse gas emissions | High | Reduced |
| Compostability | Rarely compostable, it requires special processes | Many are compostable |
| Environmental impact | High impact | Lower impact |
| Recyclability | High recyclability for some plastics (like PET), but it depends on the infrastructure | Variable |
| Production cost | Generally lower, due to the established production scale | Can be higher currently |
| Technological innovation | Mature technology, energy efficiency, and focus on recycling | Rapidly growing area |
| Consumer perception | Harmful in terms of environmental impact, but reliable in functionality | Positive |
| Social impact | The concentration of wealth in regions with petroleum reserves, negative impacts on local communities due to extraction | Can generate jobs in rural areas, stimulating local economies |
| Functional performance | Excellent barrier to oxygen and moisture | Efficient moisture and oxygen barrier |
| Compatibility with environmental regulations | Subject to strict regulations due to pollution and carbon emissions | Generally compliant with stricter environmental regulations, aligned with carbon reduction goals |
| Impact on human health | Potential migration of chemical additives to food and beverages, concerns about microplastics | Generally perceived as safer, but studies are ongoing to understand all effects |
| Waste reduction potential | Lower waste reduction potential, but recyclability in established waste management chains | Greater potential for waste reduction due to compostability and biodegradability |
| Mechanical strength | Generally superior, high durability, and strength | Good, but may be inferior compared to traditional plastics |
| Examples of products and brands | Supermarket plastic bags (PE), Coca-Cola bottles (PET), and Tupperware food containers (PE) | Biodegradable starch-based bags, packaging by "Natural," and PLA utensils |

The manufacturing license for regulated markets is revoked if, as confirmed by routine audits, a technology fails to conform to current Good Manufacturing Practice (cGMP). Therefore, meeting quality standards is usually the primary goal in these domains. Agencies, whether public or private, frequently publish recommendations in public papers; nevertheless, these suggestions are not necessarily clear. Therefore, producers are responsible for understanding the standards and proving to the auditing body, which has a deep understanding of the industry standards, that their current quality control and management systems are adequate (Verma et al., 2024b). Because the auditing entity's choices and the length of the validation process are not reliably anticipated, this adds inertia and unpredictability to process optimization attempts. There is always some wiggle room for process improvement without requalification when it comes to process and quality criteria. The Design Space is another name for this location. Official evaluations of project risks and costs could be difficult for process engineers without precise quantification. An innovative case study that quantifies the limitations imposed by the lack of room for design has been reported by earlier researchers. Several studies have shown that Good Manufacturing Compliance norms have monetary impacts on factories that produce a lot of goods (Singh et al., 2023). PSE engineers need not be experts in every field; what they do need is the ability to quickly and accurately identify areas of bio-based production that could benefit from enhancement. More study is needed in this field since optimization and decision-making in cGMP facilities are strongly affected by regulatory ramifications. To tackle this complexity, formal approaches must be developed.

## 13.9 ASSESSMENT FOR THE PRODUCTION OF BIO-BASED POLYMERS

The sustainable and circular production of bio-based chemicals and polymers should be evaluated by looking at all three pillars of sustainability: environmental, economic, and social. This evaluation needs to consider the entire life cycle of these products, including what happens to them after use (end-of-life, or EoL). To design sustainable production processes, specialized software is used to calculate the size of equipment and estimate the materials and energy needed (Shrivastava et al., 2023). This information is essential for conducting techno-economic assessments (TEA) and life-cycle assessments (LCA). These tools help identify the most sustainable way to recycle or reuse bio-based materials, such as bioplastics, after they have been used. When assessing the options for recycling or disposing of post-consumer bioplastics, it is important to use criteria and metrics that consider all aspects of sustainability—technical, economic, environmental, and social (see Figure 13.2).

Biopolymers are attracting significant attention for applications like food packaging, pharmaceuticals, and edible films because they offer sustainable, biodegradable solutions (Prasad et al., 2024). Unlike fossil-based plastics, biopolymers are non-toxic, biocompatible, and possess suitable physical properties for various uses (see Figure 13.2). They can be made from renewable resources either by producing their building blocks through fermentation or by directly synthesizing the polymers in microbial cells. Developing sustainable technologies for making and recycling

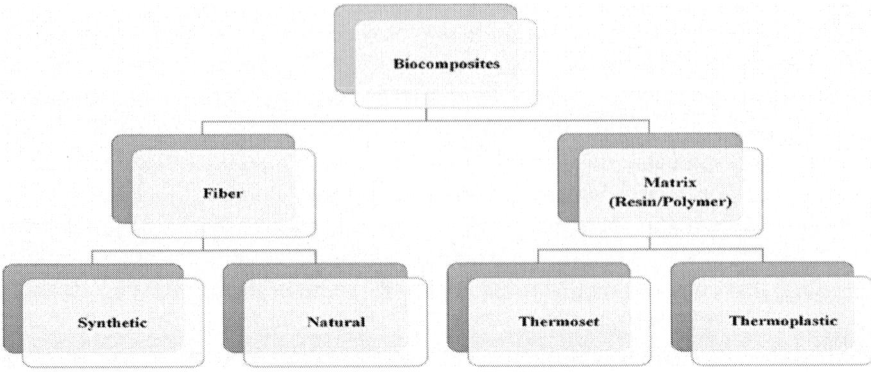


**FIGURE 13.2** Representation of various types of biocomposites utilized in sustainable development.

biopolymers requires a focus on (1) using renewable resources, (2) reducing environmental impact, and (3) improving the efficiency of conversion technologies.

This book chapter highlights advancements in methods and tools used to assess sustainable and circular processes, including recycling systems that incorporate bio-based materials into the industrial ecosystem. The LCA approach is emphasized as a proven method for evaluating the environmental, economic, and social impact of these processes. The findings can guide policymakers, entrepreneurs, and stakeholders to support research into bio-based production systems, create new standards and certifications for bio-based industries, and promote circular systems that align with sustainability goals (Singh et al., 2024a). In simpler terms, this research aims to help develop bio-based products that are sustainable from start to finish, encouraging innovation and collaboration to create a cleaner, greener future.

## 13.10 THE GLOBAL EMERGENCE OF A BIO-BASED ECONOMY

The world is witnessing a paradigm shift toward a more sustainable economic model—the bio-based economy. This transition is driven by the urgent need to mitigate climate change, reduce reliance on fossil resources, and promote environmental sustainability. A bio-based economy utilizes renewable biological resources, such as plants, animals, and microorganisms, to produce a wide range of products and services, replacing fossil-based counterparts and fostering innovation across various sectors (Semwal et al., 2023b; Majhi et al., 2023; Bugge et al., 2016).

### 13.10.1 Drivers of the Bio-Based Economy

Several key factors are propelling the global emergence of a bio-based economy:

- **Environmental concerns:** growing awareness of the environmental impacts of fossil fuels and conventional manufacturing processes has fueled the demand for sustainable alternatives. Bio-based products offer reduced carbon footprints, ameliorate environmental stresses, lower greenhouse gas emissions, and decrease reliance on non-renewable resources (Misra et al., 2024; Semwal et al., 2024; Umrao et al., 2024; von Braun, 2014).

- **Resource depletion:** the finite nature of fossil resources and geopolitical uncertainties associated with their supply have spurred the search for renewable alternatives. Bio-based resources offer a secure and sustainable supply chain for various industries (McCormick & Kautto, 2013).
- **Technological advancements:** breakthroughs in biotechnology, synthetic biology, and bioprocessing have enabled the efficient conversion of biomass into valuable products. These advancements have expanded the possibilities for bio-based materials, chemicals, and energy sources (Philp et al., 2013).
- **Economic opportunities:** the bio-based economy presents significant economic opportunities, including job creation, rural development, and new markets for bio-based products. Governments worldwide are recognizing the potential of the bio-based economy to drive economic growth and innovation (Dietz et al., 2018).
- **Consumer demand:** increasing consumer awareness and demand for sustainable products are driving the market for bio-based alternatives. Consumers are increasingly seeking products with lower environmental impacts and ethical sourcing (Scarlat et al., 2015).

### 13.10.2 Key Sectors of the Bio-Based Economy

The bio-based economy encompasses a wide range of sectors, including:

- **Bioenergy:** production of renewable energy from biomass sources, such as wood, agricultural residues, and dedicated energy crops. Bioenergy offers a sustainable alternative to fossil fuels and contributes to energy security (IEA Bioenergy, 2018).
- **Biomaterials:** development of bio-based materials for various applications, including packaging, construction, textiles, and automotive. Bio-based materials offer comparable or superior performance to conventional materials while reducing environmental impact (Mohanty et al., 2018).
- **Biochemicals:** production of chemicals and intermediates from renewable resources for use in various industries, including pharmaceuticals, cosmetics, and agriculture. Bio-based chemicals offer a sustainable alternative to petrochemicals and reduce reliance on fossil resources (Dubey & Misra, 2024; Patermann, 2017).
- **Biopharmaceuticals:** development of pharmaceuticals and healthcare products using biotechnology and bioprocessing. Biopharmaceuticals offer innovative solutions for disease treatment and prevention (Israr et al., 2024; Walsh, 2018).
- **Agriculture and food:** sustainable agricultural practices and the production of food and feed using bio-based technologies. This includes precision agriculture, organic farming, and the development of novel food sources (Kar et al., 2024; Misra et al., 2023; FAO, 2019).

## 13.11 CHALLENGES AND FUTURE OUTLOOK

While the bio-based economy holds immense promise, several challenges need to be addressed:

- **Cost competitiveness:** bio-based products often face challenges in competing with cheaper fossil-based alternatives. Continued research and development are crucial to improve the cost-effectiveness of bio-based production (Rogers et al., 2015).
- **Technological barriers:** scaling up bio-based technologies and achieving commercial viability requires overcoming technological hurdles. Further research and innovation are needed to optimize bioprocessing and improve product yields (Carus & Dammer, 2018).
- **Infrastructure and logistics:** the development of dedicated infrastructure and efficient logistics is crucial for the widespread adoption of bio-based products. This includes biomass supply chains, biorefineries, and distribution networks (IEA Bioenergy, 2020).
- **Policy and regulation:** supportive policy frameworks and regulations are essential to incentivize the development and adoption of bio-based technologies. Governments need to create a level playing field and promote sustainable practices (OECD, 2019).

The future of the bio-based economy is bright. Continued advancements in biotechnology, coupled with supportive policies and growing consumer demand, will drive the transition toward a more sustainable and resilient economic model. The bio-based economy offers a pathway to address pressing environmental challenges while fostering innovation and economic growth.

## ACKNOWLEDGMENTS

The authors acknowledge the financial assistance from the Council of Science & Technology, Uttar Pradesh (UPCST) project (Project ID 2707). The authors also thank the in-house seed grant project (File No. FY-2023–24/SMG/001 and File No. FY-2023–24/SMG/002) funded by Shri Ramswaroop Memorial University, Lucknow-Deva Road, Barabanki- 225003, U.P., India.

## REFERENCES

Bugge, M., Hansen, T., & Klitkou, A. (2016). What is the bioeconomy? A review of the literature. *Sustainability*, 8(7), 691.

Carus, M., & Dammer, L. (2018). The bioeconomy in Europe: status quo and future prospects. *Industrial Biotechnology*, 14(3), 137–140.

Dhiman, V., Singh, D., Dhiman, V., Pandey, M., Sharma, A., Pandey, H., & Verma, S. (2024). Insects as biotic stress in agriculture, pp. 193–210. https://doi.org/10.1007/978-981-97-4239-4_10

Dietz, T., Börner, J., Förster, J., & von Braun, J. (2018). Governance of the bioeconomy: a global comparative study. *Sustainability*, 10(10), 3602.

Dixit, R., Bisht, N., Misra, S., Gupta, S. C., & Chauhan, P. S. (2023). Bacillus consortia modulate transcriptional and metabolic machinery of arabidopsis plants for salt tolerance. *Current Microbiology*, 80, 77.

Dixit, V. K., Misra, S., Mishra, S. K., Joshi, N., & Chauhan, P. S. (2021). Rhizobacteria-mediated bioremediation: insights and future perspectives. In: *Soil Bioremediation: An Approach Towards Sustainable Technology* (Eds. Parray, J. A., Mahmoud, A. H. A. E., & Sayyed, R.). John Wiley & Sons, pp. 193–211.

Dubey, R., & Misra, S. (2024). Biostimulants: an eco-friendly regulator of plant stress tolerance and sustainable solution to future agriculture. *Proceedings of the National Academy of Sciences, India*. https://doi.org/10.1007/s43538-024-00328-4

FAO. (2019). *The State of the World's Biodiversity for Food and Agriculture*. Rome.

Gupta, S., Misra, S., Kumar, M., Mishra, S. K., Tiwari, S., Narayan, S., Agrawal, L., & Chauhan, P. S. (2023). Enhancement of drought tolerance in transgenic arabidopsis thaliana plants overexpressing Chickpea Ca14–3–3 Gene. *Journal of Plant Growth Regulation*, 42, 1544–1557.

IEA Bioenergy. (2018). *Bioenergy—A Key Pillar of the Clean Energy Transition.*

IEA Bioenergy. (2020). *Tracking Progress in Bioenergy: A Global Assessment.*

Israr, J., Ahamad, T., Alam, S., Misra, S., Gupta, D., Ahmad, R., Arshad, M., Khan, M. F., Maurya, P., AlShammari, L., & Ali, A. (2024). Bioactive compounds from Achyranthes aspera L. extract: a UHPLC profile and in silico study for mouth cancer. *ChemistrySelect*, 9, e202402092.

Kar, S., Mishra, S. K., Misra, S., Agarwal, R., Kumar, S., & Chauhan, P. S. (2024). Endophytic alkalotolerant plant growth-promoting bacteria render maize (Zea mays L.) growth under alkaline stress. *Current Microbiology*, 81, 43.

Majhi, B., Semwal, P., Mishra, S. K., Misra, S., & Chauhan, P. S. (2023). Arsenic stress management through arsenite and arsenate-tolerant growth-promoting bacteria in rice. *International Microbiology*, 18, 1–5.

Mani, P., Sharma, H., Gautam, D., Singh, T., Verma, S., & Hussain, R. (2017). Hydroxyapatite (ha) attenuate tio 2 toxicity in bio-system triggering E. coli and mouse bone marrow mono-nuclear cells (bmmnc's). *International Journal of Life Science and Pharma Research*, 7, 46–57.

McCormick, K., & Kautto, N. (2013). The bioeconomy in Europe: an overview. *Sustainability*, 5(6), 2589–2608.

Mishra, S. K., Misra, S., Dixit, V. K., Kar, S., & Chauhan, P. S. (2023). Ochrobactrum sp. NBRISH6 inoculation enhances Zea mays productivity, mitigating soil alkalinity and plant immune response. *Current Microbiology*, 80, 328.

Misra, S., Pandey, S., Dixit, V., Mishra, S. K., Khan, M. H., Agarwal, L., & Chauhan, P. S. (2017). Soil microbiome for enhanced crop productivity. In: *Mining of Microbial Wealth and Metagenomics* (Eds. Kalia, V. C., Shouche, Y., Purohit, H. J., & Rahi, P.). Springer Nature Singapore, pp. 227–247.

Misra, S., Prasad, P., Semwal, P., Mishra, S. K., Asif, M. H., & Chauhan, P. S. (2024). Genomic characterization of the salt-tolerant Bacillus and Jeotgalicoccus strains reveals a diverse metabolism relevant to plant growth promotion and salt stress tolerance. *3 Biotech*, 14, 1–4.

Misra, S., Semwal, P., Pandey, D. D., Mishra, S. K., & Chauhan, P. S. (2023). Siderophore-producing Spinacia oleracea bacterial endophytes enhance nutrient status and vegetative growth under iron deficit conditions. *Journal of Plant Growth Regulation*. https://doi.org/10.1007/s00344-023-11185-8.

Mohanty, A. K., Misra, M., & Drzal, L. T. (2018). Sustainable bio-composites from renewable resources: opportunities and challenges in the green materials world. *Journal of Polymers and the Environment*, 26(2), 557–575.

OECD. (2019). *The Bioeconomy to 2030: Designing a Policy Agenda.*

Patermann, C. (2017). The bioeconomy: a global perspective. *Solutions*, 8(3), 52–61.

Philp, J. C., Ritchie, R. J., & Allan, J. E. (2013). Biobased chemicals: a review. *Trends in Biotechnology*, 31(4), 219–229.

Prasad, S., Dave, A., & Misra, S. (2024). Cellulose based nanomaterials for food packaging: opportunities and challenges. In: *Organic-Based Nanomaterials in Food Packaging* (Eds. Younis, K., Yousuf, O., & Ul Islam, S.). Springer, Cham. https://doi.org/10.1007/978-3-031-63829-9_5

Rogers, M., Nielsen, D. R., & Tschirner, U. W. (2015). The role of technology in the bioeconomy. *Biofuels, Bioproducts and Biorefining*, 9(1), 1–8.

Scarlat, N., Dallemand, J. F., Monforti-Ferrario, F., & Nita, V. (2015). The role of biomass and bioenergy in a future bioeconomy: policies and facts. *Environmental Development*, 15, 3–34.

Semwal, P., Majhi, B., Pandey, D. D., Misra, S., Mishra, S. K., & Chauhan, P. S. (2023a). Agriwaste burning management through microbial intervention. In: *Recent Trends in Solid Waste Management* (Eds. Ravindran, B., Gupta, S. K., Bhat, S. A., Chauhan, P. S., & Tyagi, N.). Woodhead Publishing (Elsevier), pp. 333–347.

Semwal, P., Mishra, S. K., Majhi, B., Mishra, A., Joshi, H., Misra, S., Misra, A., Srivastava, S., & Chauhan, P. S. (2024). Bacillus australimaris protect Gloriosa superba L. against Alternaria alternata infestation. *World Journal of Microbiology and Biotechnology*, 40, 1–6.

Semwal, P., Misra, S., Misra, A., Kar, S., Majhi, B., Mishra, S. K., Srivastava, S., & Chauhan, P. S. (2023b). Endophytic Bacillus strains enhance biomass and bioactive metabolites of Gloriosa superba. *Industrial Crops and Products*, 204, 117296.

Sharma, H., & Verma, S. (2014). FemA gene in Indian isolates of methicillin resistant Staphylococcus aureus—isolation and amplification through real time PCR. *European Journal of Experimental Biology*, 4(6), 90–94.

Shrivastava, N., Singh, D., Verma, S., Singh, A., & Srivastava, A. (2023). Toxicology assessment of polymeric material for implants. In: *Assessment of Polymeric Materials for Biomedical Applications*. Taylor & Francis, pp. 33–54. https://doi.org/10.1201/9781003319139-3

Singh, D., Sharma, A., Verma, S., Pandey, H., & Pandey, M. (2024a). Impact of nanoparticles on plant physiology, nutrition, and toxicity: a short review. *Next Nanotechnology*, 6, 100081. https://doi.org/10.1016/j.nxnano.2024.100081

Singh, D., Shyam, P., Verma, S. K., & Anjali (2024b). Medical applications of systems biology. In: *Systems Biology Approaches: Prevention, Diagnosis, and Understanding Mechanisms of Complex Diseases* (Eds. Joshi, S., Ray, R. R., Nag, M., & Lahiri, D.). Springer Nature Singapore, pp. 55–70. https://doi.org/10.1007/978-981-99-9462-5_3

Singh, D., & Verma, S. (2024). Obesity in adolescents-causes and consequences. In: *Lifestyle Diseases in Adolescents: Addressing Physical, Emotional, and Behavioral Issues*. Bentham Science Publishers, pp. 107–120.

Singh, D., Verma, S., Singh, V., & Perugu, S. (2023). Green functional nanomaterials: synthesis and application, pp. 45–65. https://doi.org/10.1007/978-3-031-31104-8_3

Singh, D., Verma, S. K., & Kumar, S. (2024c). Genetic manipulation for secondary metabolite production. In: *Secondary Metabolites and Biotherapeutics*. Elsevier, pp. 161–188. https://doi.org/10.1016/B978-0-443-16158-2.00001-X

Singh, D., Verma, S. K., & Kumar, S. (2024d). Trends in secondary metabolites production from plant sources. In: *Secondary Metabolites and Biotherapeutics*. Elsevier, pp. 103–126. https://doi.org/10.1016/B978-0-443-16158-2.00006-9

Singh, D., Verma, S. K., & Shyam, P. (2024e). Identification and purification of plant secondary metabolite as medicinal raw materials. In: *Secondary Metabolites and Biotherapeutics*. Elsevier, pp. 9–38. https://doi.org/10.1016/B978-0-443-16158-2.00003-3

Singh, D., Verma, S. K., Wani, K. I., & Aftab, T. (2024f). An overview of metal and metalloid toxicity and signaling in plants. In: *Metals and Metalloids in Plant Signaling* (Ed. Aftab, T.). Springer Nature, pp. 1–17. https://doi.org/10.1007/978-3-031-59024-5_1

Srivastava, S., Singh, D., Verma, S. K., Pandey, M., Sharma, A., Pandey, H., & Mishra, A. (2024). Recent advancements in cancer vaccines: a systematic review. *Vacunas*, 25, 97–108. https://doi.org/10.1016/j.vacune.2024.02.011

Umrao, V., Yadav, S., Semwal, P., Misra, S., Mishra, S. K., Chauhan, P. S., & Shirke, P. A. (2024). Endophytic bacilli from Cyamopsis tetragonoloba (L.) Taub. induces plant growth and drought tolerance. *International Microbiology*, 12, 1–6.

Verma, S., Anjali, A., Diksha, D., & Kumar, V. (2023). Recent advancements in Skin tissue engineering in the application of Nanotechnology. *Research Journal of Biotechnology*, 18, 127–136. https://doi.org/10.25303/1802rjbt1270136

Verma, S., & Hussain, R. (2020). Comparative analysis of growth kinetics, concentration of sugar and alcohol tolerance of various yeast strains isolated from five different sources. *International Research Journal of Modernization in Engineering Technology and Science*, 2, 2582–5208.

Verma, S., Khare, R., & Singh, D. (2024a). Current perspectives on neuroendocrine tumors. *hLife*. https://doi.org/10.1016/j.hlife.2024.07.006

Verma, S., Pandey, M., Khare, R., & Singh, D. (2024b). A review on non-small cell lung cancer. *Vacunas*, 25. https://doi.org/10.1016/j.vacun.2023.12.003

Verma, S., Pandey, M., Sharma, A., & Singh, D. (2024c). Exploring ayurveda: principles and their application in modern medicine. *Bulletin of the National Research Centre*, 48. https://doi.org/10.1186/s42269-024-01231-0

Verma, S. K., Kumar, P., Mishra, A., Khare, R., & Singh, D. (2024d). Green nanotechnology: illuminating the effects of bio-based nanoparticles on plant physiology. *Biotechnology for Sustainable Materials*, 1, 1. https://doi.org/10.1186/s44316-024-00001-2

Verma, S. K., & Singh, D. (2024). Drug delivery and medical imaging using nanotechnology. In: *Applications of Nanotechnology in Biomedical Engineering*. CRC Press.

Verma, S. K., Singh, D., & Misra, S. (2024e). Innovative approaches to sustainable waste management and energy generation in developing nations. In: *Practice, Progress, and Proficiency in Sustainability* (Eds. Mehra, R., Kumar, P., & Hung, B.). IGI Global, pp. 267–300. https://doi.org/10.4018/979-8-3693-4054-7.ch010

Verma, S. K., Singh, D., & Shyam, P. (2024f). Elicitation of secondary metabolites from plants. In: *Secondary Metabolites and Biotherapeutics*. Elsevier, pp. 127–159. https://doi.org/10.1016/B978-0-443-16158-2.00009-4

Verma, S. K., Singh, V., & Singh, D. (2022). Analysis of isolated yeast strains from different sources for bio-ethanol production. *Inventum Biologicum: An International Journal of Biological Research*, 2, 3.

von Braun, J. (2014). The global bioeconomy: the role of agricultural biotechnology. *Journal of Agricultural and Food Chemistry*, 62(24), 5405–5410.

Walsh, G. (2018). Biopharmaceutical benchmarks 2018. *Nature Biotechnology*, 36(12), 1136–1145.

# 14 Innovative Processing and Functional Enhancements in Bio-Based Polymers

*Shubham Sharma, Nidhi Verma, Anmol Srivastava, Vishnu Agarwal, and Manmohit Kalia*

## 14.1 INTRODUCTION

### 14.1.1 Processing Innovations & Functional Advances in Bio-based Polymers

Bio-based polymers are polymeric biomolecules derived from natural sources such as animal, plant, and microbial sources. They have properties like biocompatibility, biodegradability, low density, reusability, moderate-to-high mechanical strength, and heat stability. These properties make bio-based polymers a valuable candidate for industrial applications. Various efforts have been made to use bio-based polymers in place of petroleum-based polymers. Recently, polysaccharide biopolymers have acquired significant attention from researchers (Jabeen and Atif, 2024). This field is just taking its first steps to explore the potential use of bio-based polymer derivatives as biosensors. Commercialization of petroleum-based polymers has made it an essentiality for everyday life, but because environmental awareness is increasing every day and there is a growing need to curb the dependency on petroleum-based polymers, much effort has been made toward the synthesis of bio-based materials by using natural polymers (Heinrich, 2019). Due to the concern for the utilization of renewable resources, eco-friendly products are given attention by researchers and also by many companies in the industry. The key parameters of sustainable manufacturing are reuse, reduce, recycle, recover, redesign, and remanufacture (Balla et al., 2019). Greenhouse gas emissions due to excess usage of fossil fuels have become a global burden, so experts are attempting to utilize it as little as possible. An amazing substitute for fossil fuels is biomass, which can be obtained from agricultural, lignocellulosic, and organic wastes that are easily available and can be used to develop sustainable biopolymers, reducing reliance on fossil-based resources. Many technologies have been developed for biomass utilization, which include drying, separation, fermentation, and pressing.

 DOI: 10.1201/9781003566205-14

## 14.2 PROCESSING PARAMETERS

Bio-based polymers like chitosan, lignin, wool, gelatin, and silk can be utilized in place of traditional plastic goods with the appropriate functional and structural features for numerous applications since they are superior to conventional petroleum-based polymers in lowering carbon dioxide emissions (Baranwal et al., 2022). Using bio-based polymers has several advantages: they are cost-effective, environmentally friendly, and readily available. However, their usage has been restricted due to their low mechanical characteristics. Fillers are being studied to improve the characteristics of bio-based polymers while lowering the total cost of the finished product.

Blends or composites of bio-based polymers provide benefits in both economic and environmental ways. By making polymer composites, we can balance performance and cost in the final polymeric composite product. The performance of these composite polymers is dependent on the processing and blend composition (Ahorsu et al., 2018). Mixing different polymers together combines the characteristics of both polymer components, and the final qualities of a blended composite material are determined by various parameters, including molecular weight, processing technique, morphology, and crystallinity. The mechanical characteristics of mixed polymers are proportional to the mixture's component ratio. For example, mixing a component with another polymer with poor mechanical qualities will not result in a composite with greater mechanical capabilities. Bio-based polymer composites are known as green products since they are combined using the melt processing method without using a solvent. The screw extruder technique, a form of melt processing, is used to combine polyamide with thermoplastic polyurethane (Rashmi et al., 2017). This mix demonstrated shape memory properties and may be utilized in sports shields as well as grip materials. Recently, nanomaterials have emerged as a transformative tool in healthcare due to their unique electrical, chemical, and mechanical properties. Carboxymethyl chitosan (CMCS), which is a water-soluble derivative of chitosan, has gained attention for its versatile properties, including biocompatibility, biodegradability, and improved solubility. CMCS-encapsulated ionic liquids are found to increase the conductivity and sensitivity of biosensors (Sharma et al., 2019).

Based on their manufacturing and origin, bio-based polymers can be classified into three major categories:

1. **Chemically synthesized polymers**: for example, lactic acid monomers polymerized into poly-(lactic acid) (PLA).
2. **Biomass-derived polymers**: for example, starch, cellulose, casein, and gluten.
3. **Microbial polymers:** for example, PHA or polyhydroxyalkanoates and bacterial cellulose.

The necessity to produce non-petroleum-based polymers drives global interest in bio-based polymers daily. Figure 14.1 illustrates how agricultural, lignocellulosic, and organic wastes serve as key feedstocks for the extraction of renewable and sustainable biopolymers, highlighting their role in reducing dependency on fossil-based

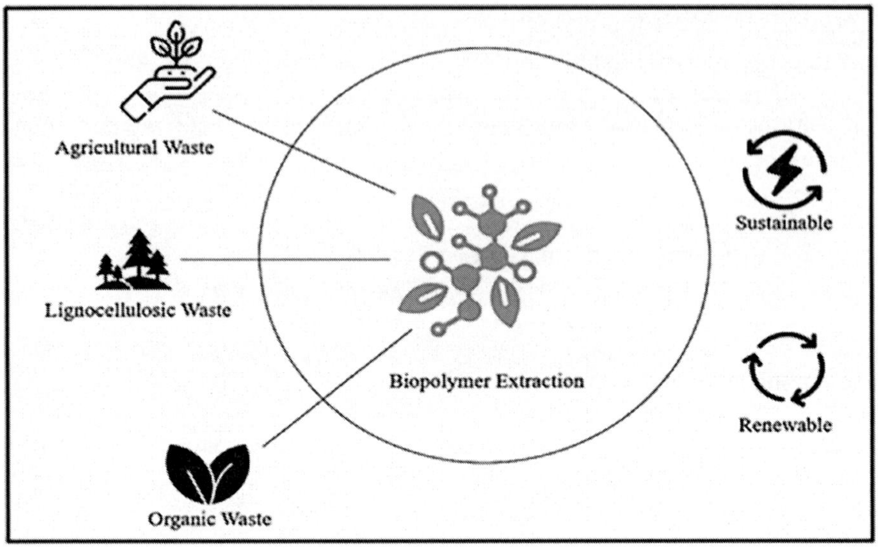


**FIGURE 14.1** Utilization of agricultural, lignocellulosic, and organic wastes for extraction of renewable and sustainable biopolymers.

resources. Continuous efforts are being made by researchers to improve energy efficiency in the production of bio-based polymers to boost environmental friendliness. There are various mechanical and physical processing methods, including milling, microwave irradiation, electron beam, spinning, extrusion, steam explosion, etc. These can be used for synthesizing bio-based polymers. Reactive extrusion is one of the cost-effective methods for producing new biopolymers with better characteristics. It is one of the fastest polymer processing technologies due to its continuity, high mixing efficiency, and high capacity, as well as it takes less time for the process to complete (Mehta et al., 2024). Furthermore, unlike other modification techniques, fibers and polymers are fed into a heated extruder during the reactive extrusion process. The heated material is then mixed to make a homogenous slurry using a screw and pushed into a mold to obtain the appropriate form (see Figure 14.2) (Formela et al., 2017).

Bio-based polymer composites are widely used in industries because they are easy to process, lightweight, chemically resistant, durable, and corrosion resistant. Melt compounding is another manufacturing procedure frequently used to create novel materials with desired qualities. Furthermore, it is commonly utilized in extruders found in the industry. It is the primary technique for producing high-molecular-weight polymers for packaging. It involves heating the material over its melting point, casting the molten polymer into a mold to get the required shape, and then cooling. The primary advantage of utilizing this method is the removal of complex processes such as polymerization reactions (Shakil, 2024).

Electrospinning is also one of the most widely used methods for bio-based polymer synthesis. It is a fiber production technique that uses electric force to draw

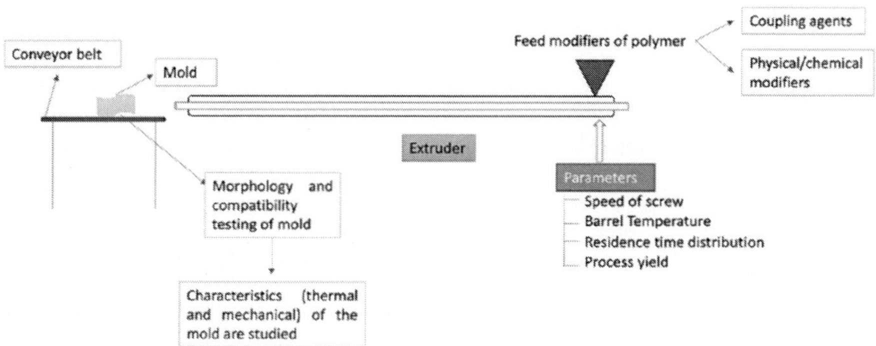


**FIGURE 14.2** Factors influencing the quality of biocomposites produced by the extrusion method.

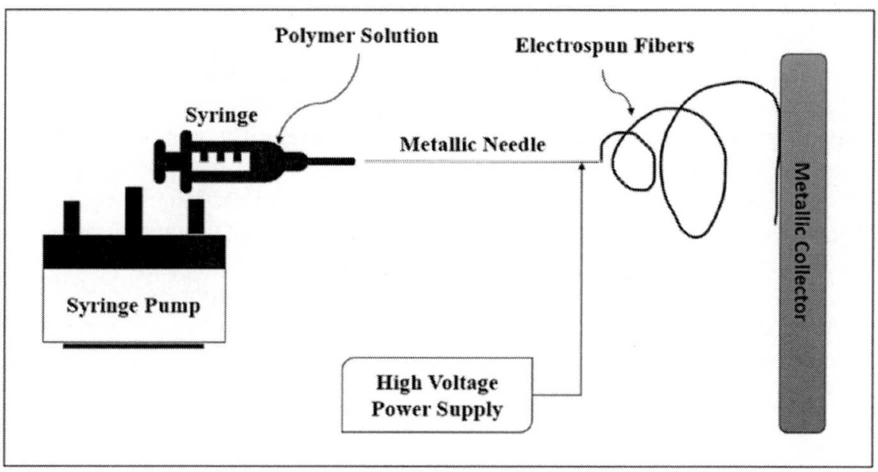


**FIGURE 14.3** Electrospinning setup containing a syringe, syringe pump, metallic needle, metallic collector, and a high voltage power supply.

charged threads of polymer solutions or polymer melts into ultra-fine fibers. This method produces fibers with diameters ranging from nanometers to micrometers, creating a web-like, fibrous material that can mimic the extracellular matrix of biological tissues. An electrospinning setup has five major components (see Figure 14.3) (Xue et al., 2019).

Electrospun fibers have a high surface-area-to-volume ratio, which makes them suitable for applications such as filtration membranes and drug administration systems that require high porosity or surface contact. Extraction and physical blending are also used to process bio-based polymers. This method allows bio-based products to be adjusted depending on the extraction procedures. Physical blending involves mixing the biopolymer components in a melt state with no chemical reactions. However, physical blending has one disadvantage: the bio-based polymer produced by this method has weak interfacial adhesion between phases, which may result in poor physical and mechanical properties for the bio-based materials being processed. For these processes, injection molding is the preferred technique, as it has very high accuracy (Ostafinska et al., 2015).

## 14.3 BIO-BASED POLYMER PROCESSING BY EXTRACTION

Solid–liquid extraction, liquid–liquid extraction, and Soxhlet extraction are some methods used to extract bioactives from solid and liquid biomass waste streams. However, these methods have limitations, such as high solvent usage and energy consumption. Ultrasound-assisted extraction and microwave-assisted extraction, on the other hand, have the potential to be faster and use less solvent. Modern extraction techniques include supercritical fluid extraction, microwave-assisted extraction, liquid extraction, and ultrasonication (Herzyk et al., 2024; Matharu et al., 2016). Choosing the right solvent throughout extraction is critical because it affects the end product and safety. Because of its low boiling point, ethanol is considered to be the safest solvent. However, ethanol is not ideal for all types of extraction; sometimes, acetone or a water–acetone mixture is a preferable solvent. Enzymatic extraction is used to recover lipids using viscozyme, proteases, and pectinase enzymes. Enzymatic extraction is generally safer than hexane-based extraction; however, lipid removal produces residues with high salt concentrations, which might be problematic when scaling up. The fundamental constraint of this enzymatic approach is its cost, which can be reduced by using immobilized enzymes. Separating substances from an extract is a costly process. The inherent qualities of the completed product are determined by the bio-based polymeric components used and the reprocessing technique (Badia et al., 2017). These qualities can be improved by adjusting thermal and other parameters. During reprocessing, various additives and fillers may be added to improve specific material properties. Reductive catalytic fractionation approach has been used to extract lignin from lignocellulose or wood (Schutyser et al., 2015). The extraction percentages in different solvents were 74% and 80% in ethylene glycol and methanol, respectively, at a reaction temperature of 250°C for 3 hours. Lignin was then transformed into phenolic mono and dimers in both solvents, which were found to develop distinct features and properties. Another method to separate the required product from its solvent is solvent extraction, which allows us to extract and recover both organic and metal ions from aqueous solutions. This approach outperforms previous extraction methods in terms of stability and selectivity. Temperature, pH, and extraction time determine the yield in solvent extraction.

## 14.4 PROCESSING BY MICROWAVE IRRADIATION

Microwave polymer processing methods are more efficient when discussing time and energy (Belkhir et al., 2022; Kabir, 2023; Khandelwal and Gupta, 2022). The microwave reactors can reduce the heating time, which plays a central role in the optimization of this method (Krishnan et al., 2022; Sharifvaghefi and Zheng, 2022). Various types of radiation are used, such as microwave, electron beam, pulse electrical field, ultrasound, and gamma rays. These radiations have been found to assist in processing methods for biomass conversion into biopolymeric materials. Microwaves are EM radiations having frequencies ranging from 0.3GHz to 30GHz. These radiations can be used to produce hydrogels. Microwave energy is absorbed by polar molecules through their polar bonds, and friction between molecules due to the polarization frequency gap generates heat, thereby heating the sample. This heating leads to the breaking of bonds. Microwave is very time-efficient compared to conventional methods of irradiation like laser heating. This method is preferred for various polymerization reactions because it has benefits like reduced heating time and homogeneity. Overall, microwave-assisted irradiation is an eco-friendly way to produce hydrogels for biomaterials and is often used to make hydrogels from polysaccharides (Duquette and Dumont, 2019; Kundu et al., 2022).

Aliphatic polyesters are a group of biodegradable bio-based polymers. An environment-friendly method to synthesize polybutylene succinate uses microwave at 200–220°C for 20 minutes, compared to conventional heating methods that take almost 5 hours. The polycondensation of polybutylene succinate was performed under atmospheric pressure in the presence of a catalyst (Ilsouk et al., 2020). Nanostructured composites of bio-based polymers with PVA under microwave radiation have been made (Riaz et al., 2011), and their properties can be compared with conventional solution mixing using TGA (thermogravimetric analysis), FTIR (Fourier transform infrared spectroscopy), SEM (scanning electron microscope), DTA (dynamic thermal field), and TEM (transmission electron microscope). The results were in favor of microwave-assisted mixing, indicating that it is an efficient method for producing composites. Synthesis of PLA from lactic acid by polymerization through ring-opening mechanism and using metals as catalysts is a standard process to produce PLA in industries. However, the problem is that the catalyst might cause toxicity because its traces might be left in the final polymer even after the polymerization process is complete. To address this problem, microwave-irradiated polymerization of lactic acid using a 2.45 GHz microwave radiation can be used, which results in polymerization in less than 1 hour (Liu et al., 2001). Similarly, soya peptone grafted with polyacrylamide can be prepared by microwave-aided polymerization along with ceric ammonium nitrate as the free radical initiator (Mahto et al., 2014).

## 14.5 PROCESSING OF BIO-BASED POLYMERS USING 3D PRINTING TECHNOLOGY

3D printing, additive manufacturing, or solid freeform fabrication is a revolutionary innovation in bioprocessing. It can be used to synthesize a wide range of object designs (see Figure 14.4.). 3D printing is a computer-driven processing method that uses 3D models created with software like AutoCAD and Blender, which aids in printing materials using bioinks. It produces 3D structures with precise shape and size (Oleksy et al., 2023).

It prints the object in a layer-by-layer fashion based on the 3D model. Many conventional manufacturing techniques can easily be replaced with 3D printing because it is much faster. Bio-based polycarbonate can be prepared using extrusion-based

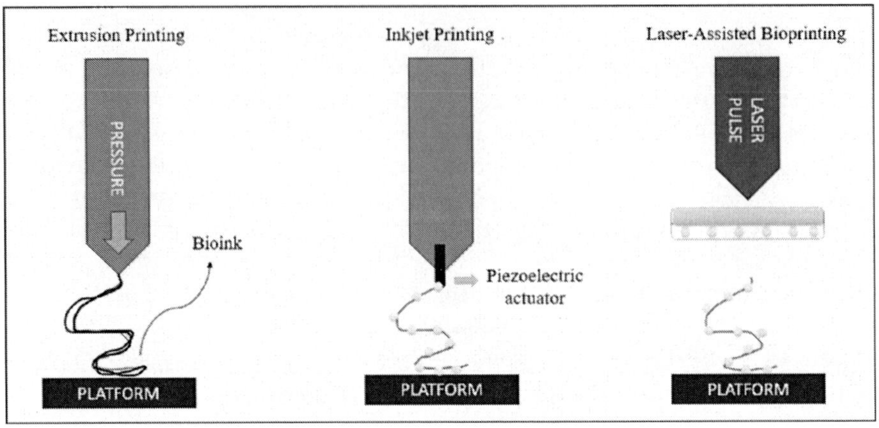


**FIGURE 14.4** Different types of 3D printing technologies.

3D printing (Park et al., 2020; Wang et al., 2017). Data are sent to the 3D printing machine via the command line, and each layer is created on top of the previous one to create the precise part shown in the 3D model. After printing, all parts are finished and ready for use. To assess the effectiveness of high reflection, they used 3D printing technology in combination with injection molding and hot processing to create a bio-based PLA-polybutylene 2,5-furan dicarboxylate composite (Leong et al., 2013). It was also examined by Long et al. using FTIR, TGA, H1-NMR (hydrogen-1 nuclear magnetic resonance), AFM (atomic force microscope), and DSC (differential scanning calorimeter). Bio-based polymers are the most suitable candidates for 3D printing scaffolds that can be used for 3D cell culture applications. 3D printing of scaffolds using bio-based polymers is an emerging area of research these days.

## 14.6 PROCESSING OF COMPOSITE FIBER-REINFORCED POLYMERS

Fiber-reinforced polymers are composite materials that contain high-strength fibers within a polymer matrix. They have a wide range of uses in daily life due to their low cost and ease of processing. Numerous processing procedures are employed for their synthesis, including injection molding, grafting, pultrusion, and extrusion compounding. Compression molding is a superior processing method for natural fibers, since injection molding might harm their characteristics (Leong et al., 2013). There are two sources of fibers, as shown in Figure 14.5.

These natural fibers have superior properties and can replace synthetic materials due to their low density, low molecular weight, fair mechanical properties, low cost, and high specific strength, and most importantly, non-toxicity or biocompatibility. Natural fibers can be used as reinforcing agents in different polymer matrices to produce composites with improved mechanical properties. Composite fiber-reinforced polymer of PLA and wool fibers can be made using the extrusion technique, and the surface of wool fibers should be enhanced before being characterized using TGA, SEM, and FTIR spectroscopy (Pawlak et al., 2019). Similarly, electrospinning can

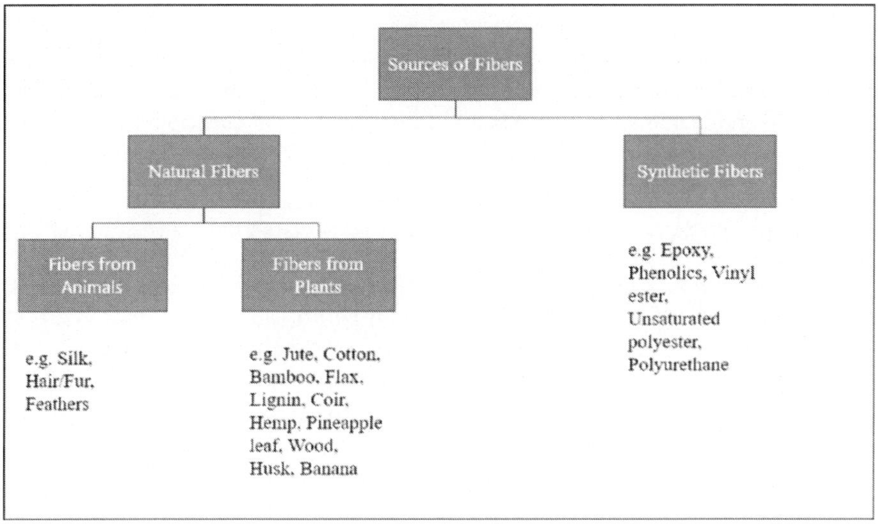


**FIGURE 14.5** Natural and synthetic sources of fibers.

be used to create a chitosan–PVA nanofibrous membrane for air filtration, which is then cured with UV light (Zhu et al., 2019). Air pollution is a global concern, and fiber-reinforced polymers can be utilized as air filters to solve it and reduce pollution. Hydrophilic silica nanoparticles can create a rough surface, increasing filtration efficiency (Hu et al., 2019). Similarly, silver nanoparticles have antibacterial characteristics. Remarkable efforts are being made all over the world to replace fossil fuel-based products with bio-based polymers. Furthermore, due to the rise in oil prices and customers' desire for environmentally friendly products, bio-based polymers are the ideal alternative.

## 14.7 CASE STUDY ON 3D PRINTED PLA

The main objective of this study was to determine the potential of PLA as a replacement for conventional plastics like acrylonitrile–butadiene–styrene (ABS) (Sakthivel et al., 2018). PLA was 3D printed using fused deposition modeling (FDM). The FDM machine used PLA wire spools to print specimens for testing. Various tests were conducted on the printed PLA specimens:

i. Mechanical testing of the specimens was done to analyze tensile strength, bending behavior, and impact resistance.
ii. Biodegradability testing was done by landfill and bacterial degradation methods. Biodegradation was monitored in the buried samples, which were kept under controlled conditions. In the bacterial degradation method, specific bacteria, which are known to degrade PLA, were used to observe the microbial degradation rate.

This study showed that PLA is very similar to ABS in terms of mechanical properties like tensile strength and impact resistance. The tensile strength of 3D printed PLA specimens was found to be 48.66 MPa (average), which indicates that the specimens have high strength and therefore, PLA can be used in biomedical or automotive applications.

## 14.8 STRUCTURAL AND FUNCTIONAL MODIFICATIONS IN BIO-BASED POLYMERS

It is one of the vital strategies to enhance the functional parameters and broaden their application potential. These alterations can be directed toward building polymers that have improved biodegradability, thermal stability, mechanical strength, and some other properties like antimicrobial activity, solvent resistance, and self-rejuvenation. From a biological standpoint, these objectives also include properties like enhanced attachment of cells and proteins to the surface and thrombogenicity. Different techniques for structural modifications include:

### 14.8.1 Chemical Modifications

These modifications include plasma treatment, wet chemical oxidation, covalent binding of polymer chains, and functionalization. Exposure of the polymer surface to wet chemicals like nitric acid, potassium permanganate, and chromic acid causes oxidation, which results in the formation of hydroxyl, carbonyl, and carboxylic acid groups on the polymer surface (Gleissner et al., 2022). In addition, when the treatment mixture includes sulfuric acid, sulfonation may occur, which in turn enhances the adhesion and wettability of the polymer as a result of an increase in the potential of hydrogen bonding. Laurichesse et al. in 2014 reported the use of lignin for the production of a novel bio-based polymer. They chemically modified lignin to form newer chemically active sites (via halogenation, amination, nitration, alkylation, and sulfomethylation), modified the hydroxyl groups, and split the lignin structure into aromatic macromers with the help of pyrolysis, hydrogenation, hydrolysis, gasification, oxidation, and enzymatic oxidation (Laurichesse et al., 2014). As a result of this modification, lignin can be applied along with polyolefins and elastomers, which impart antioxidant and flame-retardant properties. For instance, lignosulfonates show improved flowability, compressive strength, durability, and plasticity to certain polymers. On similar lines, acetylation of lignin was achieved with the help of kraft, hydrolytic lignin, and organosolv after subjecting them to a reaction containing acetic anhydride in the presence of 4-dimethylaminopyridine as a catalyst. Ethyl ether was then used to precipitate acetylated kraft lignin, while a cold methanol and water mixture was used to precipitate hydrolytic lignin and modified organosolv. It was further dried with the help of a methanol and water mixture, and then with the help of the solvent casting method, films of cellulose triacetate (CTA) composite were obtained with modified or unmodified lignin. This bio-based polymer has found applications in water treatment and is resistant to biofouling as a result of the antibacterial property of lignin. Figure 14.6 illustrates various structural modification techniques used to enhance the functional properties of bio-based polymers.

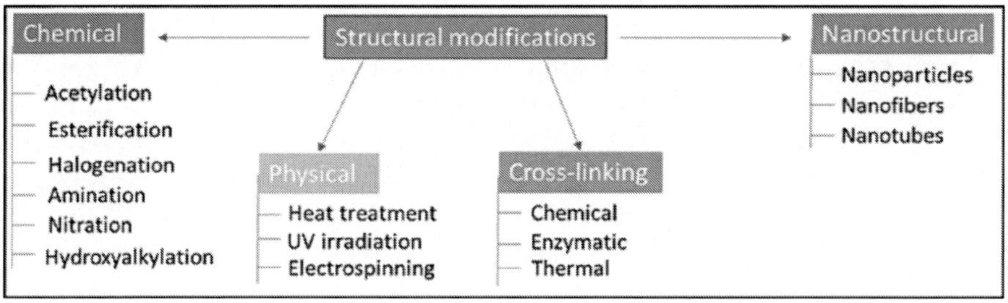


**FIGURE 14.6** Overview of structural modification techniques in bio-based polymers. Various approaches, including chemical, physical, irradiation, enzymatic, and nanostructural modifications, are illustrated.

### 14.8.2 Cross-Linking

Researchers are focusing on developing novel bio-based polymers to tackle the environmental risks associated with chemical polymers or petroleum monomers. To develop new monomers, green resources, including agricultural waste or plant biomass, are being explored. Pezzana et al. (2023) showed that furfuryl alcohol, which is derived from furfural (derived from hemicellulose after acid treatment), can be utilized to form epoxy monomer. Then, the cross-linking of this synthesized furan monofunctional epoxy monomer can be achieved in a two-step process that involves UV irradiation followed by thermal treatment. In the first step, furan monofunctional epoxy monomer was used in the presence of a cationic photoinitiator, which led to the formation of linear polymer chains due to UV-induced polymerization. In the next step, different polymer chains were cross-linked together (Pezzana et al., 2023). Lately, a lot of attention has been placed on developing films from proteins, polysaccharides, and polylactides to provide sustainable solutions for food packaging. Chemical treatments that involve acids, alkali, or cross-linking agents are being explored to enhance the functional properties of these films. It was reported that acid or alkali treatment of protein films reduced permeability and elevated tensile strength. The cross-linking agents include formaldehyde, glyceraldehyde, glyoxal, and glutaraldehyde, which further increased the resistance of these polymers against breakage and showed water-proof properties (Purewal et al., 2024). Vartiainen and Harlin (2011) found that genipin, glyoxal, and glutaraldehyde could cross-link chitosan films irrespective of the addition or absence of nanoclays. Furthermore, it was found that genipin or glutaraldehyde at a concentration of 3.3 $g/m^2$ in chitosan films effectively reduced water vapor transmission rate by 34% and 17%, respectively, in comparison to chitosan films used alone. They further reported that a significant reduction of 92% was achieved in oxygen transmission rate when glyoxal (at a concentration of 3.3 $g/m^2$) was used as a cross-linking agent between protein films (Vartiainen and Harlin, 2011). The toxicity associated with chemical cross-linking propelled the researchers to explore enzymatic cross-linking, which would be a more feasible option for food applications. In this regard, food-grade enzymes like transglutaminases are being used to cross-link protein films as they attach the γ-amine (acyl acceptor) of lysine residues to γ-carboxyamide (acyl donor) via the acyl-transferase mechanism. This enzyme can further be extended to cross-link protein and polysaccharide blends. The mechanism behind this involves oxidation of tyrosine by the tyrosinase enzyme into o-quinone residues, which later on attach to the functional groups of the chitosan polymer in the presence of transglutaminases. Both enzymatic and chemical cross-linking have led to the development of polymers with improved stability and functional enhancements.

### 14.8.3 Physical Modifications

This strategy involves altering the polymer structure without changing its chemical composition. One such method is heating, which takes into account the ability of proteins to aggregate irreversibly and form a gel network in the presence of high temperatures. An increase in temperature results in changes in intra- and intermolecular electrostatic and hydrophobic interactions in proteins, along with variations in disulfide and hydrogen bonds (Yousefi et al., 2022). This protein network's functional and physical parameters largely depend on the ionic strength and the magnitude of temperature. Thermal denaturation of different proteins occurs at different temperatures; for gluten protein, it is 90°C, while for whey protein, it is 70°C. The cross-linking of disulfide, hydrophobic, and hydrogen bonds results in elevated tensile strength. In one study, it was found that oxygen permeability was significantly reduced after high-temperature treatment of whey protein films. In addition, the protein film's rigidity and strength greatly increased due to reduced water content in the film (Benbettaïeb et al., 2016).

Another physical modification method involves ultrasound, an acoustic wave with a frequency above 20 kHz. As a result of compression and decompression waveforms of these high-energy waves, shearing effects are observed. These shear forces and the energy associated with them can easily break covalent bonds in proteins present in the aqueous solution. In addition to these forces, sonolysis of water due to these waves results in the denaturation of proteins (Zink et al., 2016). Researchers reported that UV irradiation of protein solution before the formation of films increased the tensile strength of whey-based films. In another study by Rodriguez-Turienzo et al. (2012), it was found that the whey protein coating, when treated with UV irradiation, showed lower lipid oxidation in comparison to untreated coatings, and the quality of fish was maintained as no major sensory changes were observed (Umaraw et al., 2020).

### 14.8.4 Nanostructural Modifications

This strategy involves the incorporation of nanomaterials like nanotubes, nanoparticles (NPs), and nanofibers into bio-based polymers to achieve the required functional parameters. A study by Zhou et al. found that the addition of titanium dioxide ($TiO_2$) nanoparticles to whey protein isolate films significantly increased the tensile strength of these films. In addition, this nanocomposite blocks almost

90% UV light and 70% visible light. A combination of soy protein isolates and $TiO_2$ NPs inhibited *Escherichia coli* and *Staphylococcus aureus* after treating the film with UV light for 3 hours at 365 nm. On similar lines, silver nanoparticles (AgNPs) have been incorporated into polymer nanocomposite matrices, which not only strengthen the polymer but also have inhibitory activity against a broad spectrum of bacteria, fungi, and viruses. This polymer bend was also found to be less cytotoxic due to the immobilization of $Ag^+$ ions and, hence, has found applications in food packaging industries (50). Scientists have also developed cellulose nanocrystals (CNCs) by ball milling cellulose with water and then centrifuging it. The source of cellulose used for fabrication can be bacterial, wood, algal, or hemp. CNCs, because of their rod-like morphology and high surface area to volume, have found applications in polymeric nanocomposites for reinforcement. Barhoum et al. reported that with the help of delignification and mechanical pressing techniques, anisotropic wood slices can be directly transformed into anisotropic films or anisotropic cellulose nanofiber films (ACFs), which are highly aligned. Hence, it preserves the original alignment of cellulose nanofibers (CNFs). When treated with sodium hypochlorite for lignin removal, the wood macrostructure maintains its well-aligned structure. Thereby, following delignification, CNFs undergo compression and drying to form films containing hydrogen bonds due to the presence of water molecules. During the *in situ* carbonization of wood in conjunction with other primary polymer impregnation, the arrangement of cellulose microchannels is preserved. Next, submicron-sized fiber bundles were formed after chemical cross-linking, which was further self-assembled in an aqueous solution. These fibers have a significant mechanical strength in comparison to disoriented CNFs, and these aligned CNFs have found application in nanofluidic and green electronics as they result in anisotropic transmittance and guided liquid transport (Barhoum et al., 2020).

## 14.9 BIODEGRADABILITY AND BIOCOMPATIBILITY ENHANCEMENTS IN BIO-BASED POLYMERS

### 14.9.1 Biodegradability

Petroleum-based plastics have raised environmental concerns and have facilitated the path for bio-based polymers, which are taken from plants, microorganisms, or other biomass resources. Bio-based polymers can be easily degraded in the natural environment. Figure 14.7 represents biodegradation and biocompatibility mechanisms in hydrogel and scaffold materials, outlining the role of enzymatic breakdown, microbial activity, and polymer chemistry in material degradation. Their advantages include biodegradation capacity and a capability to sustain decomposition by microorganisms into water ($H_2O$), $CO_2$, and $CH_4$ or biomass under peculiar environmental conditions, part of some bio-based materials category. They can reduce plastic pollution, lower landfill waste, and reduce dependence on non-renewable fossil fuels in a way that has put bio-based polymers at the forefront of sustainable material research (Lee and Lee, 2022; Mukherjee et al., 2023).

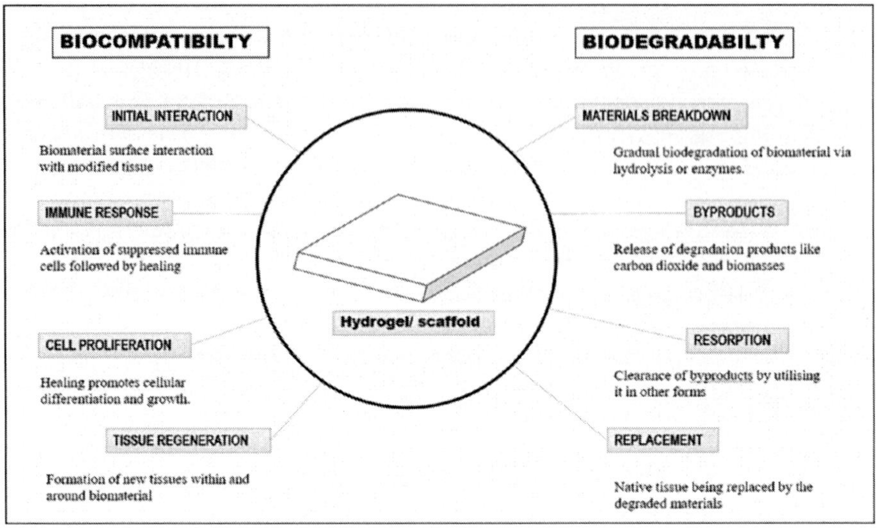


**FIGURE 14.7** Schematic representation of biocompatibility and biodegradability processes in hydrogel–scaffold materials.

#### 14.9.1.1 Biodegradability Mechanism

Biological degradation is a process driven by microbial processes, enzymatic hydrolysis, and environmental circumstances with respect to temperature, humidity, and aerobic conditions. During the process of biodegradation, organic chemicals are broken down and taken up by organotrophs, heterotrophs, algae, and fungi, releasing natural compounds in the environment, like $CO_2$ and lactic acid, and supporting biomass production.

**Chitosan**: Naturally occurring polysaccharide, which is abundant in arthropods exoskeleton, is abundant in arthropods exoskeleton, is biodegradable, and is a bio-based polymer. Chitosan is degraded by *Aeromonas hydrophila* and *Aspergillus niger* via enzymatic hydrolysis with the help of chitinase and lysozyme, which cleaves β-(1→4) glycosidic linkages. It gets degraded to form glucosamine in the monomeric form that is metabolized by the microbes. The environmental factors affect the degradation rate of chitosan (Pal et al., 2021; Priyadarshi and Rhim, 2020).

**Cellulose**: This common polymer is plentiful in the cell walls of plants and microbial cells. It has an enzyme called cellulase, which cleaves β-(1→4) glycosidic bonds, thus releasing D-glucose that is readily taken up by microorganisms (*Cellulomonas* and *Trichoderma reesei*) that can act as a precursor for the glycolysis (Entner–Doudoroff) pathway. The byproducts of the metabolized products are $CO_2$, water, and biomass. Different factors influence the biodegradation rate; amorphous cellulose is more easily degraded than crystalline one (Abe et al., 2021; Weimer, 2022).

**Silk fibroin**: Derived from the silkworm (*Bombyx mori*) cocoons, silk fibroin is a protein polymer known for its biodegradability, mechanical strength, and biocompatibility. It consists of glycine, alanine, and serine repeating units that form into β-sheets. Proteases generally degrade silk fibroin by breaking CO-NH (peptide) bonds. The biodegradation rate is influenced by the amino acid content and molecular weight. *In vivo*, it degrades over weeks or months, whereas the degradation duration varies across different environments. Silk fibroin is degraded by both microorganisms and fungi *Aspergillus niger* and *Pseudomonas putida* (Zvinavashe et al., 2022).

**Polylactic acid (PLA)**: A biodegradable polyester derived from corn starch and sugarcane. PLA biodegrades via hydrolysis followed by microbial action (*Aspergillus oryzae* and *Streptomyces* sp.). Hydrolysis creates lactic acid, which is then assimilated by other microorganisms and converted into natural byproducts. PLA degradation efficiently takes place at high temperatures (compost), while, in natural conditions, the degradation takes place slowly due to cooler temperatures (Zaaba and Jaafar, 2020).

**Polyhydroxyalkanoates (PHA):** These are synthesized by the microbes themselves as an energy storage compartment (intracellular). PHAs are fully biodegradable and are degraded by microbes and fungi (*Alcaligenes faecalis* and *Aspergillus fumigatus*) as they have a depolymerase enzyme that breaks them into hydroxyl acid monomers, which further get reduced into biomass, water, and $CO_2$; and methane is also a byproduct. Apart from environmental factors, its degradability depends on the PHA's crystallinity (Fernandes et al., 2020).

**Starch-based polymers:** Starch, a polymer composed of amylose and amylopectin, biodegrades easily due to its hydrophilic nature. Blended with other natural polymers, it enhances its strength as well as biocompatibility and biodegradation. Enzymes like amylases readily degrade starch into glucose, which can be easily metabolized by microorganisms (*Bacillus subtilis* and *Rhizopus oryzae*). The degradation of starch material is quite high in the environment (Ghasemlou et al., 2022).

### 14.9.2 Biocompatibility

Bio-based biomaterials exhibit excellent biocompatibility, making them suitable for various medical applications, which include drug delivery systems, tissue engineering, and chronic wound healing. These materials are naturally derived, biodegradable, and interact positively with biological systems, making them far superior environmentally compared to synthetic alternatives.

#### 14.9.2.1 Modifications in Biomaterials to Enhance Biocompatibility

Modification of biomaterials is necessary for better integration and performance in diverse environments. Mainly, two types of modifications are carried out: physicochemical and biological enhancement methods (Rashidi et al., 2014).

#### 14.9.2.2 Physicochemical Methods

Physicochemical alterations are focused on surface characteristics to enhance the surface interaction with surrounding tissues. Surface chemistry is a key factor in protein adsorption, cell adhesion, and biocompatibility. Techniques like surface grafting, plasma treatment, and chemical coatings are often used to enhance the surface energy, hydrophilicity, and promotion of cell adhesion and proliferation. The major findings of this method are to improve integration and to minimize hyper-immune reactions (Alba-Perez et al., 2020). The strategies include surface chemistry modifications, which are a primary strategy for improving biomaterial functions by adjusting the chemical properties of the material surfaces.

i. **Plasma treatment**: This method usually exposes the material surfaces to the plasma of reactive species (oxygen and nitrogen). During this process, functional groups like hydroxyl, carboxyl, and amine groups are added to enhance surface hydrophilicity (Milinkovic et al., 2012; Safonov et al., 2010; Weimer, 2022).
ii. **Grafting functional groups**: Functional groups like RGD (arginine, glycine, and aspartic acid) peptides, bioactive groups, and molecules can be mounted on the surface to mimic the extracellular matrix (ECM) to improve cellular interactions.
iii. **Self-assembled monolayers (SAMs):** SAMs are single-molecule thick layers formed by adsorbing specific organic molecules with tailored functional groups onto biomaterial surfaces. SAMs are used to control the hydrophilicity and hydrophobicity of a molecule, incorporating PEG (polyethylene glycol), reducing protein adsorption, and preventing microbial attachment (Ali et al., 2020; Ulman, 1996).
iv. **Biopolymer coatings:** Bioactive polymers coated surfaces with collagen, chitosan, and hyaluronic acid improve cell compatibility and tissue integration. Usually, collagen-coated surfaces have superior compatibility in bone implants (Shirzadian et al., 2012).

**Topographical modifications:** Topographical modification can alter the physical structure of the surface to improve cell attachment, orientation, and proliferation by mimicking actual ECM (Chen et al., 2013).

i. **Nano-patterning:** Creating nanostructures on the surfaces in the nanoscale architecture of natural tissues.
ii. **Micro-texturing:** This approach involves creating microscale features like grooves, ridges, and pits to guide cells for proper alignment and directional growth.

#### 14.9.2.3 Biological Methods

The major goal of biological modifications is to incorporate bioactive molecules that can be beneficial to influence cellular behaviors. Biological enhancements often include embedding growth factors, peptides, or enzymes to promote tissue regeneration and healing. Protein immobilization is an important biological approach in tissue engineering. One of the most difficult issues in this field is ensuring that the targeted protein or therapeutics are delivered to the correct location; to overcome this, the idea of protein immobilization came into the picture. Physical adsorption and entrapment are two major ways most of the techniques are associated with controlled and sustained release of desired proteins; they possess several demerits, such as rapid and unwanted absorption and desorption. However, covalent immobilization of enzymes is popular as protein functional groups usually contain many groups like carboxyl, hydroxyl, amide, and thiol groups, which facilitate covalent interactions. For example, when growth factors were added to the medium, it promoted the growth and proliferation of epithelial cells less than when they were bound to the surfaces. This area is still under research as it is a vast area in which the sensitivity and complexity of proteins are still being explored (Rashidi et al., 2014; Vasita et al., 2008).

## 14.10 CONCLUSION

Bio-based polymers offer a sustainable alternative to the fossil fuel-derived polymers. They are reducing environmental impact, enhancing biodegradability, and lowering carbon footprints. Innovative techniques like polymer blending, cross-linking, and incorporating nanocomposites are essential for improving functional properties of bio-based polymers, making them suitable for industries like automotives, textiles, medical, and pharmaceuticals. Despite advancements, the cost of production remains a barrier. Complex processing steps contribute to high cost; however, research is focused on optimizing these techniques and improving biopolymer extraction and modification to reduce expenses and increase scalability. Bio-based polymer holds significant promise as a sustainable material, provided the cost and scalability challenges are addressed. Continued research and innovation are crucial for making these materials more economically viable and achieving a wider market space.

## REFERENCES


Abe MM, Branciforti MC, Brienzo M. Biodegradation of hemicellulose-cellulose-starch-based bioplastics and microbial polyesters. *Recycling*. 2021 Mar;6(1):22.

Ahorsu R, Medina F, Constantí M. Significance and challenges of biomass as a suitable feedstock for bioenergy and biochemical production: a review. *Energies*. 2018 Dec;11(12):3366.

Alba-Perez A, Jayawarna V, Childs PG, Dalby MJ, Salmeron-Sanchez M. Plasma polymerised nanoscale coatings of controlled thickness for efficient solid-phase presentation of growth factors. *Materials Science and Engineering: C*. 2020 Aug 1;113:110966.

Ali F, Roldán-Carmona C, Sohail M, Nazeeruddin MK. Applications of self-assembled monolayers for perovskite solar cells interface engineering to address efficiency and stability. *Advanced Energy Materials*. 2020;10(48):2002989.

Badia JD, Gil-Castell O, Ribes-Greus A. Long-term properties and end-of-life of polymers from renewable resources. *Polymer Degradation and Stability*. 2017 Mar 1;137:35–57.

Balla VK, Kate KH, Satyavolu J, Singh P, Tadimeti JGD. Additive manufacturing of natural fiber reinforced polymer composites: processing and prospects. *Composites Part B: Engineering*. 2019 Oct 1;174:106956.

Baranwal J, Barse B, Fais A, Delogu GL, Kumar A. Biopolymer: a sustainable material for food and medical applications. *Polymers*. 2022 Feb 28;14(5):983.

Barhoum A, Jeevanandam J, Rastogi A, Samyn P, Boluk Y, Dufresne A, et al. Plant celluloses, hemicelluloses, lignins, and volatile oils for the synthesis of nanoparticles and nanostructured materials. *Nanoscale*. 2020 Nov 26;12(45):22845–90.

Belkhir K, Riquet G, Becquart F. Polymer processing under microwaves. *Advances in Polymer Technology*. 2022 May 6;2022:1–21.

Benbettaïeb N, Gay JP, Karbowiak T, Debeaufort F. Tuning the functional properties of polysaccharide-protein bio-based edible films by chemical, enzymatic, and physical cross-linking. *Comprehensive Reviews in Food Science and Food Safety*. 2016 Jul;15(4):739–52.

Chen CH, Ou KL, Wang WN. Variation in surface morphology and microstructure of 316L biomedical alloys immersed in artificial saliva. *Journal of Experimental & Clinical Medicine*. 2013 Feb 1;5(1):30–6.

Duquette D, Dumont MJ. Comparative studies of chemical crosslinking reactions and applications of bio-based hydrogels. *Polymer Bulletin*. 2019 May 1;76.

Fernandes M, Salvador A, Alves M, Vicente A. Factors affecting polyhydroxyalkanoates bio-Degradation in soil. *Polymer Degradation and Stability*. 2020 Dec 1;182:109408.

Formela K, Hejna A, Haponiuk J, Tercjak A. In situ processing of biocomposites via reactive extrusion. In: *Biocomposites for High-Performance Applications: Current Barriers and Future Needs Towards Industrial Development*. 2017. pp. 195–238, Woodhead Publishing.

Ghasemlou M, Daver F, Murdoch BJ, Ball AS, Ivanova EP, Adhikari B. Biodegradation of novel bioplastics made of starch, polyhydroxyurethanes and cellulose nanocrystals in soil Environment. *Science of the Total Environment*. 2022 Apr 1;815:152684.

Gleissner C, Landsiedel J, Bechtold T, Pham T. Surface activation of high performance polymer fibers: a review. *Polymer Reviews*. 2022 Aug 27;62(4):757–88.

Heinrich LA. Future opportunities for bio-based adhesives—advantages beyond renewability. *Green Chemistry*. 2019 Apr 15;21(8):1866–88.

Herzyk F, Piłakowska-Pietras D, Korzeniowska M. Supercritical extraction techniques for obtaining biologically active substances from a variety of plant byproducts. *Foods*. 2024 May 30;13(11):1713.

Hu M, Yan X, Hu X, Feng R, Zhou M. Synthesis of silver decorated silica nanoparticles with rough surfaces as adsorbent and catalyst for methylene blue removal. *Journal of Sol-Gel Science and Technology*. 2019 Mar 1;89(3):754–63.

Ilsouk M, Raihane M, Rhouta B, MeRijs Meri R, Zicans J, Vecstaudža J, et al. The relationship of structure, thermal and water vapor permeability barrier properties of poly(butylene succinate)/organomodified beidellite clay bionanocomposites prepared by in situ polycondensation. *RSC Advances*. 2020;10(61):37314–26.

Jabeen N, Atif M. Polysaccharides based biopolymers for biomedical applications: a review. *Polymers for Advanced Technologies*. 2024;35(1):e6203.

Kabir E. Application of microwave heating in polymer synthesis: a review. *Results in Chemistry*. 2023 Dec 1;6:101178.

Khandelwal P, Gupta V. Microwave joining of polymers. *IJSR*. 2022 Jul 5;11(7):627–33.

Krishnan R, Shibu SN, Poelman D, Badyal AK, Kunti AK, Swart HC, et al. Recent advances in microwave synthesis for photoluminescence and photocatalysis. *Materials Today Communications*. 2022 Aug 1;32:103890.

Kundu R, Mahada P, Chhirang B, Das B. Cellulose hydrogels: Green and sustainable soft biomaterials. *Current Research in Green and Sustainable Chemistry*. 2022 Jan 1;5:100252.

Laurichesse S, Avérous L. Chemical modification of lignins: towards biobased polymers. *Progress in Polymer Science*. 2014 Jul 1;39(7):1266–90.

Lee SM, Lee D. Effective medical waste management for sustainable green healthcare. *International Journal of Environmental Research and Public Health*. 2022 Nov 10;19(22):14820.

Leong WY, Thitithanasarn S, Yamada K, Hamada H. Compression and injection molding techniques for natural fiber composites. *Natural Fibre Composites: Materials, Processes and Applications*. 2013 Dec 1;216–32.

Liu L, Zhang C, Liao L, Wang X, Zhuo R. Microwave-assisted polymerization of D, L-lactide with stannous octanoate as catalyst. *Chinese Chemical Letters*. 2001 Aug 1;12:663–4.
Long Y, Zhang R, Huang J, Wang J, Zhang J, Rayand N, et al. Retroreflection in binary bio-based PLA/PBF blends. *Polymer*. 2017 Sep;125:138–43.
Mahto D, Rani P, MiShra S, Sen G. Microwave assisted synthesis of polyacrylamide grafted soya peptone and its application as water soluble adhesive. *Industrial Crops and Products*. 2014 Jul 1;58:251–8.
Matharu AS, Houghton JA, LuCas-Torres C, Moreno A. Acid-free microwave-assisted hydrothermal extraction of pectin and porous cellulose from mango peel waste—towards a zero waste mango biorefinery. *Green Chemistry*. 2016 Sep 26;18(19):5280–7.
Mehta J, Gupta K, Lavania S, Kumar P, Chaudhary V, Gupta P. Inherent roadmap in synthesis and applications of sustainable materials using oil based and microbial polymers. *Materials Today Sustainability*. 2024 Mar 1;25:100615.
Milinkovic I, Rudolf R, Raic K, Aleksic Z, Lazic V, Todorovic A, et al. Aspects of titanium-implant surface modification at the micro and nano levels. *Materiali in Tehnologije*. 2012 May 1;46:251–6.
Mukherjee S, Bhattacharjee S, Paul S, Nath S, Paul S. Biofilm-a syntrophic consortia of microbial cells: boon or bane? *Applied Biochemistry and Biotechnology*. 2023 Sep;195(9):5583–604.
Oleksy M, Dynarowicz K, Aebisher D. Rapid prototyping technologies: 3D printing applied in medicine. *Pharmaceutics*. 2023 Aug 21;15(8):2169.
Ostafinska A, Fortelny I, Nevoralova M, Hodan J, Kredatusova J, Slouf M. Synergistic effects in mechanical properties of PLA/PCL blends with optimized composition, processing, and morphology. *RSC Advances*. 2015 Nov 18;5(120):98971–82.
Pal P, Pal A, Nakashima K, Yadav BK. Applications of chitosan in environmental remediation: a review. *Chemosphere*. 2021 Mar;266:128934.
Park SJ, Lee JE, Lee HB, Park J, Lee NK, Son Y, et al. 3D printing of bio-based polycarbonate and its potential applications in ecofriendly indoor Manufacturing. *Additive Manufacturing*. 2020 Jan 1;31:100974.
Pawlak F, Aldas M, López-Martínez J, Samper MD. Effect of different compatibilizers on injection-molded green fiber-reinforced polymers based on poly(lactic acid)-maleinized linseed oil system and sheep wool. *Polymers*. 2019 Sep;11(9):1514.
Pezzana L, Melilli G, Guigo N, Sbirrazzuoli N, Sangermano M. Cross-linking of biobased monofunctional furan epoxy monomer by two steps process, UV irradiation and thermal treatment. *Macromolecular Chemistry and Physics*. 2023;224(1):2200012.
Priyadarshi R, Rhim JW. Chitosan-based biodegradable functional films for food packaging applications. *Innovative Food Science and Emerging Technologies* [Internet]. 2020 Jun [cited 2024 Nov 18];62(102346). Available from: www.scopus.com/inward/record.url?scp=85083645953&partnerID=8YFLogxK
Purewal SS, Kaur A, Bangar SP, Singh P, Singh H. Protein-based films and coatings: an innovative approach. *Coatings*. 2024 Jan;14(1):32.
Rashidi H, Yang J, Shakesheff KM. Surface engineering of synthetic polymer materials for tissue engineering and regenerative medicine applications. *Biomaterials Science*. 2014 Aug 26;2(10):1318–31.
Rashmi BJ, Loux C, Prashantha K. Bio-based thermoplastic polyurethane and polyamide 11 bioalloys with excellent shape memory behavior. *Journal of Applied Polymer Science* [Internet]. 2017 [cited 2024 Nov 18];134(20). Available from: https://onlinelibrary.wiley.com/doi/abs/10.1002/app.44794
Riaz U, Ashraf SM, Sharma HO. Mechanical, morphological and biodegradation studies of microwave processed nanostructured blends of some bio-based oil epoxies with poly (vinyl alcohol). *Polymer Degradation and Stability*. 2011 Jan 1;96(1):33–42.
Safonov V, Zykova A, Smolik J, Rogovska R, Donkov N, Georgieva V. The surface parameters modifications at nano scale for biomedical applications. *Journal of Physics: Conference Series*. 2010 Dec 7;253:012068.

Sakthivel A, Muthu Kumaran D, Kandasamy J. A case study of 3D printed PLA and its mechanical properties. *Materials Today: Proceedings*. 2018 Jan 1;5:11219–26.

Schutyser W, Bosch SV den, Renders T, Boe TD, Koelewijn SF, Dewaele A, et al. Influence of bio-based solvents on the catalytic reductive fractionation of birch wood. *Green Chemistry*. 2015 Nov 2;17(11):5035–45.

Shakil F. Reactive extrusion: filled polymer compounds and its applications. In: Sethi SK, Gupta HS, Verma A, editors. *Polymer Composites: From Computational to Experimental Aspects* [Internet]. Singapore: Springer Nature; 2024 [cited 2025 Feb 25]. pp. 245–68. Available from: https://doi.org/10.1007/978-981-97-0888-8_12

Sharifvaghefi S, Zheng Y. Microwave vs conventional heating in hydrogen production via catalytic dry reforming of methane. *Resources Chemicals and Materials*. 2022 Sep 1;1(3):290–307.

Sharma N, Singh D, Rani R, Sharma D, Pandey H, Agarwal V. Chitosan and its nanocarriers: Applications and opportunities. In: *Nanomaterials in Plants, Algae and Microorganisms*. 2019. pp. 267–286, Academic Press.

Shirzadian T, Bagheri S, SaeidiBorojeni H, Ghaffari P, Foroughi F, Mahboubi M. The role of biocompatible coatings of biomaterials for creation of direct and appropriate chemical bounding between bioimplant and bone tissue. *Journal of Injury and Violence Research*. 2012 Nov;4(3 Suppl 1):39.

Ulman A. Formation and structure of self-assembled monolayers. *Chemical Reviews*. 1996 Jan 1;96(4):1533–54.

Umaraw P, Munekata P, Verma A, Barba F, Singh vp, Kumar P, et al. Edible films/coating with tailored properties for active packaging of meat, fish and derived products. *Trends in Food Science & Technology*. 2020 Feb 1;98.

Vartiainen J, Harlin A. Crosslinking as an efficient tool for decreasing moisture sensitivity of biobased nanocomposite films. *Materials Sciences and Applications*. 2011 May 5;2(5):346–54.

Vasita R, Shanmugam I K, Katt DS. Improved biomaterials for tissue engineering applications: surface modification of polymers. *Current Topics in Medicinal Chemistry*. 2008;8(4):341–53.

Wang X, Jiang M, Zhou Z, Gou J, Hui D. 3D printing of polymer matrix composites: a review and prospective. *Composites Part B: Engineering*. 2017 Feb 1;110:442–58.

Weimer PJ. Degradation of cellulose and hemicellulose by ruminal microorganisms. *Microorganisms*. 2022 Dec;10(12):2345.

Xue J, Wu T, Dai Y, Xia Y. Electrospinning and electrospun nanofibers: methods, materials, and applications. *Chemical Reviews*. 2019 Apr 24;119(8):5298–415.

Yousefi N, Abbasi S. Food proteins: Solubility & thermal stability improvement techniques. *Food Chemistry Advances*. 2022 Oct 1;1:100090.

Zaaba NF, Jaafar M. A review on degradation mechanisms of polylactic acid: hydrolytic, photodegradative, microbial, and enzymatic degradation. *Polymer Engineering & Science*. 2020;60(9):2061–75.

Zhou JJ, Wang SY, Gunasekaran S. Preparation and characterization of whey protein film incorporated with TiO2 nanoparticles. *Journal of Food Science*. 2009 Sep;74(7):N50–6.

Zhu M, Xiong R, Huang C. Bio-based and photocrosslinked electrospun antibacterial nanofibrous membranes for air filtration. *Carbohydrate Polymers*. 2019 Feb 1;205:55–62.

Zink J, Wyrobnik T, Prinz T, Schmid M. Physical, chemical and biochemical modifications of protein-based films and coatings: an extensive review. *International Journal of Molecular Sciences*. 2016 Aug 23;17(9):1376.

Zvinavashe AT, Barghouti Z, Cao Y, Sun H, Kim D, Liu M, et al. *Degradation of Regenerated Silk Fibroin in Soil and Marine Environments*. Prof Benedetto Marelli [Internet]. 2022 Aug 29 [cited 2024 Nov 18]; Available from: https://dspace.mit.edu/handle/1721.1/150422

# 15 Cost and Complexity of Implementation of Bio-Based Materials

*Urvashi Srivastava, Adhishree Yadav, and Vikas Pandey*

## 15.1 SIGNIFICANCE OF BIO-BASED MATERIALS

Bio-based materials are derived from renewable biological resources. Because of their advantages for the environment, economy, and society, bio-based accoutrements that come from renewable natural resources including plants, animals and microorganisms, are becoming increasingly significant in a variety of disciplines. They vastly lessen reliance on fossil energies by furnishing a more environmentally friendly cover for goods made from petroleum; for example, dedicated use of bio-polyethylene terephthalate for packaging and polylactic acid for textiles can offer a carbon footprint reduction of up to 67% and 80%, respectively (Suarez et al., 2022). Economically speaking, bio-based products offer chances to diversify force chains and sectors, generating employment—particularly in pastoral areas where biomass husbandry is common (Moreno et al., 2020). Organic production not only improves agricultural output but also creates a bioeconomy that can stimulate local development. They also create value addition in agricultural byproducts and enhance bio-economies, which can be the drivers for regional growth. Due to health and safety benefits and reduced toxicity in products, bio-based materials find the perfect application in food packaging, healthcare, construction textile, automotive, medical, agriculture, water filtration, bio refineries, and furniture industries among others (Sillanpää & Ncibi, 2017; Arfin & Sonawane, 2018; Reichert et al., 2020; Al-Tayyar et al., 2020; Vinod et al., 2020; Wendels & Avérous, 2021; Cywar et al., 2021; Rafiqah et al., 2021; Ahmad Qamar et al., 2020; Siracusa & Blanco, 2020; Sharma et al., 2021; Ramesh Kumar et al., 2020; Arif et al., 2022; Yang et al., 2020; Deng et al., 2021; Okolie et al., 2020; Zych et al., 2020; Baranwal et al., 2022; George et al., 2020; Sheldon et al., 2021; Azimi et al., 2020; Ates et al., 2020; Walker & Rothman, 2020).

## 15.2 TECHNOLOGICAL ADVANCEMENTS

Smart manufacturing is a revolutionary strategy that uses modern technology to improve the responsiveness, flexibility, and effectiveness of manufacturing operations. It creates a smarter and more connected manufacturing sector by integrating

DOI: 10.1201/9781003566205-15

digital and physical systems. Smart manufacturing uses real-time monitoring and predictive protection to improve processes and reduce time-outs; quickly adapts to changing product specifications and needs; helps create additional specialized production; and reduces waste and energy consumption through resource optimization that guarantees constant product quality using precise control and observation, thus enhancing the green economy (Bradu et al., 2023). Smart manufacturing of biological devices is in line with Industry 4.0, also known as the Fourth Industrial Revolution. Pandey et al. (2023) illustrated the advantages of Bioprocess 4.0, and some of its advantages are summarized in Figure 15.1.

Industry 4.0 reduces costs and complexity while improving bio-based material supply chain integration and quality by design (QbD). According to various studies, the global digital manufacturing market size is anticipated to grow from USD 2.0 billion in 2024 to 12.94 billion by 2035, at a compounded annual growth rate (CAGR) of 18% during the forecast period, as shown in Figure 15.2.

Sustainable environmental behavior in Industry 4.0 can increase revenue while cutting costs and expenses. Similarly, across Industry 4.0, machinery and equipment may be shared. This comprises industries sharing their operational capacity and a means to provide services. The supplying industry may keep its machinery running and generate variable income even when it is not functioning in the industry it serves. At the same time, the industry that is using the machinery continues to operate without hindrance or loss of demand. This represents the integration of digital technology such as real-time data collection, digital twins (Rani et al., 2024), advanced robotics, big data analysis, cyber-physical systems, artificial intelligence (AI), machine learning (ML), and additive manufacturing in the production process.

### 15.2.1 The Smart Manufacturing Process for Bio-Based Materials

The smart manufacturing process is a systemized and data-driven approach to production. It can be broken down into five key stages, each enhanced by smart technologies to improve efficiency and cost-effectiveness (see Figure 15.3).

#### 15.2.1.1 Decentralization via Small Modular Plants

Due to the wide variability in the area of the distribution of raw materials, decentralization of the processing units via smaller modular plants may offer better results. Julia Riese et al. (2024) examined the effects of different degrees of decentralization in production networks of the chemical industry. It was found that mobile modular plants can reduce supply chain expenses and transportation efforts by up to 40%. The environmental assessment showed that decentralized electrified units would be responsible for only 3% of the total greenhouse gas produced by a centralized electric unit. A.D. Moreno et al. (2020) showed that by adopting small-scale processing, the return flow can be 70% as shown in Figure 15.4.

#### 15.2.1.2 Preprocessing and Treatment of Bio-Based Materials

Bio-based materials mostly require a level of preprocessing before they enter the production line. This may be in the form of physical treatments: grinding,

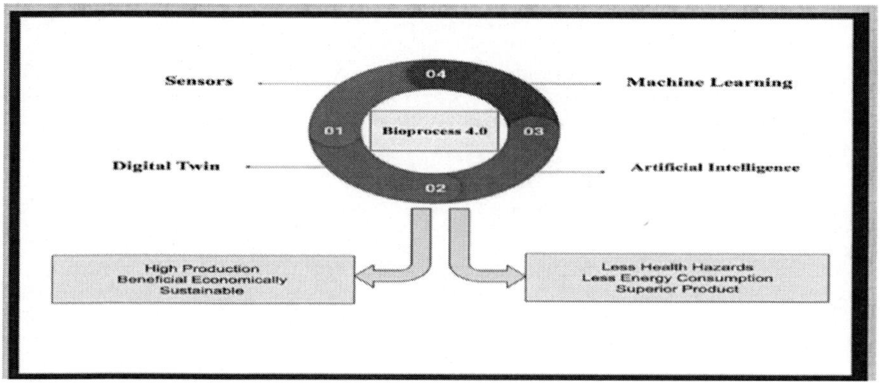


**FIGURE 15.1** Advantages of Bioprocess 4.0.

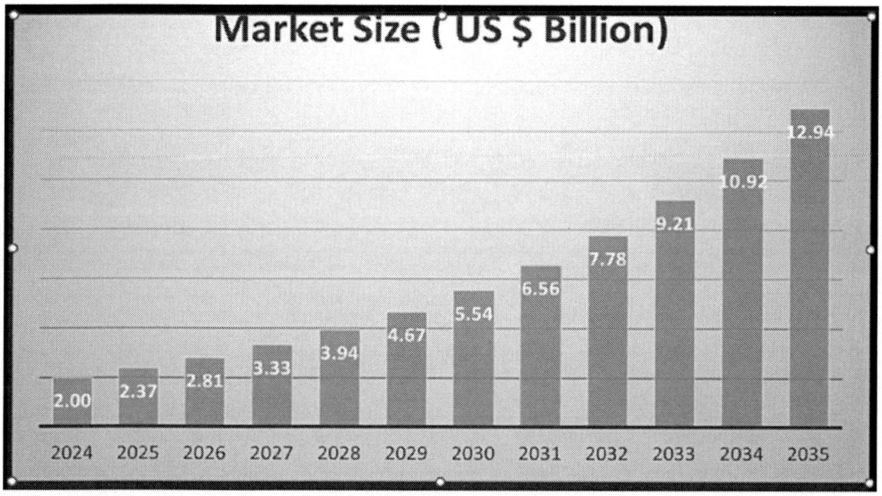


**FIGURE 15.2** Forecast on the digital manufacturing market.

drying, or fractionation. Such automated machines would be enabled with the help of algorithms and live data from sensors to vary in real time. An example would be the moisture content of biomass, detected and varied during drying to achieve an optimum level. AI-enabled systems will be able to predict the right set of preprocessing parameters from previous experience and thus ensure enhanced quality control with minimal material wastage. Nicolás M. Clauser et al. (2022) reported that researchers at the bioenergy program in Idaho National Laboratory have successfully tested an intelligence-based control system that could increase the reliability of the preprocessing equipment used in preprocessing feedstock by more than 50%.

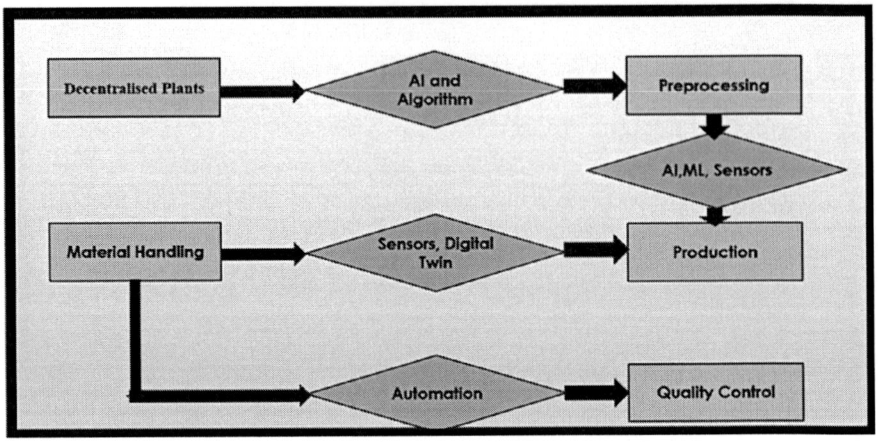


**FIGURE 15.3** Key stages in the smart manufacturing process.

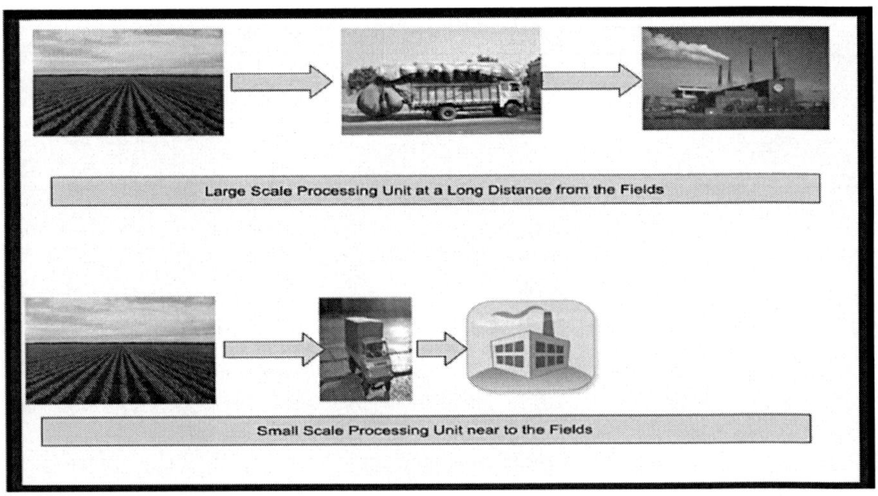


**FIGURE 15.4** Return on decentralization of network.

#### 15.2.1.3 Production and Conversion Processes

In smart manufacturing, the bioreactors or other conversion equipment will be equipped with high-end sensors that can monitor and control parameters such as temperature, pH, or pressure online. The use of digital twin technology is increasingly used to simulate the manufacturing environment and predict the ability of the production line to perform in various circumstances (Rani et al., 2024). Advanced process control (APC) systems take advantage of real-time data for automated adjustments in such production variables to ensure bio-based material processing is always at its optimum. For example, control of fermentation in bioplastic production may be kept tight to maximize yields with minimum energy consumption.

The implementation of APC and digital twin technologies typically entails an investment in the range of $1 to $5 million, depending on the complexity of the production line. However, such technologies come with huge savings regarding energy consumption (as high as 30%), reduction in raw material wastage (10–15%), and improvement in the overall production efficiency (20–25%).

#### 15.2.1.4 Automation and Robotics in Material Handling

The use of robotics in the handling of materials, such as their movement, packaging, and storage, is categorized under smart manufacturing (Taesi et al., 2023). In some smart manufacturing environments, cobots work alongside human operators, with the tasks at hand ranging from quality control to packaging. Cobots can be reprogrammed at a rapid pace to deal with different bio-based materials or changes in production demands.

The initial costs of robotic systems and AGVs range between $200,000 and $1 million for a mid-sized operation, but promise long-term savings with respect to labor—30–50% reduction in workforce strength—and also improve throughput.

#### 15.2.1.5 Real-Time Monitoring, Predictive Maintenance, and Quality Control

Throughout the smart manufacturing process, there is continuous feedback in real time concerning performance, material quality, and output through data from IoT sensors embedded in equipment. This allows for immediate adjustments to be performed that prevent deviations from target specifications. Data trends, analyzed by AI algorithms, can predict when machines are likely to fail or require maintenance, hence reducing unexpected downtimes. Predictive maintenance, in order to minimize costly repairs and maintain the production line at a functioning capacity, is among the essential practices. AI-powered cameras automate the process of defect inspection in a range of bio-based materials at different stages of production using ML models and vision systems.

Predictive maintenance systems can decrease the maintenance cost by as much as 20–30% while reducing downtime by up to 50–60%. For real-time monitoring and quality control systems, the cost saving is observed in terms of reduced waste, higher product yield, fewer recalls, and thus may potentially save anywhere from $500,000 to $2 million annually for a mid-sized facility.

### 15.2.2 Comparison of the Cost Factors in Conventional and Smart Manufacturing Processes

The primary cost factors associated with the implementation of bio-based materials include raw material costs, processing expenses, and the investment in new technologies and infrastructure. Bio-based materials often require different processing techniques compared to conventional materials, necessitating modifications to existing manufacturing systems or the development of entirely new ones. This can result in substantial capital expenditures. Additionally, the supply chain for bio-based materials is still maturing, which can lead to higher costs and variability in material availability and quality (Mehta et al., 2018; Deshpande et al., 2023). The cost analysis of conventional and smart manufacturing for bio-based materials is summarized in Table 15.1.

**TABLE 15.1**
**Cost Analysis of Conventional Manufacturing and Smart Manufacturing of Bio-Based Materials.**

| Heads | Conventional manufacturing | Smart manufacturing |
|---|---|---|
| Initial investment | Lower upfront costs, as it relies on traditional machinery and methods. However, the cost of manual labor and limited automation can increase operational expenses over time. | Higher initial investment due to advanced technologies such as automation, IoT devices, robotics, and AI-driven systems. However, this investment can result in long-term cost savings. |
| Operational efficiency | Typically less efficient, with higher material waste, energy consumption, and human error. Production can be slower, especially when scaling. | Increased efficiency with automated systems, real-time data analytics, and process optimization. Reduced material waste, energy consumption, and downtime due to predictive maintenance, ultimately lowering operational costs. |
| Labor costs | Labor-intensive, requiring more manual intervention, which leads to higher labor costs and increased risk of human error. | Reduces labor costs by automating repetitive tasks. Workforce requirements shift toward skilled roles in managing and maintaining automated systems, leading to potential savings in staffing and training. |
| Material utilization | Less precision in material usage leads to higher waste and raw material costs. | Advanced monitoring systems ensure optimal material use, reducing waste and improving yield efficiency, resulting in cost savings on raw materials. |
| Energy consumption | Higher energy consumption due to outdated equipment and less efficient processes, leading to increased energy costs over time. | Smart systems optimize energy usage, often incorporating renewable energy sources, reducing overall consumption and lowering energy bills. |
| Production flexibility | Limited flexibility in adapting to changes in demand or product customization, resulting in higher costs when reconfiguring systems or halting production. | Flexible and scalable, allowing for rapid changes in production lines, increased customization, and cost-effective handling of varied demand without significant downtime. |
| Maintenance costs | Relies on reactive maintenance, often leading to costly downtimes and unplanned repairs. | Predictive maintenance through IoT and AI reduces unplanned breakdowns and extends equipment life, cutting down on repair costs and downtime. |

**TABLE 15.1** *(Continued)*
**Cost Analysis of Conventional Manufacturing and Smart Manufacturing of Bio-Based Materials.**

| Heads | Conventional manufacturing | Smart manufacturing |
|---|---|---|
| Supply chain management | Manual tracking and inventory management can lead to inefficiencies, overstocking, or shortages, adding to inventory costs. | IoT-enabled supply chain systems allow real-time monitoring and automation, optimizing inventory levels and reducing associated costs. |
| Overall cost comparison | Short-term: conventional manufacturing may appear more cost-effective in the short term due to lower initial capital investment. | Long-term: smart manufacturing, despite higher upfront costs, yields greater long-term savings by optimizing efficiency, reducing waste, lowering labor and energy costs, and enhancing production flexibility. These benefits compound over time, making it more cost-effective and sustainable in the long run. |

### 15.2.3 Benefits of Adopting Smart Manufacturing for Bio-based Materials

In a review, P. Bradu et al. (2023) emphasized the need for adapting smart manufacturing and concluded that green technologies can help consumers save money on energy expenses and are frequently safer and healthier than their non-green equivalents. While the costs for smart manufacturing of bio-based materials might be huge, the majority of the benefits can outweigh these costs in the long run. This includes operational efficiency, sustainability, product innovation, and market competitiveness. Clauser et al. (2022) showed a comprehensive overview of adopting new technology in Industry 4.0, as shown in Figure 15.5.

#### 15.2.3.1 Operational Efficiency

##### *15.2.3.1.1 Waste Reduction and Efficient Use of Resources*

Smart manufacturing can achieve considerable benefits related to optimizing resource utilization and waste reduction. In the case of bio-based materials, this becomes even more critical because most raw materials are available seasonally and vary in their characteristics. Smart manufacturing systems utilizing real-time analytics and AI are positioned to modify the production parameters in real-time, considering any material variations, thus minimizing waste generation. This not only helps in reducing costs due to the utilization of raw materials but also assists in achieving sustainability goals.

##### *15.2.3.1.2 Predictive Maintenance*

IoT-enabled sensors and AI-driven algorithms can help enable predictive maintenance that reduces downtime and brings down the cost of routine maintenance. By way of quick identification of impending equipment failure, manufacturers can

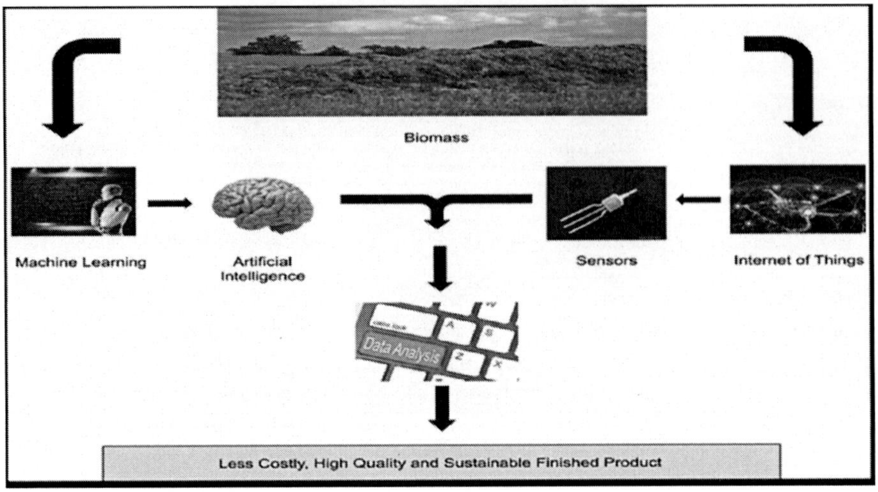


**FIGURE 15.5** Overview of adopting new technology in Industry 4.0.

conduct maintenance on their equipment during periods of low demand, thus avoiding expensive production interruptions.

##### *15.2.3.1.3 Energy Optimization*

AI algorithms predict energy demand and provide suggestions on how to reduce peak energy usage. For bio-based material production, which normally requires a tight control over temperature and humidity, smart systems can optimize the ambient conditions toward a goal of minimizing energy costs while keeping product quality.

#### 15.2.3.2 Product Innovation and Market Competitiveness

Nwakamma Ninduwezuor-Ehiobu et al. (2023) signified the integration of the innovative material for the US manufacturing market and explored the intersection of cutting-edge science, economic growth, and environmental stewardship. Smart manufacturing systems also provide more flexibility in production, hence giving manufacturers the ability to customize products with ease. Bio-based materials can have tailored solutions for particular industries, such as biodegradable packaging, sustainable textiles, or sustainable construction materials. This can make them more responsive to shifting consumer trends and changing regulatory requirements in record time. By doing so, this will give the businesses an edge over competitors in the marketplace.

#### 15.2.3.3 Long-Term Cost Economies

While the initial investment of integrating smart manufacturing might be high, long-term cost benefits will pay off. Operational efficiencies, waste reduction, energy consumption optimization, and predictive maintenance—all these factors together reduce the overall cost of production. Moreover, companies investing in smart manufacturing will be in a better position to capitalize on future market opportunities

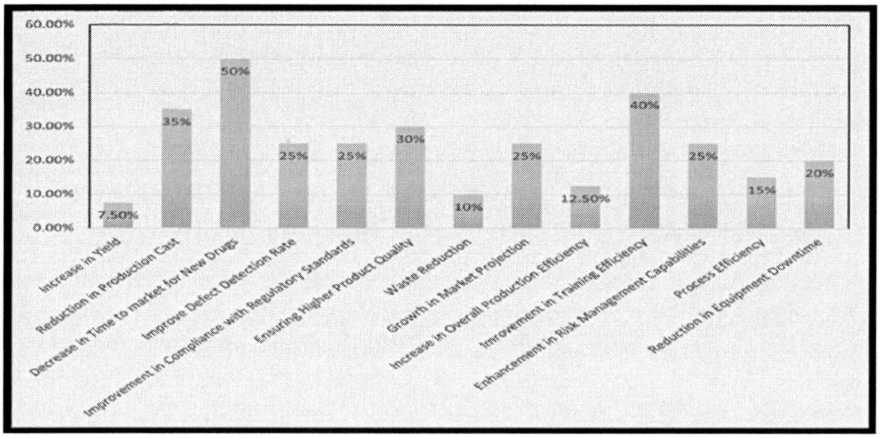


**FIGURE 15.6** Impact of adopting smart technology on manufacturing cost.

since they will have the infrastructure to quickly ramp up production efficiently. According to a report published by credenceresearch.com, the digital biomanufacturing market is projected to grow from $1157.45 million in 2024 to $6886.37 million by 3032. The graph in Figure 15.6 illustrates the reduction in manufacturing costs over time due to the implementation of smart technologies in bio-based materials production.

### 15.2.4 Weighing the Costs and Benefits

Cost–benefit analysis with respect to smart manufacturing for bio-based materials should be long term. This could be because although the initial investment in smart manufacturing may be quite high, the long-term gains in operational efficiency, sustainability, product innovation, and market competitiveness are also real. Early adopters of smart manufacturing are likely to earn first-mover advantages in lower production costs and enhanced sustainability credentials, while being quicker to market with innovative products.

#### 15.2.4.1 Break-Even Point

The break-even point to invest in smart manufacturing is relative. This is because it usually depends on the scale of operation, apart from the industry within which the venture would fall. However, in most cases, firms can realize returns on investment within 3–5 years. This is particularly so for those firms operating in highly competitive or regulated markets where efficiency and sustainability are the key factors of success.

#### 15.2.4.2 External Market Factors

Other exogenous factors that may influence the relative costs and benefits of further investment in smart manufacturing of bio-based materials concern government incentives, consumer pull for sustainable products, and rising costs for petrochemical-based feedstocks. Companies investing in smart manufacturing for bio-based materials will be best positioned to respond effectively to these external factors—both pressures and opportunities.

## 15.3 TECHNOLOGICAL COMPLEXITY IN THE ADOPTION OF SMART MANUFACTURING FOR BIO-BASED MATERIALS

The literature suggests that there is a great potential for new technologies to enable the transition toward circular economies in the plastic value chain (Aristi Capetillo et al., 2022). However, their adoption is limited by significant barriers that relate to (1) industrial lock-ins, (2) misalignment with current corporate logics, (3) production systems dynamics, (4) the maturity stage of some of the technologies with its associated growth-related concerns and economic trade-offs, and (5) lack of understanding of the technology or its effects.

Nwakamma Ninduwezuor-Ehiobu et al. (2023) studied the integration of advanced materials in smart manufacturing and recommended that collaborative steps should be taken by policymakers, researchers, and industry stakeholders. Elina Dace et al. (2024) also realized that transition requires transformative policies, purposeful innovation, access to finance, risk-taking capacity, and new and sustainable business models and markets; therefore, it is a complex process. Some of the prominent complexities are discussed next.

### 15.3.1 Variability in Bio-based Materials

Bio-based materials can only be regarded as naturally derived materials with higher variability than synthetic materials normally exhibit. There is a large variability due to climate, soil quality, and agricultural practice, which directly influence the composition and quality of raw materials. For example, lignin and cellulose in plant fibers show large variability in amount depending on the region and season in which the plants have their origin. This variability creates a huge challenge for smart manufacturing, in large part because smart manufacturing requires consistency and precision.

#### 15.3.1.1 Material Characterization and Standardization

Full material characterization will be necessary to enable the integration of bio-based materials within the framework of a smart manufacturing system. Smart systems are driven by specific data inputs that work to optimize the production process. It requires measured standards on important properties such as tensile strength, moisture content, and biodegradability. All these variables must be constantly monitored by advanced sensors and real-time monitoring systems to achieve the materials that possess the necessary threshold of quality

#### 15.3.1.2 Data Management and Analytics

Smart manufacturing systems must be able to manage high-volume datasets and utilize advanced algorithms so as to predict and correct processing parameters in real-time. For example, if a specific feedstock batch has a higher moisture content than usual, then the process should automatically modify the extrusion temperature, which otherwise might result in defects in the final product.

### 15.3.2 Process Adaptation and Flexibility

Processing techniques for bio-based materials sometimes differ quite substantially from conventional material processing. Examples include bioplastics based on PLA or PHA, which require different temperature profiles and processing conditions compared with their petroleum-based plastic counterparts. Also, when natural fibers such as hemp or flax find their way into composite applications, specific difficulties arise with respect to fiber orientation, moisture management, and matrix compatibility.

#### 15.3.2.1 Specialized Manufacturing Processes

In smart manufacturing, some of these processes require machinery that is either specially made or very adaptable. For example, 3D printing biopolymers requires special extruders that take care of the special flow characteristics of such materials. Similarly, special modifications might have to be made in injection molding equipment to ensure that, during processing, biopolymers do not degrade.

#### 15.3.2.2 Advanced Control Systems

Adoption of bio-based materials requires developing an advanced control system that will continuously monitor and alter the processing parameters in real time. This brings in the use of IoT-enabled sensors to capture temperature, pressure, humidity, and other critical data throughout the production chain. Such sensors will need to be integrated with AI-driven control algorithms that will make split-second decisions and optimize product quality.

### 15.3.3 Supply Chain Integration

Supply chain issues make the integration of bio-based materials into smart manufacturing more convoluted. Bio-based materials are derived from agricultural products. These are affected by seasonal availability and regional supply differences.

#### 15.3.3.1 Just-in-Time Manufacturing

Most of the smart manufacturing systems are set up based on just-in-time manufacturing to lower the inventory cost by enhancing production efficiency. However, for bio-based materials, the seasonal availability may badly affect just-in-time strategies. As an example, for bioplastic manufacturing, the supply of corn starch would vary seasonally with different geographical outputs of the crop. To address this challenge, the integration of supply chain real-time data and predictive analytics is very much necessary.

#### 15.3.3.2 Traceability and Certification

Most of the bio-based materials need certification regarding their sustainability credentials, such as non-GMO (genetically modified origin) origin and organic standards. Smart manufacturing systems will need to trace materials across the entire supply chain and ensure that those materials are appropriately certified. This means integrating digital tracking, such as RFID tags or blockchain-based traceability solutions. Technically complex implementations on multi-supplier ecosystems with cross-border logistics may well prove extremely difficult.

### 15.3.4 Workforce Development and Skill Building

The most important activities in the use of smart technology in the production process of bio-based materials will involve the preparation of a workforce that can work with advanced technologies like AI, data analytics, robotics, and IoT.

#### 15.3.4.1 Training and Upskilling

It is time to give way to upskilling the already existing work with the help of proper training programs. For operating, managing, and maintaining the smart manufacturing systems, using IoT sensors, AI-driven process optimization tools, and robotics, the employees require proper training.

#### 15.3.4.2 Collaboration with Educational Institutions

Specified curricula related to smart manufacturing and bio-based material science may be established with the aid of local universities and technical schools. These should include internship programs, apprenticeships, and research sponsorships, through which students gain valuable industry experience as the industry meets its increasing demand for a technology-oriented workforce.

### 15.3.5 Investment in Technological Infrastructure

This, of course, requires great investments in technological infrastructure in integrating such manufacture of bio-based materials: from smart equipment to sensors, data analytics platforms, and communication networks—all of which will link production lines for real-time data sharing and automation.

#### 15.3.5.1 IoT Adoption for Real-Time Monitoring and Data Collection

With an investment in a suitably advanced network infrastructure that supports seamless connectivity and data exchange, manufacturers can unlock IoT's potential even more fully.

#### 15.3.5.2 Integrating AI and ML for Predictive Analytics

Manufacturing companies should, therefore, invest in the development of AI-enabled data analytics platforms, integrated into IoT systems that will provide actionable insights from the data.

#### 15.3.5.3 Cybersecurity and Data Protection

Smart manufacturing infrastructure, when connected, must be guarded against cyber threats with a high level of assurance, as IoT systems may have some weaknesses, apart from the cloud-based platforms.

### 15.3.6 Regulatory Compliance and Standardization

The acceptance of bio-based materials depends on the regulatory standards of different regions and lines of business. As an example, the use of bioplastics in food packaging must meet specific standards for safety and biodegradability that come from institutions such as the US Food and Drug Administration (FDA) or European Food Safety Authority (EFSA) (Majer et al., 2018). Smart manufacturing systems will need to be able to trace whether these regulations are complied with in real time. Similarly, in manufacturing biodegradable packaging, there could be sensors detecting whether the material decomposes at a specified rate in certain conditions. The need to make sure that all regulations are complied with raises the complication of smart manufacturing systems, since such systems have to be continuously updated over time according to shifting regulations and standards. Admittedly, there are efforts at unifying standards relating to bioplastics, among other bio-based products; however, significant variations remain between regions. This is one aspect that makes standardization difficult during the implementation of smart systems by manufacturers compatible with various regulatory frameworks. The manufacturing company needs to obtain all relevant certifications, such as ISO 14001 for environmental management and ISO 50001 for energy management, showing concern for sustainability and adherence to norms. Jan Peuckert and Rainer Quitzow (2016) reported that it has been identified by a procurement survey that, for enhanced public procurement of bio-based products, effective policy measures and informational needs are required.

## 15.4 IMPACT OF ADOPTING SMART MANUFACTURING ON BIO-BASED MATERIALS

The adoption of smart manufacturing techniques for bio-based materials will leave great environmental and societal impacts, as discussed in this section.

### 15.4.1 Environmental Impact

The recent shift in global consciousness toward the environment and stewardship has prompted industries to consider alternative routes to those relying on fossil fuels. This is going to reduce greenhouse gas emissions as well as waste reduction. Nasrin Aghamohammadi and Mohsen Shahmohammadi (2023) studied the effect of adopting bio-based building materials and their contribution toward Sustainable Development Goals (SDG). The smart manufacturing equipped with predictive maintenance and sensors would be able to reduce greenhouse gas emissions by lowering energy demands. It would be able to reduce waste production also by optimizing the exact amount of ingredients required in the process. By adopting closed-loop, circular economy models, it would be possible that materials, at the end of a product's life, could be reused, recycled, or biodegraded. Raouf AliAkbari et al. (2021) mentioned that scientists have reported that recently, many advances have been made in the processing of biomass-based materials in different ways, for example, production of hydrogen from bio-based wastes, production of biofuels by raw material *Scenedesmus obliquus*, etc. Natalia A. Tarazona et al. (2022) illustrated some strategies to develop sustainable by-design polymers for the circular bio-economy.

### 15.4.2 Social Impact

Like other industries, the shift toward smart technology for bio-based materials is not without fear about job losses; however, like computers, Industry 4.0 is also expected to affect society in a positive manner.

#### 15.4.2.1 Job Creation and Skill Development

Concerning the anxiety about job replacement, the automation of smart manufacturing for bio-based materials would also open up the creation of new jobs in fields that are more technologically orientated. Hence, workforce reskilling and upskilling for smarter, higher-quality employment, possibly in technology-intensive sectors, is an opportunity availed by smart manufacturing. It trains old workers in conventional manufacturing to adapt their processes according to automation systems and provides new job opportunities in system design, process optimization, and elaboration of AI algorithms.

Farmers cultivating crops for producing bio-based materials, such as corn, sugarcane, or hemp, face supplementary demand and often an improvement in the local economic situation. Andreas Pyka (2017) introduced the idea of a dedicated innovation system (DIS) by which the compensation of decreasing jobs in traditional sectors, due to AI and robotics, is possible in the long run through the bio-economy.

#### 15.4.2.2 Community Development and Social Equity

Sourcing raw materials from rural farms positions bio-based material manufacturers in places of influence to meaningfully contribute to the economic empowerment of these communities. This is particularly important in developing countries, where sustainable agriculture and biomanufacturing can result in new sources of income and poverty alleviation.

Besides economic growth, social equity could also be cultivated along the supply chain with transparency and fairness by applying smart manufacturing technologies. Technologies such as blockchain will help verify that the bio-based materials were acquired through fair trade practices, ensuring that the farmers and workers who provided the material are adequately compensated. This could avoid cases of exploitative labor conditions, while the benefits will also spread to smallholder farmers and foster equitable benefit distribution from sustainable manufacturing. V Nikolova-Alexieva and Angelova (2019) showed that agricultural and forestry biomass can be used to produce bio-chemicals, biomaterials, and bioenergy, and hence can form low-carbon sustainable value chains. The chains will be able to create rural growth and employment, and will also meet energy and climate policy targets.

## 15.5 CASE STUDIES

### 15.5.1 Case Study 1: Hemp in Biodiesel

The hemp plant is one of the approximately 76 plants identified as useful for biodiesel production. It produces oil-bearing seeds, having good oxidation stability and one of the best kinematic viscosities. Bio-oil transesterification is a good method to convert vegetable oil into biodiesel. But it is a reversible reaction between the mono-alkyl triglycerides and alcohol in the presence of a catalyst, yielding fatty acid methyl/ethyl esters and glycerol. Therefore, it is carried out in certain environmental and molar ratio conditions to get a better yield. It has been found that reaction optimization strategies like response surface methodology (RSM), artificial neural network (ANN), and adaptive neuro-fuzzy inference system (ANFIS) considerably decrease the time duration and energy loss in attaining maximum biodiesel yield.

V. Hariram et al. (2020) employed a genetic algorithm (GA)-coupled ANN model regularized with a Bayesian ANN approach in the prediction of *Cannabis sativa* (hemp) seed biodiesel through a single-stage transesterification process. They used the central composite rotatable design (CCRD) to develop the trial experiments of up to 20 runs with variations in the input process parameters such as methanol-to-oil molar ratio, catalyst loading, and reaction duration. They claimed that by using 97.6% at a 0.225 v/v molar ratio, 1.725 w/v catalyst loading, and 60 min reaction duration, the optimal yield of hemp seed biodiesel was 97.6%. The physicochemical properties of the hemp seed oil developed using this optimizing technique were found to comply with the American Society for Testing and Materials (ASTM) standards.

### 15.5.2 Case Study 2: Bio-based Material in Packaging

Life cycle assessment (LCA) is considered a standard tool to assess the environmental impacts of PLA. The environmental performance of PLA is subjected to variation in the composition, transportation, treatment methods, and conditions involved in the process.

Junwei Li et al. (2024) developed a model obtained by integrating cradle-to-grave LCA with a rigorous process model and AI models. They used the model to conduct uncertainty and variability analyses, which are otherwise highly time-consuming. PLA (polylactic acid) produced from corn silver in the US was examined using this model. A Monte Carlo simulation was used to analyze uncertainty and variability. The results showed that the mean life cycle Global Warming Potential (GWP) of 1 kg PLA is 4.3 kg$CO_2$-eq (P5–P95 4.1–4.4) for composting PLA with natural gas combusted in the biorefinery, 3.7 kg$CO_2$-eq (P5–P95 3.4–3.9) for incinerating PLA for electricity with natural gas combusted in the biorefinery, and 1.9 kg$CO_2$-eq (P5–P95 1.6–2.1) for incinerating PLA for electricity with wood pellets combusted in the biorefinery.

## 15.6 CONCLUSION

There is great potential in the bio-based materials market. The incorporation of smart technologies into the manufacturing processes would be cost-effective in the long run. However, the said incorporation is not free from challenges like capital investments, technical upgradation, skilled worker creation, and proper regulations.

Overcoming such challenges, however, will be important in realizing the full benefit of smart manufacturing in the production of bio-based materials, with industries continuing to put more emphasis on sustainability and the development of bio-based alternatives to petrochemical products. Manufacturers will be able to exploit all the potential of smart manufacturing and drive a more sustainable and competitive bio-based materials industry with investment in the necessary skills, partnerships, and technologies.

## REFERENCES


Aghamohammadi, N., & Shahmohammadi, M. (2023). Towards sustainable development goals and role of bio-based building materials. In *Elsevier EBooks*, 243–279. https://doi.org/10.1016/b978-0-323-91149-8.00004-1

Ahmad Qamar, S., Asgher, M., Bilal, M., & Iqbal, H. M. N. (2020). Bio-based active food packaging materials: sustainable alternative to conventional petrochemical-based packaging materials. *Food Research International*, *137*, 109625. https://doi.org/10.1016/j.foodres.2020.109625

AliAkbari, R., Ghasemi, M. H., Neekzad, N., Kowsari, E., Ramakrishna, S., Mehrali, M., & Marfavi, Y. (2021). High value add bio-based low-carbon materials: conversion processes and circular economy. *Journal of Cleaner Production*, *293*, 126101. https://doi.org/10.1016/j.jclepro.2021.126101

Al-Tayyar, N. A., Youssef, A. M., & Al-hindi, R. (2020). Antimicrobial food packaging based on sustainable bio-based materials for reducing foodborne pathogens: a review. *Food Chemistry*, *310*, 125915. https://doi.org/10.1016/j.foodchem.2019.125915

Arfin, T., & Sonawane, K. (2018). Bio-based materials: past to future. In *Springer EBooks*, 1–32. https://doi.org/10.1007/978-981-13-1909-9_1

Arif, Z. U., Khalid, M. Y., Sheikh, M. F., Zolfagharian, A., & Bodaghi, M. (2022). Biopolymeric sustainable materials and their emerging applications. *Journal of Environmental Chemical Engineering*, *10*(4), 108159. https://doi.org/10.1016/j.jece.2022.108159

Aristi Capetillo, A., Bauer, F., & Chaminade, C. (2022). Emerging technologies supporting the transition to a circular economy in the plastic materials value chain. *Circular Economy and Sustainability*. https://doi.org/10.1007/s43615-022-00209-2

Ates, B., Koytepe, S., Ulu, A., Gurses, C., & Thakur, V. K. (2020). Chemistry, structures, and advanced applications of nanocomposites from biorenewable resources. *Chemical Reviews*, *120*(17), 9304–9362. https://doi.org/10.1021/acs.chemrev.9b00553

Azimi, B., Maleki, H., Zavagna, L., De la Ossa, J. G., Linari, S., Lazzeri, A., & Danti, S. (2020). Bio-based electrospun fibers for wound healing. *Journal of Functional Biomaterials*, *11*(3), 67. https://doi.org/10.3390/jfb11030067

Baranwal, J., Barse, B., Fais, A., Delogu, G. L., & Kumar, A. (2022). Biopolymer: a sustainable material for food and medical applications. *Polymers*, *14*(5), 983. https://doi.org/10.3390/polym14050983

Bradu, P., Biswas, A., Nair, C., Sreevalsakumar, S., Patil, M., Kannampuzha, S., Mukherjee, A. G., Wanjari, U. R., Renu, K., Vellingiri, B., & Gopalakrishnan, A. V. (2022). Recent advances in green technology and Industrial Revolution 4.0 for a sustainable future. *Environmental Science and Pollution Research International*, *30*. https://doi.org/10.1007/s11356-022-20024-4

Clauser, N. M., Felissia, F. E., Area, M. C., & Vallejos, M. E. (2022). Integrating the new age of bioeconomy and Industry 4.0 into biorefinery process design. *BioResources*, *17*(3), 5510–5531. https://doi.org/10.15376/biores.17.3.clauser

Cywar, R. M., Rorrer, N. A., Hoyt, C. B., Beckham, G. T., & Chen, E. Y.-X. (2021). Bio-based polymers with performance-advantaged properties. *Nature Reviews Materials*, *7*. https://doi.org/10.1038/s41578-021-00363-3

Dace, E., Cascavilla, A., Bianchi, M., Chioatto, E., Zecca, E., Ladu, L., & Yilan, G. (2024). Barriers to transitioning to a circular bio-based economy: findings from an industrial perspective. *Sustainable Production and Consumption*, *48*, 407–418. https://doi.org/10.1016/j.spc.2024.05.029

Deng, Y., Lu, T., Cui, J., Sangram Keshari Samal, Xiong, R., & Huang, C. (2021). Bio-based electrospun nanofiber as building blocks for a novel eco-friendly air filtration membrane: a review. *Separation and Purification Technology*, *277*, 119623–119623. https://doi.org/10.1016/j.seppur.2021.119623

Deshpande, Y. V., Ayer, S., Agrawal, T., Agrawal, V., Madankar, T. A., & Barve, P. S. (2023). Application of smart strategies for sustainable manufacturing of conventional machining process: a review. *Journal of the Institution of Engineers (India) Series C, 104*(6), 1267–1289. https://doi.org/10.1007/s40032-023-00995-0

George, A., Sanjay, M. R., Srisuk, R., Parameswaranpillai, J., & Siengchin, S. (2020). A comprehensive review on chemical properties and applications of biopolymers and their composites. *International Journal of Biological Macromolecules, 154*, 329–338. https://doi.org/10.1016/j.ijbiomac.2020.03.120

Hariram, V., Bharadwaj, P. M., Viswaksen, A., Surya, C. H., Ruthvin Maheej, D., Seralathan, S., & Micha Premkumar, T. (2020). Optimizing the transesterification process of hemp seed bio-oil using artificial neural network. In *Lecture Notes in Electrical Engineering*, 459–471. https://doi.org/10.1007/978-981-15-7245-6_34

Li, J., Wang, Y., Xu, C., Liu, S., Dai, J., & Lan, K. (2024). Bioplastic derived from corn stover: life cycle assessment and artificial intelligence-based analysis of uncertainty and variability. *The Science of the Total Environment*, 174349–174349. https://doi.org/10.1016/j.scitotenv.2024.174349

Majer, S., Wurster, S., Moosmann, D., Ladu, L., Sumfleth, B., & Thrän, D. (2018). Gaps and research demand for sustainability certification and standardisation in a sustainable bio-based economy in the EU. *Sustainability, 10*(7), 2455. https://doi.org/10.3390/su10072455

Mehta, P., Rao, P. K., Wu, Z., Veljko Jovanović, Wodo, O., & Kuttolamadom, M. (2018). *Smart Manufacturing: State-of-the-Art Review in Context of Conventional and Modern Manufacturing Process Modeling, Monitoring and Control*. https://doi.org/10.1115/msec2018-6658

Moreno, A. D., Susmozas, A., Oliva, J. M., & Negro, M. J. (2020). Overview of bio-based industries. In *Biobased Products and Industries*, 1–40. https://doi.org/10.1016/b978-0-12-818493-6.00001-4

Nikolova-Alexieva, V., & Angelova, M. (2019). Digital entrepreneurship: doing business for smart and sustainable bio-based economy. *IOP Conference Series: Materials Science and Engineering, 618*, 012081. https://doi.org/10.1088/1757-899x/618/1/012081

Ninduwezuor-Ehiobu, N., Tula, O. A., Daraojimba, C., Ofonagoro, K. A., Ogunjobi, O. A., Gidiagba, J. O., Egbokhaebho, B. A., & Banso, A. A. (2023). Exploring innovative material integration in modern manufacturing for advancing U.S. competitiveness in sustainable global economy. *Engineering Science & Technology Journal, 4*(3), 140–168. https://doi.org/10.51594/estj.v4i3.558

Okolie, J. A., Nanda, S., Dalai, A. K., & Kozinski, J. A. (2020). Chemistry and specialty industrial applications of lignocellulosic biomass. *Waste and Biomass Valorization*. https://doi.org/10.1007/s12649-020-01123-0

Pandey, K., Pandey, M., Kumar, V., Aggarwal, U., & Singhal, B. (2023). Bioprocessing 4.0 in biomanufacturing: paving the way for sustainable bioeconomy. *Systems Microbiology and Biomanufacturing*. https://doi.org/10.1007/s43393-023-00206-y

Peuckert, J., & Quitzow, R. (2016). Acceptance of bio-based products in the business-to-business market and public procurement: expert survey results. *Biofuels, Bioproducts and Biorefining, 11*(1), 92–109. https://doi.org/10.1002/bbb.1725

Pyka, A. (2017). Dedicated innovation systems to support the transformation towards sustainability: creating income opportunities and employment in the knowledge-based digital bioeconomy. *Journal of Open Innovation: Technology, Market, and Complexity, 3*(4), 27. https://doi.org/10.1186/s40852-017-0079-7

Rafiqah, S. A., Khalina, A., Harmaen, A. S., Tawakkal, I. A., Zaman, K., Asim, M., Nurrazi, M. N., & Lee, C. H. (2021). A review on properties and application of bio-based poly(butylene succinate). *Polymers, 13*(9), 1436. https://doi.org/10.3390/polym13091436

Ramesh Kumar, S., Shaiju, P., O'Connor, K. E., & P, R. B. (2020). Bio-based and biodegradable polymers—state-of-the-art, challenges and emerging trends. *Current Opinion in Green and Sustainable Chemistry*, *21*, 75–81. https://doi.org/10.1016/j.cogsc.2019.12.005

Rani, S., Bhambri, P., Kumar, S., Pareek, P. K., & Elngar, A. A. (Eds.). (2024). *AI-Driven Digital Twin and Industry 4.0: A Conceptual Framework with Applications*. CRC Press.

Reichert, C. L., Bugnicourt, E., Coltelli, M.-B., Cinelli, P., Lazzeri, A., Canesi, I., Braca, F., Martínez, B. M., Alonso, R., Agostinis, L., Verstichel, S., Six, L., Mets, S. D., Gómez, E. C., Ißbrücker, C., Geerinck, R., Nettleton, D. F., Campos, I., Sauter, E., & Pieczyk, P. (2020). Bio-based packaging: materials, modifications, industrial applications and sustainability. *Polymers*, *12*(7), 1558.

Riese, J., Fasel, H., Maik Pannok, & Lier, S. (2024). Decentralized production concepts for bio-based polymers—implications for supply chains, costs, and the carbon footprint. *Sustainable Production and Consumption*, *46*, 460–475. https://doi.org/10.1016/j.spc.2024.03.001

Sharma, P., Gaur, V. K., Sirohi, R., Varjani, S., Hyoun Kim, S., & Wong, J. W. C. (2021). Sustainable processing of food waste for production of bio-based products for circular bioeconomy. *Bioresource Technology*, *325*, 124684. https://doi.org/10.1016/j.biortech.2021.124684

Sheldon, R. A., Basso, A., & Brady, D. (2021). New frontiers in enzyme immobilisation: robust biocatalysts for a circular bio-based economy. *Chemical Society Reviews*. https://doi.org/10.1039/d1cs00015b

Sillanpää, M., & Ncibi, C. (2017). *Biorefineries: Industrial-Scale Production Paving the Way for Bioeconomy*, 233–270. https://doi.org/10.1007/978-3-319-55637-6_7

Siracusa, V., & Blanco, I. (2020). Bio-Polyethylene (Bio-PE), Bio-polypropylene (Bio-PP) and bio-poly(ethylene terephthalate) (Bio-PET): recent developments in bio-based polymers analogous to petroleum-derived ones for packaging and engineering applications. *Polymers*, *12*(8), 1641. https://doi.org/10.3390/polym12081641

Suarez, A., Ford, E., Venditti, R., Kelley, S., Saloni, D., & Gonzalez, R. (2022). Rethinking the use of bio-based plastics to accelerate the decarbonization of our society. *Resources, Conservation and Recycling*, *186*, 106593. https://doi.org/10.1016/j.resconrec.2022.106593

Taesi, C., Aggogeri, F., & Pellegrini, N. (2023). COBOT applications—recent advances and challenges. *Robotics*, *12*(3), 79. https://doi.org/10.3390/robotics12030079

Tarazona, N. A., Machatschek, R., Balcucho, J., Castro-Mayorga, J. L., Saldarriaga, J. F., & Lendlein, A. (2022). Opportunities and challenges for integrating the development of sustainable polymer materials within an international circular (bio)economy concept. *MRS Energy & Sustainability*, *9*(1), 28–34. https://doi.org/10.1557/s43581-021-00015-7

Vinod, A., Sanjay, M. R., Suchart, S., & Jyotishkumar, P. (2020). Renewable and sustainable biobased materials: an assessment on biofibers, biofilms, biopolymers and biocomposites. *Journal of Cleaner Production*, *258*, 120978. https://doi.org/10.1016/j.jclepro.2020.120978

Walker, S., & Rothman, R. (2020). Life cycle assessment of bio-based and fossil-based plastic: a review. *Journal of Cleaner Production*, *261*, 121158. https://doi.org/10.1016/j.jclepro.2020.121158

Wendels, S., & Avérous, L. (2021). Biobased polyurethanes for biomedical applications. *Bioactive Materials*, *6*(4), 1083–1106. https://doi.org/10.1016/j.bioactmat.2020.10.002

Yang, H., Yu, B., Xu, X., Bourbigot, S., Wang, H., & Song, P. (2020). Lignin-derived biobased flame retardants toward high-performance sustainable polymeric materials. *Green Chemistry*, *22*(7), 2129–2161. https://doi.org/10.1039/d0gc00449a

Zych, A., Tellers, J., Bertolacci, L., Ceseracciu, L., Marini, L., Mancini, G., & Athanassiou, A. (2020). Biobased, biodegradable, self-healing boronic ester vitrimers from epoxidized soybean oil acrylate. *ACS Applied Polymer Materials*, *3*(2), 1135–1144. https://doi.org/10.1021/acsapm.0c01335

# 16 Case Studies and Success Stories of Bio-Based Materials for Smart Manufacturing

*Sunil Kumar Verma, Ravi Kant Kushwaha, Akash Singh, Anshu Mishra, Sankalp Misra, Renu Khare, and Mohanned Alhussien*

## 16.1 INTRODUCTION TO CASE STUDIES

### 16.1.1 Definition of Case Studies

A case study is a qualitative research approach encompassing an in-depth examination of a definite subject, for instance, an event, group, individual, phenomenon, or organization within its real-life context. This approach allows researchers to explore complex issues in detail, providing rich insights that quantitative methods may overlook. Case studies can vary in scope and may focus on a single case (within-case research) or multiple cases (cross-case research) to derive broader generalizations about a particular phenomenon or set of phenomena. The core of a case study lies in its ability to investigate contemporary issues as they occur in their natural settings. This contrasts with experimental designs, where researchers exert control over variables. The case study method is prevalent across various fields (Crowe et al., 2011).

## 16.2 INDUSTRY APPLICATIONS OF BIO-BASED MATERIALS

Bio-based materials have found diverse applications across various industries, showcasing their versatility and potential to replace traditional fossil-based products.

### 16.2.1 Bioplastics in Consumer Goods

In terms of material qualities, polymers can be either flexible or stiff, permeable or impermeable, or hydrophilic or hydrophobic. These characteristics are dictated by the structure of the monomers, which are the repeating building units of polymers. Plastics are polymeric materials treated and shaped, usually with heat, into their final, economically relevant shape. Like those found in bottles and textiles, most plastics are thermoplastics made of linear polymer chains that can be thermally reshaped.

DOI: 10.1201/9781003566205-16

However, some polymers undergo crosslinking during processing to create thermosets, which are more durable than thermoplastics and have a shape that is largely independent of temperature, like those found in epoxies and car tires (Censi et al., 2022). The most economically significant forms of polymers processed into products are called "plastics." In contrast, these materials' physicochemical qualities and synthesis before their final shape are referred to as "polymers." The transportation, food, healthcare, and energy sectors all depend heavily on these minerals, which are essential to modern existence (Sharma & Verma, 2014). Plastics are also necessary for many elements of sustainability; plastic insulators can save energy; plastic food packaging extends shelf life, which helps minimize food waste; and lightweight plastic materials increase the fuel efficiency of automobiles and airplanes. With an annual growth rate of 4% and an annual production of about 380 million tons of plastic, 6,300 million tons of plastic garbage have been produced since 1950. Growing awareness of plastic waste's adverse effects on the environment and the greenhouse gas emissions linked to plastic drives the shift to a "circular plastic economy." Reuse and recycling rule the life cycles of materials in a circular economy, while the consumption of non-renewable resources and trash creation are reduced (Rosenboom et al., 2022).

Often called bioplastics, these materials can also be manufactured from renewable resources, even though most commercial plastics are derived from fossil fuels. The monomers, in this instance, are produced or extracted from biomass compounds (such as plant sugars) and then polymerized to create either new polymers like polyhydroxyalkanoates (PHAs) or a straight substitute for an existing plastic like polyethene (PE). Additionally, non-synthetic natural polymers, including proteins, natural rubber, and starch, can be obtained through biomass extraction. The phrase "bioplastic" is often used; however, because of its ambiguous definition, it is still misinterpreted. Bioplastics are polymers that are biodegradable, bio-based, or derived from biological processes or renewable resources (Singh et al., 2024b). Some plastics that are biodegradable but still derived from fossil fuels are also called "bioplastics"; however, this word is deceptive and should not be used. The consumer goods sector is one of the largest markets for bioplastics, encompassing a wide range of products. Key applications include packaging. Bioplastics are extensively used in food, beverages, and cosmetics packaging materials. They offer comparable mechanical properties to traditional plastics while being more environmentally friendly. For instance, bioplastics like polylactic acid (PLA) are utilized in food containers, beverage bottles, and flexible packaging solutions. Household Items such as disposable cutlery, plates, and cups are increasingly made from bioplastics (Bachleitner et al., 2023). Companies like Kaue Süd Engineering and Ramaplast S.p.A. have successfully integrated bioplastics into their product lines. Kaue Süd Engineering uses Terralene®, a sugarcane-based bioplastic, to manufacture inhalers, while Ramaplast produces high-quality cosmetic packaging using certified bio-based materials. These applications reduce dependence on fossil fuels and appeal to environmentally conscious consumers. For 15 years, Kaue Süd Engineering has relied on bioplastic from FKuR (Singh et al., 2024b). Kaue Süd Engineering has developed a variety of items using different bioplastics under the new "ecoproducts" name. Along with sand toys and household goods, the line includes a steam inhaler for customers. The inhaler's material must meet high requirements. It must be temperature resistant, safe, and easy to clean. It should also feel high-quality and be constructed from sustainable materials to reduce its environmental impact. Terralene® HD 4527, a sugarcane-based bioplastic, has been selected by Kaue Süd Engineering. This material is perfect for creating the steam inhaler because it has several benefits (Verma et al., 2023).

The steam inhaler comprises a mouthpiece, an outer and inner container, and the bioplastic Terralene® HD 4527. This material is dishwasher safe even at high temperatures, significantly lessening the customer's cleaning and maintenance burden. Terralene® HD 4527's high material density makes the inhaler feel more substantial and precious. This gives the inhaler a more upscale feel than those manufactured of traditional HDPE. The high stiffness of the Terralene® HD 4527 inhaler contributes to its exceptional resilience and durability. This guarantees a long product life and excellent dependability while in use. The inhaler needs to be able to tolerate high temperatures. After the hot water is filled, the inner container is put into the outer container. The double-walled insulation efficiently retains heat while enabling the inhaler to be carried around safely. The inhaler may be readily disassembled, washed in the dishwasher, and compactly stored afterward. Because Terralene® HD 4527 is approved for food contact, users can feel even safer. Kaue Süd Engineering has developed a steam inhaler that meets customer-specified sustainability standards and impresses with its functionality and user-friendliness, thanks to the partnership with FKuR and the use of Terralene® HD 4527. Even with PP equipment, this high-performance plastic compound has good mechanical qualities and is simple to work with in injection molding.

### 16.2.2 Bioplastics in Other Sectors of Life

Integrating bio-based materials in agriculture drives significant innovations that enhance sustainability, reduce waste, and stimulate a circular economy. These innovations leverage renewable resources derived from agricultural practices, transforming waste into valuable products and improving farming efficiency. Below are key impact areas of bio-based materials (Vigneswari et al., 2024). Bio-based fertilizers are derived from organic materials, including compost, manure, and plant residues. These fertilizers provide essential nutrients to crops and improve soil health by enhancing microbial activity and soil structure. For example, live microorganisms promote plant growth by increasing the nutrient supply to the host plant. They can be produced from agricultural waste, such as food processing byproducts, which are rich in nutrients and can enhance soil fertility. Completing organic waste from agriculture and food production to create compost reduces landfill waste while providing a sustainable source of nutrients for crops. Examples include microbial pesticides and beneficial microorganisms to control pests and diseases. For example, *Bacillus thuringiensis* (Bt) is a bacterium used to target specific insect pests without harming beneficial insects. Plant extracts such as neem or garlic can be used as natural insect repellents or fungicides, providing effective pest control while minimizing environmental impact (Atiwesh et al., 2021).

#### 16.2.2.1 Circular Economy Initiatives

The evolution of circular economy in agriculture emphasizes the importance of reusing resources while minimizing waste. Biofinery facilities transform biomass into a spectrum of bioproducts comprising bioplastics, chemicals, and biofuels (Verma & Kumar, 2022). They utilize agricultural residues as feedstock to produce high-value products while promoting sustainability. Nutrient recovery systems and technologies that recycle nutrients from agrarian waste streams back into the soil support sustainable farming practices by closing nutrient loops. Innovations in agricultural practices through bio-based materials are paving the way for sustainable development in the sector (Menossi et al., 2021). By transforming waste into valuable resources and reducing reliance on non-renewable inputs, these advancements enhance productivity and contribute to environmental conservation and economic resilience. As research continues to evolve and new technologies emerge, the potential for bio-based materials in agriculture will expand further, supporting a greener future for food production systems globally. The Bio-Based Garden Project in the Netherlands exemplifies the agricultural application of bio-based materials. Over the past decade, this initiative has tested around 80 crops for their potential as raw materials in various applications, including natural pesticides and building materials. This project emphasizes sustainable practices, such as alternative soil conditioning and biofertilization, to enhance crop resilience and productivity (see Table 16.1) (Lange et al., 2021).

### 16.2.3 Sustainable Packaging Solutions

Sustainable packaging is an essential component of modern industry, focusing on minimizing environmental impact while meeting consumer needs. It encompasses a range of materials and methods designed to decrease waste, conserve resources, and encourage a circular economy.

#### 16.2.3.1 Innovative Sustainable Packaging Solutions

Plant-based plastics (bioplastics) are biodegradable materials derived from renewable biomass sources like sugarcane or maize starch and have a reduced carbon footprint compared to conventional plastics (Kumar & Verma, 2022). Mushroom packaging made from mycelium is an innovative solution that binds agricultural waste into biodegradable packaging. It offers insulation and protection while being environmentally friendly. Single-use items are less necessary when containers such as glass jars and metal tins can be reused or refilled. This approach fosters a culture of sustainability among consumers (Leipold & Petit-Boix, 2018). Edible packaging is made from materials like seaweed or starch. Compostable materials and packaging, such as those made from PLA (starch-based) or cellulose, biodegrade in compost environments, providing an eco-friendly alternative to conventional plastics. These films dissolve in water, significantly reducing waste. Commonly used in single-dose packaging for food and detergents, they represent a novel approach to minimizing environmental impact. Upcycled packaging involves repurposing waste materials into new packaging solutions, such as

**TABLE 16.1**
**Study of Innovative Bio-Based Products.**

| Product name | Biomass platform | Technology readiness level (TRL) | Business to business (BtB) or business to consumer (BtC) |
|---|---|---|---|
| Guayule rubber | Natural rubber | 6 | BtC & BtB |
| Micro-fibrillated cellulose | Plant fibers | 8 | BtB |
| Thermoplastic biopolymers reinforced with plant fibers | Plant fibers | 5 | BtB |
| Plant fibers reinforced resin prepregs | Plant fibers | 9 | BtB |
| Self-binding composite non-woven plant fibers | Plant fibers | 7 | BtB |
| Lignin biocomposites reinforced with plant fibers | Plant fibers | 6 | BtB |
| High-Purity lignin | Lignin | 5 | BtB |
| Lignin-based carbon Nanofibers (CF) | Lignin | 5 to 6 | BtB |
| Lignin bio-oil | Lignin | 5 | BtB (maybe eventually BtC) |
| Lignin-based phenolic (LBP) resins | Lignin | 8 | BtB |
| Bio-BTX aromatic | Lignin | 6 | BtB |
| Bio-based phenol and alkylphenols | Lignin | 5 | BtB (maybe eventually BtC) |
| Bio-based polyamide 12 | Renewable oils & fats | 5 | BtB |
| Biolubricants | Renewable oils and fats | 7–8 for high-performance lubricants and 9 for loss lubrication | BtC & BtB |
| PHAs from renewable oils and fats | Renewable oils and fats | 6 | BtB |
| Limonene-based engineering polymers | Terpenes | 5 | BtB |
| Bacterial biosurfactants: sophorolipids and rhamnolipids | Natural polyelectrolytes | 7 | BtB |
| Biotechnological chitosan (third generation) | Natural polyelectrolytes | 7 | BtB |
| PHAs from urban wastes | Urban biowaste | 6 | BtB |
| Volatile fatty acids (VFAs) mixture | Urban biowaste | 7 | BtB |

reclaimed textiles or wood, thereby reducing landfill contributions (Taylor et al., 2024). Zero-waste packaging is designed to eradicate waste through its lifespan; it is reusable, compostable, or recyclable. Industry examples: FoodCraft utilizes plant-based PLA boxes that are compostable for its organic products; ITC Limited has developed biodegradable paperboards made from renewable fibers that serve as alternatives to single-use plastics in various sectors like food service and delivery; and companies like TIPA focus on creating compostable flexible packaging that mimics traditional plastic but fully biodegrades within 180 days (Tsakalova et al., 2024).

Adopting sustainable packaging practices reduces environmental impact and aligns with consumer preferences. Many consumers are enthusiastic about paying more for sustainable options, reflecting a growing demand for sustainable practices across industries. Additionally, businesses can enhance their brand image and potentially reduce costs associated with waste management. In conclusion, sustainable packaging solutions are vital for promoting environmental responsibility in industry applications. By leveraging innovative materials and practices, companies can meet consumer demands while contributing to a healthier planet. The USDA has highlighted success stories involving bio-based disposables, such as compostable food ware, which are increasingly adopted in various settings, including universities and parks. These initiatives demonstrate the growing trend toward sustainable packaging options that minimize environmental impact while maintaining functionality (Hsiao & Hu, 2024).

## 16.3 COLLABORATIVE PROJECTS AND INNOVATIONS

Collaboration among businesses, research institutions, and government entities has been pivotal in advancing bio-based innovations.

### 16.3.1 Hydrogen Production for Transportation

The H2Boost project is a significant initiative funded by the UK government to scale up biohydrogen production specifically for the transport sector. This collaborative effort involves multiple partners and aligns with the UK's broader net-zero goals, emphasizing the transition to cleaner energy sources. H2Boost focuses on developing innovative technologies to produce hydrogen from biogenic feedstocks, for instance, agricultural as well as organic waste (Babkin et al., 2023). By converting these materials into hydrogen, the project aims to create a sustainable and renewable fuel source for transportation, contributing to the decarbonization of the transport sector. The primary goal of $H_2$Boost is to enhance the scale of biohydrogen production, making it commercially viable and accessible for various transportation applications. The project seeks to integrate biohydrogen production with current energy systems and transportation networks, ensuring seamless adoption and utilization. By producing low-carbon hydrogen, H2Boost plays a crucial role in helping the UK meet its ambitious net-zero emissions targets by 2050 (Duque-Acevedo et al., 2020).

#### 16.3.1.1 Collaborative Efforts

H2Boost is a collaborative project involving various stakeholders, including the Biorenewables Development Centre (BDC), which is leading the project. BDC focuses on developing bio-based technologies and processes that efficiently convert biomass into hydrogen. Industry partners from various companies in the energy and transport sectors are involved, providing expertise and resources to facilitate technology development and implementation (Verma et al., 2024b). Government-supported UK's Net Zero Hydrogen Fund, H2Boost, benefits from substantial financial backing to foster innovation in hydrogen technologies. The project employs several cutting-edge technologies to enhance biohydrogen production. The gasification method transforms biomass into syngas (a combination of $H_2$ and CO) through extreme temperature treatment. The syngas can then be processed to extract pure hydrogen. Anaerobic digestion, utilizing microorganisms, breaks organic materials in anaerobic conditions, yielding hydrogen-rich biogas. While primarily associated with water splitting, advancements in electrolytic processes can also be applied to biogenic sources to enhance hydrogen yield (Asghar et al., 2022).

### 16.3.2 Biodegradable Solvents Development

#### 16.3.2.1 Overview of Circa Group and Cyrene™

Circa Group, an innovative cleantech company, has significantly progressed in developing biodegradable, non-toxic solvents from waste cellulose. Their flagship product, Cyrene™, is a bio-based solvent derived from levoglucosenone (LGO), a platform molecule produced through their proprietary Furacell™ process (Camp & Greatrex, 2022). This initiative addresses environmental challenges and creates commercial opportunities by providing safer alternatives to traditional petroleum-based solvents. The Furacell™ process is central to Circa's operations and involves the conversion of lignocellulosic biomass—primarily waste cellulose—into valuable chemicals. Key features of this process include feedstock utilization, which utilizes abundant non-food biomass, including agricultural residues and forestry waste, ensuring that the production does not compete with food resources. The process is designed to minimize byproducts, leading to reduced waste and lower separation costs. The primary outputs are LGO, biochar, and water. The production is nearly carbon neutral, relying on renewable energy sources and producing minimal harmful emissions (Stanfield et al., 2023).

#### 16.3.2.2 Cyrene™: A Breakthrough Solvent

Cyrene™ is a sustainable substitute for conventional solvents like DMF (dimethylformamide) and NMP (N-methyl-2-pyrrolidone), known for their toxicity and environmental impact. Cyrene™ breaks down naturally without leaving harmful residues, making it environmentally friendly. It poses significantly lower health risks compared to conventional solvents, enhancing safety for workers and consumers (Singh et al., 2024a). The solvent is suitable for various industrial applications such as pharmaceuticals, paints and coatings, textiles, and electronics (Kong et al., 2022).

The development of Cyrene™ has been bolstered by collaboration with the BDC and support from the EU Horizon 2020 research and innovation program. This partnership emphasizes the importance of collaborative research in driving innovation in sustainable chemistry. ReSolute Project: As part of this initiative, Circa is constructing the ReSolute plant in France, which will scale up Cyrene™ production to 1,200 tons annually. This facility represents a significant investment in clean technology and aims to establish a robust supply chain for bio-based solvents (Yu et al., 2021).

The shift toward biodegradable solvents like Cyrene™ opens up numerous commercial opportunities. There is an increasing need for sustainable chemicals compelled by regulatory pressures and consumer preferences for environmentally friendly products. Industry Cyrene™ has been trialed in over 1,000 applications, demonstrating its effectiveness across various sectors. Its performance often surpasses traditional solvents, making it an attractive option for manufacturers looking to improve sustainability. The establishment of production facilities like ReSolute not only contributes to environmental goals but also creates jobs in clean technology sectors. The development of biodegradable solvents by Circa Group exemplifies how collaboration can lead to innovative solutions that address pressing environmental challenges. By transforming waste cellulose into high-performance solvents like Cyrene™, this initiative promotes sustainability and paves the way for commercial success in various industries. As the project progresses, it holds the potential to significantly reduce reliance on toxic solvents while supporting the transition toward a greener economy (Li et al., 2022).

### 16.3.3 Research and Development in Agri-Tech

#### 16.3.3.1 Overview of Fontus Environmental Limited

Fontus Environmental Limited is an agri-tech consultancy dedicated to advancing sustainable agricultural practices through innovative research and development. The company collaborates with various stakeholders, including academic institutions, farming organizations, and industry partners, to investigate novel plant protection products. This interdisciplinary approach focuses on collaboration significance for referring environmental tests while creating commercial opportunities. The focus is on developing effective alternatives to conventional chemical pesticides to mitigate pest and disease pressures while minimizing ecological impact. By researching novel plant protection solutions, Fontus aims to improve the resilience of crops against pests, diseases, and climate-related stresses. The initiative supports adopting integrated pest management promotion strategies that combine physical, chemical, and biological tools to manage pest populations sustainably. A key goal is to decrease reliance on synthetic chemicals in agriculture, thereby reducing potential harm to ecosystems and human health (Chauhan et al., 2023).

#### 16.3.3.2 Collaborative Partnerships

Fontus Environmental collaborates with diverse stakeholders, including partnering with universities and research organizations, allowing access to cutting-edge research and expertise in plant sciences and agronomy. Collaborations with industry bodies facilitate knowledge transfer and help align research efforts with practical agricultural needs. Government agencies engaging with government entities ensure compliance with regulatory frameworks and support initiatives to promote sustainable agriculture. Private sector companies' partnerships with agribusinesses enable the commercialization of developed products, facilitating their adoption by farmers. The interdisciplinary nature of Fontus Environmental's projects underscores the significance of collaboration in agri-tech. Combining expertise from various fields—such as biology, chemistry, agronomy, and environmental science—leads to more comprehensive solutions that address complex agricultural challenges. Collaborative efforts foster innovation by bringing together diverse perspectives and resources, accelerating the development of new technologies and practices. Partnerships accelerate knowledge exchange between industry and academia, ensuring that research findings effectively translate into practical applications for farmers (D'Almeida & de Albuquerque, 2024).

With increasing consumer awareness regarding sustainability, there is an increasing need for sustainable agricultural inputs that minimize environmental impact. Governments are increasingly promoting sustainable farming practices through policies and incentives, creating an encouraging environment for implementing innovative solutions. As sustainability becomes a global priority, there is potential for Fontus Environmental's innovations to be marketed internationally, expanding their impact beyond local markets. By reducing reliance on expensive synthetic inputs and improving crop yields through sustainable practices, farmers can enhance their profitability while contributing to environmental stewardship. Fontus Environmental Limited exemplifies how collaborative research and development in agri-tech can lead to innovative solutions for sustainable agriculture. The company addresses critical ecological challenges by focusing on novel plant protection products through interdisciplinary partnerships while creating commercial opportunities within the agricultural sector. As these initiatives progress, they promise to transform farming practices toward more sustainable models that benefit producers and the planet (Bomtempo & Alves, 2014).

## 16.4 IMPACT OF BIO-BASED MATERIALS ON SUSTAINABLE MANUFACTURING PRACTICES

The integration of bio-based materials into manufacturing processes has significant implications for sustainability. Integrating bio-based materials into manufacturing has emerged as a transformative approach to enhancing resource efficiency and reducing waste (see Figure 16.1). Companies that adopt bio-based materials often report significant environmental benefits, particularly in sustainable production practices. This discussion focuses on how bioplastics and other bio-based materials contribute to these goals (Verma et al., 2024a). Bio-based materials like bioplastics result from renewable resources like plants, agricultural residues, and waste biomass. This change from fossil fuels toward renewable feedstocks reduces dependency on finite resources and promotes a more sustainable supply chain (Ladu & Morone, 2024). For instance, bioplastics can be produced from corn or sugarcane, which sequesters

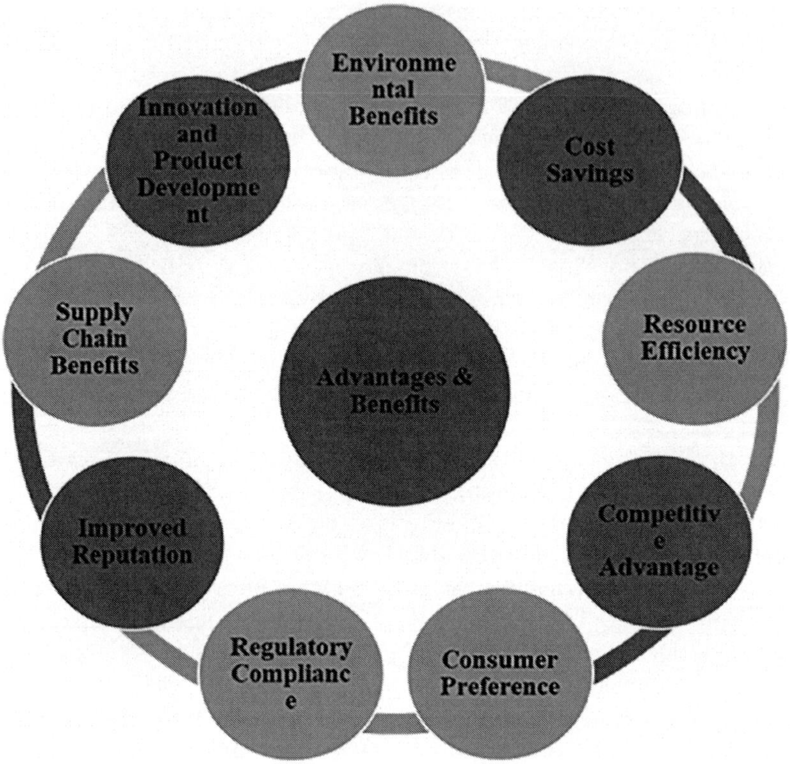


**FIGURE 16.1** Advantages and benefits of smart and sustainable manufacturing practices.

carbon dioxide during growth and replenishes annually, unlike petroleum-based plastics that deplete non-renewable resources. The production processes for bio-based materials typically require less energy when paralleled with conventional plastics. For example, bioplastic synthesis often involves lower temperature processes and fewer energy-intensive steps, reducing greenhouse gas emissions. Studies indicate that switching to bioplastics reduces greenhouse gas emissions by up to 25%, which is compared to conventional plastics. Bio-based materials generally have a smaller carbon footprint throughout their life cycle. The carbon dioxide absorbed by the plants used as feedstock offsets the emissions generated during production and disposal. For example, substituting fossil-based polyethene with bio-based polyethene could prevent over 73 million tons of $CO_2$ emissions annually (see Figure 16.1) (Ioannidou et al., 2020).

Numerous bioplastics are fabricated to be compostable or biodegradable, providing a sustainable end-of-life option that traditional plastics lack. For instance, polylactic acid (PLA) can biodegrade in industrial composting facilities, diverting waste from landfills and reducing methane emissions associated with plastic degradation. This characteristic minimizes environmental pollution and contributes to nutrient recycling in soil. The use of bio-based materials supports the circular economy principles by facilitating recycling in addition to composting pathways. Few bioplastics are compatible with prevailing recycling infrastructures, allowing them to be processed alongside traditional plastics without compromising quality. This integration helps reduce overall waste generation and promotes resource recovery. Innovative processes are being developed to convert agricultural and industrial waste into valuable bio-based products. For example, researchers are exploring methods to produce biodegradable plastics from wastewater or solid organic waste using microbial fermentation. This decreases waste but generates economic opportunities in waste management (Okolie et al., 2023).

### 16.4.1 Case Studies Illustrating Impact

#### 16.4.1.1 Coca-Cola's PlantBottle™

Coca-Cola's PlantBottle™ represents a significant advancement in sustainable packaging, being the world's first fully recyclable PET plastic bottle made partially from renewable plant materials. Launched in 2009, the PlantBottle uses up to 30% plant-based materials, primarily from sugarcane, which allows Coca-Cola to reduce its reliance on fossil fuels and lowers its carbon footprint while maintaining the functionality and recyclability of traditional PET bottles. The innovation behind the PlantBottle lies in its unique manufacturing process:

**Biomaterial sourcing:** The PlantBottle is produced using natural sugars extracted from plants, specifically sugarcane. This process converts these sugars into monoethylene glycol (MEG), a key constituent of PET plastic. The leftover component, terephthalic acid (PTA), has traditionally been derived from petroleum sources.

**Partnerships for innovation:** Coca-Cola collaborated with Virent, a company specializing in biofuels and biochemicals, to develop the technology that enables this conversion. This partnership exemplifies how cross-industry collaboration can drive sustainable innovations.

**Environmental impact:** Since its launch, the PlantBottle initiative has reportedly helped avoid more than 315,000 metric tons of $CO_2$ emissions—the equivalent of removing nearly one million vehicles from the road. This reduction highlights the potential of bio-based materials in mitigating climate change impacts associated with conventional plastics (Tullo, 2009).

## 16.5 COST SAVINGS AND COMPETITIVE ADVANTAGE

Integrating bio-based materials into manufacturing practices is increasingly recognized as a viable strategy for achieving cost savings and gaining a competitive edge. By leveraging renewable resources, companies can reduce material costs, minimize waste disposal fees, and fulfil the increasing demands of consumers for sustainable products. This discussion explores how bio-based materials contribute to cost efficiency and competitive advantage in various industries. Bio-based materials, such as bioplastics made from agricultural feedstocks like corn or sugarcane, often have stable prices compared to petroleum-based materials. This price stability arises from the renewable nature of farm products, which are less susceptible to the volatile fluctuations of fossil fuel markets. For example, bioplastics produced from agricultural waste lead to lower raw material costs and provide farmers with additional revenue streams (Berger & Heath, 2007).

The production processes for many bio-based materials typically require less energy than traditional plastics. For instance, the synthesis of polylactic acid (PLA), a common bioplastic, can be achieved using less energy-intensive methods compared to the petrochemical processes used for conventional plastics. This reduction in energy consumption not only lowers production costs but also minimizes greenhouse gas emissions associated with manufacturing. Utilizing bio-based materials can lead to significant reductions in waste disposal fees. Additionally, using agricultural residues or waste in bioplastic production can help minimize overall waste generation by turning potential waste into valuable products. A life cycle assessment (LCA) approach reveals that bio-based materials can lead to long-term cost savings. For example, buildings constructed with bio-based materials have been reported to cost two-thirds less to run than those made with conventional materials due to improved energy efficiency and reduced maintenance costs (Brick & Lai, 2018).

## 16.6 REGULATORY COMPLIANCE AND MARKET DEMAND

The growing focus on sustainability has led to stringent industry regulations, compelling companies to adopt bio-based materials as part of their manufacturing processes. This shift not only aids in regulatory compliance but also aligns with the developing market needs for eco-friendly products. Integrating bio-based materials into manufacturing practices presents significant opportunities for businesses to enhance their competitive advantage while mitigating risks associated with environmental regulations (McCright et al., 2016).

### 16.6.1 Regulatory Acquiescence

Governments and regulatory agencies worldwide are implementing stricter laws to lessen the adverse environmental effects of conventional materials, especially plastics. For instance, the European Union has introduced directives such as the Single-Use Plastics Directive and Waste Framework Directive, which set ambitious targets for waste reduction and recyclability. Companies that adopt bio-based materials are better positioned to comply with these evolving regulations, as many bio-based products are designed to be biodegradable or recyclable. Specific regulations provide exemptions for bio-based materials, particularly those that demonstrate biodegradability or are derived from renewable resources (Park & Lin, 2020). For example, under the REACH regulation, specific naturally occurring substances may be exempt from certain requirements if they meet strict conditions. This regulatory landscape encourages companies to invest in bio-based alternatives, as compliance can reduce administrative burdens and costs associated with traditional material regulations.

Compliance with environmental standards is often validated through certifications and ecolabels that indicate a product's sustainability credentials. These certifications help businesses demonstrate their commitment to sustainability, making it easier for consumers and procurement officers to identify environmentally friendly options. For example, the EU Ecolabel and other multi-issue ecolabels postulate a frame for evaluating the environmental performance of bio-based products, further enhancing their marketability. Conducting LCAs is crucial for understanding bio-based materials' environmental impact during their lifespan—from raw material extraction to end-of-life disposal. Regulatory bodies increasingly require companies to perform LCAs to ensure compliance with sustainability standards. Companies can fulfil regulatory requirements while promoting eco-friendly practices by adopting bio-based materials that demonstrate lower carbon footprints and reduced environmental impacts (van Harreveld et al., 2015).

## 16.7 FUTURE ASPECTS OF BIO-BASED MATERIALS FOR SMART MANUFACTURING

The future of bio-based materials in smart manufacturing is perched for considerable growth, driven by technological developments, increasing regulatory pressures, and rising consumer demand for sustainable products. As industries transition toward greener practices, bio-based materials offer a viable solution to address environmental concerns while enhancing operational efficiency. This chapter explores the future potential of bio-based materials within smart manufacturing, highlighting key trends, innovations, and challenges that will shape their adoption. The advancements of new technologies are crucial for successfully integrating bio-based materials in manufacturing processes. Innovations in biotechnology, such as fermentation and enzymatic processes, enable the production of high-performance bioplastics and composites that can compete with traditional materials. For instance, companies like Genomatica are pioneering technologies that enhance feedstock conversion efficiency, resulting in lower production costs and improved material properties. Moreover, advancements in material science are leading to the creation of bio-based materials with enhanced functionalities. These include biodegradable polymers that maintain structural integrity while offering additional properties such as UV resistance or antimicrobial effects (Gargalo et al., 2020). Such innovations will expand the applicability of bio-based materials across various sectors, including packaging, automotive, and construction. As governments worldwide tighten regulations around sustainability and plastic usage, companies adopting bio-based materials will be better positioned to comply with environmental standards. Initiatives like national policies to reduce plastic waste drive demand for sustainable alternatives. Companies that proactively integrate bio-based materials into their product lines can mitigate risks associated with regulatory changes while enhancing their market appeal.

Furthermore, compliance with emerging sustainability certifications will become increasingly important. Businesses investing in bio-based solutions will likely benefit from favorable regulatory treatment and access to new markets prioritizing eco-friendly products. Consumer preferences shift toward sustainable products, creating a robust market demand for bio-based materials (Tyagi et al., 2023). Research specifies which environmentally conscious consumers are ready to pay a premium for products made from renewable resources. This trend is pronounced in sectors where sustainability is becoming a key purchasing criterion. To capitalize on this demand, manufacturers must focus on transparency and communication regarding the environmental benefits of their bio-based products. Companies that effectively market their commitment to sustainability will likely gain a competitive advantage in an increasingly crowded marketplace. The future of bio-based materials is closely linked to the principles of a circular economy. By utilizing renewable resources and promoting recyclability or biodegradability, bio-based materials contribute to waste reduction and resource efficiency. Companies are exploring innovative approaches to integrate bio-based materials into closed-loop systems where waste is minimized and materials are reused or composted. For example, startups like Biome Bioplastics are developing biodegradable plastics from lignin—a byproduct of the paper industry—demonstrating how waste could be turned into valuable resources (Products, 2000). Such initiatives foster sustainable manufacturing practices that align with circular economy goals.

## 16.8 CHALLENGES AHEAD AND CONCLUSION OF BIO-BASED MATERIALS FOR SMART MANUFACTURING

As the adoption of bio-based materials in smart manufacturing continues to grow, several challenges remain. These include technical, economic, regulatory, and market dynamics that could impact the successful integration of bio-based materials into mainstream manufacturing practices. The availability of high-quality biomass feedstocks is critical for producing bio-based materials (Verma et al., 2025). Fluctuations in agricultural yields due to climate change, pests, or market demand can disrupt supply chains. Ensuring consistent quality across different sources is essential for maintaining product performance. The transportation and storage of bio-based raw materials can be more complex than traditional fossil-based materials. Companies must develop efficient logistics systems to manage the supply chain effectively, which may involve additional costs. Bio-based materials can save long-term costs through reduced waste disposal fees and regulatory compliance. Achieving economies of scale is crucial for making bio-based options more economically viable (Mujtaba et al., 2022).

Companies must balance these investments with short-term profitability pressures. As governments implement stricter environmental regulations, companies must stay informed about changing compliance requirements related to bio-based materials. Navigating this complex regulatory landscape can be challenging, particularly for smaller businesses without dedicated compliance teams. Establishing clear certification standards for bio-based materials is essential for building consumer trust. However, the lack of universally accepted standards can create confusion in the market. Despite growing interest in sustainability, some consumers remain unaware of the benefits of bio-based materials. Effective marketing strategies are needed to educate consumers about the advantages of these products and encourage responsible disposal methods. There may be misconceptions regarding the performance and reliability of bio-based materials compared to traditional options. Addressing these concerns through transparent communication and product testing will be vital for gaining consumer acceptance. While advancements in material science are ongoing, some bio-based alternatives may not yet match the performance characteristics of conventional materials in specific applications. Continued innovation is necessary to enhance their properties and broaden their applicability. Adapting current manufacturing processes to incorporate bio-based materials may require significant changes in equipment or workflows (Halonen et al., 2020). Companies must assess the feasibility and costs associated with these adaptations.

## 16.9 CONCLUSION

The future of bio-based materials in smart manufacturing holds immense promise as industries strive to align with sustainability goals and meet growing consumer demand for eco-friendly products. While supply chain complexities, economic viability, regulatory compliance, consumer awareness, and technological limitations exist, proactive strategies can mitigate these hurdles. Investing in research and development will be crucial for enhancing the performance of bio-based materials and achieving cost competitiveness. Collaboration among stakeholders—including manufacturers, suppliers, policymakers, and consumers—will foster innovation and drive the successful integration of bio-based solutions into mainstream manufacturing practices. As companies increasingly adopt bio-based materials, they will contribute to environmental sustainability and position themselves favorably within a rapidly evolving market landscape. By embracing these innovative solutions, businesses can enhance their competitiveness. In summary, while the path forward presents challenges, the potential benefits of bio-based materials—ranging from reduced carbon footprints to improved brand loyalty—make them an essential component of innovative manufacturing strategies to foster a sustainable future.

## REFERENCES


Asghar, A., Sairash, S., Hussain, N., Baqar, Z., Sumrin, A., & Bilal, M. (2022). Current challenges of biomass refinery and prospects of emerging technologies for sustainable bioproducts and bioeconomy. *Biofuels, Bioproducts and Biorefining*, *16*(6), 1478–1494. https://doi.org/10.1002/bbb.2403

Atiwesh, G., Mikhael, A., Parrish, C. C., Banoub, J., & Le, T.-A. T. (2021). Environmental impact of bioplastic use: a review. *Heliyon*, *7*(9), e07918. https://doi.org/10.1016/j.heliyon.2021.e07918

Babkin, A., Shkarupeta, E., Tashenova, L., Malevskaia-Malevich, E., & Shchegoleva, T. (2023). Framework for assessing the sustainability of ESG performance in industrial cluster ecosystems in a circular economy. *Journal of Open Innovation: Technology, Market, and Complexity*, *9*(2), 100071. https://doi.org/10.1016/j.joitmc.2023.100071

Bachleitner, S., Ata, Ö., & Mattanovich, D. (2023). The potential of CO2-based production cycles in biotechnology to fight the climate crisis. *Nature Communications*, *14*(1), 6978. https://doi.org/10.1038/s41467-023-42790-6

Berger, J., & Heath, C. (2007). Where consumers diverge from others: identity signaling and product domains. *Journal of Consumer Research*, *34*(2), 121–134. https://doi.org/10.1086/519142

Bomtempo, J. V., & Alves, F. C. (2014). Innovation dynamics in the biobased industry. *Chemical and Biological Technologies in Agriculture*, *1*(1), 19. https://doi.org/10.1186/s40538-014-0019-8

Brick, C., & Lai, C. K. (2018). Explicit (but not implicit) environmentalist identity predicts pro-environmental behaviour and policy preferences. *Journal of Environmental Psychology*, *58*, 8–17. https://doi.org/10.1016/j.jenvp.2018.07.003

Camp, J. E., & Greatrex, B. W. (2022). Levoglucosenone: bio-based platform for drug discovery. *Frontiers in Chemistry*, *10*. https://doi.org/10.3389/fchem.2022.902239

Censi, V., Saiano, F., Bongiorno, D., Indelicato, S., Napoli, A., & Piazzese, D. (2022). Bioplastics: a new analytical challenge. *Frontiers in Chemistry*, *10*. https://doi.org/10.3389/fchem.2022.971792

Chauhan, D., Singh, D., Pandey, H., Khan, Z., Srivastava, R., Dhiman, V. K., & Dhiman, V. K. (2023). Impact of transcription factors in plant abiotic stress: a recent advancement for crop improvement. In *Plant Transcription Factors* (pp. 271–286). Elsevier. https://doi.org/10.1016/B978-0-323-90613-5.00005-4

Crowe, S., Cresswell, K., Robertson, A., Huby, G., Avery, A., & Sheikh, A. (2011). The case study approach. *BMC Medical Research Methodology*, *11*(1), 100. https://doi.org/10.1186/1471-2288-11-100

D'Almeida, A. P., & de Albuquerque, T. L. (2024). Innovations in food packaging: from bio-based materials to smart packaging systems. *Processes*, *12*(10), Article 10. https://doi.org/10.3390/pr12102085

Duque-Acevedo, M., Belmonte-Ureña, L. J., Yakovleva, N., & Camacho-Ferre, F. (2020). Analysis of the circular economic production models and their approach in agriculture and agricultural waste biomass management. *International Journal of Environmental Research and Public Health*, *17*(24), Article 24. https://doi.org/10.3390/ijerph17249549

Gargalo, C. L., Udugama, I., Pontius, K., Lopez, P. C., Nielsen, R. F., Hasanzadeh, A., Mansouri, S. S., Bayer, C., Junicke, H., & Gernaey, K. V. (2020). Towards smart biomanufacturing: a perspective on recent developments in industrial measurement and monitoring technologies for bio-based production processes. *Journal of Industrial Microbiology & Biotechnology*, *47*(11), 947–964. https://doi.org/10.1007/s10295-020-02308-1

Halonen, N., Pálvölgyi, P. S., Bassani, A., Fiorentini, C., Nair, R., Spigno, G., & Kordas, K. (2020). Bio-based smart materials for food packaging and sensors—a review. *Frontiers in Materials*, *7*. https://doi.org/10.3389/fmats.2020.00082

Hsiao, C.-J., & Hu, J.-L. (2024). Biomass and circular economy: now and the future. *Biomass*, *4*(3), Article 3. https://doi.org/10.3390/biomass4030040

Ioannidou, S. M., Pateraki, C., Ladakis, D., Papapostolou, H., Tsakona, M., Vlysidis, A., Kookos, I. K., & Koutinas, A. (2020). Sustainable production of bio-based chemicals and polymers via integrated biomass refining and bioprocessing in a circular bioeconomy context. *Bioresource Technology*, *307*, 123093. https://doi.org/10.1016/j.biortech.2020.123093

Kong, Z., Xue, Y., Hao, T., Wu, J., Zhang, Y., Rong, C., Chen, H., Song, L., Li, D., Pan, Y., & Li, Y.-Y. (2022). Carbon-neutral treatment of *N*, *N*-dimethylformamide-containing wastewater by up-flow anaerobic sludge blanket: CO2 reduction and bio-energy cleaner production. *Journal of Cleaner Production*, *380*, 134880. https://doi.org/10.1016/j.jclepro.2022.134880

Kumar, P., & Verma, S. (2022). *Entrepreneurial Opportunities in Bioenergy* (pp. 32–43). https://doi.org/10.4018/978-1-6684-5269-1.ch003

Ladu, L., & Morone, P. (2024). Sustainability assessments of bio-based products: from research to practice (and standards). *Societal Impacts*, *3*, 100041. https://doi.org/10.1016/j.socimp.2024.100041

Lange, L., Connor, K. O., Arason, S., Bundgård-Jørgensen, U., Canalis, A., Carrez, D., Gallagher, J., Gøtke, N., Huyghe, C., Jarry, B., Llorente, P., Marinova, M., Martins, L. O., Mengal, P., Paiano, P., Panoutsou, C., Rodrigues, L., Stengel, D. B., van der Meer, Y., &

Vieira, H. (2021). Developing a sustainable and circular bio-based economy in EU: by partnering across sectors, upscaling and using new knowledge faster, and for the benefit of climate, environment & biodiversity, and people & business. *Frontiers in Bioengineering and Biotechnology*, *8*. https://doi.org/10.3389/fbioe.2020.619066

Leipold, S., & Petit-Boix, A. (2018). The circular economy and the bio-based sector—perspectives of European and German stakeholders. *Journal of Cleaner Production*, *201*, 1125–1137. https://doi.org/10.1016/j.jclepro.2018.08.019

Li, L., Zhang, F., Tu, R., Yu, H., Wang, H., Sun, Y., Jiang, E., & Xu, X. (2022). N,N-Dimethylformamide solvent assisted hydrothermal pretreatment of Chlorella for coproduction of sugar, nitrogenous compounds and carbon dots. *Bioresource Technology*, *344*(Pt A), 126143. https://doi.org/10.1016/j.biortech.2021.126143

McCright, A. M., Dunlap, R. E., & Marquart-Pyatt, S. T. (2016). Political ideology and views about climate change in the European Union. *Environmental Politics*, *25*(2), 338–358. https://doi.org/10.1080/09644016.2015.1090371

Menossi, M., Cisneros, M., Alvarez, V. A., & Casalongué, C. (2021). Current and emerging biodegradable mulch films based on polysaccharide bio-composites. A review. *Agronomy for Sustainable Development*, *41*(4), 53. https://doi.org/10.1007/s13593-021-00685-0

Mujtaba, M., Lipponen, J., Ojanen, M., Puttonen, S., & Vaittinen, H. (2022). Trends and challenges in the development of bio-based barrier coating materials for paper/cardboard food packaging; *a review*. *Science of The Total Environment*, *851*, 158328. https://doi.org/10.1016/j.scitotenv.2022.158328

Okolie, O., Kumar, A., Edwards, C., Lawton, L. A., Oke, A., McDonald, S., Thakur, V. K., & Njuguna, J. (2023). Bio-based sustainable polymers and materials: from processing to biodegradation. *Journal of Composites Science*, *7*(6), Article 6. https://doi.org/10.3390/jcs7060213

Park, H. J., & Lin, L. M. (2020). Exploring attitude–behavior gap in sustainable consumption: comparison of recycled and upcycled fashion products. *Journal of Business Research*, *117*, 623–628. https://doi.org/10.1016/j.jbusres.2018.08.025

Products, N. R. C. (US) C. on B. I. (2000). Processing technologies. In *Biobased Industrial Products: Priorities for Research and Commercialization*. National Academies Press (US). www.ncbi.nlm.nih.gov/books/NBK232954/

Rosenboom, J.-G., Langer, R., & Traverso, G. (2022). Bioplastics for a circular economy. *Nature Reviews Materials*, *7*(2), 117–137. https://doi.org/10.1038/s41578-021-00407-8

Sharma, H., & Verma, S. (2014). FemA gene in Indian isolates of methicillin resistant Staphylococcus aureus—isolation and amplification through real time PCR. *European Journal of Experimental Biology*, *4*(6), 90–94.

Singh, D., Sharma, A., Verma, S. K., Pandey, H., & Pandey, M. (2024a). Impact of nanoparticles on plant physiology, nutrition, and toxicity: a short review. *Next Nanotechnology*, *6*, 100081. https://doi.org/10.1016/j.nxnano.2024.100081

Singh, D., Shyam, P., Verma, S. K., & Anjali. (2024b). Medical applications of systems biology. In S. Joshi, R. R. Ray, M. Nag, & D. Lahiri (Eds.), *Systems Biology Approaches: Prevention, Diagnosis, and Understanding Mechanisms of Complex Diseases* (pp. 55–70). Springer Nature Singapore. https://doi.org/10.1007/978-981-99-9462-5_3

Stanfield, M. K., Terry, R. S., Smith, J. A., & Thickett, S. C. (2023). Levoglucosan and levoglucosenone as bio-based platforms for polymer synthesis. *Polymer Chemistry*, *14*(44), 4949–4956. https://doi.org/10.1039/D3PY01019H

Taylor, D., Chong, K., & Röder, M. (2024). Designing biomass policy: the political economy of renewable energy for net zero. *WIREs Energy and Environment*, *13*(2), e512. https://doi.org/10.1002/wene.512

Tsakalova, M., Onorato, C., Karathanasis, A. Z., & Deligkiozi, I. (2024). A circular economy framework for the assessment of bio-based value chains. *Sustainable Chemistry for the Environment*, *6*, 100099. https://doi.org/10.1016/j.scenv.2024.100099

Tullo, A. (2009). MATERIALS Coca-Cola to introduce biobased content for bottle plastic. *Chemical & Engineering News*, *87*, 9. https://doi.org/10.1021/cen-v087n021.p009a
Tyagi, P., Ahmed, S., & Lai, C. w. (2023). *Future Prospects of Biobased Materials* (pp. 749–764). https://doi.org/10.1016/B978-0-323-91677-6.00013-1
van Harreveld, F., Nohlen, H. U., & Schneider, I. K. (2015). Chapter five—the ABC of ambivalence: affective, behavioral, and cognitive consequences of attitudinal conflict. In J. M. Olson & M. P. Zanna (Eds.), *Advances in Experimental Social Psychology* (Vol. 52, pp. 285–324). Academic Press. https://doi.org/10.1016/bs.aesp.2015.01.002
Verma, S., Anjali, A., Diksha, D., & Kumar, V. (2023). Recent advancements in skin tissue engineering in the application of nanotechnology. *Research Journal of Biotechnology*, *18*, 127–136. https://doi.org/10.25303/1802rjbt1270136
Verma, S., & Kumar, P. (2022). *Biofuel Policies in India: An Assessment of Policy Barriers* (pp. 44–64). https://doi.org/10.4018/978-1-6684-5269-1.ch004
Verma, S. K., Kumar, P., Mishra, A., Khare, R., & Singh, D. (2024a). Green nanotechnology: illuminating the effects of bio-based nanoparticles on plant physiology. *Biotechnology for Sustainable Materials*, *1*(1), 1. https://doi.org/10.1186/s44316-024-00001-2
Verma, S. K., Singh, D., & Misra, S. (2024b). Innovative approaches to sustainable waste management and energy generation in developing nations. In R. Mehra, P. Kumar, & B. Hung (Eds.), *Practice, Progress, and Proficiency in Sustainability* (pp. 267–300). IGI Global. https://doi.org/10.4018/979-8-3693-4054-7.ch010
Verma, S. K., Singh, D., Sharan, A., Khare, R., Rai, S., Kumar, P., & Singh, V. (2025). Innovations in biomass-derived materials for efficient $CO_2$ sequestration. *Chemistry Africa*. https://doi.org/10.1007/s42250-024-01179-8
Vigneswari, S., Kee, S. H., Hazwan, M. H., Ganeson, K., Tamilselvan, K., Bhubalan, K., Amirul, A.-A., & Ramakrishna, S. (2024). Turning agricultural waste streams into biodegradable plastic: a step forward into adopting sustainable carbon neutrality. *Journal of Environmental Chemical Engineering*, *12*(2), 112135. https://doi.org/10.1016/j.jece.2024.112135
Yu, S., Sharma, R., Morose, G., & Nagarajan, R. (2021). Identifying sustainable alternatives to dimethyl formamide for coating applications using Hansen Solubility Parameters. *Journal of Cleaner Production*, *322*, 129011. https://doi.org/10.1016/j.jclepro.2021.129011

# 17 Challenges of Predictive Analytics Adoption in Supply Chain Management

## *A Case from the Healthcare Industry*

*Bhavna Pandey, Manoj Kumar Khurana, and Rohit Rampal*

## 17.1 INTRODUCTION

Over the past few decades, there has been a huge rise in available data. Big Data emerged as a result of advancements in worldwide communication, internet use, sensor technology, and social media platforms (Brandtner et al., 2021). Big data is a substantial contributor to data accumulation due to its complex nature, making it difficult to manage through traditional information programming. It presents issues in data storage, analysis, sharing, and security. Big data serves as a repository of information that makes predictive analytics possible, which is essential in the competitive and evolving world around (Galli & Ocampo, 2019). It is often used for predictive analytics, which uses statistical expertise to forecast events for the future based on historical data. Data scientists frequently employ the following data prediction techniques: neural networks, Bayesian statistics, decision trees, regression modeling, support vector machines, etc. To prepare data for additional analysis using the proper prediction methodologies, the raw data are first obtained from many sources and then cleaned, integrated, and other procedures are carried out (Dubey et al., 2019). Figure 17.1 displays a detailed taxonomy for predictive analytics (Nugroho, 2023). Taxonomy organizes and classifies predictive analytics methodologies, giving a base for their application in big data contexts.

Today's business conditions have become more uncertain. Leveraging the exponential rise of data volume to apply PA and develop reliable forecasting models is appealing for many sectors. One application domain is the global supply chain, where interactions among multiple stakeholders, including manufacturers, distributors, and retailers, produce vast volumes of data that have enormous potential for PA use (Birkel et al., 2020). The foundation of our economy is the supply chain

DOI: 10.1201/9781003566205-17

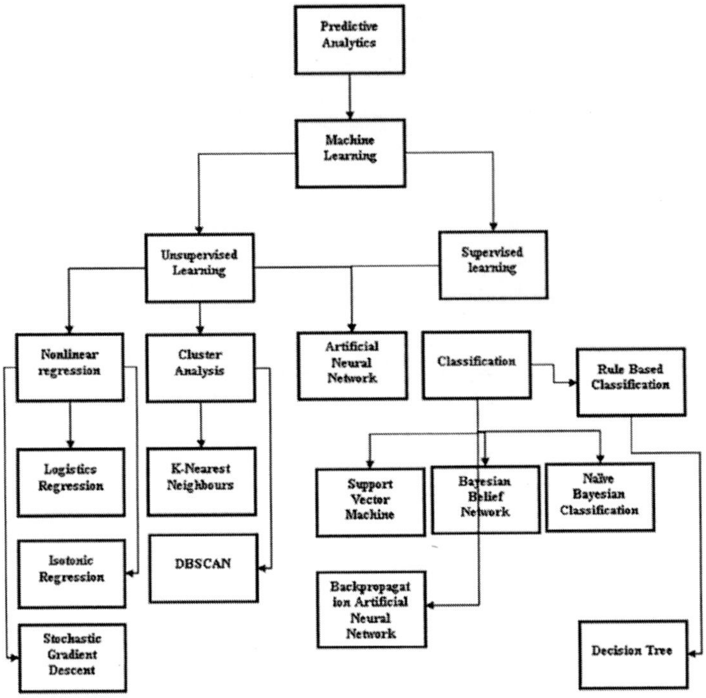


**FIGURE 17.1** Taxonomy of PA.

management (SCM). Supply chains (SC) are the preferred method of value creation for producing and distributing goods and services, as well as for providing value to consumers, regardless of sector or industry. SCM must use new strategies to handle and manage complicated SC networks because of the many challenges and high levels of uncertainty. Simultaneously, more and more data are created and shared within the SC and among the SCM tasks and activities (Darbanian et al., 2024). SC face numerous disruptions, high turnover rates, and shifting market challenges, which make them particularly vulnerable to future uncertainties. Reliable PA models might assist in managing these dynamics and building robust SC networks. Some research has examined PA in SCM, but it is at a generic SCM level, and there are no insights about the analysis of challenges of PA adoption in SCM. Additionally, there is very little discussion in the literature regarding the implementation procedure and specific challenges associated with implementing PA in the field of SCM. Thus, the goal of this research is to examine the framework and challenges that companies face when implementing PA. The objectives of this study are as follows:

RO1: to figure out the challenges of implementing PA tools in SCM.
RO2: to evaluate the rank and interrelationships among the challenges using the hybrid F-AHP and DEMATEL methods.

The structure of this chapter is as follows: an overview of the recent research in these areas is provided in Chapter 2. The methodology used to carry out the study is presented in Chapter 3, and the outcomes of the study and analysis are presented in Chapter 4. Chapter 5 concludes by outlining the research study's broad results, limitations, and implications of the findings for practitioners as well as recommendations for additional research.

## 17.2 LITERATURE REVIEW

The study contains a thorough literature review to analyze the PA and its challenges resulting from the adoption in order to achieve the research objectives. Several electronic databases, including SCOPUS, Web of Science, and Google Scholar, were also searched for articles published by major publishers. A pair of search keywords including "barrier" OR "challenge" OR "critical success factor" AND "data analytics" OR "predictive analytics" AND "supply chain management" OR "SCM" were used. This section presents an overview about PA in SCM.

Several studies have examined various aspects of implementing PA in businesses. Kache and Seuring (2017) presented the first insights in the field of SCM and big data analytics research. They conducted a Delphi study with 20 international experts and identified four primary challenges when introducing big data analytics: information and cybersecurity, IT capabilities and infrastructure, integration and collaboration, and governance and compliance. Schoenherr and Speier-Pero (2015) surveyed logistics, purchasing, and operations management professionals and identified the following primary issues when introducing PA: untrained and inexperienced individuals, time constraints, and integration with existing systems. However, Hazen et al. (2014) highlighted the significance of data quality in SCM operations. With an emphasis on the interactions among technology, human behavior, and organizational context that take place during the post-adoption phases of technology in retail supply chain (RSC) organizations, Sodero et al. (2019) aimed to investigate the social process of big data and predictive analytics use for logistics and SCM. Galli and Ocampo (2019) examined the relationship between PA tools and continuous improvement. Nugroho (2023) performed an extensive literature study to address challenges related to data quality in predictive analytics. The study by Birkel et al. (2020), which used a multiple case technique, provided detailed insights into six real transport logistics cases. The findings reveal challenges that are both internal and external. From an internal standpoint, the technical implementation presents the biggest obstacles. External challenges include relational and external circumstances that can hinder implementation success. A study by Bag et al. (2022) found that firms using big data and predictive analytics are more likely to share information about the circular economy with stakeholders and have a strong relationship with them, leading to increased trust and engagement. Table 17.1 shows the detailed list of finalized challenges of PA in SCM.

## 17.3 RESEARCH METHODOLOGY

### 17.3.1 F-AHP Method

Analytic hierarchy process (AHP), developed by Thomas L. Satty in 1980, is a quantitative technique to deal with complex MCDM problems. It systematically breaks down the complex problem into a hierarchical structure. In this hierarchy, the goal is

**TABLE 17.1**
**List of Challenges of PA Implementation in SCM.**

| Challenges | Description | References |
|---|---|---|
| Data integration complexity (CPA1) | Diversity in data from multiple sources might make data integration more challenging | Moktadir et al. (2019) |
| Low quality of information sharing (CPA2) | This hinders the ability of SC to operate effectively, which sometimes results in misleading conclusions and ineffective decision-making | Bag et al. (2022); Nugroho (2023); Hazen et al. (2014) |
| Data scalability (CPA3) | Scalability becomes a challenge when data volumes increase, affecting operational speed and performance | Alharthi et al. (2017) |
| Ineffective handling of data produced from multiple sources (CPA4) | Both organized and unstructured data are present in the data produced from many sources, ensuring data quality is a key challenge | Schoenherr and Speier-Pero (2015); Bag et al. (2022) |
| Lack of policies regarding privacy, data security, and intellectual property (CPA5) | Implementing PA in SCM can provide considerable challenges in terms of protecting sensitive data and ensuring compliance with data governance policies and regulations | Omar et al. (2019); Tambuskar et al. (2024) |
| Resilient infrastructure for storing massive amounts of supply chain data (CPA6) | To handle massive volumes of data without crashing or slowing down, and to scale up or down as needed, organizations require a resilient infrastructure | Dehkhodaei et al. (2023) |
| Change management (CPA7) | Any major system upgrade faces challenges in gaining support from stakeholders and overcoming organizational resistance to change | Brandtner et al. (2021) |
| High cost of available solutions and uncertainty about ROI (CPA8) | Sophisticated software tools are needed to implement these analytics, and maintaining and acquiring these tools can be expensive. This might pose a serious challenge for smaller companies | Moktadir et al. (2019) |
| Lack of skilled data professionals (CPA9) | The current workforce lacks appropriate training and education in PDA, a lack of development and training leads to poor performance among data analysts | Brandtner et al. (2021) |
| Time constraints (CPA10) | The process of developing, testing, and customizing predictive analytics initiatives takes time and involves multiple stages | Birkel et al. (2020); Schoenherr and Speier-Pero (2015) |

**TABLE 17.2**
**Linguistic Scale for F-AHP.**

| Relative importance | Linguistic variables | TFN scale |
|---|---|---|
| 1 | Equal importance | (1,1,1) |
| 2 | Moderately equal importance | (1,2,3) |
| 3 | Moderately importance | (1,3,5) |
| 4 | Moderately strongly importance | (2,4,6) |
| 5 | Strongly importance | (3,5,7) |
| 6 | Strongly very importance | (4,6,8) |
| 7 | Very strongly importance | (5,7,9) |
| 8 | Very strongly extremely importance | (6,8,9) |
| 9 | Extremely importance | (7,9,9) |

at the top, criteria and sub-criteria at the levels and sub-levels, while alternatives are at the bottom. AHP is regarded as the most potent and successful decision-making approach for resolving complicated issues having several conflicting or clashing criteria. AHP is easily understandable, flexible, and simple in application, which makes this technique more prevalent to apply. But vagueness or uncertainty in the decision-making of individuals brings limitations to Satty's AHP. To deal with such problems, AHP was integrated with the fuzzy set theory. It enables the decision maker's judgment to account for uncertainties and vagueness. The following steps summarize the F-AHP method based on Goyal et al. (2021).

Step 1: Construct a pairwise comparison matrix (Ć) by taking the inputs from the decision-makers using the linguistic scale given in Table 17.2. If there are "m" criteria, then there will be a (m × m) matrix.

$$\acute{C} = \begin{bmatrix} \acute{a}_{11}^{k} & \acute{a}_{12}^{k} & \cdots & \acute{a}_{1m}^{k} \\ \acute{a}_{21}^{k} & \cdots & \cdots & \acute{a}_{2m}^{k} \\ \cdots & \cdots & \cdots & \cdots \\ \acute{a}_{m1}^{k} & \acute{a}_{m1}^{k} & \cdots & \acute{a}_{mm}^{k} \end{bmatrix}$$

Here, $\acute{a}_{ij}^{k}$ shows the preference for i$^{th}$ criteria over j$^{th}$ criteria by k$^{th}$ decision maker.

Step 2: If there are many decision-makers, their preferences can be aggregated using the weighted geometric mean of their judgments, which can be determined as shown below.

$$\acute{a}_{i,j} = \prod_{k=1}^{K} \left( \acute{a}_{ij}^{k} \right)^{\omega_k} \tag{17.1}$$

where $\omega_k$ is the weight given to k[th] decision maker and $\sum_{k=1}^{K}\omega_k = 1$.

Step 3: The pairwise comparison matrix is updated in accordance with the total preferences.

$$\acute{C} = \begin{bmatrix} \acute{a}_{11} & \acute{a}_{11} & \cdots & \acute{a}_{1m} \\ \acute{a}_{21} & \cdots & \cdots & \acute{a}_{2m} \\ \cdots & \cdots & \cdots & \cdots \\ \acute{a}_{m1} & \acute{a}_{m2} & \cdots & \acute{a}_{mm} \end{bmatrix}$$

Step 4: The given equation determines the geometric mean of the fuzzy comparison values for each criterion.

$$v_i = \left(\prod_{j=1}^{m}\acute{a}_{i.j}\right)^{1/m} \tag{17.2}$$

Step 5: The given equation determines the fuzzy weights.

$$F_i = v_i \times \left(\sum\nolimits_{j=1}^{m} v_i\right)^{-1} \tag{17.3}$$

Step 6: Obtained weights are fuzzy sets. These have to be defuzzified using the center of area method.

$$F_i = \frac{\alpha_i + \beta_i + \gamma_i}{3} \tag{17.4}$$

Step 7: As the sum of all these defuzzified weights is not equal to 1. Hence, it needs to be normalized to get the final weight.

$$w_i = \frac{\mathrm{F}_i}{\sum\nolimits_{j=1}^{m}\mathrm{F}_j} \tag{17.5}$$

By using the above steps, the weights of each challenge are computed.

### 17.3.2 DEMATEL Method

The DEMATEL approach has gained popularity for its ability to visually represent complex causal relationships. This approach uses digraphs to organize the involved factors into cause-and-effect groups. The digraph might illustrate a fundamental idea of the contextual relationship between system components, where values stand for the degree of influence. The relationship between cause-and-effect elements can, therefore, be transformed into a comprehensible structural model of the system using the DEMATEL. The DEMATEL can suggest the most crucial criterion that influences other criteria. The following summarizes the DEMATEL method following Tzeng et al. (2007).

**TABLE 17.3**
**Linguistic Scale for DEMATEL.**

| Linguistic variables | Score |
|---|---|
| Degree of influence | 4 |
| Very high influence (VH) | 3 |
| High influence (H) | 2 |
| Low influence (L) | 1 |
| Very low influence (VL) | 0 |

Step 1: Formulate the direct-relationship matrix.
Using the scale provided in Table 17.3, the decision maker first evaluates the correlation between the attribute groups, considering the influence of each criterion.

$$A = \begin{bmatrix} 0 & \acute{a}_{11} & \dots & \acute{a}_{1j} & \acute{a}_{1n} \\ \acute{a}_{21} & 0 & \dots & \acute{a}_{2j} & \acute{a}_{2n} \\ \dots & \dots & \dots\dots & & \dots \\ \acute{a}_{n1} & \acute{a}_{n2} & \dots\dots & & 0 \end{bmatrix} \tag{17.6}$$

Step 2: Normalization of direct relation matrix.
The above-formulated matrix needs to be normalized using the following formulas based on the pairwise comparison.

$$Y = \lambda.\ A \tag{17.7}$$

where $\lambda = \frac{1}{\max_{1 \le i \le n}\left(\sum_{j=1}^{n} \acute{a}_{ij}\right)}$ (i, j = 1, 2, . . . , n)

Step 3: Compute the total–relation matrix.
Here, the estimation is done using Equation (17.8) and the identity matrix.

$$T = Y(I - Y)^{-1} \tag{17.8}$$

where $I$ is an identity matrix.

Step 4: Calculate the sum of rows and columns in the total relation matrix.
Equation (17.9) is used to estimate the total relation sum of the rows and columns.

$$R_i = \left[\sum_{j=1}^{n} t_{ij}\right]_{n\times 1} = [t_i]_{n\times 1} \quad i = 1, 2, \dots\dots n \tag{17.9}$$

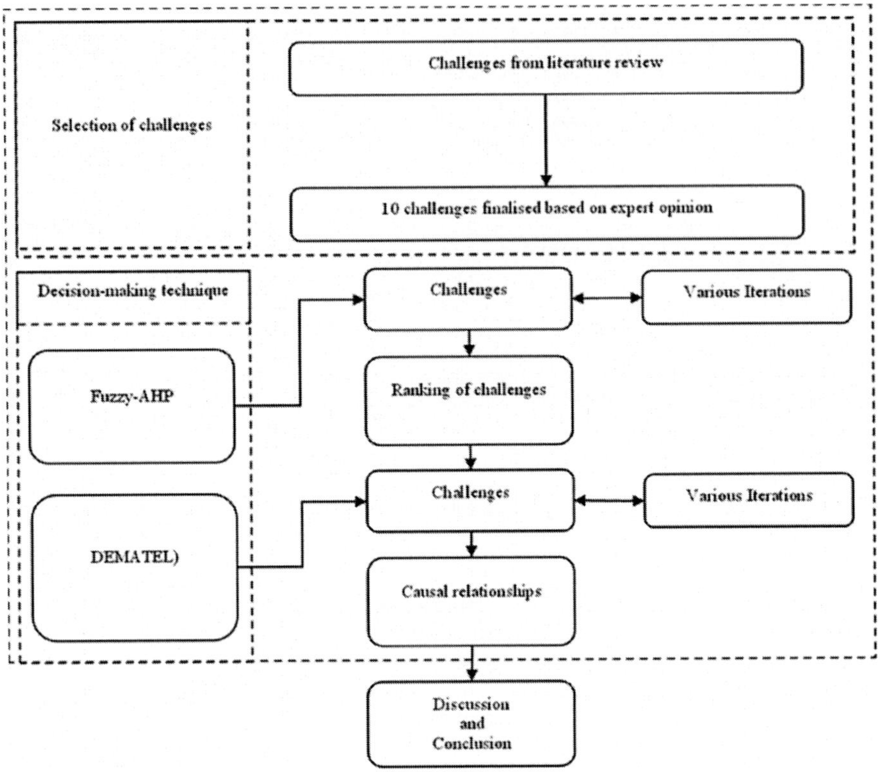


**FIGURE 17.2** Research methodology.

$$C_j = \left[ \sum_{i=1}^{n} t_{ij} \right]_{1 \times n} = \left[ t_j \right]_{n \times 1} \quad j = 1, 2, \ldots\ldots n$$

Step 5: Set up a threshold value to obtain the digraph.

By calculating the average of the elements in matrix T, the threshold value is established. The dataset of (R + C, R – C) can be mapped to obtain the digraph. Figure 17.2 shows the proposed research methodology.

## 17.4 ANALYSIS OF RESULTS AND DISCUSSION

### 17.4.1 Case Study

To demonstrate the practical relevance of the proposed framework, an empirical case study was conducted at a healthcare company that specializes in generic drugs, clinical research, and manufacturing. ABC (anonymized for confidentiality) has played a significant role in transforming healthcare services across the country for over a decade. ABC experienced a rise in supply chain interruptions following the COVID-19 pandemic. It became essential to maintain optimal levels of inventory in order to reduce such risks and deliver goods to customers on schedule. So ABC decided to combine onto a single global chart of accounts using data analytical tools in order to address the challenges and create an agile, data-driven supply chain at the foundational level. It is certain that its business services will be accessible anytime, anywhere because it is currently advancing its operations on a strong IT infrastructure. The business is also cutting its overall infrastructure expenses and the duration of scheduled maintenance downtime.

The case organization, thus, decided to implement PA tools in its SC for a smooth transformation by conducting a study intended to rank and interrelate different challenges for its successful implementation. The proposed F-AHP and DEMATEL framework would help practitioners in the case organization to handle the PA implementation challenges more effectively.

### 17.4.2 Discussion of Results

Developing countries face challenges in incorporating PA into their industrial businesses and SCM. Industry managers in emerging nations aim to identify and address these challenges. Countries like India are also facing significant obstacles in implementing PA tools in their manufacturing sector. This study uses an integrated fuzzy AHP and DEMATEL method to identify and prioritize the challenges concerning PA in SCM based on the current production environment in India. The ten major barriers are identified and included in this study based on a literature review. The fuzzy AHP approach is utilized to calculate the weights and ranks of the predictive analytics challenges in SCM. The F-AHP pair comparison questionnaires were prepared and consulted by multiple experts from industry and academia. Additionally, the experts were requested to use Table 17.2 to compare the challenges pairwise. Table 17.4 shows the pair comparison table of challenges. Using Table 17.2 and the F-AHP steps presented in Equations (17.1)–(17.5), the weights and ranks of the challenges were calculated using Excel software. Table 17.5 shows the weights and ranks of the challenges.

The results of F-AHP show that low quality of information sharing (CPA2) is the most significant challenge of PA adoption in SCM. The degree of success attained by an organization is strongly correlated with how well it uses its data to extract pertinent knowledge and information sharing to guide decision-making. High-quality data are required to facilitate data analysis. The correct analytical tools combined with high-quality data will yield significant insights and support rationale decision-making (Nugroho, 2023). Frequent data review and cleaning practices are essential for ensuring excellent data quality. The challenge of ineffective handling of data produced from multiple sources (CPA4) was identified as the second major challenge. Businesses struggle to effectively aggregate and synchronize data due to the growth of data produced by multiple IoT devices, supplier networks, and third-party logistics. Data integration complexity (CPA1) ranked third, and data scalability (CPA3) ranked fourth. Complex integration procedures may lead to more operational expenses and

**TABLE 17.4**
**Comparison Matrix for Challenges Based on F-AHP Scale.**

| Challenges | CPA1 | CPA2 | CPA3 | CPA4 | CPA5 | CPA6 | CPA7 | CPA8 | CPA9 | CPA10 |
|---|---|---|---|---|---|---|---|---|---|---|
| CPA1 | 1 | 1/3 | 2 | 1/2 | 2 | 6 | 8 | 4 | 5 | 8 |
| CPA2 | 3 | 1 | 3 | 2 | 8 | 9 | 9 | 4 | 8 | 9 |
| CPA3 | 1/2 | 1/3 | 1 | 1/2 | 3 | 4 | 6 | 3 | 7 | 7 |
| CPA4 | 2 | 1/2 | 2 | 1 | 2 | 8 | 6 | 7 | 8 | 9 |
| CPA5 | 1/2 | 1/8 | 1/3 | 1/2 | 1 | 2 | 2 | 2 | 8 | 4 |
| CPA6 | 1/6 | 1/9 | 1/4 | 1/8 | 1/2 | 1 | 1/2 | 1/2 | 4 | 3 |
| CPA7 | 1/8 | 1/9 | 1/6 | 1/6 | 1/2 | 2 | 1 | 1/2 | 2 | 7 |
| CPA8 | 1/4 | 1/4 | 1/3 | 1/7 | 1/2 | 2 | 2 | 1 | 7 | 5 |
| CPA9 | 1/5 | 1/8 | 1/7 | 1/8 | 1/8 | 1/4 | 1/2 | 1/7 | 1 | 2 |
| CPA10 | 1/8 | 1/9 | 1/7 | 1/9 | 1/4 | 1/3 | 1/7 | 1/5 | 1/2 | 1 |

**TABLE 17.5**
**Results of F-AHP Analysis.**

| Challenges | Weights | Rank |
|---|---|---|
| CPA1 | 0.15 | 3 |
| CPA2 | 0.29 | 1 |
| CPA3 | 0.13 | 4 |
| CPA4 | 0.20 | 2 |
| CPA5 | 0.07 | 5 |
| CPA6 | 0.03 | 8 |
| CPA7 | 0.04 | 7 |
| CPA8 | 0.05 | 6 |
| CPA9 | 0.02 | 9 |
| CPA10 | 0.01 | 10 |

redundant data management efforts. To enhance overall data quality and facilitate the scalability of supply chain operations, strengthening the efficiency of data integration is essential. Lack of policies regarding privacy, data security, and intellectual property (CPA5) ranked fifth. Reliable data governance frameworks are required to safeguard sensitive information, enforce legal requirements, and foster confidence among supply chain partners. The high cost of available solutions and uncertainty about ROI (CPA6) ranked sixth. It can be challenging to measure the expenses and return on investment (ROI) of predictive analytics initiatives, especially when the benefits are indirect or long-term. Without precise measures, it can be hard to justify the investment in analytics initiatives. It should have well-defined goals and key performance indicators, which can be regularly monitored and reported on to show the benefit of analytics efforts. Change management (CPA7) ranked seventh. Resilient infrastructure for storing massive amounts of supply chain data (CPA6) ranked eighth. Organizations can organize vast amounts of data, adopt analytics, make access easier, and guarantee data integrity with the help of an efficient data infrastructure, which eventually produces information and knowledge that is useful for making decisions. It also includes the policies, regulations, and standards to guarantee the accuracy and usability of the data. Lack of skilled data professionals (CPA9) ranked ninth. The lack of qualified data experts is acknowledged as a challenge to the adoption of data-driven solutions in SCM. Time constraints (CPA10) ranked tenth.

DEMATEL is used to determine the inter-relationships between the challenges. The direct-influence and total-influence matrices were first obtained. The direct-influence matrix is shown in Table 17.6, and the total-influence matrix is shown in Table 17.7. Using the scale in Table 17.3 and the DEMATEL steps presented in Equations (17.6)–(17.9), the analysis was conducted. Table 17.8 shows the final results of the DEMATEL analysis for the challenges of PA adoption in SCM.

**TABLE 17.6**
**Initial Direct Relationship Matrix.**

| Challenges | CPA1 | CPA2 | CPA3 | CPA4 | CPA5 | CPA6 | CPA7 | CPA8 | CPA9 | CPA10 |
|---|---|---|---|---|---|---|---|---|---|---|
| CPA1 | 0 | 3 | 2 | 3 | 2 | 1 | 2 | 2 | 1 | 4 |
| CPA2 | 2 | 0 | 2 | 2 | 3 | 1 | 2 | 2 | 1 | 3 |
| CPA3 | 1 | 2 | 0 | 3 | 1 | 2 | 1 | 2 | 1 | 2 |
| CPA4 | 2 | 2 | 3 | 0 | 2 | 1 | 1 | 1 | 2 | 2 |
| CPA5 | 1 | 3 | 2 | 4 | 0 | 3 | 2 | 2 | 3 | 1 |
| CPA6 | 2 | 2 | 3 | 2 | 3 | 0 | 1 | 2 | 3 | 1 |
| CPA7 | 2 | 1 | 1 | 3 | 4 | 2 | 0 | 3 | 2 | 1 |
| CPA8 | 1 | 2 | 2 | 2 | 2 | 2 | 1 | 0 | 3 | 2 |
| CPA9 | 2 | 2 | 1 | 4 | 2 | 3 | 2 | 4 | 0 | 2 |
| CPA10 | 1 | 2 | 1 | 2 | 4 | 2 | 1 | 2 | 1 | 0 |

**TABLE 17.7**
**Total Relationship Matrix.**

| Challenges | CPA1 | CPA2 | CPA3 | CPA4 | CPA5 | CPA6 | CPA7 | CPA8 | CPA9 | CPA10 |
|---|---|---|---|---|---|---|---|---|---|---|
| CPA1 | 0.1114 | 0.2766 | 0.2402 | 0.3048 | 0.257 | 0.1635 | 0.1859 | 0.2144 | –0.096 | 0.0706 |
| CPA2 | 0.1842 | 0.1418 | 0.2257 | 0.2559 | 0.2732 | 0.1566 | 0.1784 | 0.2054 | –0.059 | 0.0007 |
| CPA3 | 0.1285 | 0.1989 | 0.1229 | 0.2558 | 0.1725 | 0.1707 | 0.1179 | 0.1786 | –0.083 | 0.033 |
| CPA4 | 0.1943 | 0.2356 | 0.2719 | 0.1897 | 0.2417 | 0.168 | 0.1478 | 0.1839 | 0.045 | 0.0047 |
| CPA5 | 0.1762 | 0.2969 | 0.2671 | 0.3701 | 0.1957 | 0.2606 | 0.1997 | 0.2366 | –0.053 | 0.0267 |
| CPA6 | 0.1965 | 0.2398 | 0.2807 | 0.2767 | 0.2783 | 0.1254 | 0.1497 | 0.2221 | 0.0133 | –0.067 |
| CPA7 | 0.1848 | 0.1855 | 0.192 | 0.2933 | 0.3008 | 0.1938 | 0.0937 | 0.2403 | –0.046 | –0.108 |
| CPA8 | 0.1483 | 0.2246 | 0.2269 | 0.257 | 0.234 | 0.1987 | 0.1391 | 0.1278 | 0.019 | –0.002 |
| CPA9 | 0.2078 | 0.2523 | 0.2235 | 0.3585 | 0.2678 | 0.2524 | 0.1935 | 0.3033 | –0.165 | –0.007 |
| CPA10 | 0.1295 | 0.2031 | 0.1696 | 0.2282 | 0.2808 | 0.1775 | 0.1243 | 0.1829 | –0.048 | –0.155 |

**TABLE 17.8**
**Results of DEMATEL Analysis.**

| Challenges | Ri | Ci | Ri+Ci | Ri – Ci | Identity |
|---|---|---|---|---|---|
| CPA1 | 1.73 | 1.66 | 3.39 | 0.07 | Cause |
| CPA2 | 1.56 | 2.26 | 3.82 | –0.69 | Effect |
| CPA3 | 1.30 | 2.22 | 3.52 | –0.92 | Effect |
| CPA4 | 1.68 | 2.79 | 4.47 | –1.11 | Effect |
| CPA5 | 1.98 | 2.50 | 4.48 | –0.52 | Effect |
| CPA6 | 1.72 | 1.87 | 3.58 | –0.15 | Effect |
| CPA7 | 1.53 | 1.53 | 3.06 | 0.00 | Effect |
| CPA8 | 1.57 | 2.10 | 3.67 | –0.52 | Effect |
| CPA9 | 1.89 | -0.47 | 1.41 | 2.36 | Cause |
| CPA10 | 1.29 | -0.20 | 1.09 | 1.50 | Cause |

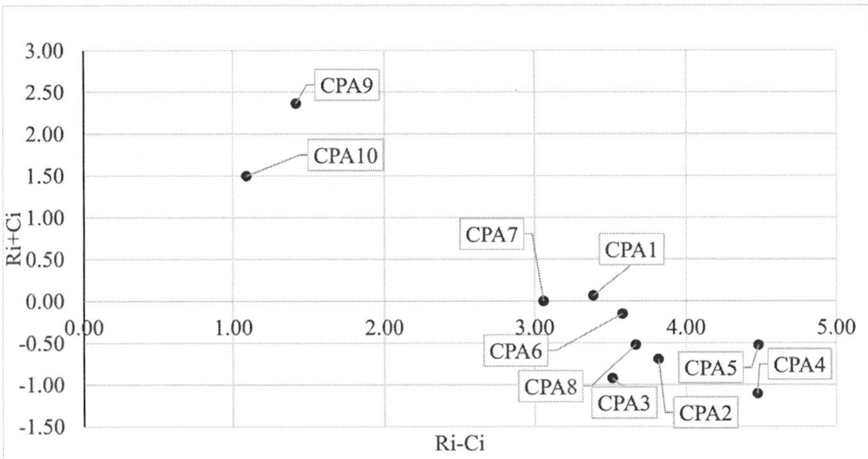


**FIGURE 17.3** The causal diagram for challenges.

A comprehensive causal relationship map has been created based on the Ri + Ci and Ri – Ci data from Table 17.8, and shown in Figure 17.3. The Ri value indicates the impact on other challenges, while the Ci component indicates the effect from other challenges. Ri + Ci indicates the degree of relationship among challenges, while Ri – Ci reflects the degree of effect. Ri – Ci determines whether a challenge is a cause or effect. A positive value indicates a causal factor, whereas a negative number indicates an effect. Based on these findings, we may divide the challenges into causal and effect groups, allowing for a more systematic approach to resolving every issue. Among the most significant causal challenges are data integration complexity (CPA1), lack of skilled data professionals (CPA9), and time constraints (CPA10).

Improving workforce skills through focused training programs might reduce skill gaps, improving management of other dependent elements such as data integration complexity and scalability. Similarly, addressing time restrictions by allocating personnel and prioritizing tasks based on impact can eliminate bottlenecks and allow for more proactive data management methods. Low quality of information sharing (CPA2), data scalability (CPA3), ineffective handling of data produced from multiple sources (CPA4), lack of policies regarding privacy, data security, and intellectual property (CPA5), resilient infrastructure for storing massive amounts of supply chain data (CPA6), change management (CPA7), and high cost of available solutions and uncertainty about ROI (CPA8) are in the effect group.

## 17.5 CONCLUSIONS

The current study aimed to investigate and analyze the key challenges for PA adoption in SCM by utilizing fuzzy analytical hierarchical process (F-AHP) and decision-making trial and evaluation laboratory (DEMATEL). After identifying the key challenges through a comprehensive literature review, this study employed F-AHP to rank them and DEMATEL to analyze them. The findings led to the proposal of a framework for analyzing PA adoption. These studies highlight challenges in adopting and implementing Industry 4.0 and offer broad approaches to address them. The study's outcomes identified five critical challenges using F-AHP: low quality of information sharing (CPA2), ineffective handling of data produced from multiple sources (CPA4), data integration complexity (CPA1), data scalability (CPA3), and lack of policies regarding privacy, data security, and intellectual property (CPA5).

The result emphasizes the need for practices that include frequent data validations and audits to maintain the highest standards for data quality. Effective data governance practices protect against misleading and insecure data while ensuring consistency in quality. Enterprise data come from a wide variety of sources. It is understandable why extracting, converting, and making data suitable for a destination system takes time and resources since each source has its own access rules. Therefore, to strengthen an ethical and successful predictive analytics operation in SCM, organizations must create clear data governance rules, enforce data quality protocols, and carry out frequent audits to guarantee their compliance with internal and external legislation. This strategy includes classifying data, establishing responsibilities for ownership and stewardship, and putting access controls in place to limit who has the ability to see and alter particular datasets. It also entails creating precise standards for data quality, managing metadata, and putting policies in place for data destruction and retention.

The DEMATEL analysis of challenges in PA adoption within SCM highlights critical insights into the interdependencies among these challenges. By calculating direct and total influences, and through the use of causal–effect relationships, the analysis identifies that three challenges act as causes, while others are influenced downstream. The three key causal challenges include data integration complexity (CPA1), lack of skilled data professionals (CPA9), and time constraints (CPA10). Addressing these causes can mitigate cascading effects on other challenges, as improvements in workforce skills, streamlined data integration, and optimized time management create a foundation for successful PA adoption. Conversely, challenges, such as low quality of information sharing (CPA2), data scalability (CPA3), and ineffective handling of data produced from multiple sources (CPA4), fall in the effect category. These effects can be alleviated through interventions targeting causal challenges. By prioritizing resources to improve training, set clear integration protocols, and manage time effectively, organizations can systematically enhance data quality, scalability, and infrastructure resilience. This structured approach not only facilitates the resolution of interconnected challenges but also supports a sustainable, efficient PA ecosystem in SCM. Ultimately, targeting root causes in a structured sequence offers a strategic roadmap to overcoming barriers in SCM PA adoption.

The study and its findings also have some limitations that should be addressed. The major limitation was that the study included only ten significant challenges and four experts for decision-making. The study's findings and conclusions may be biased due to limited challenges and expertise. To mitigate this, greater numbers of respondents should be employed. The study's focus on major aspects of the healthcare industry limits its results and analysis to specific scenarios. These conclusions may not apply to other aspects of management, such as operations or strategy. Future research could explore the relationship between these challenges in different industries and managerial scenarios.

## REFERENCES


Alharthi, A., Krotov, V., & Bowman, M. (2017). Addressing barriers to big data. *Business Horizons*, *60*(3), 285–292.

Bag, S., Gupta, S., & Wood, L. (2022). Big data analytics in sustainable humanitarian supply chain: barriers and their interactions. *Annals of Operations Research*, *319*(1), 721–760.

Birkel, H., Kopyto, M., & Lutz, C. (2020, June). Challenges of applying predictive analytics in transport logistics. *Proceedings of the 2020 on Computers and People Research Conference* (pp. 144–151).

Brandtner, P., Udokwu, C., Darbanian, F., & Falatouri, T. (2021, January). Applications of big data analytics in supply chain management: findings from expert interviews. *Proceedings of the 2021 4th International Conference on Computers in Management and Business* (pp. 77–82).

Darbanian, F., Brandtner, P., Falatouri, T., & Nasseri, M. (2024). Data analytics in supply chain management: a state-of-the-art literature review. *Operations and Supply Chain Management: An International Journal*, *17*(1), 1–31.

Dehkhodaei, A., Amiri, B., Farsijani, H., & Raad, A. (2023). Barriers to big data analytics (BDA) implementation in manufacturing supply chains. *Journal of Management Analytics*, *10*(1), 191–222.

Dubey, R., Gunasekaran, A., Childe, S. J., Papadopoulos, T., Luo, Z., Wamba, S. F., & Roubaud, D. (2019). Can big data and predictive analytics improve social and environmental sustainability?. *Technological Forecasting and Social Change*, *144*, 534–545.

Galli, B. J., & Ocampo, L. A. (2019). Exploring the relationships between continuous improvement and predictive analytics. *International Journal of Advanced Operations Management*, *11*(3), 189–210.

Goyal, S., Garg, D., & Luthra, S. (2021). Sustainable production and consumption: analysing barriers and solutions for maintaining green tomorrow by using fuzzy-AHP–fuzzy-TOPSIS hybrid framework. *Environment, Development and Sustainability*, *23*, 16934–16980.

Hazen, B. T., Boone, C. A., Ezell, J. D., & Jones-Farmer, L. A. (2014). Data quality for data science, predictive analytics, and big data in supply chain management: an introduction to the problem and suggestions for research and applications. *International Journal of Production Economics*, *154*, 72–80.

Kache, F., & Seuring, S. (2017). Challenges and opportunities of digital information at the intersection of Big Data Analytics and supply chain management. *International journal of operations & production management*, *37*(1), 10–36.

Moktadir, M. A., Ali, S. M., Paul, S. K., & Shukla, N. (2019). Barriers to big data analytics in manufacturing supply chains: a case study from Bangladesh. *Computers & Industrial Engineering*, *128*, 1063–1075.

Nugroho, H. (2023). A review: data quality problem in predictive analytics. *IJAIT (International Journal of Applied Information Technology)*, 79–91.

Omar, Y. M., Minoufekr, M., & Plapper, P. (2019). Business analytics in manufacturing: current trends, challenges and pathway to market leadership. *Operations Research Perspectives*, *6*, 100127.

Schoenherr, T., & Speier-Pero, C. (2015). Data science, predictive analytics, and big data in supply chain management: current state and future potential. *Journal of Business Logistics*, *36*(1), 120–132.

Sodero, A., Jin, Y. H., & Barratt, M. (2019). The social process of Big Data and predictive analytics use for logistics and supply chain management. *International Journal of Physical Distribution & Logistics Management*, *49*(7), 706–726.

Tambuskar, D. P., Jain, P., & Narwane, V. S. (2024). An exploration into the factors influencing the implementation of big data analytics in sustainable supply chain management. *Kybernetes*, *53*(5), 1710–1739.

Tzeng, G. H., Chiang, C. H., & Li, C. W. (2007). Evaluating intertwined effects in e-learning programs: a novel hybrid MCDM model based on factor analysis and DEMATEL. *Expert Systems with Applications*, *32*(4), 1028–1044.

# 18 Nanotechnology and Biomaterials in Smart Manufacturing

*Vivek Kumar Dhiman, Vinay Kumar Dhiman, Naveen Mani Tripathi, and Rupendra Shrestha*

## 18.1 INTRODUCTION TO NANOTECHNOLOGY

### 18.1.1 Overview of Nanotechnology and Biomaterials

The term "nanotechnology" was introduced by Norio Taniguchi in 1974. He described it as the "processing of separation, consolidation, and deformation of materials at the atomic or molecular level" (Sandhu 2006). In a foundational paper titled "On the Basic Concept of Nano-Technology," the authors stated, "In the processing of materials, the minimal size of stock removal, accretion, or flow of materials is likely one atom or one molecule, specifically 0.1–0.2 nm in length" (Sandhu 2006). Since then, nanotechnology has been broadly defined as the study of materials on a scale of 1 billionth of a meter ($10^{-9}$ m). This field focuses on manipulating matter at the atomic and molecular levels, and uses principles from physics, chemistry, and materials science to understand the unique properties of nanomaterials (Binns 2021).

The prefix "nano-" and even "nanobio-" (or bionano-) have emerged as a result of the development of nanotechnology (Omran 2020). We believe that the highest level of success in the field of nanoscale research can be achieved via a combination of rigorous discipline and pragmatic approaches. As an example, using the phrase "nanomaterial," which is used rather often, what structural characteristic of a material should the prefix "nano-" allude to? Are there any additional features provided, such as crystal size, grain size, domain size, or surface topology? What is the point at which a standard material transforms into a nanomaterial? If it were not for the fact that the term is used in a consistent and contemporary manner, its very existence ought to be called into question, since, from an etymological standpoint, the word is a complete construct. It is far more rational to refer to a material as a "nanostructured material," but a pragmatic approach would imply that a nanomaterial is "any form of a material that is composed of discrete functional parts, many of which have one or more dimensions of the order of 100 nm or less." This approach is logically followed by other definitions, such as the definition of a nanocrystalline material, which is "a material that is comprised of many crystals, the majority of which have one or more dimensions of the order of 100 nm or less," and the definition of a nanocomposite, which is a "multi-phase material in which the majority of the dispersed phase


DOI: 10.1201/9781003566205-18

components have one or more dimensions of the order of 100 nm or less" (Borisenko and Ossicini 2013). When we reach this point, we are confronted with the issue of the relationship between nanotechnology and biomaterials research.

Biomaterials in nanotechnology are substances, either created or naturally sourced, intended to engage with biological systems at the nanoscale, therefore enhancing areas such as medical treatment, diagnostics, and tissue engineering (Kovářík et al. 2017). These materials are designed to execute specific roles inside the body, including tissue healing, targeted medication administration, and enhancement of cell proliferation. At the nanoscale, biomaterials exhibit distinctive features such as augmented surface area, heightened reactivity, and capacity for meticulous adjustment, facilitating precise interactions with cellular and molecular structures (Kumarage et al. 2022). The prevalent biomaterials in nanotechnology include nanoparticles, hydrogels, nanofibers, liposomes, dendrimers, and nanostructured coatings. Nanoparticles composed of metals, polymers, or ceramics are often used in drug delivery and medical imaging, while hydrogels may encapsulate medications or cells for regulated release, usually assisting in tissue scaffolding. Nanofibers created by electrospinning replicate the extracellular matrix, making them appropriate for wound healing and tissue regeneration. Liposomes, lipid-derived vesicles, enable targeted medicine delivery to particular anatomical regions, minimize adverse effects, and enhance therapeutic efficacy. Dendrimers, which are extensively branched polymers, may transport several therapeutic substances, thereby augmenting therapy versatility. Nanostructured coatings are used in medical equipment to enhance biocompatibility, diminish infections, and facilitate cellular adhesion and integration (Kumarage et al. 2022). Utilizing these materials, nanotechnology enables more efficient, targeted, and less invasive therapies to improve healthcare results.

### 18.1.2 Role of Nanotechnology and Biomaterials

Nanotechnology spans multiple domains, including biology, physics, chemistry, and engineering, making it a highly interdisciplinary field. Defined as the design, synthesis, modification, characterization, and production of materials, devices, and systems of 1–100 nm, nanotechnology leverages the unique physical and chemical properties of nanoparticles, such as melting point, electrical and thermal conductivity, magnetism, and catalytic activity, which differ from those of their bulk counterparts because of their high surface-to-volume ratio. This phenomenon, known as the "size effect," describes how nanoparticles' physical characteristics vary with size, while the "surface effect" pertains to their chemical properties. For instance, reducing the material size increases the surface area and enhances reactivity. Techniques such as coating, doping, and compositing nanoparticles with biomaterials can transform the properties of engineered biomaterials, enhancing features such as biocompatibility, nontoxicity, and durability (Basova et al. 2021).

A "biomaterial," defined as a substance engineered to interact with biological systems for medical purposes, can be natural or synthetic and is used to support, repair, or replace damaged tissues or biological functions (Hudecki et al. 2019). Historically, biomaterials have been primarily derived from animal tissues. Currently, synthetic biomaterials are developed using contributions from biology, physics, chemistry,

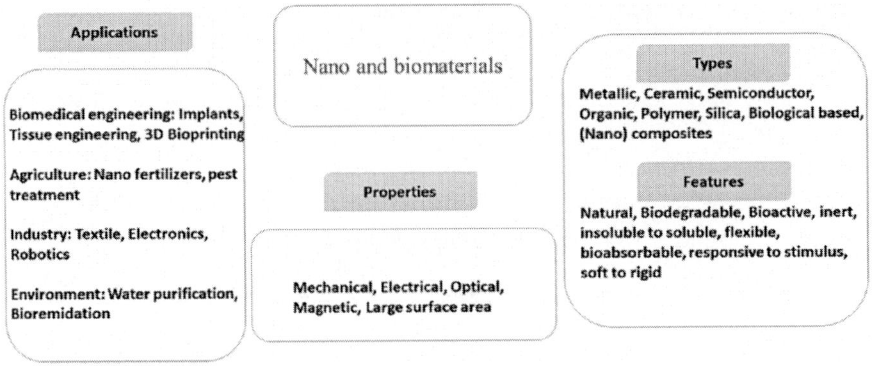


**FIGURE 18.1** Nano and biomaterials: types, applications, features, and properties.

computational science, and materials engineering. Early bioengineered materials relied on ceramics, metals, polymers, glass, and composites, which are still foundational in medical implants, tissue repair, cancer treatment, biosensors, and drug delivery systems. In biomaterials and biocomposites, crucial design considerations include biocompatibility, mechanical integrity with the surrounding bone tissue, non-toxicity, and biodegradability of byproducts. Nanotechnology has introduced advanced methodologies that serve as sophisticated templates for creating innovative biomaterials with enhanced functionality. Applications include orthopedics, nanofilm biomaterials, wound healing, tissue engineering, and cancer nanotechnology (Kumarage et al. 2022). This study emphasizes the origins and properties of the biomaterials and nanomaterials used in modern biomaterials and biocomposites (see Figure 18.1).

## 18.2 FUNDAMENTALS OF NANOTECHNOLOGY AND BIOMATERIALS

### 18.2.1 Historical Development and Milestones

The use of nanostructured materials has not been discovered recently. As early as the fourth century CE, the Romans incorporated nanosized metals into decorative glasses and cups. A well-known example of this is the Lycurgus Cup (Raut et al. 2024). Under light, the cup appears green; however, when it is lit from within, it becomes lively red. This color-changing effect is due to the presence of metal nanoparticles, similar to the technique used in famous stained-glass windows from the 10th–12th centuries. To understand the color shift seen in the Lycurgus Cup—from gold's typical yellowish hue to a striking red—one can compare the absorption spectra of the bulk gold metal film to that of a gold colloidal film. A thin bulk gold metal sheet absorbs most of the visible light spectrum and is strongly in the infrared range, showing a slight dip between 400 and 500 nm, which gives it a blue appearance when backlit (Raut et al. 2024). The Crusaders utilized Damascus steel sword blades in the seventh century during the medieval period. The material exhibited a distinctive wavy banding pattern, referred to as a damask, characterized by enhanced mechanical properties and atypically sharp cutting edges. High-resolution transmission electron microscopy (HRTEM) images of the Damascus steel samples demonstrated the presence of iron carbide nanowires (cementite) and carbon nanotubes within the blades. These intricate structures may account for the significant hardness and exceptional sharpness of the blades (Reibold et al. 2006).

Although relatively modern, the field of biomaterials has roots that can be traced back thousands of years, making a formal history challenging to compile. Nevertheless, historical insights have revealed both ancient uses and the rapid evolution of biomaterials in recent decades. Artificial glass eyes and wooden dentures have been widely used throughout much of the documented history. Synthetic plastics emerged in the early 20th century, sparking a wave of implantation experiments. With hindsight into biomaterial toxicology, it is evident that many of these early efforts were destined for failure. Poly (methyl methacrylate) (PMMA), however, marked a breakthrough when it was introduced to dentistry in 1937. During World War II, fragments of PMMA from shattered aircraft gunnery turrets were inadvertently lodged in pilots' eyes, demonstrating the material's mild foreign body response (Ratner and Zhang 2020).

Shortly after the war, Voorhees experimented with the Vinyon N parachute cloth as a vascular prosthesis. By 1958, Rob's cardiovascular surgery textbook suggested that surgeons could use Dacron fabric from drapery shops as arterial prostheses. In the early 1960s, Charnley pioneered the use of PMMA, ultrahigh-molecular-weight polyethylene, and stainless steel for total hip replacement. Despite these milestones in synthetic materials for medical use, the term "biomaterial" has not yet been used (Ratner and Zhang 2020). The origins of the term "biomaterial" remain uncertain, but the field likely took shape through symposia held at Clemson University in the late 1960s and the early 1970s. These symposia laid the groundwork for the Society for Biomaterials, which was established in 1975 (Ratner and Zhang 2020). With society's formation, the field shifted from ad hoc material use by individual physicians addressing urgent medical needs to researchers and engineers designing materials with precise specifications and scientists studying biocompatibility. This new discipline of biomaterials grew alongside the Society for Biomaterials, which became a forum for advancements and collaborations that influenced fields such as drug delivery, biosensors, and bioseparation. Today, the field of biomaterials has grown extensively, with dedicated academic departments, specialized programs, and research institutes. Many companies are now integrating biomaterials into devices, supported by educational and research initiatives that shape the science and engineering of biomaterials. This growth is documented in the Society for Biomaterials Educational Directory (1992), reflecting the expansive role that biomaterials now play in technology and healthcare.

## 18.3 SMART MANUFACTURING: THE INTEGRATION OF NANOTECHNOLOGY AND BIOMATERIALS

### 18.3.1 How Nanotechnology Enhances Smart Manufacturing

Advances in nanotechnology have led to the development of a vast array of nanomaterials that are now widely used across numerous fields. The evolution of these materials began with the first generation, which focused on passive nanostructures, such as nanoparticles. The second generation introduced active nanostructures, whereas the third generation saw the emergence of nanosystems with hierarchical 3D designs. The fourth generation is characterized by molecular nanosystems (see Table 18.1).

**TABLE 18.1**
**Highlights of Nanotechnology and Biomaterials Use in Biomedicine.**

| Application | Role of nanomaterials | Biomaterials |
|---|---|---|
| Drug delivery | Nanoparticles enable targeted drug delivery with controlled release, reducing side effects and enhancing efficacy. | Biocompatible materials (e.g., liposomes, dendrimers) encapsulate drugs, providing a stable and responsive delivery system. |
| Tissue engineering | Nanofibers and scaffolds mimic natural tissue structures, promoting cell growth and tissue regeneration. | Biomaterials such as hydrogels and nanofibrous scaffolds support cell adhesion and tissue repair, mimicking the extracellular matrix. |
| Orthopedic implants | Nanocoatings improve osseointegration and reduce infection rates. | Strong, biocompatible metals, ceramics, and composites provide structural support and mechanical compatibility with bone. |
| Cancer therapy | Nanoparticles (e.g., gold, iron oxide) enable targeted drug delivery and hyperthermia treatment. | Biocompatible carriers enhance the selective targeting of cancer cells while minimizing harm to surrounding tissues. |
| Wound healing | Nanotechnology enhances antibacterial properties in wound dressings and accelerates tissue repair. | Biomaterials like hydrogels and nanofibrous mats provide a moist environment, supporting healing and reducing scarring. |
| Diagnostics and imaging | Quantum dots and magnetic nanoparticles improve the accuracy of medical imaging and diagnostics. | Biocompatible materials ensure stable and safe use of contrast agents for enhanced imaging of specific tissues or organs. |
| Cardiovascular applications | Nanoparticles in stents reduce plaque buildup and promote arterial healing. | Biomaterials with anti-inflammatory properties are used in stents and grafts, promoting vascular integration and healing. |
| Bone regeneration | Nanocoatings on implants enhance mineralization and bone tissue interaction. | Bioceramics and composite scaffolds encourage bone cell proliferation and faster healing. |
| Biosensors | Nanoscale sensors detect biomolecules and pathogens with high sensitivity and specificity. | Biomaterials provide biocompatible surfaces for effective analyte detection, ensuring reliable and safe biosensing devices. |
| Antimicrobial coatings | Nanoparticles (e.g., silver, copper) provide antimicrobial properties, preventing infections on surfaces. | Biocompatible coatings help prevent biofilm formation on medical devices, reducing hospital-acquired infections. |
| Gene therapy | Nanocarriers, like liposomes and polymer nanoparticles, deliver genes to target cells efficiently. | Biomaterials stabilize genetic material and ensure precise delivery to target cells for gene therapy applications. |
| Dental applications | Nano-hydroxyapatite particles are used in dental fillers and coatings, improving remineralization and strength. | Biocompatible composites are used in restorations, providing strength and durability for dental fillings and implants. |

**TABLE 18.1** ***(Continued)***
**Highlights of Nanotechnology and Biomaterials Use in Biomedicine.**

| Application | Role of nanomaterials | Biomaterials |
|---|---|---|
| Vaccines and immunotherapy | Nanoparticles serve as adjuvants and carriers, enhancing immune response and enabling targeted delivery. | Biomaterials like liposomes and hydrogels encapsulate antigens, providing controlled release for vaccine applications. |
| Ophthalmology | Nanoparticles allow for targeted drug delivery in eye treatments, improving efficacy and reducing side effects. | Biocompatible hydrogels and films are used in contact lenses and eye implants, supporting controlled drug release. |
| Neuroengineering | Nanoscale electrodes are used in neural implants, improving signal accuracy and interface with nerve cells. | Biocompatible polymers in neural interfaces support cell adhesion and reduce inflammatory response in the nervous system. |
| Anti-cancer photothermal therapy | Gold nanoparticles absorb light and convert it into heat to target and destroy cancer cells selectively. | Biocompatible carriers stabilize the nanoparticles and ensure safety in photothermal applications. |
| Antioxidant delivery | Nanoparticles deliver antioxidants to target sites, protecting cells from oxidative damage. | Biomaterials help encapsulate antioxidants, ensuring gradual release and enhanced therapeutic effects. |
| Environmental sensors | Nanomaterials detect environmental toxins with high sensitivity in water, air, and soil. | Biodegradable biomaterials are used in sensor design, minimizing environmental impact and enhancing sensor lifespan. |
| Blood purification | Magnetic nanoparticles capture and remove toxins and pathogens from blood in extracorporeal systems. | Biocompatible materials in purification systems ensure safety for blood-contact applications, reducing risks of clotting. |
| Artificial skin and wound dressings | Nanofibers mimic the skin's structure and promote faster wound healing. | Biocompatible hydrogels and polymers support skin regeneration, moisture retention, and antibacterial protection. |
| Antiviral coatings | Nanoparticles disrupt viral structures on surfaces, helping prevent viral spread in healthcare settings. | Biomaterials provide a stable and safe surface for antiviral coatings on high-contact medical equipment and surfaces. |
| Controlled tissue scaffolding | Nanofiber scaffolds with controlled degradation rates support specific tissue growth rates and structures. | Natural polymers (e.g., collagen, chitosan) promote cell adhesion and compatibility in tissue engineering scaffolds. |
| Organ transplants | Nanocoatings reduce rejection and improve organ preservation pre-transplant. | Biomaterials with immune-modulating properties minimize rejection and support graft integration post-transplantation. |
| Stem cell engineering | Nanoparticles facilitate stem cell differentiation and homing to target tissues for regenerative therapies. | Biomaterials support cell growth, directing stem cell differentiation for tissue repair and regeneration. |
| Cancer hyperthermia therapy | Iron oxide nanoparticles generate heat under magnetic fields, targeting cancer cells while sparing healthy tissue. | Biomaterials act as carriers for magnetic nanoparticles, enhancing precision and biocompatibility in hyperthermia treatments. |

## 18.4 NANO- AND BIOMATERIALS IN SMART MANUFACTURING

### 18.4.1 Synthesis, Characterization, and Properties of Nanomaterials for Smart Manufacturing Applications

In terms of synthesis, the three primary routes are physical, chemical, and biological routes (Chauhan et al. 2023; Singh et al. 2024; Budhiraja et al. 2024). Physical methods, such as laser ablation and mechanical milling, enable the formation of nanoparticles through the manipulation of bulk materials, achieving precise particle sizes with controlled morphology. Chemical approaches, including sol–gel synthesis and chemical vapor deposition (CVD), are crucial for producing high-quality nanoparticles, particularly metal oxides and carbon-based materials. Biological synthesis techniques that leverage microorganisms, plants, and algae represent an emerging eco-friendly paradigm. These biosynthetic methods exploit the reductive capacity of biomolecules to produce nanoparticles, often yielding products with enhanced biocompatibility and reduced toxicity. Notably, fungal and plant-mediated synthesis of nanoparticles, which use enzymes and plant phytochemicals as reducing agents, circumvents the environmental hazards associated with conventional chemical synthesis and aligns with sustainability goals by minimizing waste and hazardous byproducts.

Applications of nanomaterials span multiple sectors, with notable impacts on the environmental, biomedical, and agricultural fields. The high surface reactivity and tunable properties of nanomaterials, such as metal oxides and carbon-based nanoparticles, make them potent agents for environmental remediation. For example, $TiO_2$ and ZnO nanoparticles have been extensively employed in photocatalysis for water and air purification, where they facilitate the degradation of organic pollutants and heavy metals through oxidative mechanisms (Reddy et al. 2019). In biomedical applications, nanoparticles enable advancements in drug delivery, where their small size and modifiable surfaces allow for the encapsulation of therapeutic agents that can be directed to specific tissues or cells, thus minimizing systemic side effects. For example, iron oxide nanoparticles are used as MRI contrast agents, while gold and silver nanoparticles show potential in photothermal cancer treatments, where localized heating of malignant cells is achieved through targeted irradiation (Kharey et al. 2023). Moreover, antimicrobial applications benefit from the bactericidal properties of metal nanoparticles, which can be integrated into coatings for medical devices to provide enhanced protection against pathogens.

In the agricultural domain, nanotechnology offers solutions that enhance crop yield and reduce environmental impacts through the development of nano-encapsulated pesticides and fertilizers that ensure controlled release. Such technologies minimize the leaching and runoff associated with traditional agrochemicals, thus improving their efficacy while mitigating ecological harm. Additionally, nanosensors embedded in the soil or integrated within plants enable real-time monitoring of moisture, nutrient levels, and pathogen presence, thus facilitating precision agriculture and optimizing resource usage (Bharti et al. 2024). There are several challenges related to the stability and optimization of nanoparticles, emphasizing the role of capping agents and polymeric coatings in preventing agglomeration and enhancing biocompatibility. Stabilizers, such as polyethylene glycol (PEG) and chitosan, are used extensively to maintain nanoparticle dispersion and functionalize surfaces for targeted biomedical applications, such as drug delivery and diagnostic imaging (Ahmad et al. 2011). Significant emphasis has been placed on the potential environmental and health implications of nanomaterials. Although the benefits of nanotechnology have been extensively reported, concerns regarding nanoparticle toxicity, persistence in the environment, and interaction with biological systems are recognized as critical research areas. Sustainable synthesis approaches, such as green biosynthesis using plant extracts and microbial processes, are advocated as viable solutions to these concerns as they reduce reliance on toxic chemicals and align with environmental sustainability objectives (see Table 18.2).

**TABLE 18.2**
**Application of Nano- and Biomaterials in Smart Manufacturing.**

| Nanomaterial | Synthesis method | Particle size (nm) | Properties | Applications in smart manufacturing | Unique characteristics |
|---|---|---|---|---|---|
| Metal nanoparticles | Chemical reduction, green synthesis | 1–100 | High surface area, catalytic activity, optical properties | Catalysts in chemical processes, antibacterial coatings | Optical tunability, biocompatibility |
| Nanowires | Electrospinning, Vapor-Liquid-Solid (VLS) growth | 10–1000 | High electrical conductivity, High aspect ratio | Flexible and stretchable electronics, Smart sensors | Unique conductivity pathways |
| Carbon nanotubes (CNTs) | Arc discharge, chemical vapor deposition (CVD) | 1–50 | High strength, electrical and thermal conductivity | Structural reinforcement, conductive inks, lightweight composites | Exceptional mechanical properties |
| Quantum dots | Colloidal synthesis, hydrothermal methods | 2–10 | Tunable photoluminescence, quantum confinement effects | Display technology, bioimaging, quantum computing | Size-dependent optical properties |
| Graphene | Exfoliation, chemical vapor deposition (CVD) | 0.5–10 | High electrical and thermal conductivity, mechanical strength | Flexible electronics, heat sinks, protective coatings | Two-dimensional structure, high flexibility |
| Metal oxide nanoparticles | Sol–gel, hydrothermal, co-precipitation | 10–500 | Magnetic properties, catalytic activity | Environmental sensors, photocatalysis, energy storage | Stability under harsh conditions |
| Nanocomposites | In-situ polymerization, sol–gel process | 1–500 | Enhanced mechanical strength, thermal stability | Lightweight structures, automotive components, aerospace | Hybrid functionality |

*(Continued)*

**TABLE 18.2** *(Continued)*
**Application of Nano- and Biomaterials in Smart Manufacturing.**

| Nanomaterial | Synthesis method | Particle size (nm) | Properties | Applications in smart manufacturing | Unique characteristics |
|---|---|---|---|---|---|
| Dendrimers | Divergent and convergent Synthesis | 1–10 | Tunable surface functionality, monodispersity | Drug delivery, nanofiltration, chemical sensing | Branching structure, high surface functionality |
| Nanofibers | Electrospinning, phase Separation | 50–1000 | High tensile strength, largesurface-to-volume ratio | Filtration membranes, tissue engineering, wearables | Porous structure |
| Polymeric nanoparticles | Emulsion polymerization, self-assembly | 1–500 | Biodegradable, controlled drug release | Biomedical devices, targeted drug delivery, coatings | Biocompatibility, controlled release capabilities |
| Silica nanoparticles | Stöber process, sol–gel | 10–200 | Chemical stability, biocompatibility | Drug delivery systems, cosmetic additives, catalysts | Tunable pore size, inertness |
| Fullerenes | Arc discharge, laser ablation | ~1 | High electron affinity, chemical stability | Organic photovoltaics, medical diagnostics, superconductors | Spherical carbon structure, electron acceptor |
| Ceramic nanoparticles | Sol–gel, flame spray pyrolysis | 10–100 | Hardness, heat resistance | Thermal barrier coatings, Abrasive materials, electronics | High melting point, wear resistance |
| Magnetic nanoparticles | Co-precipitation, thermal decomposition | 5–100 | Superparamagnetic properties, high magnetic saturation | Magnetic storage media, MRI contrast agents, targeted drug delivery | Responsiveness to magnetic fields |

### 18.4.2 Nanostructured Materials for Enhanced Mechanical, Thermal, and Electrical Properties

The unique properties of nanostructured materials have positioned them as transformative components across various applications that require exceptional mechanical, thermal, and electrical capabilities. By engineering materials at the nanoscale, scientists and engineers can overcome certain limitations of conventional materials, often achieving properties that were previously unattainable. Nanostructured materials exhibit superior mechanical properties such as high tensile strength, improved hardness, and enhanced wear resistance. The key to these improvements lies in their fine grain sizes, which reduce defect accumulation and restrict the movement of dislocations, thereby leading to enhanced strength and toughness. Carbon nanotubes (CNTs), graphene, and nanocrystalline alloys are prominent examples

(Ahmad et al. 2015). Incorporating nanomaterials into metal matrices (to form nanocomposites) has shown promising results, significantly increasing mechanical strength without sacrificing ductility. Such materials are utilized in fields requiring durable and lightweight materials, such as the aerospace and automotive industries. Heat management is crucial for high-performance electronics, automotive systems, and energy applications. Nanostructured materials can offer superior thermal conductivity or act as insulators, depending on their structural configuration. For example, nanomaterials such as graphene and boron nitride nanosheets have high thermal conductivities, making them ideal for heat dissipation in electronic components. Conversely, materials such as aerogels have nanoscale pores that trap air and provide excellent thermal insulation. Nanostructuring allows precise control over these properties, creating tailored solutions for various applications, from electronic cooling systems to barrier coatings. The electrical conductivity of materials can be significantly improved by nanostructuring. For example, CNTs and graphene exhibit exceptional electrical conductivity owing to their high carrier mobility and low resistivity at the nanoscale (Wang and Weng, 2018). Integrating these nanomaterials into conductive composites can enhance their overall conductivity, making them suitable for applications in flexible electronics, batteries, and supercapacitors. Additionally, nanostructured semiconductors such as silicon nanowires and quantum dots display unique electronic behaviors, enabling innovations in optoelectronic devices, photovoltaics, and transistors (see Table 18.3).

Nanostructured materials have potential applications in various sectors, including aerospace, electronics, healthcare, and energy. By strengthening the structural components to enhance energy storage systems and enable more efficient electronic devices, the versatility of nanostructured materials can be transformed. Ongoing research is focused on refining manufacturing processes, improving scalability, and ensuring environmental and economic feasibility. As synthesis techniques advance, we expect the scope of applications of these enhanced nanostructured materials to expand further.

### 18.4.3 Nanotechnology-Based Bioprinting for Biomaterial Fabrication

Nanotechnology plays a crucial role in the synthesis of nanophase biomaterials, which are used in various biomedical fields, including 3D printing, drug delivery, orthopedic implants, and diagnostics. Nanoscale biomaterials can be used as bioinks for 3D printing applications. Bioinks are essential for the bioprinting process and consist of liquid cell suspensions housed in printing cartridges that create an optimal environment for cell growth and development (Bhatt and Anbarasu 2017). The cells were suspended until they were transferred to a cartridge containing liquid matrix material. This process facilitates the adhesion of cells to the extracellular matrix through proteins and biological growth factors, leading to the formation of a bioscaffold that initiates tissue regeneration. In addition, bioinks serve as a protective barrier for cells during printing, ensuring their preservation as they are transformed into biological structures. The primary function of bioinks is to aid in the regeneration of tissues on synthetic substrates. A wide variety of inks are commercially available and are sourced from both natural and synthetic materials. An ideal nanoscale bioink should incorporate multiple biomaterials at the nanometer scale to enhance the printing efficiency. Recent technological advancements have enabled the synthesis of

**TABLE 18.3**
**Properties of Nanomaterials and Their Uses.**

| Material | Nanostructure type | Enhanced property | Applications | Properties |
|---|---|---|---|---|
| Carbon nanotubes (CNTs) | Tubular nanostructures | Mechanical, electrical | Electronics, composite materials, drug delivery, sensors, energy storage devices | High strength-to-weight ratio, excellent conductivity |
| Graphene | 2D nanostructure | Electrical, thermal | Batteries, sensors, flexible electronics, coatings, EMI shielding, supercapacitors | Superior conductivity, high surface area |
| Nanostructured ceramics | Nanograined or layered | Mechanical, thermal | Aerospace, armor, cutting tools, biomedical implants, thermoelectric devices | High hardness, improved fracture toughness |
| Metallic nanoparticles | Spherical or other shapes | Electrical, optical | Catalysts, biomedical imaging, drug delivery, electronics, water treatment | High surface-to-volume ratio, tunable properties |
| Nanocomposites | Matrix with nanofillers | Mechanical, Electrical | Automotive, aerospace, structural materials, sports equipment, coatings, electromagnetic interference (EMI) shielding | Tailored properties based on filler and matrix choice |
| Metal Nanowires | 1D nanostructure | Electrical | Conductive films, sensors, nanoelectronics, transparent electrodes, flexible electronics | High conductivity, flexibility in thin films |
| Silicon Nanowires | 1D nanostructure | Electrical, thermal | Thermoelectrics, photovoltaics, lithium-ion batteries, biosensors | Enhanced thermoelectric efficiency, high conductivity |
| Quantum dots | Zero-dimensional | Optical, electrical | Display technology, medical imaging, photodetectors, solar cells, quantum computing | Unique optical properties, size-dependent tunability |
| Nanostructured polymers | Nanoporous or layered | Mechanical, thermal | Filtration, packaging, insulation, drug delivery, tissue engineering | Lightweight, customizable thermal and mechanical properties |
| Metal oxide nanostructures | Nanorods, nanosheets | Electrical, catalytic | Sensors, catalysts, energy storage devices, environmental remediation, supercapacitors | Enhanced chemical reactivity, improved sensing accuracy |

nanoscale biomaterials by integrating nanoparticles with existing materials, resulting in the development of more effective bioinks, which are transforming the bioprinting industry (Chandra et al. 2024). A suitable nano-bioink candidate must demonstrate sufficient mechanical strength, rigidity, and shape fidelity for scaffold formation; precise tuning for optimal extrusion; the ability to create scaffolds that mimic the native cellular microenvironment; biocompatibility; the potential for chemical modifications to tailor the cellular environment; and the capacity for large-scale synthesis with minimal batch-to-batch variation. This technology has significant applications in skin repair, making it possible to treat minor injuries, such as lacerations and burns. Initially, bioink development focused on skin regeneration; however, the technology has since expanded to include bone and cartilage regeneration, establishing bioinks as a versatile regenerative technology.

### 18.4.4 Biodegradable Biomaterials for Sustainable Production and Waste Reduction

Materials engineered to interact with biological systems are called biomaterials. They are bioactive materials that can be readily compacted into human tissue. They exhibit a significant level of biodegradability. They are often used in pharmaceuticals, tissue engineering, the production of human body components, and many industrial applications. Sustainability and biomaterials mutually reinforce and facilitate the production and development of innovative sustainable materials using modern technical methods. Biomaterials are environmentally friendly. They are either derived from various biological sources or are produced using green technologies (Hosseini et al. 2020). Several biomaterials have been produced and effectively employed in various biomedical disciplines as alternatives to conventional materials. They are effectively utilized in medical treatments, such as cancer therapy, ligament and tendon repair, orthopedic applications, ophthalmic applications for contact lens design, wound healing, nerve regeneration therapy, breast implants, and the production of various surgical devices. In addition, they have some non-biomedical uses. In addition to their various features and applications, biomaterials are also used in biological systems and medical devices. Common biomaterials include ceramics, metals, and their alloys, particularly bioactive glass; polymeric substances and their composites; biopolymers and their composites; and materials synthesized from animal-based sources (Biswal et al. 2020). Sustainable biomaterials represent an interplay between renewable natural resources and biomaterials, offering promising prospects for the advancement of novel sustainable development methods over the next few years. Currently, considerable efforts and emphasis are directed toward the creation and design of materials derived from sustainable resources, gradually supplanting traditional materials. Currently, we are at a pivotal juncture when sustainable energy is essential, which is achievable via advancements in green technology. The use of biomaterials is progressively increasing compared to that of conventional materials. A variety of sophisticated approaches are used to synthesize biomaterials with specific features, including thermal or hydrothermal synthesis and surfactant templates that facilitate the controlled creation of bionanomaterials. Numerous processes are used for the fabrication of nanomaterials for biological purposes; however, hydrothermal carbonization is the predominant method, conducted at moderate pressures and temperatures in an aqueous medium. The product produced using this process may be used in several applications, including medicine delivery, $CO_2$ sequestration, catalysis in diverse organic syntheses, sensors, and chemical vapor deposition (Nizamuddin et al. 2018).

### 18.4.5 Nanoparticle-Modified Biomaterials for Biodegradable Scaffolds and Tissue Engineering

Nanomaterials offer promising alternatives owing to their comparable size to natural tissue components such as proteins. Tissue regeneration in the body is affected by nanoscale events. Surface modification of implants with physiologically inspired nanoscale features is essential for the development of advanced orthopedic materials. Research indicates that implant surfaces with nanoscale roughness promote consistent and superior bone growth compared with those with micron-scale roughness (Shirazi 2023). Numerous studies have demonstrated enhanced osteogenic integration on nanostructured surfaces made of various materials, including ceramics, metals, polymers, and composites, outperforming conventional materials with micron grain sizes (Liang et al. 2024). Porous scaffolds are critical for tissue engineering and serve as templates for cell proliferation and tissue development. Although these scaffolds are made from both natural and synthetic polymers, problems such as low cell density and active cell development still exist. Nanoscale hydrogel scaffolds present a potential solution to these challenges. Researchers have enhanced the properties and applicability of hydrogels by incorporating nanoparticles into them. A bioactive nanoparticle-stimulated three-dimensional scaffold was developed to facilitate bone tissue formation using human mesenchymal stem cells (Gao et al. 2017). They tested bioactive ceramic nanoparticles for osteogenesis within PEGDMA scaffolds using bone marrow-derived mesenchymal stem cells (hMSCs). During simultaneous polymerization, stem cells in PEGDMA are co-printed with bioactive glass (BG) and hydroxyapatite (HA) nanoparticles, enabling precise 3D placement. After 21 days, cell viability was 86%, and the compressive modulus was 358.91 kPa across all groups. The PEG-HA group exhibited the highest alkaline phosphatase activity, as confirmed by quantitative PCR gene expression analysis (Gao et al. 2014). Additionally, Masson's trichrome staining revealed the most substantial collagen deposition in the PEG-HA scaffolds, indicating that hydroxyapatite was more effective than BG in promoting hMSC osteogenesis in bioprinted bone structures. These biomaterials can be used as nanocomposite implants in tissue engineering applications.

## 18.5 CHALLENGES AND SOLUTIONS

### 18.5.1 Technical Challenges and Engineering Solutions

Nanotechnology is transforming smart manufacturing across industries such as healthcare and electronics by enhancing precision, efficiency, and flexibility. However, challenges, such as system integration, cybersecurity, scalability, quality control, and environmental sustainability, must be addressed. Ensuring interoperability between traditional systems and IoT devices is vital for avoiding inefficiencies. Technologies such as IoT and cyber-physical systems enable connectivity, whereas communication protocols such as OPC UA and MQTT enhance machine-to-machine interaction, improving data flow. Cloud platforms and digital twins facilitate predictive maintenance and support legacy systems. However, IoT connectivity also increases the vulnerability to cyber threats, making data protection crucial. Scaling production while maintaining precision in nanomanufacturing is complex and requires flexible and reconfigurable manufacturing systems for quick adaptation. Additive manufacturing fosters rapid prototyping and on-demand production, whereas AI-driven scheduling improves market responsiveness. Maintaining consistent quality at the nanoscale remains challenging because minor deviations can cause performance issues. Advanced metrology and AI quality control enhance reliability, and non-destructive testing safeguards material integrity. Machine learning detects anomalies to prevent errors. Nanomanufacturing processes may generate environmentally hazardous byproducts, particularly reactive nanoparticles. Closed-loop systems aim to recycle waste and reduce emissions, whereas effective filtration minimizes nanoparticle release. Life cycle assessment tools evaluate the environmental impact of nanomaterials and low-energy methods, along with renewable sources, to promote sustainable manufacturing.

### 18.5.2 Regulatory and Safety Considerations

Global efforts have been made to establish standards and regulations for nanotechnology, with a focus on health, safety, and environmental concerns. With the rapid evolution of nanotechnology, international bodies, such as the International Organization for Standardization, and related organizations, such as IEC and CEN, have become instrumental in setting technical standards. ISO established Technical Committee 229 specifically for nanotechnology, which developed 64 standards by 2018 (Soltani and Pouypouy 2019). Although this is a significant start, the number of standards is not yet sufficient to fully encompass a variety of applications and complexities within nanotechnology, leading to gaps that affect product safety and quality control. At the regulatory level, countries and regions approach nanotechnology governance in different ways. The European Union has implemented robust frameworks, particularly the REACH regulation, which requires the registration, evaluation, and restriction of chemicals and nanomaterials. This regulation mandates stringent safety evaluation procedures for nanomaterials, particularly concerning consumer and environmental safety. In contrast, the US employs a more case-specific approach, with the Toxic Substances Control Act serving as the primary regulatory framework (Allan et al. 2021). This act treats nanomaterials as new chemicals that require specific risk assessment, but generally integrates them into existing chemical regulatory structures. These frameworks represent two distinct approaches: the EU's comprehensive and consumer-centered policies versus the US approach, which is more flexible and aimed at accommodating business growth while managing safety concerns on a case-by-case basis.

Many countries have implemented their own national standards and certification systems to address the unique risks of nanotechnology and support the commercialization of safe products. For example, China, the US, and Iran have established national systems that address consumer safety and environmental impact. Certification programs such as NanoTrust in Iran, NanoQ in Thailand, and NanoMark in Taiwan seek to enhance consumer trust by ensuring that certified nanotechnology products meet specified safety and quality standards. These certifications play a crucial role in countries in which comprehensive regulatory frameworks may not yet be fully developed (Pogany et al. 2024). Although India does not have comprehensive nanotechnology-specific HSE regulations, they are increasingly aligning with international best practices. Key agencies, such as the Indian Council of Medical Research and the Central Pollution Control Board, monitor environmental and health impacts, but detailed safety protocols specific to nanomaterials are still under development. India's nanotechnology regulatory framework is still developing, with challenges in enforcing nanospecific safety and labeling standards. Efforts are underway to align with international standards; however, there is a need for a dedicated policy to address the unique risks of nanomaterials. Increased collaboration with international regulatory bodies, clear HSE policies, and stronger public awareness measures will likely form the foundation of India's approach to nanotechnology regulation in the coming years (see Table 18.4).

### 18.5.3 Challenges in Scaling Up Nanotechnology and Biomaterial Production for Smart Manufacturing

The large-scale adoption of nanotechnology in manufacturing presents significant challenges that must be addressed to unlock its transformation potential. One major issue is the high cost of nano-engineered materials, which arises from complex synthesis processes and expensive raw materials, limiting accessibility for small and medium-sized enterprises. To mitigate this, companies can focus on optimizing production methods, exploring alternative materials, and creating larger manufacturing plants to achieve economies of scale (Grumezescu, 2017). Another challenge is the uniform dispersion of nanomaterials, such as carbon nanotubes, in composites; inconsistent distribution can lead to weak mechanical performance. Advanced dispersion techniques such as high-energy ball milling and chemical functionalization, along with improved characterization methods, can help achieve better integration of nanomaterials.

Before biomaterials can be applied clinically, they must meet a series of stringent requirements. Therapeutic biomaterials must be non-toxic, non-carcinogenic, non-antigenic, or non-mutagenic. In addition, those intended for interaction with the blood must exhibit thromboresistance; however, some biomaterials used in non-cardiovascular applications can be exempted if the device containing them is designed to act as a hemostatic agent. Safety and efficacy testing for biomaterial-based devices involves evaluating physical and biological characteristics, considering the entire device, its intended duration of use, and interactions with target tissues (Hench and Thompson 2010). Various standards and guidelines,

**TABLE 18.4**
**Nanotechnology Regulations and Authorities around the World.**

| Category | Considerations | Details |
|---|---|---|
| **International standards** | Standardization gaps | ISO, IEC, and CEN lead efforts, but only 64 standards have been published by ISO (TC229) by 2018, which is insufficient for rapid market growth. |
| | Key organizations | ISO's TC229 focuses on defining nanotechnology terms, measurements, and HSE standards; IEC/TC113 and CEN/TC352 contribute to regional and technical standards. |
| | Global harmonization needs | Differences between ISO and EU definitions create challenges; a globally accepted definition for "nanomaterials" is crucial. |
| **EU regulations** | REACH (registration, evaluation, authorization, and restriction of chemicals) | Requires thorough safety evaluations for nanomaterials, covering manufacturing, environmental, and consumer safety. Focused on high consumer protection. |
| | Specific product rules | Biocidal products regulation (BPR), classification, labeling, and packaging (CLP) regulation for hazardous nanomaterials, and directives for food and cosmetic products containing nanomaterials. |
| **US regulations** | Toxic Substances Control Act (TSCA) | Requires nanomaterials to be registered as new chemicals, focusing on risk assessments on a case-by-case basis. Offers flexibility for businesses but is less stringent on consumer labeling. |
| | EPA Programs | EPA launched initiatives like the Nanoscale Materials Stewardship Program to evaluate nanomaterial risks, focusing on environmental impacts. |
| **National certifications** | NanoTrust (Iran) | Voluntary certification system, emphasizing nanoscale conformity, quality assurance, and safety for consumer products. |
| | NanoQ (Thailand) | First Thai certification for nanotechnology products, covering textiles, paints, and household items. Jointly managed by the National Nanotechnology Center. |
| | NanoMark (Taiwan) | World's first nanotechnology certification system; aims to protect consumers and enhance the global competitiveness of Taiwanese nanotech products. |
| **Health, safety, and environmental (HSE) considerations** | HSE Guidelines and Occupational Safety | International and regional regulations focus on HSE through WHO guidelines, NIOSH exposure limits, and country-specific occupational safety standards. |
| | Life cycle assessment (LCA) | Suggested as a critical component for assessing environmental impact, waste management, and long-term effects of nanomaterials, although comprehensive LCA standards are still developing. |

such as ISO 10993–1 and ASTM International F748, provide essential protocols for assessing cytotoxicity, mutagenicity, and hemolysis. Moreover, assessing known or potentially toxic chemical entities is crucial, and the contents of biocompatibility test reports must be carefully defined. Numerous biomaterials are available for toxicity testing; however, the design for acute or subchronic toxicity assessments must comply with regulatory standards and the intended use of the product. Safety testing for biologics, particularly when biomaterials are combined with biological molecules or cells, must align with the FDA regulations outlined in 21 CFR parts 600. The U.S. The Pharmacopeial Convention provides methods for conducting both *in vitro* and *in vivo* biological reactivity tests, ensuring a comprehensive approach for assessing the safety and biocompatibility of biomaterial-based products (Hench and Thompson 2010). Scalability is critical because lab processes often struggle at the industrial level, impacting material properties and consistency. Pilot projects and computational modeling can identify challenges early, whereas continuous manufacturing techniques may ensure consistent production. Environmental issues arise from the potential toxicity of nanomaterials and their production chemicals, necessitating green chemistry, life cycle analyses, and advanced wastewater treatment to minimize their ecological impacts. Moreover, the integration of nanomaterials into existing workflows requires equipment adjustment and training. Comprehensive training, collaboration with manufacturers, and standardized quality control can facilitate this transition.

## 18.6 FUTURE TRENDS AND INNOVATIONS

The future of nanotechnology and biomaterials in smart manufacturing promises to revolutionize production through nanoscale material enhancements, thereby improving efficacy and sustainability. Nanotechnology modifies materials at the molecular level, resulting in biomaterials with superior mechanical, thermal, and electrical properties for aerospace, automotive, and healthcare applications. Combining nanotechnology with biomaterials may lead to intelligent materials capable of responding to environmental changes, facilitating adaptive manufacturing, and self-repairing products. This synergy promotes sustainable techniques by sourcing biomaterials from renewables, enhancing bioplastics, and enabling advanced methods, such as nanoscale 3D printing, for precise structures. Sensitive nanotechnology-based sensors can improve real-time monitoring, quality control, and efficiency. The increasing use of biomaterials in medical devices, implants, and tissue engineering has met the demand for sustainable healthcare. Collaborative robots (cobots) enhance production safety and accuracy. Data analytics and AI refine the material selection and manufacturing processes. Integrating nanotechnology and biomaterials in manufacturing fosters innovation, transforms industries, and enhances the quality of life. Intelligent medicine that can sense and administer therapies is essential for medical progress. Despite significant research on responsive hydrogels and polymers, the effectiveness of patient treatment relies on biological interactions. Hydrogel-based materials integrated into micro- and nanosystems are promising for drug delivery and biosensor technologies. Micro- and nanotechnologies present new opportunities for simulating biological systems and impacting therapeutic procedures via tissue engineering, drug delivery, and cellular biosensors. These advancements will broaden the application of these technologies in therapy and medication delivery.

Nanoscale engineering is a powerful tool for biological applications driven by advancements in lithography, such as nanoimprints. Many polymers, such as responsive hydrogels, can be shaped using micro- and nano-fabrication, thereby facilitating high-throughput production for clinical use. Recent research has highlighted how responsive hydrogel biomaterials can aid in drug delivery, tissue engineering, and biosensing through improved nanoimprint methods. Over a million patients benefit from controlled release systems annually, highlighting their impact on medicine. However, challenges remain in the development of drug delivery systems that target specific conditions. Interest in biomedicine has increased because of these challenges. Recent evaluations of hydrogels sensitive to environmental factors have shown promise as smart biomaterials for drug delivery and biosensors. Hydrogel materials are adaptable and can be modified with biomolecules, enabling the development of targeted therapies for specific diseases. The evolution of intelligent therapeutic and diagnostic systems has transformed tissue engineering and controlled drug delivery systems. Many innovative systems for FDA approval will begin preclinical research in the coming years. Integrating multiple components into nanoscale features allows for efficient and economical production necessary for clinical application. We suggest that a new drug administration field will evolve, creating devices that deliver precise therapeutic doses in response to biomarkers or stimuli and provide treatment only when needed.

## 18.7 CONCLUSION

In recent decades, nanotechnology has significantly impacted biomaterials by introducing innovative materials that enhance their application efficiency, selectivity, and cost-effectiveness. Novel biomaterials developed using nanotechnology exhibit unique properties and functions, which were previously unattainable. However, owing to safety and toxicity concerns, only certain types of nanomaterials are currently used, as nanomaterials vary widely in their characteristics. When incorporating nanoparticles into biomaterials, it is crucial to consider their potential toxicity and immune response. Nanomaterial toxicity often depends on factors such as surface characteristics, size, and dosage. Therefore, further research is essential for understanding thc environmental and toxicological effects of nanobiomaterials. Many nanomaterials can respond to various stimuli such as temperature, magnetism, light, electricity, and sound, allowing for the future development of intelligent and automated nanobiomaterials. Instead of single-function applications, multifunctional nanobiomaterials have potential, such as those that promote bone and tissue regeneration while reducing the immune response. Additionally, nanobiomaterials can be integrated with electronic devices such as nanoelectrochemical systems and advanced nanobioelectronics. Characterization methods and models are being developed to better understand the properties of nanobiomaterials, supporting the future applications of nanotechnology in biomaterials.

## REFERENCES

Ahmad I, Yazdani B, Zhu Y (2015) Recent advances on carbon nanotubes and graphene reinforced ceramics nanocomposites. *Nanomaterials* 5 (1):90–114

Ahmad MB, Tay MY, Shameli K, Hussein MZ, Lim JJ (2011) Green synthesis and characterization of silver/chitosan/polyethylene glycol nanocomposites without any reducing agent. *International Journal of Molecular Sciences* 12 (8):4872–4884

Allan J, Belz S, Hoeveler A, Hugas M, Okuda H, Patri A, RauscHer H, Silva P, Slikker W, Sokull-Kluettgen B (2021) Regulatory landscape of nanotechnology and nanoplastics from a global perspective. *Regulatory Toxicology and Pharmacology* 122:104885

Basova TV, Vikulova ES, Dorovskikh SI, HAssan A, Morozova NB (2021) The use of noble metal coatings and nanoparticles for the modification of medical implant materials. *Materials & Design* 204:109672

Bharti A, Jain U, Chauhan N (2024) From lab to field: nano-biosensors for real-time plant nutrient tracking. *Plant Nano Biology* 100079

Bhatt A, Anbarasu A (2017) Nanoscale biomaterials for 3D printing. *IOSR Journal of Pharmacy and Biological Sciences* 12:80–86

Binns C (2021) *Introduction to nanoscience and nanotechnology*. John Wiley & Sons

Biswal T, BadJena SK, Pradhan D (2020) Sustainable biomaterials and their applications: a short review. *Materials Today: Proceedings* 30:274–282

Borisenko VE, Ossicini S (2013) *What is what in the nanoworld: a handbook on nanoscience and nanotechnology*. John Wiley & Sons

Budhiraja N, Peta S, Singh D (2024) Antibacterial, antioxidant, and cytotoxicity analysis of green synthesis silver nanoparticles from Oregano, Rosemary, and Thyme leaf extract. *Ecology Environment and Conservation* 30:102–106. doi:10.53550/EEC.2024.v30i06s.016

Chandra DK, Reis RL, Kundu SC, KumAr A, Mahapatra C (2024) Nanomaterials-based hybrid bioink platforms in advancing 3D bioprinting technologies for regenerative medicine. *ACS Biomaterials Science & Engineering* 10 (7):4145–4174

Chauhan V, Dhiman VK, Mahajan G, Pandey A, Kanwar SS (2023) Synthesis and characterization of silver nanoparticles developed using a novel lipopeptide (s) biosurfactant and evaluating its antimicrobial and cytotoxic efficacy. *Process Biochemistry* 124:51–62

Gao G, Schilling AF, Yonezawa T, Wang J, Dai G, Cui X (2014) Bioactive nanoparticles stimulate bone tissue formation in bioprinted three-dimensional scaffold and human mesenchymal stem cells. *Biotechnol J* 9 (10):1304–1311. doi:10.1002/biot.201400305

Gao G, Zhang XF, Hubbell K, Cui X (2017) NR2F2 regulates chondrogenesis of human mesenchymal stem cells in bioprinted cartilage. *Biotechnology and Bioengineering* 114 (1):208–216

Grumezescu AM (2017) *Nanoscale fabrication, optimization, scale-up and biological aspects of pharmaceutical nanotechnology*. William Andrew

Hench LL, Thompson I (2010) Twenty-first century challenges for biomaterials. *Journal of the Royal Society Interface* 7 (suppl_4):S379–S391

Hosseini ES, Dervin S, Ganguly P, Dahiya R (2020) Biodegradable materials for sustainable health monitoring devices. *ACS Applied Bio Materials* 4 (1):163–194

Hudecki A, Kiryczyński G, Łos MJ (2019) Biomaterials, definition, overview. In: *Stem cells and biomaterials for regenerative medicine*. Elsevier, pp 85–98

Kharey P, Goel M, Husain Z, Gupta R, Sharma D, ManIkandan M, Palani I, Gupta S (2023) Green synthesis of biocompatible superparamagnetic iron oxide-gold composite nanoparticles for magnetic resonance imaging, hyperthermia and photothermal therapeutic applications. *Materials Chemistry and Physics* 293:126859

Kovářík T, Křenek T, Bělský P, Šesták J (2017) Biomaterials and nanotechnology approach to medical enhancement. *Thermal physics and thermal analysis: from macro to micro, highlighting thermodynamics, kinetics and nanomaterials*. Springer, pp 449–470

Kumarage V, Siriwardane I, Sandaruwan C, Kandanapitiya M, Kottegoda N (2022) Nanotechnology applications in biomaterials; A review. *Journal of Materials Research and Technology* 3:32–54

Liang W, Zhou C, Bai J, Zhang H, Long H, Jiang B, Liu L, Xia L, Jiang C, Zhang H (2024) Nanotechnology-based bone regeneration in orthopedics: a review of recent trends. *Nanomedicine* 19 (3):255–275

Nizamuddin S, Siddiqui MTH, Mubarak N, Baloch HA, Mazari SA, Tunio M, Griffin G, Srinivasan M, Tanksale A, Riaz S (2018) Advanced nanomaterials synthesis from pyrolysis and hydrothermal carbonization: a review. *Current Organic Chemistry* 22 (5):446–461

Omran BA (2020) *Nanobiotechnology: a multidisciplinary field of science*. Springer

Pogany A, Aungkavattana P, Bañares M, BeitollAhi A, Bim V, Briffa S, Bochon A, Cassee F, Doridot F, Exner T (2024) *INISS-Nano: revised concept and action plan (International Network Initiative on Safe and Sustainable Nanotechnologies), ICAM—Catholic Institute of Arts and Crafts*

Ratner BD, Zhang G (2020) A history of biomaterials. In: *Biomaterials science*. Elsevier, pp 21–34

Raut SS, Singh R, Lekhak UM (2024) Naturally occurring nanoparticles (NONPs): a review. *Next Sustainability* 3:100037

Reddy CV, Reddy KR, Shetti NP, MishrA A, Basu S (2019) Recent progress in TiO2-and ZnO-based nanostructured hybrid photocatalysts for water purification and hydrogen generation. In: *Nanoscale materials in water purification*. Elsevier, pp 815–843

Reibold M, Paufler P, Levin AA, Kochmann W, Pätzke N, Meyer D (2006) Carbon nanotubes in an ancient Damascus sabre. *Nature* 444 (7117):286–286

Sandhu A (2006) Who invented nano? *Nature Nanotechnology* 1 (2):87–87. doi:10.1038/nnano.2006.115

Shirazi S (2023) *Mesenchymal stem cell and immune cell modulation by nano-scale surface topography of dental implants*. University of Illinois at Chicago

Singh D, SHarma A, Verma S, Pandey H, Pandey M (2024) Impact of nanoparticles on plant physiology, nutrition, and toxicity: a short review. *Next Nanotechnology* 6:100081. doi:10.1016/j.nxnano.2024.100081

Soltani AM, Pouypouy H (2019) Chapter 19—Standardization and regulations of nanotechnology and recent government policies across the world on nanomaterials. In: Ghorbanpour M, Wani SH (eds) *Advances in phytonanotechnology*. Academic Press, pp 419–446. doi:10.1016/B978-0-12-815322-2.00020-1

Wang Y, Weng GJ (2018) Electrical conductivity of carbon nanotube-and graphene-based nanocomposites. In: *Micromechanics and nanomechanics of composite solids*. Springer, pp 123–156

# 19 Integration of Artificial Intelligence and Biomaterials

*Vivek Kumar Dhiman, Devendra Singh, Vinay Kumar Dhiman, Rajasekaran Subbarayan, and Rupendra Shrestha*

## 19.1 INTRODUCTION

Biomaterials are necessary because of their essential for various medical applications, including tissue engineering, drug delivery systems, implanted medical devices, and radiological phantoms. Biomaterials require a detailed analysis and understanding in relation to living tissues for interactions based on their properties. Enhancement of the design and optimization of biomaterials is required for performance (Kuusisto, 2024). A critical aspect of biomaterials is their interaction with biological systems, which is fundamental to their performance and efficacy. Biomaterials are foundational for creating medical devices, frameworks for tissue engineering, and sophisticated drug delivery systems (Chauhan et al., 2024a, 2024b). Artificial intelligence (AI)-powered algorithms are transforming this process by predicting material properties and quality by analyzing vast material databases and identifying them for further research. This data-based approach has significantly accelerated the development of new materials, surpassing the capacity of traditional methods feasible for human researchers. Beyond expediting discovery, AI enhances our understanding of fundamental interactions at the biomaterial interface. Accelerating the discovery of materials with desirable requirements has enabled the optimal design of biomaterials for specific applications. AI can interpret mechanical properties, biocompatibility, and degradation rates to generate different designs with precise specifications for implants as medical devices. Assurance of biomaterial development for biocompatibility to prevent adverse reactions while interacting with living tissues is paramount (Rajitha et al., 2024). AI models predict biocompatibility by analyzing the chemical composition and structure of materials, significantly reducing their dependence on extensive *in vitro* and *in vivo* tests. AI also facilitates the customization of biomaterials per individual need, particularly in tissue engineering, where personalized implants can be designed for unique anatomy. The information used by AI is used to support decision-making and hidden patterns using databases with information on experimental data on biological and material aspects of research. AI also enhances the biomaterial manufacturing processes by ensuring consistency and quality.


DOI: 10.1201/9781003566205-19

Automation and real-time monitoring of AI are used to reduce error rates and improve the production efficiency of more reliable and scalable biomaterials. Machine learning (ML) algorithms also efficiently detect patterns and predict outcomes from complex datasets that integrate physicochemical material properties and omics data. This capability improves the selection of materials for specific applications and provides insight into their bioactive mechanisms. AI is crucial for biomaterial personalization (Jeznach et al., 2024). As biomedicine actively moves toward accurate fabrication systems, disease diagnosis, tissue engineering, drug development and dosage, and other complex diagnoses, AI-assisted predictive algorithms can be increasingly used to create or develop biomaterials. This customized approach reduces side effects, maximizes therapeutic outcomes, and promises significantly improved patient care. Integrating AI into biomaterial research exemplifies the synergistic enhancement of both fields. The complexity of biological systems presents a rich landscape for AI to refine its algorithms, whereas AI enables biomaterials to explore an expansive realm of possibilities.

Recent advancements in computational technologies have led to the development of AI, and its components, such as ML, deep learning (DL), and neural networks (NN), have revolutionized the field of science (see Table 19.1). These cutting-edge technologies have opened new avenues for the design, optimization, and predictive modeling of biomaterials. This research explores the incorporation of ML and AI in developing biomaterials, emphasizing their roles in improving material characteristics, promoting creative design methods, and precisely forecasting material performance in various conditions. AI-driven methodologies were examined for their ability to accelerate the creation of novel biomaterials through data-informed decision-making and advanced modeling techniques. The potential of these technologies to accelerate the development of novel biomaterials is exciting. (Gokcekuyu et al., 2024). Additionally, ML approaches have been analyzed for their crucial role in predicting biomaterial responses to varying environmental stimuli, thereby significantly reducing trial-and-error experimentation and enhancing material reliability. This confirmed the efficiency and reliability of these technologies.

Specifically, we delve into how ML and other AI components can be harnessed to enhance the performance and functionality of biomaterials, thereby enabling more precise tailoring of their properties to meet specific medical needs. To further advance this field, future research should focus on expanding the application areas of these technologies and conducting more in-depth investigations into their specific contributions to biomaterial science. By doing so, we can fully realize the transformative potential of computational intelligence in revolutionizing biomaterial research and production, paving the way for groundbreaking advancements in medical sciences.

## 19.2 AI AND ML METHODS

AI is the computational imitation of human intelligence, which works with human-like cognition. These activities encompass visual perception, speech recognition, decision-making, and language translation. AI operates through multiple components, including ML, DL, and others, allowing machines to function with greater sophistication. AI and ML encompass a broad range of methods and techniques for creating systems that can learn from data, make decisions, and solve complex problems. Supervised learning is among the most popular techniques in ML, where models learn from labeled datasets that include both known input data and their corresponding outputs (Tyagi & Chahal, 2020). This approach works well for both classification and regression tasks. For instance, linear regression is commonly used to predict continuous outcomes, such as stock prices, housing values, and material properties in biomaterials, by identifying the relationships between variables. Similarly, logistic regression serves as a technique for binary classification problems, predicting discrete outcomes like "yes" or "no," such as whether a biomaterial is biocompatible. Advanced methods like decision trees model decision-making in a hierarchical way, while random forests enhance accuracy by merging several decision trees to minimize overfitting. Support vector machines (SVMs) classify complex biomaterial datasets by finding an optimal hyperplane to separate data classes, such as by predicting tissue adhesion or cellular responses to a given material. For simpler problems, k-nearest neighbors (k-NN) is often applied to classify biomaterials based on their similarity to known materials in a database (Rickert & Lieleg, 2022).

Inspired by the human brain, NN play a critical role in the analysis of vast biomaterial datasets. They are particularly effective in predicting the mechanical properties of polymers, biodegradation rates, and performance of tissue scaffolds. DL approaches, such as convolutional neural networks (CNNs), efficiently process medical imaging data, enabling tasks such as analyzing biomaterial–tissue interactions and identifying defects in implant structures (Tiwari et al., 2023). These AI and ML methods are increasingly utilized in biomaterial design and optimization, aiding the development of smart implants, tissue-engineered scaffolds, and drug delivery systems, thereby contributing significantly to the advancement of medical science (Kayadurmus et al., 2024). By leveraging these tools, researchers can analyze complex biological responses, accelerate biomaterial discovery, and predict the material performance more accurately. As a result, AI and ML have revolutionized the field of biomaterials, enabling the creation of highly personalized and efficient solutions in regenerative medicine, biomedical engineering, and healthcare (see Table 19.1).

In the development of biomaterials, AI and ML have been increasingly utilized to enhance the design, optimization, and prediction of material properties and biological interactions (Han & Wu, 2024). The process begins with data collection, in which various data points are gathered from experimental studies on biomaterial properties, biological responses, and environmental conditions. This can include mechanical properties, such as tensile strength, chemical composition, biological compatibility (e.g., cell adhesion), and factors such as temperature or pH. Data were then preprocessed, noise was removed, missing values were handled, and key features were extracted from raw experimental datasets, such as material composition or biological performance metrics. Data are also often normalized or standardized to ensure consistency across diverse datasets.

Once the data were prepared, exploratory data analysis was performed to identify trends, correlations, or outliers. Visualizations such as scatter plots or histograms are used to analyze the relationships between materials and their biological responses,

**TABLE 19.1**
**AI Components and Their Application (Gokcekuyu et al., 2024; Suwardi et al., 2022).**

| AI components | Type | Algorithms | Description | Application in biomaterials |
|---|---|---|---|---|
| **Machine learning** | **Supervised learning** | Linear Regression | Estimates ongoing values according to input variables. | Predicting mechanical properties of biomaterials (e.g., tensile strength). |
| | | Logistic egression | Predicts binary outcomes (classification). | Determining biocompatibility of materials (e.g., yes/no). |
| | | Decision trees | Model decisions in a tree-like structure. | Classifying biomaterials based on toxicity or material composition. |
| | | Random forest | A collection of decision trees designed to enhance accuracy and minimize overfitting. | Predicting biodegradability or material response to specific conditions. |
| | | Support vector machines (SVMs) | Data are classified by finding an optimal hyperplane to separate data classes. | Classifying tissue responses to implants or biomaterial interaction. |
| | | k-Nearest neighbors (k-NN) | Categorizes data points according to their closeness to existing data points. | Identifying similar biomaterials based on known properties. |
| | **Unsupervised learning** | Clustering (e.g., K-means) | Group data into clusters based on similarity. | Grouping biomaterials with similar chemical or mechanical properties. |
| | | Principal component analysis | Reduces the dimensionality of data while retaining significant features. | Reducing large biomaterial datasets for analysis and visualization. |
| | | Autoencoders | Neural networks that learn compressed data representations. | Identifying anomalies in biomaterial datasets or structural defects. |
| | **Reinforcement learning** | Q-Learning | Learning through rewards and penalties to optimize actions. | Optimizing biomaterial design for specific performance goals. |
| | | Deep Q-networks (DQN) | Integrates neural networks and reinforcement learning to tackle complex decisions. | Designing adaptive biomaterials with dynamic responses to environmental conditions. |

*(Continued)*

**TABLE 19.1** *(Continued)*
**AI Components and Their Application (Gokcekuyu et al., 2024; Suwardi et al., 2022).**

| AI components | Type | Algorithms | Description | Application in biomaterials |
|---|---|---|---|---|
| **Deep learning** | | Artificial neural networks (ANNs) | Models inspired by the human brain are effective for complex patterns. | Predicting biological responses like cell adhesion to biomaterials. |
| | | Convolutional neural networks | Processes image data to recognize spatial features. | Analyzing medical images for biomaterial–tissue interactions or defects. |
| | | Recurrent neural networks (RNNs) | Analyses sequential data for time-based predictions. | Monitoring changes in biomaterial degradation rates over time. |
| **Hybrid methods** | | Ensemble methods | Combines multiple algorithms to enhance performance. | Predicting material performance under varying biological environments. |
| | | Neuro-fuzzy systems | Integrates neural networks with fuzzy logic for decision-making. | Optimizing biomaterial properties based on uncertain biological data. |

and techniques for clustering, like k-means can help group materials with similar characteristics. Subsequently, the AI/ML models were trained using the data. To predict material properties, supervised learning techniques such as NN, SVMs, and random forests are commonly applied to predict the mechanical strength, degradation rates, and cell–material interactions (Sinha et al., 2023). For unsupervised learning tasks, techniques such as clustering or dimensionality reduction help discover hidden patterns in data or group biomaterials with similar properties. The AI/ML pipeline has accelerated the discovery of new biomaterials. This allows researchers to optimize material properties for specific applications, such as implants, scaffolds for tissue regeneration, and drug delivery systems (see Figure 19.1). By leveraging these advanced methods, the development of biocompatible and effective biomaterials is more efficient, with the potential to predict and optimize material performance and biological response before physical testing.

## 19.3 MACHINE LEARNING

ML trains computers to learn patterns and relationships from data without explicit programming. It has revolutionized industry by creating intelligent systems capable of performing cognitive tasks akin to human reasoning. Recent advancements in ML have resulted in the development of systems with human-like intelligence,

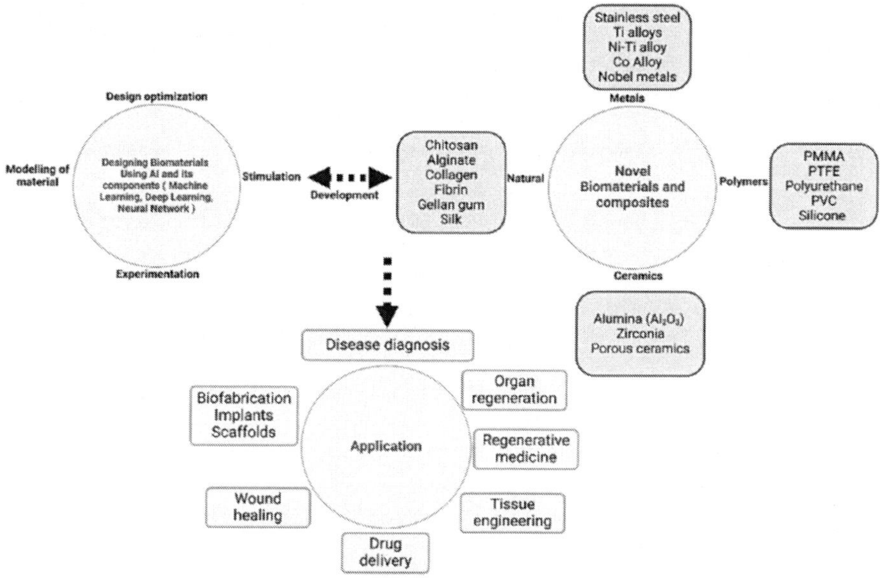


**FIGURE 19.1** Pipeline for the development of biomaterials using AI and its components.

profoundly transforming industries and improving daily life. ML-powered AI systems use analytical models to generate forecasts, make recommendations, and provide insights. These systems minimize the need for manual encoding of decision rules, relying instead on data-driven approaches for model development. A key application of ML in biomaterials is predictive modeling. Researchers can build models that forecast the behavior, properties, and performance of biomaterials by training algorithms on extensive and varied datasets (Gokcekuyu et al., 2024). For example, ML algorithms can predict biocompatibility or mechanical properties, minimizing reliance on labor-intensive and costly trial-and-error experiments. This approach significantly reduces development time and resource consumption while improving outcomes. ML facilitates accelerated design and synthesis of biomaterials.

Conventional material discovery techniques typically require extensive iterative experimentation. In comparison, ML algorithms can process large datasets, uncover significant patterns, and suggest enhanced material designs. This not only simplifies the development process but also encourages the creation of innovative materials that possess improved biocompatibility, mechanical properties, and overall performance. ML is driving innovation in the field by enabling tailored material design and opening doors for new applications. Another important field where ML is advancing is in the analysis and interpretation of medical imaging data related to biomaterials. Sophisticated medical imaging techniques produce large amounts of intricate data that can be difficult to process and analyze quickly. ML algorithms excel at analyzing and classifying medical images, assisting with disease diagnosis, patient monitoring, and treatment planning. Moreover, image-based ML tools can be used to assess biomaterial performance and integrity, providing rapid feedback for researchers and clinicians. (Patel & Webb, 2023). Despite its significant advantages,

**TABLE 19.2**
**Major Applications of AI and Biomaterials.**

| Field | AI applications | Biomaterials applications | Examples |
|---|---|---|---|
| Healthcare | AI enhances diagnosis by analyzing medical imaging such as MRI and X-rays, predicting treatment outcomes, and monitoring chronic diseases using data from wearable devices. | Biomaterials combine biocompatibility with functionality in implants, surgical instruments, wound dressings, and drug delivery systems. | AI models analyze CT scans to detect tumors with remarkable accuracy, often surpassing traditional diagnostic methods.<br>Biodegradable sutures: crafted from polylactic acid (PLA), these sutures naturally dissolve as the wound heals. |
| Regenerative medicine | AI predicts scaffold designs that replicate natural tissue architecture and optimizes protocols for efficient cell seeding and growth. | Biomaterials provide a foundation for tissue regeneration, supporting cell adhesion and integrating harmoniously with native tissues. | 3D-printed AI-designed scaffolds: tailored for repairing bone defects with precision.<br>Hydrogel scaffolds: engineered to deliver growth factors for effective cartilage regeneration. |
| Medical devices | AI enables real-time monitoring and feedback in wearables and implants, such as smart pacemakers. | Biomaterials create durable and biocompatible devices, such as stents, pacemakers, and orthopedic implants. | AI-enhanced smart pacemaker: continuously monitors heart activity and autonomously adjusts pacing. Titanium alloy hip implants: durable, long-lasting replacements for enhanced joint function. |
| Pharmaceuticals | AI speeds up drug discovery by simulating molecular interactions and refining drug designs. | Biomaterials provide controlled drug delivery, enhancing efficacy while minimizing side effects. | AI-driven drug discovery: identifying drug candidates for Alzheimer's disease.<br>Biodegradable microparticles: release drugs in a timed fashion for chemotherapy. |
| Surgical assistance | AI-driven robotic systems conduct precise surgeries, minimizing complications and shortening recovery time. | Biomaterials are utilized in surgical meshes, sutures, and wound dressings to enhance the healing process. | Da Vinci robotic system: AI-enhanced robotic surgeries for accurate and precise procedures.<br>Biomaterial meshes: help prevent hernia recurrence following surgery. |

**TABLE 19.2** *(Continued)*
**Major Applications of AI and Biomaterials.**

| Field | AI applications | Biomaterials applications | Examples |
|---|---|---|---|
| Diagnostics | AI processes data from imaging, blood tests, and genetic profiles to facilitate early disease detection. | Biosensors use biomaterials to detect specific biomarkers, such as glucose or proteins. | AI-driven radiology: detects early-stage pneumonia through X-ray analysis.<br>Glucose biosensors: utilizes nanomaterials for effective diabetes monitoring. |
| Prosthetics | AI improves prosthetic performance by analyzing user movement patterns and enhancing control. | Biomaterials enable the creation of prosthetic components that are lightweight, durable, and biocompatible. | AI-powered bionic arms: offer accurate movement and sensory feedback.<br>Silicone prosthetics: feature comfortable, skin-friendly designs. |
| Environmental Health | AI tracks air and water quality, analyses pollution sources, and forecasts environmental impacts. | Biomaterials replace harmful plastics, minimize waste, and aid in ecosystem restoration. | AI pollution mapping models: track $CO_2$ levels in cities.<br>Biodegradable plastics: replace traditional plastics in packaging. |
| Nanotechnology | AI designs nanoparticles with specific properties for diagnostics, drug delivery, or imaging. | Biomaterials at the nanoscale target specific cells or tissues, improving therapeutic efficiency. | AI-enhanced nanoparticles: target and deliver chemotherapy drugs to tumors.<br>Nanocoatings for implants: protect against bacterial infections. |
| Personalized medicine | AI analyzes patient data to tailor treatments, including drug dosage, therapy type, and schedules. | Biomaterials enable custom implants, drug delivery systems, or devices tailored to individual patient needs. | AI-driven personalized treatments: tailored for oncology patients.<br>Customized hydrogel implants: conform to individual joint structures. |
| Smart materials | AI creates adaptive materials that react to environmental changes, such as variations in temperature or pH. | Biomaterials incorporate sensors and adaptive systems for real-time adjustments. | AI-powered smart contact lenses: track glucose levels in tears.<br>pH-responsive hydrogels: release drugs exclusively in acidic environments, such as tumor sites. |

ML in biomaterial research has several challenges. Data availability and quality are key obstacles. Creating reliable ML models depends on standardized, high-quality datasets that are often hard to access and sufficiently large. Moreover, interpreting and validating these models can be complicated, requiring expertise and a thoughtful approach to their limitations.

Ethical considerations are vital in healthcare applications of ML. It is essential to protect data privacy, eliminate algorithmic biases, and uphold transparency in decision-making to foster trust and guarantee fair outcomes. Numerous ML approaches are now widely applied in biomaterial research, offering innovative solutions to complex challenges. These solutions empower researchers to tackle insurmountable obstacles, making them feel capable and confident. Below, we discuss the foundational principles of AI and ML methods, their advancements, and their integration into various domains, including biomaterials. Supervised learning is a part of ML in which models are trained on labeled datasets where the input and associated output (target) values are known. The main objective is to understand the correlation between the inputs and outputs, allowing the model to provide precise predictions of unfamiliar data. Supervised learning problems are often categorized into regression (forecasting continuous outcomes) and classification (identifying discrete groups). Common techniques consist of linear regression to forecast continuous variables, logistic regression for binary classification, decision trees, random forests for structured decision-making, and SVMs for categorizing data into separate groups (Gokcekuyu et al., 2024). NN, particularly DL models, are widely used to discern intricate correlations among labeled data. Supervised learning has many applications, including predicting the mechanical characteristics of biomaterials, detecting illnesses in medical imaging, assessing material biocompatibility, and classifying emails as spam or non-spam.

Conversely, unsupervised learning addresses unlabeled data to reveal hidden patterns, connections, or structures within a dataset, without specified outputs. In contrast to supervised learning, it does not require labeled examples, rendering it particularly advantageous for tasks such as clustering, wherein analogous data points are aggregated, and dimensionality reduction, which streamlines extensive datasets by diminishing their features while maintaining critical information (Gokcekuyu et al., 2024). Prominent approaches include k-means clustering for data segmentation into clusters, hierarchical clustering for tree-structured groupings, and principal component analysis for dimensionality reduction of intricate datasets. For biomaterials, unsupervised learning may uncover novel material classifications with analogous qualities, evaluate trends in biological reactions, or streamline extensive biomaterial information for enhanced displays. Supervised and unsupervised learning are essential for addressing intricate issues in healthcare, engineering, and data analysis.

Reinforcement learning (RL) is a type of ML where actions and decisions are based on learning during interaction with an environment. This trial and error aims to maximize a reward signal by adapting the actions to achieve long-term goals. RL has been applied in robotics, gaming, and healthcare (Fayaz et al., 2022). The objective is to map the actions during the learning process or through the learning policy and maximize the reward during the outcome. Value functions, including the state-value function (V) and action-value function (Q), denote the anticipated total reward associated with a particular policy. These functions assist in the decision-making process to maximize long-term rewards. The RL must encounter the exploration phase to find better ways to obtain strategies or techniques. The training utilizes algorithms like deep Q-networks, SARSA, and Q-learning to refine the agent's value estimates based on the rewards received and state transitions observed. Challenges with RL include exploration, dealing, and delays in results. Ethical and safe learning process reinforcement for real-world applications, autonomous complex behavior, adaptation of tasks of a wide variety, and feedback are all required to improve RL.

DL, a subset of AI and ML, utilizes artificial neural networks (ANNs) to tackle intricate problems by analyzing large datasets. Drawing inspiration from the human brain's architecture, DL algorithms are made up of several layers of interconnected nodes (neurons), which allow them to autonomously identify and learn high-level features from unprocessed input data (Kumar & Thakur, 2012). These layers, input, hidden, and output, allow the network to process data hierarchically and progressively identify patterns and relationships. DL excels in tasks that require the recognition of intricate patterns, such as image classification, speech recognition, natural language processing, and time-series forecasting. CNNs are widely used for image and video analysis, such as for detecting biomaterial defects or analyzing tissue interactions in medical imaging. Recurrent neural networks (RNNs) and their advanced forms, such as long short-term memory (LSTM) networks, are ideal for sequential data analysis, such as predicting biomaterial degradation over time or analyzing biological signals. DL methods, including autoencoders and generative adversarial networks (GANs), are applied to highly unstructured and complex data to identify anomalies or generate new biomaterial designs. DL requires massive computational power and large datasets for training, and often leverages graphics processing units and cloud computing to accelerate performance (Kumar & Thakur, 2012). Its ability to deliver high accuracy and adapt to diverse domains makes it a cornerstone in modern AI applications such as personalized medicine, biomaterial optimization, autonomous systems, and medical diagnostics.

DL utilizes NN with multiple layers to learn the hierarchical representations of data. This enables the handling of high-dimensional data, making DL particularly effective in fields such as image recognition, speech processing, and predictive modeling. The advantages of DL include its ability to outperform humans in controlled scenarios, particularly in terms of its precision and consistency. It reduces biases and adapts to data drift, ensuring robust performance over time (Ahmed et al., 2023). Facilitates the reuse of pre-trained models, accelerating deployment in new applications. AI and ML methods are increasingly being adopted in biomaterial research, leveraging their advanced problem-solving capabilities to tackle critical challenges. By analyzing large datasets, ML algorithms can precisely predict biomaterial behaviors, such as biocompatibility and degradation rates. This reduces the reliance on trial-and-error experimentation. DL models have enabled the development of custom-designed biomaterials with improved properties. These models analyze the relationships between material composition and performance, and offer innovative insights for optimizing material synthesis. ML models are instrumental in analyzing imaging data from biomaterial experiments, such as microstructural analysis or surface characterization. DL tools can classify images, identify defects, and assess the material integrity (You et al., 2024). Selecting the correct ML or DL model for specific applications remains challenging, particularly when transitioning from controlled research settings to real-world applications. Model interpretability and validation require domain-specific expertise to ensure a reliable output.

## 19.4 FUTURE DIRECTIONS

As programming frameworks, computational resources, and data accessibility improve, ML methods will become increasingly integral to biomaterial research. Future advancements will likely include combining AI, ML, and physics-based models for more accurate and interpretable predictions, addressing ethical concerns like data privacy and algorithmic fairness to ensure broader acceptance of AI-driven solutions, and leveraging AI with nanotechnology and bioinformatics for groundbreaking innovations in biomaterials. The future of ML in biomaterials research looks bright. With datasets increasing in size and quality, the precision and trustworthiness of ML models are on the rise. The combination of ML with innovative technologies like nanotechnology and bioinformatics presents significant opportunities for enhancing biomaterial development. Predictive models could anticipate degradation rates, optimize material selection for specific applications, or simulate long-term interactions with biological systems. This potential for advancement is fascinating and should inspire optimism within the scientific community.

## 19.5 CONCLUSION

Integrating AI and ML into biomaterial research represents a paradigm shift, enabling faster, more efficient, and more precise advancements. By addressing the current challenges and expanding interdisciplinary collaborations, this field can harness the full potential of ML to shape the next generation of biomaterials. The continued evolution of these technologies will likely drive significant breakthroughs in the development and application of biomaterials in medicine, industry, and other fields.

## REFERENCES

Ahmed, S. F., Alam, M. S. B., Hassan, M., Rozbu, M. R., Ishtiak, T., Rafa, N., Mofijur, M., Shawkat Ali, A., & Gandomi, A. H. (2023). Deep learning modelling techniques: current progress, applications, advantages, and challenges. *Artificial Intelligence Review*, *56*(11), 13521–13617.

Chauhan, A., Rana, G., Dutta, V., Kumari, A., Krishna Rao, S., Subbarayan, R., Ravi, K., Selvaraj, S., & Ghotekar, S. (2024a). Recent trends in phyto-mediated iron-based nanomaterials for environmental remediation and biomedical applications. *Inorganic Chemistry Communications*, *160*, 111976. https://doi.org/10.1016/j.inoche.2023.111976

Chauhan, V., Pandey, A., Mahajan, G., Dhiman, V., & Kanwar, S. S. (2024b). Synergistic exploration of surfactin-capped silver nanoparticles: bioinformatics insights, antibacterial potency, and anticancer activity. *3 Biotech*, *15*(1), 13. https://doi.org/10.1007/s13205-024-04174-5

Fayaz, S. A., Jahangeer Sidiq, S., Zaman, M., & Butt, M. A. (2022). Machine learning: an introduction to reinforcement learning. In *Machine Learning and Data Science: Fundamentals and Applications* (pp. 1–22). Wiley Online Library.

Gokcekuyu, Y., Ekinci, F., Guzel, M. S., Acici, K., Aydin, S., & Asuroglu, T. (2024). Artificial intelligence in biomaterials: a comprehensive review. *Applied Sciences*, *14*(15), 6590.

Han, S., & Wu, J. (2024). Artificial intelligence (AI) meets biomaterials and biomedicine. *Smart Materials in Medicine*, *5*(2), 251–255. https://doi.org/10.1016/j.smaim.2024.03.001

Jeznach, O., Tabakoglu, S., Zaszczyńska, A., & Sajkiewicz, P. (2024). Review on machine learning application in tissue engineering: what has been done so far? Application areas, challenges, and perspectives. *Journal of Materials Science*, *59*(46), 21222–21250. https://doi.org/10.1007/s10853-024-10449-2

Kayadurmus, H. M., Ayran, M., Goktug, S., Gunduz, O., & Dogan, C. (2024). Spectroscopic characterization of biomaterials for tissue engineering applications. In *Biomedical Materials & Devices* (pp. 1–17). Springer.

Kumar, K., & Thakur, G. S. M. (2012). Advanced applications of neural networks and artificial intelligence: a review. *International Journal of Information Technology and Computer Science*, *4*(6), 57–68.

Kuusisto, M. (2024). *Artificial Intelligence Solutions for the Recycling and Utilization of Biomaterials for Industrial Applications* (Bachelor thesis). HAMK Häme University of Applied Sciences, Hämeenlinna, Finland.

Patel, R. A., & Webb, M. A. (2023). Data-driven design of polymer-based biomaterials: high-throughput simulation, experimentation, and machine learning. *ACS Applied Bio Materials*, *7*(2), 510–527.

Rajitha, A., Kansal, L., Raj, G., Kalra, R., Dhamija, K., & Abdul-Zahra, D. S. (2024). Biomaterials and artificial intelligence: predictive modeling and design. *E3S Web of Conferences*.

Rickert, C. A., & Lieleg, O. (2022). Machine learning approaches for biomolecular, biophysical, and biomaterials research. *Biophysics Reviews*, *3*(2).

Sinha, P., Roshini, D., Daoo, V., Abraham, B. M., & Singh, J. K. (2023). Integrating machine learning and molecular simulation for material design and discovery. *Transactions of the Indian National Academy of Engineering*, *8*(3), 325–340. https://doi.org/10.1007/s41403-023-00412-z

Suwardi, A., Wang, F., Xue, K., Han, M. Y., Teo, P., Wang, P., Wang, S., Liu, Y., Ye, E., & Li, Z. (2022). Machine learning-driven biomaterials evolution. *Advanced Materials*, *34*(1), 2102703.

Tiwari, N., Gupta, B. K., Prakash, A., Husain, M., & Singh, D. (2023). Applying deep learning and machine learning algorithms for the identification of medicinal plant leaves based on their spectral characteristics. *Journal of Advanced Zoology*, *44*.

Tyagi, A. K., & Chahal, P. (2020). Artificial intelligence and machine learning algorithms. In *Challenges and Applications for Implementing Machine Learning in Computer Vision* (pp. 188–219). IGI Global.

You, S., Fan, Y., Chen, Y., Jiang, X., Liu, W., Zhou, X., Zhang, J., Zheng, J., Yang, H., & Hou, X. (2024). Advancements and prospects of deep learning in biomaterials evolution. *Cell Reports Physical Science*, *5*(9).

# 20 Future Trends in Smart Biomanufacturing

*Adhishree Yadav, Pramod Yadav, Urvashi Srivastava, and Manash K. Paul*

## 20.1 INTRODUCTION

Smart biomanufacturing involves integrating advanced technologies and intelligent systems into the production of biological products like pharmaceuticals, biofuels, and food. This approach utilizes innovations such as artificial intelligence (AI), the internet of things (IoT), big data analytics, and automation to enhance efficiency, sustainability, and product quality (Gargalo et al., 2020). Key components include automation and robotics, which streamline processes and reduce human error; data analytics that inform decision-making and predictive maintenance; IoT integration for real-time monitoring of production conditions; and AI, which optimizes processes and improves product yields. The focus on sustainability aims to minimize waste and energy consumption, promoting a circular economy. Benefits include increased efficiency, improved quality, cost reductions, and faster time-to-market. Applications span pharmaceuticals, food and beverages, and biofuels, though challenges such as integration complexity, data security, and skill gaps remain (Veloso & Ferreira, 2017).

The significance of future trends in smart biomanufacturing lies in their potential to transform production processes, enhance efficiency, and promote sustainability across various industries. As technology continues to evolve, several key trends are expected to shape the future landscape. The increasing use of automations, robotics, AI, and machine learning (ML) will streamline production, reduce human error, and allow for more flexible manufacturing systems, such as personalized medicine and tailored food products, enabling companies to respond quickly to market demands. A growing emphasis on environmental responsibility will lead to more sustainable practices, including waste reduction, resource optimization, and the use of renewable materials, aligning with global sustainability goals (Veloso & Ferreira, 2017). Future trends will focus on enhancing supply chain transparency and resilience through IoT and blockchain technologies, ensuring that disruptions are minimized and that quality standards are maintained. Increased collaboration among industry, academia, and research institutions will foster innovation, leading to the development of new processes and technologies that enhance biomanufacturing capabilities. Overall, these trends will not only improve operational efficiencies and product quality but also address pressing global challenges such as sustainability and health, positioning smart biomanufacturing as a crucial player in the future of manufacturing (Gargalo et al., 2020).

 DOI: 10.1201/9781003566205-20

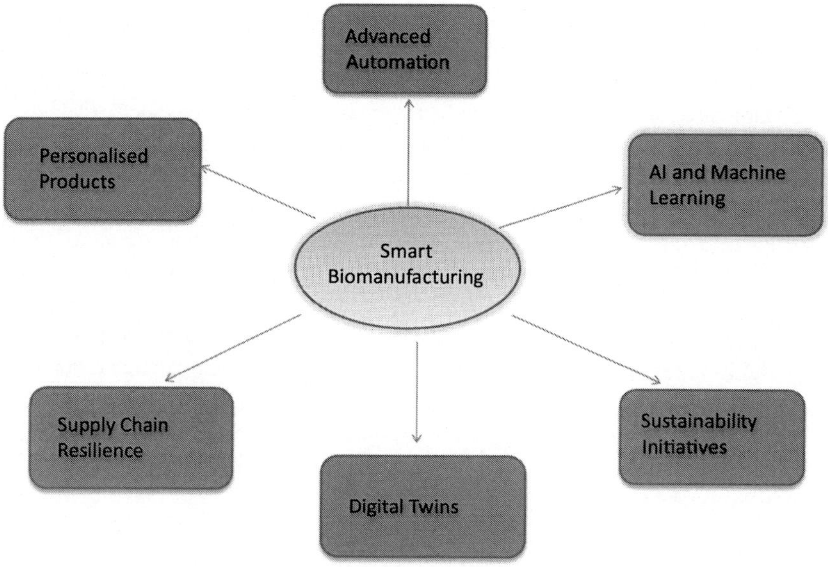


**FIGURE 20.1** Schematic diagram of key trends of smart biomanufacturing.

## 20.2 TECHNOLOGICAL INNOVATIONS

Technological innovations in smart biomanufacturing are revolutionizing the production of biological products, enhancing efficiency, quality, and sustainability. Here are some key innovations shaping the field (Figure 20.1).

### 20.2.1 Integration of AI and ML

AI refers to the development of computer systems that can perform tasks typically requiring human intelligence. These tasks include things like problem-solving, understanding natural language, recognizing patterns, and making decisions. ML is a subset of AI that focuses on creating algorithms and models that allow computers to learn from data and improve their performance over time without being explicitly programmed for each task (Veloso & Ferreira, 2017). The integration of AI and ML refers to using ML techniques to enhance AI systems. By embedding ML into AI, we give AI systems the ability to learn, adapt, and become more accurate over time. AI and ML algorithms analyze vast datasets from bioprocesses to identify optimal conditions for fermentation, cell culture, and purification. By predicting outcomes based on historical data, these technologies can fine-tune parameters like temperature, pH, and nutrient concentrations, leading to higher yields and reduced cycle times. ML models predict equipment failures before they occur by analyzing sensor data (Helleckes et al., n.d.). This proactive approach minimizes downtime, reduces maintenance costs, and ensures uninterrupted production processes. AI-powered analytics can monitor critical quality attributes in real-time during production. By detecting deviations early, these systems enable immediate corrective actions, ensuring consistent product quality and compliance with regulatory standards. AI enhances supply chain management by forecasting demand and optimizing inventory levels. AI tools can integrate and analyze data from various sources—such as laboratory experiments, production logs, and environmental sensors—providing a comprehensive view of the manufacturing process (Praveen Chakravarthi et al., n.d.). In drug discovery, AI and ML accelerate the identification of promising drug candidates by analyzing large datasets from high-throughput screening experiments. This leads to faster development timelines and a more efficient R&D process and customized of biological products, such as personalized medicine or tailored food products. AI models help design and optimize metabolic pathways in microorganisms for producing specific biochemicals. This approach allows for the engineering of strains that maximize yield and minimize byproducts. Digital twin technology, powered by AI, creates virtual representations of biomanufacturing processes. This allows for simulation, scenario analysis, and real-time monitoring, facilitating process improvements and predictive analytics (Praveen Chakravarthi et al., n.d.; Saha et al., 2024).

### 20.2.2 Automation and Robotics in Biomanufacturing

Automation and robotics are essential for advancing biomanufacturing, significantly enhancing efficiency, precision, and consistency in production processes. These technologies streamline workflows, reduce cycle times, and facilitate continuous operation, which boosts overall productivity. By performing repetitive tasks with high accuracy, robotics minimizes human error and ensures consistent product quality. Additionally, automation decreases reliance on manual labor for routine tasks, allowing skilled workers to focus on more complex processes (Lee et al., 2023). Safety is also improved, as robots can handle hazardous materials and work in environments that may be unsafe for humans. Moreover, automated systems can be easily scaled to meet changing production demands, enabling manufacturers to adapt swiftly to market fluctuations (Tang et al., 2020; Wolf et al., 2022). There are several applications of this technology, such as robotics are used to manage and monitor fermentation processes, ensuring optimal conditions for microbial growth and product yield. In drug discovery, automated systems facilitate rapid screening of compounds, accelerating the identification of potential drug candidates. Automation streamlines cell culture processes, including inoculation and media exchange, improving overall efficiency and consistency. Automated systems equipped with sensors conduct real-time monitoring of critical quality attributes, enabling immediate corrections and reducing waste (Figure 20.2). Robotics enhances the packaging process by ensuring accurate packing and labeling, which is essential for regulatory compliance (Lee et al., 2023).

### 20.2.3 Advanced Monitoring and Sensor Technology

Advanced monitoring and sensor technology are revolutionizing biomanufacturing by providing real-time data that enhances process culture medium exchange (CEME) control, quality assurance, and overall operational efficiency (Konishi et al., 2021).

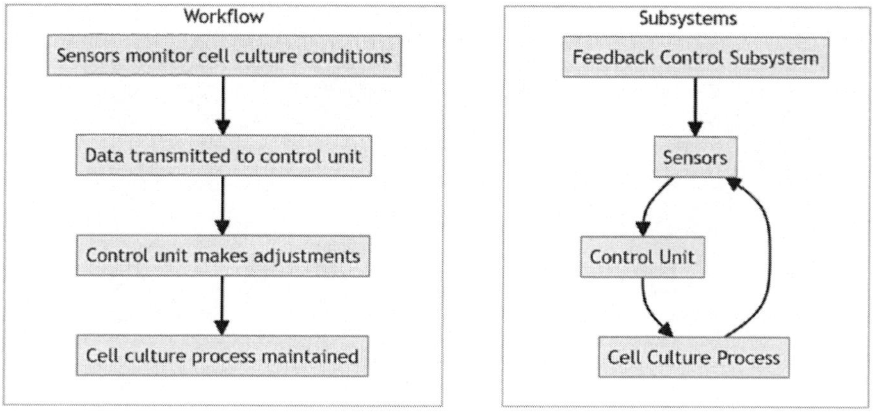


**FIGURE 20.2** Flowchart of an automated cell culture system composed of subsystems for feedback control.

Advanced sensors are used in bioreactors to monitor conditions like temperature, pH, and gas concentrations, ensuring optimal growth conditions for microorganisms or cells. Automated systems can take samples at set intervals, reducing the need for manual intervention while maintaining data integrity and accuracy. Sensors integrated directly into the production line provide continuous analysis of product quality attributes, enabling real-time adjustments to maintain consistency. Advanced monitoring systems track environmental conditions within production facilities, ensuring compliance with safety and regulatory standards. Sensors can monitor inventory levels and environmental conditions during transportation and storage, enhancing traceability and quality assurance throughout the supply chain (Figure 20.3). Furthermore, historical data analysis allows for predictive analytics, which helps identify potential issues proactively, reducing downtime (Konishi et al., 2021).

### 20.2.4 Case Studies

Amgen, a leading biotechnology company, implemented ML algorithms to analyze large datasets from its bioprocessing operations. By using predictive analytics, Amgen was able to identify optimal conditions for cell growth and product yield in its bioreactors. This approach not only increased production efficiency but also reduced the time and resources required for process development (Heyndrickx et al., 2024). Genentech employs advanced robotic platforms for the purification of biopharmaceuticals. Their automation systems handle the repetitive tasks involved in chromatography and filtration, ensuring high precision and reducing the risk of contamination (Menéndez, 2004). Merck implemented advanced sensor technologies in their fermentation processes for vaccine production. By using a combination of multi-parameter sensors, they monitored critical variables such as biomass, nutrient levels, and byproduct concentrations in real time. This approach allowed Merck to optimize fermentation conditions, resulting in a 30% increase in product yield and

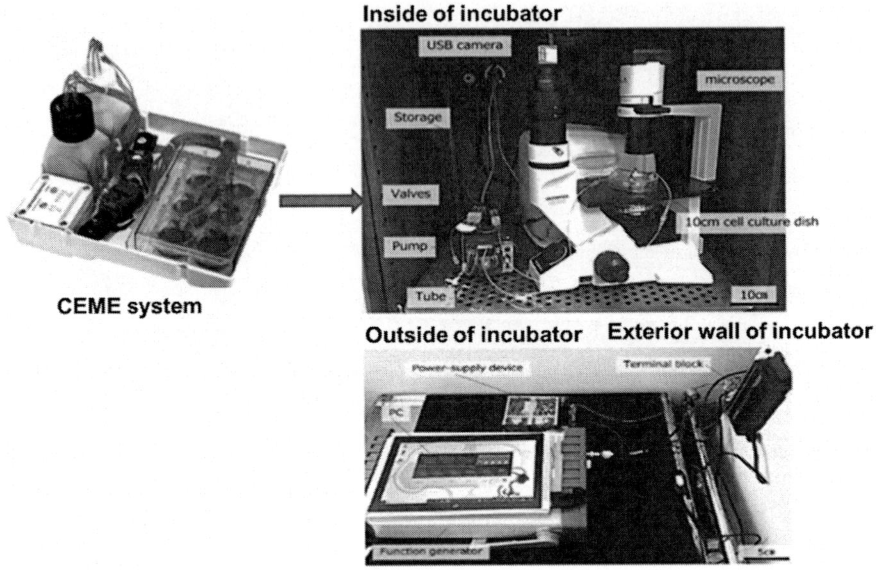


**FIGURE 20.3** The CEME system is a portable device for the automatic replacement of cell culture media, which is used in applications for the continuous change of cell culture media.

enhanced batch consistency (*Merck Study: MRNA Technology a 'Game Changer' for Asia-Pacific Vaccine Manufacturers | Merck*, n.d.).

## 20.3 DIGITIZATION IN BIOMANUFACTURING

Digitization in biomanufacturing involves integrating digital technologies into the production processes of biological products, such as vaccines, enzymes, and therapeutic proteins, leading to greater efficiency, improved quality, and enhanced regulatory compliance. Key aspects include real-time monitoring through advanced sensors, in-line analytics using technologies like Raman spectroscopy, and automation in bioreactors that reduces manual intervention and variability. Centralized data platforms integrate information from production and supply chains, often leveraging cloud computing for enhanced accessibility (Udugama et al., 2021). Advanced analytics and AI facilitate predictive analytics, enabling the identification of potential quality issues before they occur, while digital twins allow for real-time simulation and optimization of processes. The use of blockchain technology enhances traceability and transparency in the supply chain. The benefits of digitization are substantial, including increased efficiency through reduced downtime, improved product quality due to enhanced monitoring, and streamlined compliance with regulatory requirements (Figure 20.4). Furthermore, digitization leads to cost reductions by minimizing waste and optimizing resource utilization, while providing greater flexibility and scalability in production. Notable case studies highlight these advancements: Amgen

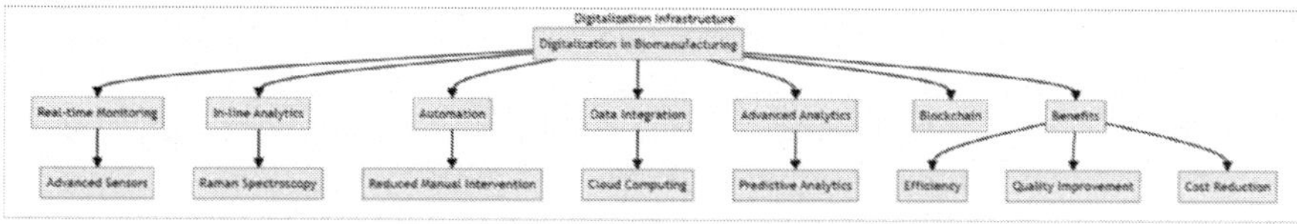


**FIGURE 20.4** Digitalization infrastructure in biomanufacturing.

has integrated digital platforms to enhance production agility, Pfizer accelerated vaccine development during the COVID-19 pandemic using digital technologies, Genentech adopted AI-driven quality control systems to improve product consistency, and Bristol Myers Squibb implemented a digital supply chain strategy for enhanced resilience (Udugama et al., 2021).

### 20.3.1 Advancements in the Role of Digital Twins

Digital twins provide a virtual representation of physical assets and systems, allowing for continuous monitoring of biomanufacturing processes. By integrating data from sensors and IoT devices, they enable real-time tracking of critical parameters such as temperature, pH, and cell density, facilitating immediate adjustments to maintain optimal conditions. Advancements in ML algorithms allow digital twins to analyze historical data and simulate various production scenarios. This capability helps identify the most efficient operating conditions and streamline workflows, resulting in increased yields and reduced production times (Udugama et al., 2021). Digital twins facilitate predictive maintenance by continuously analyzing the health of equipment and processes. They can forecast potential failures or maintenance needs before they occur, minimizing downtime and reducing operational costs. Digital twins create a shared platform for stakeholders across different departments, such as R&D, production, and quality assurance. This collaboration fosters knowledge sharing and accelerates decision-making, improving overall operational efficiency. By maintaining a comprehensive digital record of the production process, digital twins help ensure compliance with regulatory requirements (Chen et al., 2020). They enable easy data access for audits and inspections, streamlining the compliance process. Digital twins serve as valuable training tools for employees, providing a safe environment for learning about process operations and troubleshooting without disrupting actual production. This simulation capability enhances workforce readiness and skill development. Digital twins support the entire product life cycle, from development to commercialization. They enable companies to analyze performance data post-launch, helping to inform future improvements and adaptations.

### 20.3.2 Impact on Design and Operational Strategies

The integration of digital twins in biomanufacturing profoundly impacts both design and operational strategies. In terms of design, digital twins enhance process simulation, allowing engineers and scientists to visualize and optimize entire manufacturing workflows before physical implementation (Zhao et al., 2023). This capability facilitates iterative design improvements, enabling rapid prototyping and refinement of bioreactor configurations and process flows. Additionally, digital twins support scalability and flexibility, ensuring that systems can adapt to varying production demands. They also streamline regulatory compliance by incorporating necessary checks into the design phase, reducing approval delays. On the operational side, digital twins provide real-time monitoring and control of critical parameters, empowering operators to make data-driven adjustments that enhance process stability. Predictive analytics embedded in digital twins allow for proactive maintenance, minimizing downtime and disruptions. Furthermore, the data generated supports informed decision-making across the organization, fostering collaboration among cross-functional teams, including R&D, production, and quality assurance (Canzoneri et al., 2024). This shared understanding enhances innovation and problem-solving capabilities. Lastly, by simulating various operational scenarios, digital twins identify opportunities for resource optimization and waste reduction, thereby supporting sustainability initiatives. Overall, the influence of digital twins is significant, driving improvements in efficiency, quality, and adaptability in biomanufacturing processes (Canzoneri et al., 2024; Zhao et al., 2023).

### 20.3.3 Case Study

Bristol Myers Squibb implemented digital twin technology to optimize their biomanufacturing processes, particularly in the production of monoclonal antibodies. The company created a digital twin of their bioreactor systems, allowing for real-time monitoring of critical process parameters. This included integrating data from sensors that tracked temperature, pH, and nutrient levels. The use of digital twins enabled more precise control over the production process, resulting in improved yield and reduced batch variability. The ability to simulate various scenarios also allowed for rapid adjustments in response to unexpected changes, enhancing overall efficiency (*Revolutionizing Healthcare through Data Science and Predictive Technology—Bristol Myers Squibb*, n.d.).

## 20.4 SINGLE-USE TECHNOLOGIES IN BIOMANUFACTURING

Single-use technologies (SUT) in biomanufacturing refer to disposable systems and components that facilitate the production of biopharmaceuticals. These technologies have gained widespread adoption due to their ability to enhance flexibility, minimize contamination risks, and streamline manufacturing processes. This detailed overview explores the various components, advantages, challenges, and notable case studies associated with SUT in biomanufacturing (Figure 20.5). There are many key components of SUT, such as disposable bioreactor bags (Mitra & Murthy, 2022).

These are typically made from biocompatible materials such as polyethylene or other polymers, are utilized for cell culture and fermentation processes. These bioreactors allow for the growth of cells or microorganisms without the need for traditional stainless-steel reactors. Each bioreactor is used for a single batch, significantly reducing the time required for cleaning and sterilization between runs. Single-use filtration systems incorporate disposable filters designed for applications

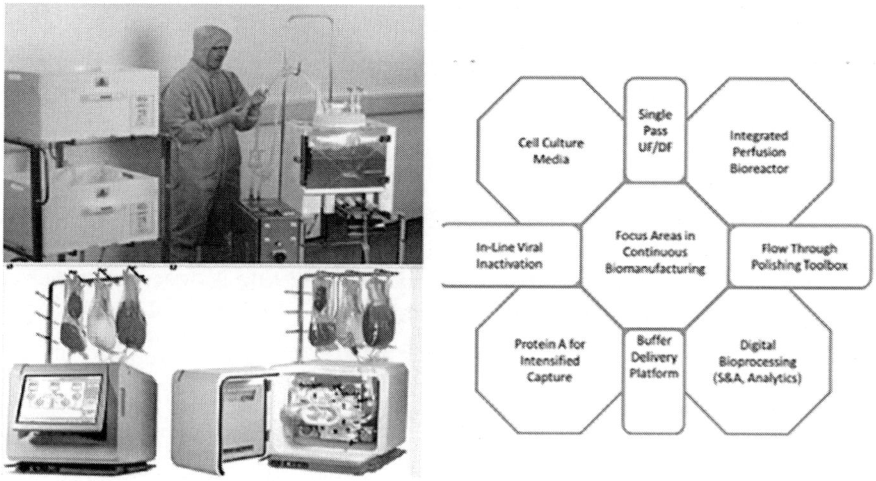


**FIGURE 20.5** Single-use bioreactor and disposable biopharmaceutical technology for smart biomanufacturing.

such as cell retention, clarification, and sterilization of the final product. Single-use filtration helps maintain sterility throughout the production process while ensuring that only desired materials pass through (*Large-Scale, Single-Use Depth Filtration Systems*, n.d.; Love et al., n.d.). They are particularly important in the upstream and downstream processing stages. Disposable mixing bags and systems facilitate the mixing of components, such as cell culture media and other ingredients. These systems minimize the risk of contamination and allow for easy handling. They are especially useful for small-batch productions where traditional mixing equipment would require extensive cleaning. Disposable sampling systems enable aseptic collection of samples during the production process. These devices are designed to maintain sterility while providing easy access for quality control testing at various stages of production, ensuring that product quality is consistently monitored. Disposable connectors and tubing systems are employed for the transfer of fluids between various stages of biomanufacturing. These components facilitate fluid transfer without the need for cleaning or sterilization, which enhances operational efficiency and reduces the risk of contamination (Love et al., n.d.; Zhao et al., 2023).

These SUT offer various advantages over traditional such as the use of disposable components minimizes the chances of cross-contamination between batches. Each component is used only once, ensuring that no residual material from previous batches can affect the new production run. While the initial investment in single-use systems can be higher than traditional systems, SUT often lead to lower operational costs. This is due to reduced requirements for cleaning, validation, and maintenance, resulting in significant savings over time. SUT enable manufacturers to quickly adapt to changing production needs and market demands (Samaras et al., 2022). This flexibility is crucial in an industry where product specifications and quantities can shift rapidly. The complexity of cleaning and sterilization protocols is greatly reduced with SUT, leading to shorter production cycles and increased throughput. This efficiency allows companies to respond more rapidly to customer needs and market changes. Digital twins and modeling of single-use systems facilitate easier scale-up from laboratory to commercial production. Companies can quickly adjust processes based on virtual simulations, making it easier to adapt to different batch sizes (Langer & Rader, 2014).

Despite the various advantages, there are still many challenges that need to be in considerations. Selecting materials for single-use systems that do not leach harmful substances into the product is critical. Extensive testing is necessary to ensure that materials maintain product integrity and safety. The increased use of disposable components raises concerns regarding environmental sustainability and waste management (Dhiman et al., 2022; Verma et al., 2024a, 2024b). Manufacturers must develop strategies for responsible disposal and consider options for recycling or repurposing materials. Single-use technologies must comply with stringent regulatory standards. Thorough validation and documentation processes are necessary to demonstrate that these systems meet safety and efficacy requirements set by regulatory bodies (Langer & Rader, 2014; Samaras et al., 2022).

**Case studies:** Genentech integrated single-use bioreactors into its production lines to enhance operational efficiency. The switch to disposable bioreactors led to a significant reduction in cleaning times and increased flexibility, allowing Genentech to respond quickly to changes in production demands. Amgen adopted single-use systems across its biomanufacturing processes to improve efficiency and reduce contamination risks. This initiative resulted in a 25% reduction in production turnaround times, enabling faster delivery of products to market while maintaining high-quality standards (*Revolutionizing Healthcare through Data Science and Predictive Technology—Bristol Myers Squibb*, n.d.).

## 20.5 ADVANCEMENTS IN BIOMANUFACTURING TECHNOLOGY

Biomanufacturing technology has advanced significantly due to innovations in several areas. Synthetic biology allows for the efficient design of biological systems for producing biochemicals and biofuels, while CRISPR and gene editing enable precise modifications in microorganisms, enhancing yields and product quality. Metabolic engineering optimizes production pathways for pharmaceuticals and bio-based materials, and microbial consortia improve efficiency and resilience in production processes. Automation and AI enhance precision and speed in manufacturing, and cell-free systems offer flexibility by utilizing extracted cellular machinery. Sustainability initiatives focus on renewable feedstocks and waste reduction, making biomanufacturing more eco-friendly (Gargalo et al., 2020).

### 20.5.1 Genome Engineering and Synthetic Biology

Genome engineering encompasses a suite of techniques that enable precise alterations to an organism's DNA. The advent of CRISPR-Cas9 technology has marked a revolutionary shift in this field, providing researchers with a powerful tool for gene editing. CRISPR, short for clustered regularly interspaced short palindromic repeats, allows for targeted modifications at specific genomic locations, significantly enhancing the efficiency and accessibility of genetic manipulation. CRISPR-Cas9 technology utilizes a guide RNA to direct the Cas9 enzyme to a specific DNA sequence, where it creates a double-strand break (Hillary & Ceasar, 2023). The cell's natural repair mechanisms then introduce changes at the target site. This method has broad applications, from creating genetically modified organisms (GMOs) to potential therapeutic strategies for genetic disorders. An advancement on CRISPR, base editing enables the conversion of one DNA base pair into another without causing double-strand breaks. This precision minimizes errors and off-target effects, making it a promising tool for correcting point mutations associated with various diseases. Often described as "the next generation of CRISPR," prime editing allows for more versatile and accurate DNA modifications (Li et al., 2023). It can insert, delete, or replace sequences with minimal collateral damage, representing a significant leap in gene editing capabilities. Standardized biological parts, known as BioBricks, standardized genetic components can be assembled in various combinations to create new biological functions. This modular approach simplifies the engineering of complex biological systems. Metabolic engineering involves reprogramming the metabolic pathways of organisms to enhance the production of specific compounds. For example, microorganisms can be engineered to produce biofuels, pharmaceuticals, and specialty chemicals, offering sustainable alternatives to traditional production methods. Advances in gene synthesis technology allow for the rapid and cost-effective production of synthetic DNA. This capability is crucial for constructing complex genetic circuits and novel organisms (Hillary & Ceasar, 2023).

Genome engineering is used to develop crops with enhanced traits, such as disease resistance, improved nutritional profiles, and increased yield. For example, scientists have successfully created rice varieties that are resistant to specific pathogens, reducing reliance on chemical pesticides. Gene therapy is a rapidly advancing application of genome engineering. Techniques like CRISPR are being tested in clinical trials to treat genetic disorders such as sickle cell disease and muscular dystrophy by correcting mutations at their source (Boti et al., 2023; Li et al., 2023). Genome engineering is being employed to develop microbial strains capable of bioremediation—using engineered microbes to clean up environmental pollutants. This application highlights the potential for genome engineering to address ecological challenges. Synthetic biology integrates principles from engineering and biology to design and construct new biological parts, devices, and systems. This multidisciplinary approach aims to create organisms with novel functionalities, expanding the boundaries of what is possible in biotechnology. Synthetic biology is revolutionizing the production of various products, from biofuels to food ingredients. Engineered microorganisms can convert renewable resources into high-value products, reducing dependence on fossil fuels and minimizing environmental impact (Boti et al., 2023). In the realm of medicine, synthetic biology is paving the way for personalized therapies, including tailored vaccines and engineered probiotics that can target specific health issues. These innovations hold promise for more effective treatments and disease prevention strategies. Synthetic biology is also being utilized in environmental conservation efforts. For instance, engineered plants can be designed to sequester carbon more efficiently, contributing to climate change mitigation (Boti et al., 2023; Li et al., 2023).

### 20.5.2 Personalized Medicine and Future Trends

Personalized medicine and biomanufacturing are rapidly evolving fields that are increasingly intertwined. Advances in genomic sequencing are enabling more precise identification of individual genetic variations, leading to tailored treatment plans, especially in oncology and rare diseases. The use of biomarkers to predict treatment responses is becoming standard. This helps in selecting the most effective therapies and monitoring disease progression (Ghebrehiwet et al., 2024). Innovations in CRISPR and other gene-editing technologies allow for targeted therapies that can correct genetic disorders at the source. Wearables and mobile health apps are being used to collect patient data in real-time, enhancing personalization through continuous monitoring and feedback. There is a growing emphasis on involving patients in treatment decisions, leading to more personalized care based on individual preferences and lifestyles.

In current biomanufacturing trends, the traditional batch manufacturing is being replaced by continuous processes, which improve efficiency, reduce costs, and allow for more consistent product quality. The integration of AI and ML in biomanufacturing is optimizing production processes, predicting equipment failures, and enhancing quality control. Biomanufacturers are increasingly adopting modular systems that can be quickly adapted for different products, improving responsiveness to market demands. There's a strong push toward greener biomanufacturing practices, including the use of renewable resources, waste minimization, and energy-efficient processes. As personalized medicine grows, biomanufacturing will need to adapt to produce smaller batches of tailored therapies, requiring more flexible and agile manufacturing systems. Overall, the future of personalized medicine and biomanufacturing is poised for significant transformation, driven by technological advancements and a deeper understanding of individual patient needs (Boti et al., 2023; Ghebrehiwet et al., 2024; Vicente et al., 2020).

### 20.5.3 Next-Generation Biopharmaceutical Production

Biopharmaceuticals, or biologics, are a type of medicine produced using living organisms, such as bacteria, yeast, or mammalian cells. Unlike traditional small-molecule drugs, which are chemically synthesized, biopharmaceuticals are derived from biological sources, making them more complex in structure and function. Advanced techniques like genetic engineering, recombinant DNA technology, and monoclonal antibody production are at the forefront of this revolution. These methods allow scientists to design and produce drugs that can target specific cells or proteins in the body, offering more effective treatments with fewer side effects (Vicente et al., 2020).

One of the key innovations in next-generation biopharmaceutical production is the use of recombinant DNA technology. This involves inserting a gene of interest into a host organism's DNA, enabling it to produce large quantities of the desired protein. For example, insulin used to treat diabetes was once extracted from animal pancreases. Another groundbreaking area is the production of monoclonal antibodies. Monoclonal antibodies are identical immune cells that are clones of a unique parent cell and can bind to specific targets in the body. This technology has been transformative in treating diseases like cancer, autoimmune disorders, and infectious diseases. For instance, Herceptin, a monoclonal antibody used to treat certain breast cancers, specifically targets the HER2 protein on cancer cells, thereby blocking their growth (Cinti et al., 2024; Ghebrehiwet et al., 2024). Gene therapy is also a promising frontier. By delivering new or modified genes into a patient's cells, gene therapy can potentially correct genetic disorders at their source. This approach offers hope for treating diseases that were once thought incurable, such as cystic fibrosis, muscular dystrophy, and certain types of cancer. Another exciting development is the advent of biosimilars. Biosimilars are highly similar to already approved biopharmaceuticals but are typically offered at a lower cost. They provide more treatment options and foster competition, potentially lowering healthcare costs without compromising quality (Cinti et al., 2024).

### 20.5.4 Adoption of Industry 4.0 Technology

Industry 4.0 is the term used to describe the fourth industrial revolution, a period marked by the fusion of digital, physical, and biological worlds. Imagine factories where machines can communicate with each other, make decisions, and carry out tasks with minimal human intervention. It is like the factory has a brain, making it more intelligent and efficient. One of the core components of Industry 4.0 is the IoT, robotics, AI, and ML, which enable it to handle complex tasks, adapt to new situations, and even collaborate with human workers (Figure 20.6) (Nargund et al., 2023).

Big data and analytics also play a significant role in Industry 4.0. Manufacturers now have access to vast amounts of data from various sources, such as production lines, supply chains, and customer feedback. Cloud computing is another pillar of Industry 4.0. By storing and processing data in the cloud, manufacturers can access information from anywhere at any time. This enables real-time collaboration, flexibility, and scalability. For instance, design teams in different locations can work on a project simultaneously, speeding up the development process and bringing products to market faster (Pandey et al., 2024). Augmented reality (AR) and virtual

**FIGURE 20.6** Robotic packaging done by pick and pack machines for accurate and smart biomanufacturing.

reality (VR) are also making their mark on Industry 4.0. These technologies allow workers to visualize complex data, simulate production processes, and train in a virtual environment. For example, AR can provide maintenance workers with real-time, step-by-step instructions, overlaid onto their view of the equipment, ensuring accurate and efficient repairs. VR, on the other hand, can be used for training purposes, allowing workers to practice their skills in a safe, controlled environment before applying them in the real world. One of the most transformative aspects of Industry 4.0 is the concept of the smart factory. A smart factory integrates all these technologies to create a highly automated and flexible production environment. In a smart factory, machines, devices, and systems are interconnected, communicating and coordinating with each other to optimize production processes. This results in faster production times, higher-quality products, and lower costs. However, adopting Industry 4.0 technologies is not without its challenges (Pandey et al., 2024). One major hurdle is the need for a skilled workforce that can manage and operate these advanced technologies. Companies must invest in training and education to ensure their employees have the necessary skills to thrive in this new environment. Additionally, the initial investment in Industry 4.0 technologies can be significant, but the long-term benefits, such as increased efficiency and cost savings, often outweigh the costs (Isoko et al., 2024).

### 20.5.5 Case Studies

Pfizer has been at the forefront of adopting smart biomanufacturing technologies. They have integrated advanced sensors and IoT devices into their production lines to monitor and control bioprocesses in real-time. This has led to improved productivity, reduced waste, and enhanced product quality. By leveraging big data analytics, Pfizer can predict maintenance needs and optimize production schedules, ensuring a seamless manufacturing process (*Pfizer Applys Latest Digital Technology to Optimize COVID-19 Vaccine Development | Pfizer 2021 Annual Report*, n.d.).

## 20.6 SUSTAINABLE BIOMANUFACTURING PRACTICES

As we look to the future, it is critical to integrate sustainability into biomanufacturing. By doing so, we ensure that the advancements in biotechnology contribute to a healthier planet. This chapter delves into strategies for achieving environmental sustainability and enhancing energy efficiency while reducing waste in biomanufacturing.

### 20.6.1 Strategies for Environmental Sustainability

The biomanufacturing industry is uniquely positioned to spearhead initiatives that foster environmental sustainability. These strategies are multifaceted, focusing on reducing the environmental impact of production processes, minimizing waste, and optimizing resource use. One pivotal strategy is the adoption of green chemistry principles. For biomanufacturers, this means developing processes that use fewer toxic solvents, biodegradable materials, and renewable feedstocks. Another key approach is the implementation of closed-loop systems. In a closed-loop system, waste products are recycled back into the production process, reducing the need for raw materials and minimizing waste. This can be achieved through on-site waste treatment facilities that convert waste into reusable materials. Biomanufacturers should prioritize using raw materials that are sustainably sourced and have minimal environmental impact. This includes sourcing materials from suppliers that follow environmentally friendly practices and ensuring that materials are harvested or produced in a manner that does not deplete natural resources. The development of eco-friendly packaging is another important strategy. By using biodegradable or recyclable packaging materials, biomanufacturers can significantly reduce the environmental impact of their products. By understanding the full environmental impact of their products, biomanufacturers can identify areas for improvement and implement strategies to reduce their ecological footprint. Lastly, investing in employee training and education on sustainability can foster a culture of environmental responsibility within the organization.

### 20.6.2 Energy Efficiency and Waste Reduction

Energy efficiency is a cornerstone of sustainable biomanufacturing. By optimizing energy use and implementing energy-efficient technologies, biomanufacturers can reduce their energy consumption, lower operational costs, and decrease their carbon footprint. One effective strategy is the use of high-efficiency bioreactors. Modern bioreactors are designed to maximize energy efficiency by optimizing conditions for cell growth and product formation. These bioreactors often incorporate advanced control systems that monitor and adjust parameters such as temperature, pH, and oxygen levels in real-time, ensuring optimal performance with minimal energy use. By using LED lighting and advanced controls, biomanufacturers can reduce energy consumption and improve the working environment. LED lights are more energy-efficient and have a longer lifespan compared to traditional lighting, leading to cost savings and reduced environmental impact. The implementation of smart HVAC (heating, ventilation, and air conditioning) systems is another critical step. These systems use sensors and automation to adjust heating and cooling based on real-time conditions, ensuring energy is used efficiently.

Waste reduction is equally important in sustainable biomanufacturing. By minimizing waste generation and maximizing resource efficiency, biomanufacturers can create a more sustainable and cost-effective production process. Process intensification is one approach to waste reduction. This involves designing processes that achieve higher productivity with fewer resources and less waste. By optimizing the entire production process, biomanufacturers can reduce waste and improve overall efficiency. Advanced recycling techniques can also play a significant role in waste reduction. On-site recycling facilities can process waste materials and convert them into valuable resources that can be reused in the production process. The re-purposing of byproducts is another effective waste reduction strategy. By finding new applications for byproducts, biomanufacturers can turn waste into valuable resources. For example, waste biomass from fermentation processes can be used to produce biofuels, feed additives, or fertilizers. This not only reduces waste but also creates additional revenue streams for the company.

## 20.7 CHALLENGES AND FUTURE OUTLOOK

### 20.7.1 Regulatory Considerations and Compliance

Navigating the regulatory landscape is one of the most significant challenges in biomanufacturing. Given the complexity of biopharmaceuticals and the potential risks associated with their production and use, stringent regulatory frameworks are in place to ensure safety, efficacy, and quality. However, these regulations can also pose obstacles to innovation and efficiency (Cinti et al., 2024; Vicente et al., 2020). One of the primary regulatory challenges is the need for rigorous testing and validation. Biopharmaceuticals, unlike small-molecule drugs, are highly complex and variable. Regulatory agencies, such as the FDA (Food and Drug Administration) in the US and the EMA (European Medicines Agency) in Europe, require extensive testing to ensure that biopharmaceuticals are safe and effective. The time and cost associated with these processes can be substantial, potentially delaying the introduction of new therapies to the market. Additionally, regulatory requirements can vary significantly between countries. Companies must navigate different sets of rules, guidelines, and documentation requirements, which can be both time-consuming and costly (Nargund et al., 2023). Harmonizing regulatory standards across regions could help streamline this process, but achieving such alignment is challenging due to differing national priorities and regulatory philosophies. Biomanufacturers must implement comprehensive quality management systems (QMS) to ensure that their products consistently meet regulatory standards. This includes stringent controls on raw materials, manufacturing processes, and final product testing. Regulatory agencies conduct regular inspections and audits to verify compliance with these standards, and any deviations can result in significant penalties, product recalls, or even facility shutdowns (Pandey et al., 2024).

With the emergence of advanced biomanufacturing, regulatory agencies must adapt to keep pace with technological advancements, developing new guidelines and standards to ensure the safe and effective use of these cutting-edge therapies (Isoko et al., 2024). By fostering collaboration between regulatory agencies, industry stakeholders, and academic institutions, the biomanufacturing industry can work toward a more streamlined and adaptive regulatory environment that supports innovation while maintaining high standards for safety and quality (Nargund et al., 2023; Vicente et al., 2020).

### 20.7.2 Technological and Economic Challenges

The adoption of advanced biomanufacturing technologies brings numerous benefits, but it also presents significant technological and economic challenges. One of the primary technological challenges is the complexity of biopharmaceutical production. This complexity requires sophisticated production methods, advanced control systems, and highly skilled personnel to ensure consistent and high-quality products. Scaling up biopharmaceutical production from laboratory research to commercial manufacturing is another critical challenge (Pandey et al., 2024). The transition from small- to large-scale production can be fraught with difficulties, as the processes that work well in the lab may not translate smoothly to larger volumes. Factors such as cell growth, nutrient supply, and waste removal become more complicated at scale, requiring extensive process optimization and validation to ensure success. The high cost of biomanufacturing technologies is also a significant economic challenge. Advanced equipment, such as high-efficiency bioreactors, automated control systems, and analytical tools, can be prohibitively expensive, especially for small- and medium-sized enterprises (SMEs). Additionally, the cost of raw materials, such as specialized media and reagents, can be substantial. These economic barriers can hinder the entry of new players into the market and limit the ability of existing companies to invest in innovative technologies (Udugama et al., 2021).

Intellectual property (IP) considerations add another layer of complexity to the technological and economic landscape. Biomanufacturers must navigate a complex web of patents and proprietary technologies, which can result in costly licensing fees and potential legal disputes. Workforce development is another critical challenge. The advanced nature of biomanufacturing technologies requires a highly skilled workforce with expertise in areas such as molecular biology, bioengineering, and data analytics. Investing in education and training programs is essential to building a skilled workforce that can support the growth and innovation of the biomanufacturing industry (Cinti et al., 2024; Pandey et al., 2024).

### 20.7.3 Emerging Trends and Opportunities

The future of biomanufacturing is bright, with several emerging trends and opportunities poised to reshape the industry. One of the most promising trends is the rise of personalized medicine. Advances in genomics, proteomics, and bioinformatics are enabling the development of tailored therapies that target specific genetic and molecular profiles (Vicente et al., 2020). Another exciting trend is the integration of digital technologies into biomanufacturing processes. Industry 4.0 concepts, such as the IoT, AI, and big data analytics, are transforming the way biopharmaceuticals are produced (Pandey et al., 2024). Big data analytics can provide valuable insights into production performance, helping to identify areas for improvement and drive continuous optimization. The development of advanced biomanufacturing platforms, such as continuous bioprocessing and modular production systems, is also opening up new opportunities (Isoko et al., 2024). Continuous bioprocessing offers several advantages over traditional batch processes, including higher productivity, reduced downtime, and more consistent product quality. Modular production systems, which consist of standardized and interchangeable units, provide greater flexibility and scalability. The convergence of biomanufacturing and synthetic biology is another exciting area of opportunity. Synthetic biology involves the design and construction of new biological parts, devices, and systems, as well as the redesign of existing biological systems for useful purposes (Vicente et al., 2020). As the industry grows, there is increasing awareness of the need to minimize environmental impact and promote sustainable practices. Companies are investing in green technologies, such as renewable energy sources, biodegradable materials, and eco-friendly production processes. By prioritizing sustainability, biomanufacturers can reduce their environmental footprint, enhance their corporate social responsibility, and meet the growing demand for environmentally conscious products. Finally, collaboration and partnerships are driving innovation and growth in the biomanufacturing industry. By working together, biomanufacturers, academic institutions, regulatory agencies, and other stakeholders can share knowledge, resources, and expertise. Public–private partnerships and consortia are also playing a crucial role in advancing biomanufacturing technologies and fostering a supportive environment (Ghebrehiwet et al., 2024; Pandey et al., 2024; Vicente et al., 2020).

## 20.8 CONCLUSIONS

Smart biomanufacturing is rapidly evolving, pushing the boundaries of technology and redefining industry standards. This chapter has explored the various transformative trends—such as AI integration, digital twin technology, and Industry 4.0—that are not only enhancing efficiency and quality but also paving the way for sustainable practices in biomanufacturing. Moreover, advancements in genome engineering and synthetic biology are enabling breakthroughs in personalized medicine and tailored biopharmaceutical production, demonstrating the sector's pivotal role in addressing modern healthcare needs. Environmental responsibility and sustainability are also emerging as core elements in the future of biomanufacturing. Green chemistry principles, single-use technologies, and energy-efficient processes reflect a shift toward eco-friendly production. These practices are essential as the industry strives to balance innovation with environmental impact, ensuring that future growth aligns with global sustainability goals. In the coming years, regulatory frameworks will play a crucial role in the adoption and standardization of advanced technologies within biomanufacturing. As innovation accelerates, regulatory agencies face the challenge of adapting to new technologies like CRISPR, personalized medicine, and continuous bioprocessing. This holistic approach—combining technology, sustainability, and regulatory foresight—will undoubtedly shape the future of biomanufacturing, ensuring its relevance and resilience in an ever-evolving world.

## REFERENCES

A Comprehensive Strategy to Evaluate Single-use Bioreactors for Pilot-scale Cell Culture Production | American Pharmaceutical Review. (n.d.). Retrieved October 27, 2024, from www.americanpharmaceuticalreview.com/Featured-Articles/111768-A-Comprehensive-Strategy-to-Evaluate-Single-use-Bioreactors-for-Pilot-scale-Cell-Culture-Production/

Boti, M. A., Athanasopoulou, K., Adamopoulos, P. G., Sideris, D. C., & Scorilas, A. (2023). Recent advances in genome-engineering strategies. *Genes*, *14*(1), 129. https://doi.org/10.3390/GENES14010129

Canzoneri, M., Horner, M., & Banerjee, P. (2024). Digital twins in biomanufacturing. *Bioconvergence Revolution*, 59–81. https://doi.org/10.1007/978-3-031-61593-1_4

Chen, Y., Yang, O., Sampat, C., Bhalode, P., Ramachandran, R., & Ierapetritou, M. (2020). Digital twins in pharmaceutical and biopharmaceutical manufacturing: a literature review. *Processes*, *8*(9), 1088. https://doi.org/10.3390/PR8091088

Cinti, C., Trivella, M. G., Joulie, M., Ayoub, H., & Frenzel, M. (2024). The roadmap toward personalized medicine: challenges and opportunities. *Journal of Personalized Medicine*, *14*(6), 546. https://doi.org/10.3390/JPM14060546

Dhiman, V. K., Singh, D., Pandey, H., Chauhan, D., Dhiman, V. K., & Pandey, D. (2022). Mycoremediation of agricultural waste for the cultivation of edible mushroom. In: Aftab, T. (Ed.), *Sustainable Management of Environmental Contaminants. Environmental Contamination Remediation and Management*. Cham: Springer. https://doi.org/10.1007/978-3-031-08446-1_18

Gargalo, C. L., Udugama, I., Pontius, K., Lopez, P. C., Nielsen, R. F., Hasanzadeh, A., Mansouri, S. S., Bayer, C., Junicke, H., & Gernaey, K. V. (2020). Towards smart biomanufacturing: a perspective on recent developments in industrial measurement and monitoring technologies for bio-based production processes. *Journal of Industrial Microbiology and Biotechnology*, *47*(11), 947–964. https://doi.org/10.1007/S10295-020-02308-1

Ghebrehiwet, I., Zaki, N., Damseh, R., & Mohamad, M. S. (2024). Revolutionizing personalized medicine with generative AI: a systematic Review. *Artificial Intelligence Review*, *57*(5), 1–41. https://doi.org/10.1007/S10462-024-10768-5/TABLES/6

Helleckes, L. M., Hemmerich, J., Wiechert, W., Von Lieres, E., & Grünberger, A. (2023). Machine learning in bioprocess development: From promise to practice. *Trends in biotechnology*, *41*(6), 817–835.

Heyndrickx, W., Mervin, L., Morawietz, T., Sturm, N., Friedrich, L., Zalewski, A., . . . & Ceulemans, H. (2024). MELLODDY: cross-pharma federated learning at unprecedented scale unlocks benefits in QSAR without compromising proprietary information. *Journal of Chemical Information and Modeling*, *64*(7), 2331–2344. https://doi.org/10.1021/ACS.JCIM.3C00799

Hillary, V. E., & Ceasar, S. A. (2023). A review on the mechanism and applications of CRISPR/Cas9/Cas12/Cas13/Cas14 proteins utilized for genome engineering. *Molecular Biotechnology*, *65*(3), 311–325. https://doi.org/10.1007/S12033-022-00567-0/FIGURES/8

Isoko, K., Cordiner, J. L., Kis, Z., & Moghadam, P. Z. (2024). Bioprocessing 4.0: a pragmatic review and future perspectives. *Digital Discovery*, *3*(9), 1662–1681. https://doi.org/10.1039/D4DD00127C

Konishi, S., Hashimoto, T., Nakabuchi, T., Ozeki, T., & Kajita, H. (2021). Cell and tissue system capable of automated culture, stimulation, and monitor with the aim of feedback control of organs-on-a-chip. *Scientific Reports*, *11*(1), 1–14. https://doi.org/10.1038/s41598-020-80447-2

Langer, E. S., & Rader, R. A. (2014). Single-use technologies in biopharmaceutical manufacturing: a 10-year review of trends and the future. *Engineering in Life Sciences*, *14*(3), 238–243. https://doi.org/10.1002/ELSC.201300090

*Large-Scale, Single-Use Depth Filtration Systems. (n.d.). Retrieved October 27, 2024, from www.bioprocessintl.com/filtration/large-scale-single-use-depth-filtration-system*s

Lee, D. H., Kim, H., Sung, B. H., Cho, B. K., & Lee, S. G. (2023). Biofoundries: bridging automation and biomanufacturing in synthetic biology. *Biotechnology and Bioprocess Engineering*, *28*(6), 892–904. https://doi.org/10.1007/S12257-023-0226-X/METRICS

Li, Z. H., Wang, J., Xu, J. P., Wang, J., & Yang, X. (2023). Recent advances in CRISPR-based genome editing technology and its applications in cardiovascular research. *Military Medical Research*, *10*(1), 1–20. https://doi.org/10.1186/S40779-023-00447-X

Love, T., Cherradi, Y., & Process, S. (n.d.). *Adoption Of Single-Use For Final Filtration*. Retrieved October 27, 2024, from www.bioprocessonline.com/doc/adoption-of-single-use-for-final-filtration-0001

Menéndez, R. E. (2004). Process automation a new suite of automated bioreactors at genentech: A case study employing standards, guidelines, and software: From requirements to validation. *Pharmaceutical Engineering*, *24*(6), 8–16.

*Merck Study: mRNA Technology a 'Game Changer' for Asia-Pacific Vaccine Manufacturers | Merck. (n.d.). Retrieved October 27, 2024, from www.merckgroup.com/en/news/future-of-vaccine-manufacturing-31-05-2023.htm*l

Mitra, S., & Murthy, G. S. (2022). Bioreactor control systems in the biopharmaceutical industry: a critical perspective. *Systems Microbiology and Biomanufacturing*, *2*(1), 91–112. https://doi.org/10.1007/S43393-021-00048-6

Nargund, S., Mauch, K., Esswein, S., Niebel, B., Rafigh, M., & Pendse, P. (2023). *Biopharma 4.0: Digital Technologies Driving Smart Manufacturing* (pp. 319–335). https://doi.org/10.1007/978-3-031-45669-5_11

Pandey, K., Pandey, M., Kumar, V., Aggarwal, U., & Singhal, B. (2024). Bioprocessing 4.0 in biomanufacturing: paving the way for sustainable bioeconomy. *Systems Microbiology and Biomanufacturing*, *4*(2), 407–424. https://doi.org/10.1007/S43393-023-00206-Y/METRICS

*Pfizer Applys Latest Digital Technology to Optimize COVID-19 Vaccine Development | Pfizer 2021 Annual Report. (n.d.). Retrieved October 27, 2024, from www.pfizer.com/sites/default/files/investors/financial_reports/annual_reports/2021/story/latest-digital-technology-to-covid-vaccine-efforts/*

Praveen Chakravarthi, G., Rambabu, V., Ramamurty DSVNM, & Rahul, G. (n.d.). AI and machine learning in biotechnology: a paradigm shift in biochemical innovation. *International Journal of Plant, Animal and Environmental Sciences*, *14*(4), 70–80. Retrieved October 27, 2024, from www.fotunejournals.com/ai-and-machine-learning-in-biotechnology-a-paradigm-shift-in-biochemical-innovation.html

*Revolutionizing Healthcare Through Data Science and Predictive Technology—Bristol Myers Squibb. (n.d.). Retrieved October 27, 2024, from www.bms.com/media/media-library/scientific-media-resources/revolutionizing-healthcare-through-data-science-and-predictive-t.htm*l

Saha, R., Chauhan, A., & Rastogi Verma, S. (2024). Machine learning: an advancement in biochemical engineering. *Biotechnology Letters*, *46*(4), 497–519. https://doi.org/10.1007/S10529-024-03499-8/METRICS

Samaras, J. J., Micheletti, M., & Ding, W. (2022). Transformation of biopharmaceutical manufacturing through single-use technologies: current state, remaining challenges, and future development. *Annual Review of Chemical and Biomolecular Engineering*, *13*, 73–97. https://doi.org/10.1146/ANNUREV-CHEMBIOENG-092220-030223

Tang, T., Fu, L., Guo, E., Zhang, Z., Wang, Z., Ma, C., Zhang, Z., Zhang, J., Huang, J., & Si, T. (2020). Automation in synthetic biology using biological foundries. *Chinese Science Bulletin*, *66*(3), 300–309. https://doi.org/10.1360/TB-2020-0498

Udugama, I. A., Lopez, P. C., Gargalo, C. L., Li, X., Bayer, C., & Gernaey, K. V. (2021). Digital Twin in biomanufacturing: challenges and opportunities towards its implementation. *Systems Microbiology and Biomanufacturing*, *1*(3), 257–274. https://doi.org/10.1007/S43393-021-00024-0

Veloso, A. C., & Ferreira, E. C. (2017). Online analysis for industrial bioprocesses: broth analysis. In *Current Developments in Biotechnology and Bioengineering: Bioprocesses, Bioreactors and Controls* (pp. 679–704). https://doi.org/10.1016/B978-0-444-63663-8.00023-9

Verma, S. K., Kumar, P., Mishra, A., et al. (2024a). Green nanotechnology: illuminating the effects of bio-based nanoparticles on plant physiology. *Biotechnology for Sustainable Materials*, *1*, 1. https://doi.org/10.1186/s44316-024-00001-2

Verma, S. K., Singh, D., & Misra, S. (2024b). Innovative approaches to sustainable waste management and energy generation in developing nations. In R. Mehra, P. Kumar, & B. Hung (Eds.), *Municipal Solid Waste Management and Recycling Technologies* (pp. 267–300). IGI Global Scientific Publishing. https://doi.org/10.4018/979-8-3693-4054-7.ch010

Vicente, A. M., Ballensiefen, W., & Jönsson, J. I. (2020). How personalised medicine will transform healthcare by 2030: the ICPerMed vision. *Journal of Translational Medicine*, *18*(1), 1–4. https://doi.org/10.1186/S12967-020-02316-W/FIGURES/1

Wolf, Á., Wolton, D., Trapl, J., Janda, J., Romeder-Finger, S., Gatternig, T., Farcet, J. B., Galambos, P., & Széll, K. (2022). Towards robotic laboratory automation Plug & Play: the "LAPP" framework. *SLAS Technology*, *27*(1), 18–25. https://doi.org/10.1016/j.slast.2021.11.003

Zhao, B., Li, X., Sun, W., Qian, J., Liu, J., Gao, M., Guan, X., Ma, Z., & Li, J. (2023). BioDT: an integrated digital-twin-based framework for intelligent biomanufacturing. *Processes*, *11*(4), 1213. https://doi.org/10.3390/PR11041213

# 21 Advanced Manufacturing and Industry 4.0

## *Transforming Manufacturing Systems*

*Pushpendra Tyagi, Naveen Mani Tripathi, Nikhil Gupta, Dhruv Kumar, Virat Mani Tripathi, Ambrish Singh, Sachchidanand Sharan Das, Rahutosh Ranjan, and Ankit Sharma*

### 21.1 INTRODUCTION

From steam-driven mechanisms to computer- and digital-based fully automated production systems, manufacturing has undergone continuous and rigorous revolutions since the first Industrial Revolution. These changes have made manufacturing more complex, automated, and sustainable, allowing people to operate machines with ease, efficiency, and persistence (Qin et al., 2016). The fourth major industrial uprising, or "Industry 4.0," is defined as a different degree of organization and command over the entire value chain of a product's life path, with a tendency toward more customized customer needs (Rüßmann et al., 2015). Industry 4.0's primary goal is to fulfill the needs and expectations of individual customer, which has a direct impact on areas such as order management, R&D, manufacturing systems, delivery, and product utilization and recycling (Neugebauer et al., 2016). The examination of the function of humans in the production environment is the real distinction between industrial uprising 4.0 and computer-integrated assembling and manufacturing. While CIM emphasizes workerless production systems, Industry 4.0 emphasizes the dominant role of human workers in carrying out production activities (Thoben et al., 2017). Physical objects, including sensors, gadgets, and enterprise assets, are encouraged to be connected to the internet and one another by the Industry 4.0 model (Sipsas et al., 2016). In order to transfer to a common, unified, and multifaceted method, process, and IT solution, design and drafting methods across all sectors should be evaluated and their acceptability for a new, modern, multidisciplinary approach blueprint for product development to be checked (Rennung et al., 2016). The production process is broken up into tiny, distinct units that only exchange information about the subsequent phases in the process, which increases flexibility and productivity and most likely reduces coordination complexity (Brettel et al., 2014). Industry 4.0 calls for replacing conventional machinery with intelligent, automated, and self-learning

DOI: 10.1201/9781003566205-21

devices in order to enhance overall performance and efficiency, as well as maintenance management in conjunction with environmental interaction (Lee et al., 2014). Building an open, high-quality manufacturing platform for organizational-networked information applications is the prime goal of Industry 4.0 (Bahrin et al., 2016). Industry 4.0's primary goals include real-time data observing, data analysis, product status in process and position tracking, and the ability to store instructions for production process control (Almada-Lobo, 2015). The idea of maintenance management is also seen as one of the crucial factors to be taken into account in an Industry 4.0 setting in order to benefit both technically and financially (Mosyurchak et al., 2017). Industry 4.0 involves the use of these technologies. Such technologies are given following:

(i) Connected manufacturing or IIoT
(ii) Big data
(iii) System integration (horizontal and vertical).
(iv) Simulations
(v) Clouds
(vi) Augmented reality
(vii) Autonomous robots
(viii) 3D printing
(ix) Electronic information security

The advent of smart and connected physical systems or physical computing, which are interconnected systems with one another via the network, is the basic foundation of connected manufacturing (IIoT). They do not require human interaction and permit the swapping of data on a national or international scale. Kaisler et al. (2013) define it as follows: Big data is defined as "the amount of data that is just too large for our technology to effectively store, manage, and process." A significant amount of electronic data is gathered in real time from CPSs in an Industry 4.0 culture. In order to create a highly dynamic and effective system, an integrated system (horizontal and vertical) entails connecting all of the major participants in the complete supply chain (Peres et al., 2018). The method of manufacturing process can benefit from the use of simulations as a digital tool, and they can be used to optimize real-time data from CPSs (Chong et al., 2018). The technical foundation for connecting and communicating different Application Center Industry 4.0 components is a cloud-dependent IT platform (Landherr et al., 2016). By superimposing digital data on reality, augmented reality (AR) technology can logically integrate human–CPS interaction (Figueiredo et al., 2014). In intelligent manufacturing facilities, robotic applications complement a wide range of services by assisting operators with their tasks and collaborating with different robots in the system (Sadik & Urban, 2017). Data from different inspection activities are also sent to robots. Physical objects may now be 3D printed thanks to additive manufacturing. For example, 3D CAD digital designs are used to depict complex shapes (Chua et al., 2010). According to Wells et al. (2013), cybersecurity is the amalgamation of intellectual systems to safeguard shared data against cyber threats. Shafiq et al. (2015) state that the aim of Industry 4.0 uprising are to apply the human–machine interaction (HMI) advanced model, attain IoT-sanctioned production optimization technique in smart industries, give new and advanced levels of services and value chain interaction business models, monitor products and parts, and help machines, products, and parts communicate with one another. Numerous nations have lately established a number of initiatives to promote the advancement and utilization of Industry 4.0 technology. Some other researchers also aimed at the cutting-edge and sustainable development of the bio-based additive manufacturing processes (Agrawal et al., 2024; Atal et al., 2023; Bhadouria et al., 2022; Ranjan et al., 2024a, 2024b; Ranjan & Tripathi, 2024; Tripathi & Sharma, 2024; Satishkumar et al., 2024a, 2024b; Jafar et al., 2024; Singh et al., 2024; Tripathi et al., 2024; Sharma et al., 2024; Ratsimba et al., 2021; Neveu et al., 2020b, 2020c, 2020d).

## 21.2 INDUSTRY 4.0 PRESENT STATUS AND TECHNOLOGIES

Humanity has advanced significantly with the fourth most advanced industrial revolution, which has been fueled by the technical advancements that people currently possess. It primarily relies on the automation or digitization of industries via the internet of things (IoT); it is also integrated with the modern, smart, and connected physical systems. These systems will be autonomous, meaning they will use machine learning technologies to make their own judgments. Additionally, this system will gather data in real-time situations, analyze it, and store it on the cloud. The IoT, big data, data examination, augmented reality, cybersecurity, robotics, layer manufacturing, 3D printing, cloud-based computing, artificial intelligence (AI), and 5G networks are just a few of the pillars that Industry 4.0 will enable. Furthermore, Industry 4.0 can reduce manufacturing costs by 10–30%, logistics expenses by approximately 10–30%, and quality management costs by approximately 10–20%, according to Rojko (2017). Because Intel industry equipment has network-connected sensors (IoT) that continuously collect and store data in the cloud system for further processing, the information has a direct connection to big data and data analytics. Additionally, because of the frequent communication and interaction between internet-connected devices, which render them extremely vulnerable to computer attacks, the IoT is associated with cybersecurity.

### 21.2.1 Major Technologies (Industry 4.0)

The major technologies used in Industry 4.0 nowadays include:

#### 21.2.1.1 Industrial Robots

Nowadays, industrial robots are mostly preprogrammed to do certain tasks, and the majority of commercially used robot applications have extremely limited human interaction and physical engagement capabilities. Full robot liberty for difficult tasks in shapeless environments will not be possible for some time. Advanced robot programming requires a highly competent programmer, and the amount of work needed for constantly changing assignments is impossible to accomplish. Human validation has been encouraged as a flexible framework for programming robots because it reduces the complexity of robot activities and enables end users to command robots naturally and easily without the requirement for exact programming. The traditional method involves having humans direct the object's movements while the robot acts as a humble and obedient agent in the interaction. For human–robot cooperation to be as symmetric as human–human collaboration, a robot must be able to perform both the active and passive aspects of the relationship, like a person would. The purpose of human expansion mapping in Kruijff et al. (2006) provides an illustration of this. More sophisticated compared to their predecessors, collaborative robots will act alongside humans in industries to complete a significant number of tasks efficiently. They will also enable a significant reduction in the costs associated with constructing safety cells that previously kept the robots out of the way of humans (Mikalef et al., 2018).

#### 21.2.1.2 Augmented Reality (AR)

Several services are supported by augmented reality (AR)-based systems, including mobile device maintenance instructions and part selection in a warehouse. Industries can use augmented reality to give employees real-time data to enhance decision-making and workflow. As they examine the real system that requires maintenance, workers may be given maintenance-related instructions on how to replace a specific part (Rüßmann et al., 2015). AR is a computer illustration style that uses virtual items in the actual world to transform the physical industrial environment into a digital one (Ceruti et al., 2019).

#### 21.2.1.3 Additive Manufacturing (AM)

Using data from 3D models, Layer manufacturing is a methodology that combines many types of intricate geometries and structures. This process consists of printing sequential layers of materials that are produced one on top of another. Layer manufacturing, which has evolved over decades and makes use of a range of advanced techniques, materials, and equipment, makes it possible to rebuild production and logistical systems. Kinesiology, prototyping, and construction are just a few of the industries that have made extensive use of additive manufacturing (Ngo et al., 2018). Powder rheology of additive manufacturing powders were also tested by different researchers to investigate the angle of repose, flowability, and electrostatic charge (Tripathi et al., 2019a, 2019b, 2019c, 2019d; Neveu et al., 2020a; Lumay et al., 2019; Kalman et al., 2019a, 2019b, 2019c).

#### 21.2.1.4 System Integration (Horizontal and Vertical)

Horizontal type integration is the cornerstone of a close, high-level collaboration between several organizations. It is associated with inter-company totality and leverages information technology to improve the product life cycle by creating a sophisticated, synchronized ecosystem within the same value creation network. A structured manufacturing and production system, vertical integration is linked to the intravenous business totality and serves as the foundation for data sharing and cooperation among the various tiers of the organization's power system, including smart management planning, good production programming, and excellent business planning.

#### 21.2.1.5 Cloud Computing

Another crucial subject for the role that networked system integration plays in the development of Industry 4.0 is cloud-dependent operating systems. The word "cloud" refers to cloud-dependent design and production as well as cloud computing.

Cloud-dependent manufacturing suggested a coordinated and networked production system that represents "available on-demand"-based manufacturing. Demand-based manufacturing develops and runs adaptable cyber-physical manufacturing processes by pooling distributed manufacturing resources. Here, the primary goal is to increase efficiency by lowering the costs associated with the product life cycle and enabling the best possible use of resources by managing customer-focused, variable-demand projects (Thames and Schaefer, 2017a, 2017b). Cloud-dependent design and production processes, according to Thames and Schaefer (2017a), show integrated and collaborative product creation techniques that depend on open modification through platforms based on social dealing and crowd-based sourcing.

#### 21.2.1.6 Smart and Connected Physical Systems

Technology that makes it easier to organize and integrate industrial networking structure between their computational power and physical infrastructure is called "smart and connected physical systems," or embedded systems. To carry out decentralized acts in this situation, digital and physical tools need to be coordinated and linked with other devices. Stated differently, embedded systems typically integrate novel practicality, such as computing and communication infrastructures, with physical reality (Bagheri et al., 2015).

#### 21.2.1.7 Artificial Intelligence (AI)

These days, collaborative robotics and AI are two different domains. AI is beginning to be recognized as another crucial Industry 4.0 technology. The employment of computer-advanced and modern programs or software to carry out challenging tasks is known as AI. These algorithms enable robots of all kinds to carry out specified tasks or jobs, allowing them to quickly and efficiently adapt to changing circumstances. Robots with AI are able to learn and interact with one another automatically. In the modern industrial complex, collaborative robots with AI provide a vital instrument for the most successful, efficient, and excellent operation of industrial processes (Benotsmane & Kov, 2019).

#### 21.2.1.8 Cybersecurity

Cybersecurity's primary duty is to safeguard data stored on a device or a network. The technology can safeguard information from both physical and cyber threats. It is essential to realize that there is always a danger involved and that no system available in the world is totally safe. Along with local security (cybersecurity), physical security (steel doors with good technology, safes, and biometric system) is a key method that helps preserve data in a data center, for instance, and should be considered. Cybersecurity concerns focus on the main key areas: data privacy, data completeness, and data openness (Isaca, 2017).

#### 21.2.1.9 Big Data

The term "big data" describes the enormous amounts of partially structured and raw data that are challenging to handle, examine, and process due to their complexity, the speed at which these databases are expanding, and the variety of sources from which they come. What an organization can do with its data is more important than its quantity. These figures can help a company, enterprise, or organization make critical decisions that reinforce their balance sheet. In addition to information from internet searches and social networks, the analysis of whole the information gathered by the latest techniques includes information from blogs, contactless technology sensors built into cars, trucks, and other devices, and more (Alcácer & Cruz-Machado, 2019; Vaidya et al., 2018).

#### 21.2.1.10 Simulation

The simulation is the newest, greatest, and most advanced tool available in terms of cost reduction. Simulation modeling is a complex technique that uses a real or hypothetical system replica to enable running some tests to evaluate system design, according to Alcácer & Cruz-Machado (2019). Simulation modeling makes it possible to create complex and flexible products that allow for the gathering of data and knowledge without intervening with the system program, allowing for the testing of various concepts and systems, resource policies, and new actions before their actual implementation.

#### 21.2.1.11 Internet of Things

The main purpose of IoT technology is to connect "objects" or gadgets to the network. Real-time data gathering on the production process of a product and the behavior of the organizational environment would be made possible by the integration of detectors and controllers into industrial technology. These data would then be analyzed using the big data technique. This significant volume of data will be stored in the cloud with the help of cloud-based computing technology (Rojko, 2017).

#### 21.2.1.12 5G Network

Important programming is made possible by the software-defined architecture of the 5G network, which creates distinct layers for various applications. New and unique commercial use cases will be made possible by this. Since 5G technologies will be largely powered by the Connected Network development mechanism, which will provide more power and allow for an impressive increase in the number of instruments linked to the 5G network, it is imperative that we talk about them. 5G networks offer better real-time application capability and are approximately 100 times quicker than 4G networks. It will provide users with continuous exposure by supporting billions of 5G devices and several billion applications at a rate of 10 gigabytes per second with a time interval of less than one millisecond. Along with the principles of Industry 4.0, this most recent technology can be a major factor in the integration of existing and cutting-edge technologies. These technologies can offer a great platform for integrating advanced vehicles, industrial processes, robotics, machines, and self-guiding machinery, among other things. 5G is a significant technological development that will enable the rapid and secure communication and interconnection of other technologies, supporting the amazing expansion of IOT in a smart company. Any manufacturing and production company planning to transition to a smart, modern, and intelligent industry in the near future will need a 5G network (Rao & Prasad, 2018).

## 21.3 INDUSTRY 4.0 CASE STUDY

The manufacturing process of DVDs is the best example of the utilization of technologies that use Industry 4.0.

### 21.3.1 Manufacturing Process of DVD

A large-scale manufacturing facility was established in Noida to produce CDs, DVD+Rs, DVD-Rs, and other products. High-quality products will be produced by the newest, fully automated DVD disc production lines. The production lines, which have a cycle time of two seconds per disc and are currently the fastest and most sophisticated for DVD discs, were acquired from the firm Origin, Taurus. The company employs a contemporary Guann Way offset printer for disc printing, which can produce full-color prints of the best quality at a pace of less than one second per disc. The following procedures are involved in the production of DVD discs:

#### 21.3.1.1 Injection

From the large bag to the Sumitomo injection molding machine, polycarbonate raisins are transported via stainless steel tubing. In the Sumitomo injection molding machine, the polycarbonate is melted and then injected into the mold containing the stamper. While a CD manufacturer requires a single injection machine, the DVD production line contains two.

#### 21.3.1.2 Cooling and Dye Coating

An incredibly quick robotic arm removes the discs from the instruments after injection, and they transfer to the conveyor belt to cool. One (active) disc has Azo dye applied to it, while the other (dummy) disc in the coating chamber is left uncovered. Figure 21.1 depicts the DVDR Origin line's cooling and dye-coating sections.

**FIGURE 21.1** Cooling section on DVDR Origin line.

#### 21.3.1.3 Sputtering

In a sputtering coater, cooled discs are covered with a thin layer of metal. One disc has a 40 nm-thick layer of silver placed on it, while the other disc is left clear and free of metal coatings. Sputter deposition is one method of thin film deposition using physical vapor deposition (PVD). On a "substrate" like a polycarbonate blank substrate, material is ejected from a "target" that serves as a source. Resputtering occurs when an ion or atom bombardment causes the deposited material to re-emit during the deposition process.

#### 21.3.1.4 Bonding and Inline Testing

At this point, the discs are joined together and strengthened by UV light. Discs are subjected to specialized scanners that test their optical characteristics and separate those that do not adhere to tight quality standards. The graphic depicts the UV and bonded sections of the DVDR Origin line.

#### 21.3.1.5 TE/FE Tester and Offline Testing

Disc tracking and focusing error are verified in the TE/FE tester. A DVD drive is a very complex product that needs extensive offline testing. Of 200–300 discs, we select one and test it on particular devices. The figures depict TE/FE tester on DVDR origin line and offline testing line (Figures 21.5 and 21.6).

#### 21.3.1.6 Printing and Packing

Following testing, an offset machine is used to print the surface of the DVD disc. The final stage of production includes packaging the discs and covers into DVD boxes and sealing them with cello wrap. Except for the packaging stage, every stage of production processes is carried out in chambers with extremely high air purity, maintained through rigorous incoming air filtration and strict hygienic regulations.

**FIGURE 21.2** Dye coating section on DVDR Origin line.

**FIGURE 21.3** Bonding section on DVDR Origin line.

**FIGURE 21.4** UV section on DVDR Origin line.

## 21.4 INDUSTRY 4.0 FUTURE TRENDS

A significant shift in viewpoint is brought about by Industry 5.0, which centers on people as the primary pole of the manufacturing sector. The integration of production, manufacturing, and marketing in Society 5.0 is considered more important than in Industry 5.0. In Society 5.0, the given goods and services will be tailored to the demands of the client. The objective is to create a merger of humankind and technological growth where humans and robots enhance rather than replace one another's tasks. The use of cobots and robots has significantly changed how repetitive, dangerous, and destructive occupations are performed. Furthermore, in such a society, one needs to be qualified to take the initiative because human labor requires intellectual

FIGURE 21.5 TE/FE tester on DVDR Origin line.

FIGURE 21.6 TE/FE tester on DVDR offline testing line.

output. Instead of robots replacing people, the main objective is for humans and machines to enhance one another's activities. According to this disruptive approach, education must depart from its conventional framework to create an intelligent society. Cobots need to be combined with new hardware, software, and tools, and people need to be trained in sophisticated technology for Society 5.0. One major dissimilarity between Industry 4.0 and Society 5.0 is that the former employs technology to support or improve human work, while the latter bases the industrial revolution on robotics and other technological foundations. It is crucial to emphasize that Industry

4.0's technological advancements boost Society 5.0's productivity and efficiency by emphasizing the importance of human creativity and craftsmanship in the production and manufacturing process (Fukuyama, 2018). The objective is to build a civilization with smart cities, driverless cars, integrated intelligent industries, etc. The objective is to lessen any negative impacts on management systems and the business sector, in addition to addressing climate change. More trust between people, less corruption in public administration, more equitable resource distribution, and greater collaboration between individuals, communities, and nations can all be ideal components of a more supportive society. Increased credibility and trust between consumers and enterprises would serve as the cornerstone of the manufacturing process in terms of markets. It is also critical to keep in mind that, in Industry 4.0, personalization of products and services must be done under actual market demands to prevent overstock and unsalable inventory. The fundamental notion is to create only what the population truly needs in order to minimize waste and the costs associated with overproduction. By doing this, the environmental impact and $CO_2$ emissions are also reduced (Savanevičienė et al., 2019). It is anticipated that Society 5.0 will be more ecologically conscious and inclusive, as well as more adept at managing individualism and simultaneously integrating the environment and society. There is a general expectation for a more sustainable future with interrelated and integrated environmental, social, and economic impact issues. Industry 5.0, a recent advancement that has enhanced human–machine collaboration and communication, is the future. Industry 4.0, which encapsulated the change in cyber-physical networks, has given way to Industry 5.0, which has radically changed the way people communicate, collaborate, and live. The principles regulating human–machine interaction and our collaborative processes must be defined by the next wave of technological development. As automation, machine intelligence, and even robotics become more ubiquitous and assist workers by assuming large sections of production, distribution, and operations, the degree of collaboration between humans and equipment will change. Because of the rapid pace of development, states, legislators, regulators, and all personnel face challenges. The necessity of including people in the process has been acknowledged by the industry.

## 21.5 LEGAL ISSUES WITH INDUSTRY 4.0

In India, the legal aspects of Industry 4.0 primarily focus on data privacy, cybersecurity, intellectual property rights, and labor laws related to automation. Additionally, there are regulatory frameworks designed to address these issues and ensure the safe and ethical implementation of emerging technologies such as robotics, AI, and big data analytics in manufacturing processes. The "SAMARTH Udyog Bharat 4.0" initiative is the key policy framework guiding the adoption of Industry 4.0 technologies. The main legal concerns associated with Industry 4.0 in India include the significant impact of legal technology on legal informatics systems. Cybersecurity and data governance are recognized as major legal and ethical issues. Furthermore, the considerations for sustainable legal technology within the legal domain remain limited.

## 21.6 CONCLUSION

The major focus of this chapter was the thought of Industry 4.0, the fourth industrial uprising, which makes it possible for intelligent, effective, efficient, personalized, and customized production at reasonable costs. Smaller sensors, more intelligent gadgets, faster processors, and less costly data transmission and storage could all contribute to the development of smarter machines and goods that can communicate with and learn from one another which is shown in the case study of DVDR production. This document presents a perspective on the current and future state of Industry 4.0. In addition to the strengths of Industry 4.0—AI, 5G technology, and systematic reviews, which are added as extra strength since some specialists think they are essential to the smart factory—this chapter also introduces the history of the industrial uprising.

## REFERENCES

Agrawal, A. P., Ali, S., Tripathi, N. M., Kumar, R., & Srivastava, P. (2024). Optimizing structural steel spur gear design for reduced stress concentrations in Industry 4.0 using finite element analysis. *Recent Patents on Mechanical Engineering*, *17*(1), e050724231667. https://doi.org/10.2174/0122127976309163240620080241

Alcácer, V., & Cruz-Machado, V. (2019). Scanning the Industry 4.0: a literature review on technologies for manufacturing systems. *International Journal of Engineering, Science and Technology*, Xxxx.

Almada-Lobo, F. (2015). The Industry 4.0 revolution and the future of Manufacturing Execution Systems (MES). *Journal of Innovation Management JIM*, *3*(4), 16–21.

Atal, S., Dwivedi, D., & Tripathi, N. M. (2023). Corrosion under insulation (CUI) in oil and gas industries. In *Functional Materials for the Oil and Gas Industry* (pp. 127–138). CRC Press. https://doi.org/10.1201/9781003242550-8

Bagheri, B., Yang, S., Kao, H., & Lee, J. (2015). Cyber-physical systems architecture for self-aware machines in Industry 4.0 environment. *IFAC-Pap Online, 48*(3), 1622–1627.

Bahrin, M. A. K., Othman, M. F., Nor, N. H., & Azli, M. F. T. (2016). Industry 4.0: a review on industrial automation and robotic. *Jurnal Teknologi (Sciences & Engineering)*, 137–143. eISSN 2180–3722.

Benotsmane, R., & Kov, G. (2019). Economic, social impacts and operation of smart factories in Industry 4.0 focusing on simulation and artificial intelligence of collaborating robots. *Social Sciences*, *8*(5), 143.

Bhadouria, A. S., Kumar, A., Raj, D., Verma, A., Singh, S., Tripathi, P., Kumar, Y., Sinha, A. S. K., Tripathi, N. M., Yeneneh, A. M., & Dwivedi, D. (2022). Corrosion mitigation in oil reservoirs during CO injection. In *Nanotechnology for CO2 Utilization in Oilfield Applications* (pp. 127–145). Elsevier. https://doi.org/10.1016/B978-0-323-90540-4.00014-4

Brettel, M., Friederichsen, N., & Keller, M. (2014). How virtualization, decentralization and network building change the manufacturing landscape: an Industry 4.0 perspective. *International Journal of Mechanical, Aerospace, Industrial, Mechatronic and Manufacturing Engineering*, *8*(1), 37–36.

Ceruti, A., Marzocca, P., Liverani, A., & Bil, C. (2019). Maintenance in aeronautics in an Industry 4.0 context: the role of augmented reality and additive manufacturing. *Journal of Computational Design and Engineering*, *6*, 516–526. https://doi.org/10.1016/j.jcde.2019.02.001

Chong, S., Pan, G.-T., Chin, J., et al. (2018). Integration of 3D printing and Industry 4.0 into engineering teaching. *Sustainability* 10: https://doi.org/10.3390/su10113960

Chua, C. K., Leong, K. F., & Lim, C. S. (2010). *Rapid Prototyping: Principles and Applications, 3rd edn,* World Scientific.

Figueiredo, M. J. G., Cardoso, P. J. S., Gonçalves, C. D. F., & Rodrigues, J. M. F. (2014). Augmented reality and holograms for the visualization of mechanical engineering parts. *2014 18th International Conference on Information Visualisation* (pp. 368–373).

Fukuyama, M. (2018). Society 5.0: aiming for a new human-centered society. *Japan SPOTLIGHT, 27*(Society 5.0), 47–50.

Isaca. (2017). *Cyber Security Fundamentals Study Guide, 2nd edn,* ISACA.

Jafar, M. R., Tripathi, N. M., Yadav, M., & Nasato, D. S. (2024). Additive manufacturing in the age of Industry 4.0 and beyond. In *Advances in Pre-and Post-Additive Manufacturing Processes* (pp. 213–230). CRC Press. https://doi.org/10.1201/9781003428862-11

Kaisler, S., Armour, F., Espinosa, J., & Money, W. (2013). Big data: Issues and challenges moving forward. *2013 46th Hawaii International Conference on System Sciences* (pp. 995–1004). IEEE.

Kalman, H., Portnikov, D., Gabrieli, O. G., & Tripathi, N. M. (2019a). What do pneumatic conveying and hydraulic conveying have in common?. *Powder Technology, 354*, 485–495. https://doi.org/10.1016/j.powtec.2019.06.022

Kalman, H., Portnikov, D., Gabrieli, O. G., & Tripathi, N. M. (2019b). What do pneumatic conveying and hydraulic conveying have in common? In J. Sobota & R. Haldenwan (Eds.), *19th International Conference on Transport and Sedimentation of Solid Particles, 2019* (pp. 203–210). (International Conferences on Transport and Sedimentation of Solid Particles; Vol. TS 19). Wydawnictwo Uniwersytetu Przyrodniczego we Wroclawiu. https://doi.org/10.30825/4.12-23.2019

Kalman, H., Santo, N., & Tripathi, N. M. (2019c). Particle velocity reduction in horizontal-horizontal bends of dilute phase pneumatic conveying. *Powder Technology, 356*, 808–817. https://doi.org/10.1016/j.powtec.2019.08.081

Kruijff, G.-J., Zender, H., Jensfelt, P., & Christensen, H. I. (2006). Clarification dialogues in human-augmented mapping. Proc. ACM/IEEE Int. Conf. Human-robot Int., Salt Lake City, UT, March 2006.

Landherr, M., Schneider, U., Bauernhansl, T. (2016). The Application Centre Industrie 4.0—Industry-driven manufacturing, research and development. *49th CIRP Conference on Manufacturing Systems (CIRP-CMS 2016), Procedia CIRP, 57*, 26–31.

Lee, J., Kao, H. A., & Yang, S. (2014). Service innovation and smart analytics for Industry 4.0 and Big Data environment, product services systems and value creation. *Proceedings of the 6th CIRP Conference on Industrial Product-Service Systems, Procedia CIRP, 16*, 3–8.

Lumay, G., Tripathi, N. M., & Francqui, F. (2019). How to gain a full understanding of powder flow properties, and the benefits of doing so. *ONdrugDelivery*, 42–47.

Mikalef, P., Pappas, I., Krogstie, J., & Giannakos, M. (2018). Big Data analytics capabilities: a systematic literature review and research agenda. *Information Systems and e-Business Management, 16*, 547–578. https://doi.org/10.1007/s10257–017–0362-y

Mosyurchak, A., Veselkov, V., Turygin, A., & Hammer, M. (2017). Prognosis of behaviour of machine tool spindles, their diagnostics and maintenance. *MM Science Journal,* 2100–2104.

Neugebauer, R., Hippmann, S., Leis, M., & Landherr, M. (2016). Industrie 4.0- Form the perspective of applied research. *49th CIRP conference on Manufacturing systems (CIRP-CMS 2016)* (pp. 2–7).

Neveu, A., Francqui, F., & Tripathi, N. M. (2020a). Granular material characterization at the macroscopic scale for the validation and the calibration of numerical models. *The Proceedings of 2020 Virtual AIChE Annual Meeting, (Particle Technology Forum)*, San Francisco, CA, USA, ISBN: 978-0-8169-1114-1.

Neveu, A., Lumay, G., Pillitteri, S., Monsuur, F., Pauly, T., Ribeyre, Q., Francqui, F., Vandewalle, N., & Tripathi, N. M. (2020b). Physical characterization of blends containing mesoporous particles with a focus on electrostatic properties. In *2020 AIChE Virtual Spring Meeting and 16th Global Congress on Process Safety (2020 AIChE Virtual Spring Meeting and 16th Global Congress on Process Safety)*. AIChE. ISBN: 978-0-8169-1113-4.

Neveu, A., Tripathi, N. M., Francqui, F., Lumay, G., & Rigo, O. (2020c). Characterization of powders spreadability in the framework of additive manufacturing processes. *The Proceedings of 2020 Virtual AIChE Annual Meeting, (Particle Technology Forum)*, San Francisco, CA, USA, ISBN: 978-0-8169-1114-1.

Neveu, A., Tripathi, N. M., Rigo, O., Francqui, F., & Lumay, G. (2020d). Experimental investigation of spreadability of metal powders in recoating process. *Proceedings—Euro PM2020 Congress and Exhibition (Proceedings—Euro PM2020 Congress and Exhibition)*. European Powder Metallurgy Association (EPMA). ISBN: 978-1-8990-7251-4.

Ngo, T., Kashani, A., Imbalzano, G., Nguyen, K., & Hui, D. (2018). Additive manufacturing (3D printing): a review of materials, methods, applications and challenges. *Composites Part B Engineering*, *143*, 172–196. https://doi.org/10.1016/j.compositesb.2018.02.012

Peres, R. S., Dionisio Rocha, A., Leitao, P., & Barata, J. (2018). IDARTS—Towards intelligent data analysis and real-time supervision for INDUSTRY 4.0. *Computers in Industry*, *101*, 138–146. https://doi.org/10.1016/j.compind.2018.07.004

Qin, J., Liu, Y., & Grosvenor, R. (2016). A categorical framework of manufacturing for Industry 4.0 and Beyond, Changeable, Agile, Reconfigurable & Virtual Production. *Procedia CIRP, 52,* 173–178.

Ranjan, R., & Tripathi, N. M. (2024). Optimizing solar power: engineering efficiency in CdTe/MAsnI3 double absorber solar cell. In *2nd International Conference on Emerging Materials, Smart Manufacturing & Computational Intelligence (ICEMSMCI-2024)*. Punjab: Chitkara University.

Ranjan, R., Singh, D., Pandey, V., Sharma, A., Sinha, A. S. K., & Tripathi, N. M. (2024a). A comprehensive review on the machining process: unconventional as an alternative to conventional machining. In *Advances in Pre-and Post-Additive Manufacturing Processes* (pp. 163–195). CRC Press. https://doi.org/10.1201/9781003428862-9

Ranjan, R., Verma, S., & Tripathi, N. M. (2024b). The organic HTL in highly efficient germanium-alloyed CsSnI3 perovskite solar cell: a modelling and computational analysis. *Physica Scripta*, *99*(6), 066010. https://doi.org/10.1088/1402-4896/ad4e15

Rao, S. K., & Prasad, R. (2018). Impact of 5G technologies on smart city. *Wireless Personal Communications*, *100*(1), 161–176.

Ratsimba, A., Zerrouki, A., Tessier-Doyen, N., Nait-Ali, B., André, D., Duport, P., Neveu, A., Tripathi, N., Francqui, F., & Delaizir, G. (2021). Densification behaviour and three-dimensional printing of Y2O3 ceramic powder by selective laser sintering. *Ceramics International*, *47*(6), 7465–7474. https://doi.org/10.1016/j.ceramint.2020.11.087

Rennung, F., Luminosu, C. T., & Draghici, A. (2016). Service provision in the framework of Industry 4.0, SIM 2015/13th International Symposium in Management. *Procedia—Social and Behavioral Sciences*, *221*, 372–377.

Rojko, A. (2017). Industry 4.0 concept: background and overview. *International Journal of Interactive Mobile Technologies*, *11*(5), 77–91.

Rüßmann, M., Lorenz, M., Gerbert, P., & Waldner, M. (2015). Industry 4.0: The Future of Productivity and Growth in Manufacturing Industries. *Boston Consulting Group, 9*(1), 54–89.

Sadik, A. R., & Urban, B. (2017). An ontology-based approach to enable knowledge representation and reasoning in Worker-Cobot agile manufacturing. *Future Internet*, 9. https://doi.org/10.3390/fi9040090

Satishkumar, P., Haldar, B., Tripathi, N. M., Sharma, A., Desai, D. J., Sharma, V. K., Seenivasan., S., & Babbar, A. (2024a). Effect of interlayer temperatures and heat inputs on porosity and hydrogen solubility in wire arc additive manufactured AA2618 aluminium: a comparative study between pulsed-MIG and CMT methods. In *Advances in Pre-and Post-Additive Manufacturing Processes* (pp. 137–162). CRC Press. https://doi.org/10.1201/9781003428862-8

Satishkumar, P., Kumar, R., Tripathi, N. M., Sharma, A., Desai, D. J., Verma, A., Natarajan., N., & Babbar, A. (2024b). Integrating titanium coating with 3D-printed GFRP panels: an innovative approach to harnessing composite strengths in engineering applications. In *Advances in Pre-and Post-Additive Manufacturing Processes* (pp. 18–38). CRC Press. https://doi.org/10.1201/9781003428862-2

Savanevičienė, A., Statnickė, G., & Vaitkevičius, S. (2019). Individual innovativeness of different generations in the context of the forthcoming society 5.0 in Lithuania. *The Engineering Economist, 30*(2), 211–222.

Shafiq, S. I., Sanin, C., Toro, C., & Szczerbicki, E. (2015). Virtual engineering object (VEO): toward experience-based design and manufacturing for Industry 4.0. *Cybernetics and Systems, 46*, 35–50.

Sharma, A., Babbar, A., Singh, K., Singh, A. K., Tripathi, N. M., & Desai, D. J. K. (2024). Effect of minimum quantity lubrication (MQL) method on machining characteristics for ductile substrates: a future direction. In *Modern Hybrid Machining and Super Finishing Processes* (pp. 35–59). CRC Press. https://doi.org/10.1201/9781003327905-3

Singh, A. K., Shakir, M. H., Yadav, S., Yadav, M., Tripathi, N. M., & Fidan, I. (2024). Investigations on the mechanical characterization of additively manufactured polymer composite materials fabricated via MEX. In *Advances in Pre-and Post-Additive Manufacturing Processes* (pp. 64–77). CRC Press. https://doi.org/10.1201/9781003428862-4

Sipsas, K., Alexopoulos, K., Xanthakis, V., & Chryssolouris, G. (2016). Collaborative maintenance in flow-line manufacturing environments: an Industry 4.0 approach. *5th CIRP Global Web Conference Research and Innovation for Future Production, Procedia CIRP, 55*, 236–241.

Thames, J. L., & Schaefer, D. (2017a). Cybersecurity for Industry 4.0 and advanced manufacturing environments with ensemble intelligence. In L. Thames & D. Schaefer (Eds.), *Cybersecurity for Industry 4.0.1* (pp. 243–265). Berlin: Springer (Springer Series in Advanced Manufacturing).

Thames, J. L., & Schaefer, D. (2017b). Industry 4.0: an overview of key benefits, technologies, and challenges. In L. Thames & D. Schaefer (Eds.), *Cybersecurity for Industry 4.0.1* (pp. 1–33). Berlin: Springer (Springer Series in Advanced Manufacturing).

Thoben, K. D., Wiesner, S., & Wuest, T. (2017). Industrie 4.0 and smart manufacturing- a review of research issues and application examples. *International Journal of Automation and Technology, 11*(1), 4–16.

Tofail, S. A. M., Koumoulos, E. P., Bandyopadhyay, A., Bose, S., O'Donoghue, L., & Charitidis, C. (2018). Additive manufacturing: scientific and technological challenges, market uptake and opportunities. *Materials Today, 21*(1), 22–37.

Tripathi, N. M., & Sharma, A. (Eds.). (2024). *Advances in Pre-and Post-Additive Manufacturing Processes: Innovations and Applications*. CRC Press. https://doi.org/10.1201/9781003428862

Tripathi, N. M., Francqui, F., & Lumay, G. (2020). Influence of relative air humidity on the flow property of fine powders. *The Proceeding of Third International Conference on Powder, Granule and Bulk Solids: Innovations and Applications PGBSIA 2020 February 26–28, 2020* (pp. 63–68).

Tripathi, N. M., Kalman, H., & Levy, A. (2019a). Investigation of energy loss in horizontal bends in dilute phase pneumatic conveying systems. *International Congress on Particle Technology, PARTEC-2019*, Nurnberg, Germany.

Tripathi, N. M., Levy, A., & Kalman, H. (2024). Pressure drop through bends at different orientations in dilute phase pneumatic conveying systems. In *Powder Technology Conclave & Exhibition*. Patiala, Punjab, India: PTC.

Tripathi, N. M., Portnikov, D., Levy, A., & Kalman, H. (2019b). Bend pressure drop in horizontal and vertical dilute phase pneumatic conveying systems. *Chemical Engineering Science*, *209*, 115228. https://doi.org/10.1016/j.ces.2019.115228

Tripathi, N. M., Bocquet, S., Francqui, F., Brochu, M., & Lumay, G. (2019c). Metallic powders thermal degradation: influence on spreadability, packing dynamics and electrostatics. *International Conference on Additive Manufacturing with Powder Metallurgy (AMPM2019)*, Phoenix, AZ, USA (pp. 262–277).

Tripathi, N. M., Lumay, G. & Francqui, F. (2019d). How to assess cohesion of unconsolidated powders at high flow fields? *9th International Granulation Workshop*, Lausanne, Switzerland.

Vaidya, S., Ambad, P., & Bhosle, S. (2018). Industry 4.0—a glimpse. *Procedia Manufacturing*, *20*, 233–238.

Wells, L. J., Camelio, J. A., Williams, C. B., & White, J. (2013). Cyber-physical security challenges in manufacturing systems. *Manufacturing Letters*, *2*, 74–77. https://doi.org/10.1016/j.mfglet.2014.01.005

# Index

## U



## V



## W



## X



## Y



## Z